More Food: Road to Survival

Edited by

Roberto Pilu

&

Giuseppe Gavazzi

Università degli Studi di Milano, Department of Agricultural and Environmental Sciences - Production, Landscapes, Agroenergy - Milano, Italy

General:

1. Any dispute or claim arising out of or in connection with this License Agreement or the Work (including non-contractual disputes or claims) will be governed by and construed in accordance with the laws of the U.A.E. as applied in the Emirate of Dubai. Each party agrees that the courts of the Emirate of Dubai shall have exclusive jurisdiction to settle any dispute or claim arising out of or in connection with this License Agreement or the Work (including non-contractual disputes or claims).
2. Your rights under this License Agreement will automatically terminate without notice and without the need for a court order if at any point you breach any terms of this License Agreement. In no event will any delay or failure by Bentham Science Publishers in enforcing your compliance with this License Agreement constitute a waiver of any of its rights.
3. You acknowledge that you have read this License Agreement, and agree to be bound by its terms and conditions. To the extent that any other terms and conditions presented on any website of Bentham Science Publishers conflict with, or are inconsistent with, the terms and conditions set out in this License Agreement, you acknowledge that the terms and conditions set out in this License Agreement shall prevail.

Bentham Science Publishers Ltd.
Executive Suite Y - 2
PO Box 7917, Saif Zone
Sharjah, U.A.E.
Email: subscriptions@benthamscience.org

BENTHAM SCIENCE

CONTENTS

FOREWORD

In a recent seminar on agriculture sustainability, an old Professor of agronomy, present in the audience, commented that all what was said was correct, but that he was grown in an age where the increase of agriculture productivity was a central dogma, today, apparently, correctly sacrificed in favour of a better interaction between environment and food production. His feeling is probably motivated by robust reasons that several agronomists and geneticists, me included, share: we feel almost out of place in a cultural landscape which has forgotten promises and achievements of the green revolution, effective in nourishing a planet at a time of rapid population growth, a condition persisting still today. In 1960 the global yield of cereal grains was around 10 billions tons, while in the next 40 years it doubled. The green revolution had to do with the creation of high yielding plant varieties capable to efficiently utilize increasing agro-technical inputs. This was achieved by a rational use of genetics to recombine in single genotypes the resistance to pests, insects and abiotic stresses. In Italy, genetic and agronomic progress based on improved varieties was impressive: from 1945 to 1995 average yields increased from 0.68 to 2.57 t/ha for durum wheat, from 1.04 to 4.51 for common wheat, from 1.14 to 9.01 for maize, from 3.68 to 5.74 for rice and from 22.9 to 47.2 t/ha for sugar beet.

One can ask what are the reasonings behind the adoption of new paradigms and priorities by scientists and institutions dedicated to the future of agriculture. The first consideration is that while the south of the planet still has to solve problems of true rural development, in the western societies the social implications of agriculture are part of the more general problem of finding a balance between food and feed production and care of the environment [1]. As a consequence, the theme is debated with contributions of social scientists and politicians not familiar with agricultural specificities; as a result, their genuine good intentions frequently ignore the technical consequences of the solutions they suggest. The second point to be raised concerns our poor capacity to adopt proper parameters defining agriculture sustainability, particularly in terms of maintenance of soil fertility. On the issue, two contrasting visions have emerged. The first supports the adoption of mild systems with decreased productivity, a *wildlife-friendly* agriculture reducing its impact on the environment while minimizing the negative effects of fertilizers and pesticides. The second suggests more *intensive* agricultural systems avoiding the necessity to plogh new virgin soils which, frequently, for climatic and edaphic conditions are marginal lands hosting peculiar sources of biodiversity.

In any case, a general agreement exists on the need to consider with priority the problem of future sustainability of agriculture. A possible evolution is that agriculture will remain intensive, but being based on methods and principles derived more from biology than from chemistry. Along this line of thought, new suggestions are currently emerging concerning the radical modifications of our agricultural systems [2]. This will imply that defining a possible future should reconsider the difficulties inherent both to the practical use of the biology of the living components of agricultural processes, and to the need that such components will be properly managed in terms of sustainability. This is the core issue of this volume dedicated to the breeding of tomorrow crops.

The first group of contributions introduces macro-agronomic and economic topics, related also to the comparison between industrial and subsistence agriculture. This part includes a discussion on the role and impact of genetics in support of future yield gains. The next four chapters take into account the biological-genetic components responsible for the interaction among plants and the environment: seed germination and plant nutrition; plant development; photosynthesis. Four contributions follow, grouped under the title *Tools*. This is the most evocative part of the volume: it illustrates the methodological revolution linking genomic resources and the capacity to predict plant phenotype and behaviour based on molecular markers; the adoption of new crops adapted to sustainable agricultural systems (one example is perennial cereal grains); molecular approaches to heterosis and apomixis; the role of epigenetics in determining the yield capacity of superior varieties. The volume ends with a chapter on quality and security of field-produced commodities and with a discussion on the state of art of the breeding of minor cereal grains.

The consideration of what the volume offers, allows to anticipate, at different levels, a vision on principles, methods and conclusions on the future sustainability of food production. A first level is the attention here dedicated to reappraise relevance and role of genetics in the sustainability context. Particularly in terms of resources dedicated, the possibility of future food crises should, in fact, suggest to stress the central role of the breeding of conventional and future-tailored varieties, once the social role of this activity is recognized, as done in the past with the peace Nobel prize assigned to Norman Borlaug.

A different level of discussion sees the future as interpreted in terms of targets to be assigned to plant breeding. Two cases are topical. The first regards the hybrid varieties in terms of contribution to yield increase. The adoption of hybrid crops as a final outcome of genetic selection, indeed, is becoming obligatory even for plants where autogamy does not favour an easy production of hybrid seeds. In this respect, molecular breeding, boosted by genomics, has contributed to bring again the phenomenon of heterosis to the attention to plant breeders, considering the possibility of revealing its molecular bases and of using effective prediction methods of hybrid value [3]. The second case has to do with perennialism. Compared to annual plants, perennials reduce the need of energy and agrochemicals, as well as of soil and nitrogen losses and of irrigation water.

A last consideration is proper to mitigate the impression that in the future food production may represent a problem of difficult solution. In the past plant breeders have successfully used genetics, but their approach to yield increase was essentially empirical. The incoming century, however, has already shown that varieties resistant to biotic and abiotic stresses can be developed using rational predictive methods based on molecular markers and exploiting genomics and transgenosis [4]. More recent molecular technologies allow to generate mutations, with positive phenotypic effects, at very precise nucleotide positions in genes with a known sequence. It can be concluded that the road to survival will be largely dependent on the accumulation of knowledge and on the evolution of methods capable to meet our future food needs.

REFERENCES

[1] Salamini F. Conference delivered at the Adunanza Solenne di chiusura dell'Anno Accademico dell'Accademia Nazionale dei Lincei, 26 giugno.

[2] USA Toward sustainable agricultural systems in the 21st Century. 2010. p. Washington, DCThe National Academy Press

[3] Wallace JG, Larsson SJ, Buckler ES. Entering the second century of maize quantitative genetics. Heredity (Edinb) 2014; 112(1): 30-8. [http://dx.doi.org/10.1038/hdy.2013.6] [PMID: 23462502]

[4] Kempken F, Jung C. Genetic modification of plants: agriculture, horticulture and forestry. 2009. p. BerlinSpringer Verlag

Francesco Salamini
Onorary Professor
University of Köln
Germany

PREFACE

The need for more food is rapidly increasing as the world population is growing at a fast pace. The scientific community involved in crop production and its improvement is being called upon to find solutions to the expanding global demand for crop plants and their products. Two options are available to increase yields: the first consists of an increase in the areas under production, and the second, an improvement of productivity on existing farmland. Of the two options the second seems preferable, as it avoids the disruption of existing ecosystems as well as an increase in greenhouse gas emissions. In both cases two strategies can be adopted, one classical, of extensive industrial agriculture largely adopted in developed countries to produce major crops like corn, rice, wheat and soybean and another one, known as sustainable agriculture, characterized by a production more respectful of the ecosystem. The two strategies are not mutually exclusive and which one of the two should be adopted depends on the characteristics of local production methods and on economic and political considerations, as well as the choice of global *versus* local food production and consumption. In this context, to really improve the situation, we should focus our efforts on the areas of the world where the nutrition of the population should be improved, like Africa, India, Bangladesh, Pakistan, China and some regions of South America. In these areas the solutions to solve the problem of hunger should be local and specific, related to the real needs of the population and respectful of local traditions. The knowledge, experience and know-how available to the western world could be invaluable tools for improving their agricultural production. Application of our model of industrial agriculture should be avoided. Only by trying to understand what are the needs of these populations and exporting our knowledge to improve their situation can we hope to contribute to solve their problems. These considerations are developed in the first two chapters of the book. The following chapter will deal with genetic variability as an essential source of plant improvement. The following chapters will analyse basic physiological processes which represent bottlenecks for productivity and the efforts that could be directed to increase the efficiency of these processes. The topics analysed will be the genetic control of seed size; germination and seedling elongation, representing crucial steps in plant development; photomorphogenesis and the effects of light on aspects related to yield, such as photoperiod and shade avoidance, photosynthesis and the sink-source flux; and mineral nutrition. These topics will be covered in chapters 5 to 9. We will concentrate on factors that are directly related to yield, omitting those indirectly affecting productivity like herbicide- and pest-resistance, drought tolerance and cold resistance. In the last part of the book, attention will be given to some of the tools available to the researcher to achieve plant improvement. We will focus attention on available tools such as molecularly assisted breeding, gene editing, domestication of new species, heterosis and apomixis.

IMPORTANCE OF THE FIELD

The importance of increasing productivity of the major crops to meet the demand of an expanding population is self-evident. What is not so obvious is how to achieve a significant improvement in a short time, and what tools we can rely upon to accomplish a second "green revolution". The great majority of the contributors to the chapters of the book are teachers of advanced courses to graduate students in Biotechnology or to post-graduate students in Ph.D. programs and they feel that this book could be of interest for their students.

Roberto Pilu &
Giuseppe Gavazzi
Università degli Studi di Milano
Department of Agricultural and Environmental Sciences
- Production, Landscapes, Agroenergy –
Milano, Italy

List of Contributors

A. Michele Stanca Department of Life Sciences, Università degli Studi di Modena e Reggio Emilia, Modena, Italy

Alessandro Tondelli Consiglio per la ricerca e la sperimentazione in agricoltura (CREA), Genomics Research Centre, Fiorenzuola d'Arda, Italy

Carlo Pozzi Department of Agricultural and Environmental Sciences – Production, Landscapes, Università degli Studi di Milano, Agroenergy - Milano, Italy

Damiano Martignago Department of Biosciences, Università degli Studi di Milano, Milano, Italy

Emidio Albertini Department of Agricultural, Food and Environmental Sciences, Università degli Studi di Perugia, Perugia, Italy

Elahe Tavakol Department of Agricultural and Environmental Sciences – Production, Landscapes, Università degli Studi di Milano, Agroenergy - Milano, Italy
Department of Crop Production and Plant Breeding, College of Agriculture - Shiraz, Shiraz University, Iran

Enrico Francia Department of Life Sciences, Università degli Studi di Modena e Reggio Emilia, Modena, Italy

Fabio Veronesi Department of Agricultural, Food and Environmental Sciences, Università degli Studi di Perugia, Perugia, Italy

Fabio Fornara Department of Biosciences, Università degli Studi di Milano, Milano, Italy

Filippo Geuna Department of Agricultural and Environmental Sciences – Production, Landscapes, Università degli Studi di Milano, Agroenergy - Milano, Italy

Francesca Galbiati Department of Biosciences, Università degli Studi di Milano, Milano, Italy

Franz W. Badeck Consiglio per la ricerca e la sperimentazione in agricoltura (CREA), Genomics Research Centre, Fiorenzuola d'Arda, Italy

Daniele Rosellini Department of Agricultural, Food and Environmental Sciences, Università degli Studi di Perugia, Perugia, Italy

Gabriella Consonni Department of Agricultural and Environmental Sciences – Production, Landscapes, Università degli Studi di Milano, Agroenergy - Milano, Italy

Gaia Pigna Department of Biosciences, Università degli Studi di Milano, Milano, Italy
Department of Plant Sciences, The University of Tennessee, Knoxville, TN, USA

Gianluca Bretani Department of Agricultural and Environmental Sciences – Production, Landscapes, Università degli Studi di Milano, Agroenergy - Milano, Italy

Giuseppe Gavazzi Department of Agricultural and Environmental Sciences- Production, Landscapes, Università degli Studi di Milano, Agroenergy - Milano, Italy

Jorge Gomez-Ariza Department of Biosciences, Università degli Studi di Milano, Milano, Italy

Laura Rossini Department of Agricultural and Environmental Sciences – Production, Landscapes, Università degli Studi di Milano, Agroenergy - Milano, Italy
Parco Tecnologico Padano - Lodi, Italy

Mario Motto Fondazione Istituto Tecnico Superiore per le nuove Tecnologie della Vita, Bergamo, Italy

Martina Persico	Department of Agricultural and Environmental Sciences – Production, Landscapes, Università degli Studi di Milano, Agroenergy - Milano, Italy
Martina Landini	Department of Biosciences, Università degli Studi di Milano, Milano, Italy
Massimo Galbiati	Department of Biosciences, Università degli Studi di Milano, Milano, Italy
Michela Landoni	Department of Biosciences, Università degli Studi di Milano, Milano, Italy
Monica Colombo	Research and Innovation Centre, Fondazione Edmund Mach, San Michele all'Adige (Trento), Italy
Piero Morandini	Department of Biosciences, Università degli Studi di Milano, Milano, Italy
Paolo Pesaresi	Department of Biosciences, Università degli Studi di Milano, Milano, Italy
Roberto Pilu	Department of Agricultural and Environmental Sciences – Production, Landscapes, Università degli Studi di Milano, Agroenergy - Milano, Italy
Salvatore Ceccarelli	Consultant, Rete Semi Rurali, Scandicci, Firenze, Italy
Stefano Bocchi	Università degli Studi di Milano,, Department of Agricultural and Environmental Sciences – Production, Landscapes, Agroenergy - Milano, Italy
Stefano Sangiorgio	Department of Agricultural and Environmental Sciences- Production, Landscapes, Università degli Studi di Milano, Agroenergy - Milano, Italy
Vittoria Brambilla	Department of Biosciences, Università degli Studi di Milano, Milano, Italy
Valeria Terzi	Consiglio per la ricerca e la sperimentazione in agricoltura (CREA), Genomics Research Centre, Fiorenzuola d'Arda, Italy

CHAPTER 1

The Yield in the Context of Industrial *Versus* Sustainable Agriculture

Stefano Bocchi*

Università degli Studi di Milano, Department of Agricultural and Environmental Sciences – Production, Landscapes, Agroenergy - Milano, Italy

Abstract: During the Green Revolution both the yield and the global production significantly increased. The yield increase was achieved, for some main crops, thanks to the so called high yielding varieties. Higher global production was also due to the increase of the crop production surface which took place especially in some areas of the planet. In the current scenario of rapid human population increase, with a sharp increase of livestock, the challenge is to achieve efficient, productive, sustainable and resilient land use, while conserving biodiversity and assuring, everywhere, food security inside a framework of sustainable diets. The paper, after a discussion on the meanings of such concepts as yield, yield gap, production and global production describes some of the main issues related to increased intensification of food security and global productivity in the current discussions on the potential of the Green Revolution approach and the agro-ecological paradigm.

Keywords: Agro-food system, Diet gap, Food security, Global production, Sustainability, Yield, Yield gap.

BASIC TERMS

The issue related to crop yield, despite its fundamental importance for our future, even though extensively studied, has been poorly defined and discussed on a sufficiently broad time-space scale.

The role of the technology in yield change has often been confounded by other influences [1].

During the Green Revolution the crop yield has been the main, if not the only, goal to be considered and the farms have been viewed for decades as industries where input is converted in output thanks to an industrial-like production process.

* **Corresponding author Stefano Bocchi:** Department of Agricultural and Environmental Sciences – Production, Landscapes, Agroenergy, Università degli Studi di Milano, Milano, Italy; Tel: +39 02 50316588; E-mail: stefano.bocchi@unimi.it

Roberto Pilu and Giuseppe Gavazzi (Eds.)

Few studies have been carried out by referring to theoretical and practical analysis influences of the Green Revolution approach on the innovation in agriculture, including both the positive and negative consequences on the natural resources.

Crop yield is the weight of the so called economic products (*i.e.* grain, root vegetables, and fruits, *etc.*) at standard moisture content, referred to unit of land area cultivated per crop, conventionally and generally referred to in metric tons per hectare (t/ha). Energy, protein, oil, vitamin, micronutrients contents in the total weight are of fundamental importance in yield analysis taking into account the influence on the final utilization of the value chain (human diet and nutrition) when the nutritive, energetic, economic balances, also have to be considered.

As underlined by Fischer *et al.* [1] the energy contents reflect the cost of biosynthesis of the major constituents of the product. Cereals for examples are characterized by a total energy content of around 15 MJ/kg, whereas soybean contains about 24 MJ/kg, the comparison of the yield obtained from these crops must consider these different energy costs.

For agriculture the main figure is average yield in terms of t/ha, not only referred to field and farm, but also to different levels of the territorial systems *i.e.* districts, regional, and national. **Farm Yield** (FY), reported from yield measurements, or more often from surveys, are part of the local and national statistics annually collected without considering the cases where, for various reasons, the district is not planted to its full potential.

The possibility is not always considered, such as in warm climates, to have different crops/harvests, per year, in the same field. Nevertheless this FY is generally indicated as **Real Yield**, different from the so-called **Potential Yield** (PY), which is the uppermost end of the yield scale, which is reached with the combination of some important factors. When the most appropriate varieties are cultivated with the best agronomic management, there are no manageable abiotic and biotic stresses [2].

"PY defines what might be obtained for particular plants species when not limited in technology, *i.e.* when the best cultivars, fertilizer, machinery, labor, and knowledge are all available and applied in the best possible ways" [3].

The concept is close to the so-called **Attainable Yield** corresponding to the best yields achieved through skillful use of available technology. It is usually achieved in experiment centers or by the best farmers [3]. This simple theoretical definition does not have an easy method that actually measures it. The sowing date can be a complication. The optimal sowing date may be constrained in a multiple cropping system [4]. PY is usually determined with direct measurements or indirect estimates in plots, in two types of experiment: comparative variety ones and in plot/field experiments carried out by crop physiologists or agronomists. In this type of PY determination sampling errors occur. Crop modeling can be used to predict PY in different environments and their accuracy has significantly improved. Integrated methods, *i.e.* direct measurements, modeling and expert opinion can be used [5]. The integrated methods are particularly useful when the so-called water-limited potential yield (PYw) has to be determined. The crop yield depends on the quantity of available water and the PYw is generally calculated as a linear function of the water supply, but variation in rainfall during the development stages can create a more complex picture and modify this linearity.

Current yield in a given agricultural area is usually a poor indicator of potential performance, falling on a continuum between crop failure and potential yield. FAO defines **Actual Yield** (AY) as the average yield of a district.

The concepts of the actual attainable potential yields are useful for defining the agronomic concept of **intensification** of the farming system: where actual yields are close to the projected attainable ones. The farming system and the agriculture of the area can be described as intensive. The **intensification** of farming systems increases when the available technology is appropriately adopted and as the proportion of time in crop is relative to fallow increases. The yield can be referred to both the total biomass obtained from the growth/development process and the part of this biomass. The term biomass indicates the total dry biomass accumulated by the crop, where the term **Economic Yield** (**EY** or simply **yield**) indicates the portion useful to humans as food/fiber/fuel or as feed. The fraction yield/total biomass is defined as coefficient of economic yield, the **Harvest Index** (HI) is calculated as the useful fractions/above-ground biomass.

If we compare PY, AY (or FY), EY and calculate the differences (*i.e.* (PY-AY) we have a better knowledge on what is defined as **Yield Gap** (YG). It can be expressed in percentage on PY or on FY. The latter is more appropriate since it indicates how much is the possible, desirable increase in actual grain yields that is achievable by farmers. Scientific literature supports the notion of a minimum yield gap (FY equals EY depending strongly on prices). If the future prices will be favorable for the farmers it is suggested [1] that the minimum yield gap is 30% of FY; that is to say EY is 23% below PY [4]. The yield gap across 40 agricultural regions around the world was calculated to range between 25 and 400%. (For more information and more recent data refer to both [4] and http://www.yieldgap.org/). Many of the countries with the highest YG have the poorest access to technology, infrastructure and capital required for the model of Green Revolution agricultural development.

The so-called **Global Crop Production (GCP)** is referred to the global amount of commercial biomasses or products (grains, fruits, roots, tubers *etc.*) obtained with cultivation practices on a given total area. It is obviously dependent on cultivated area (total cultivated surface expressed in ha) and yield (t/ha of part of the obtained biomass). As observed recently [6], changes in land use for global crop production have been strongly driven by increases in land area devoted to the three major cereals (wheat, rice, maize). During last century's two decades of the 1960s and 1970s, the area expansion for the three cereals represented more than 70% of land use increase for all crops, followed by two decades during which both total crop area and area in major cereals remained relatively constant. This stability period came to a sharp end in 2002, when the crop production area starts to increase at nearly ten million hectares per year, 60% due to increased production of wheat, rice, and maize. An additional 25% can be attributable to the enlargement of soybean area. This trend of crop-area increase has occurred in South America, Asia, and Africa [6]. It is worthwhile to note and remember that two crops out of four are strongly related to meat production.

At a global scale, rates of yield increase have been clearly linear for most major cereal crops since the beginning of the '60s of the last century, the second phase of Green Revolution, when the trends were driven by rapid adoption of Green Revolution technologies that were largely a one-time innovation, including the development of High Yielding Varieties (HYV), and due to wide spread use of commercial fertilizers and pesticides, investments to expand irrigation infra-structures, and dependence on fossil-fuel energy. More recently a decline of this increase rate has been observed and strong evidence of upper yield plateaus in some of the world's most intensive cropping systems has been observed. Despite the increase of around 60% in investment in agricultural Research and Development in USA "the rate of maize yield gain has remained linear, implying that the marginal yield increase per unit of research investment has decreased substantially over time. Approaches that rely on compound rates of yield increase or constant linear rates with no upper limit to yield growth are not supported by the analysis of historical yield trends and current understanding of crop physiology and they are likely to overestimate future increases in crop yields by a large margin" [6].

Moreover, some recent studies [7 - 10] on the phenomenon of the "**paradox of the scale**" analyzed "the inverse farm size-productivity relationship". These studies established that small and diversified farming systems show higher productivity per area rather than the big monoculture farm.

The Green Revolution: Crop Innovation For Yield Increase

The analysis of the main features of the Green Revolution can be useful for better understanding our common future after recalling some basic concepts, definitions, trends related to the yield, production, and productivity,

In the 1970s, E. Rogers described as "Diffusion of Innovation" the introduction and diffusion of hybrid corn in the Corn Belt of United States of America (mainly in Iowa) from the 1940s [11] up to the 60s. The peculiar conditions that were within the USA system of production, the existing research centers and the network of farmers in the Corn Belt allowed to development and test: (a) an innovation scheme driven mainly by genetic (gene revolution) manipulation and agronomic intensification; and (b) a dissemination process useful for multiplying the research results (*i.e.* extension service). The corn hybrid has been developed in some U.S.A. research centers based on the concept of plant ideotype and on innovative genetic improvement techniques, crop physiology and agronomy. Several farmers by adopting corn hybrids were able to increase yields, specializing the farm and simplifying the agronomic schemes to meet the requirements of a new growing market of a new *commodity*.

Green Revolution, at this first step, occurred under the following conditions [12]:

- most of the farmers of the area were producing for the same market the same commodity;
- each of them, too small for affecting the price of the commodities, was in competition for the current price, trying to increase the income through yield increases, possibly at a lower cost;
- due to inelasticity of the demand, every try to increase yield exerted a downward pressure on prices;
- most of the farmers had access to credit, fertilizers, information (radio, newspapers, extension agents) and were members of organizations.

The new technology could produce a process of innovation, represented by an S-curve efficiently describing the trend: slow increase at first, then much more quick and finally stabilized or decreasing. Cochrane, proposed the so called *Agricultural Treadmill* theory: in the early stages, the first farmers able to adopt the new technology, achieve good results in a market condition characterized by pre-technology status. If the number of farmers modifying the production system increases, the **total production** increases, the state of the market changes with prices falling. Market forces, defined as the treadmill, propel these phases of the diffusion process. Elder farmers and small farms with intrinsic weaknesses tend to sell or lease the land, causing a growth towards an economy process of scale for the entire sector, an increase of the farm average size.

The diffusion of this type of innovation based on treadmill system determines new social and political conditions with strong consequences both at the micro and macro-level, such as:

- strong increase of migration from rural to urban areas (farmers in industrialized countries currently account for only 4-5% of the total workforce; in the less industrialized countries farmers can be more than 70%);
- farmers do not maintain for a long period benefits of the adopted technological innovation;
- countries can improve their competitiveness on the global market only ***if and when*** their food policy is well defined and the industries become more efficient;
- not all the farmers have the same behavior; only the early adopters take advantages;
- the innovation is focused on a new ideotype of crop, more specifically new cultivars (**High Yielding Variety**) with expected/potential higher yield, requiring higher amount of macroelements, especially nitrogen, phosphorus and potassium during specific phases of the cycle. The macro-element uptake dynamics of HYV maintain a generally relatively low rate during the first vegetative phase. During the elongation phase the crop requirements sharply increase, so that the mineralizing patterns are not usually sufficient. The HYV-technological package include a significantly increasing use of chemical fertilizers (increased costs).

The Green Revolution was enabled by the invention of the Haber-Bosch process, based on the production of **high volumes of industrial nitrogen** fertilizers, but also on the strong changes of many farms, extension services in the world. Before Green Revolution, the crop nitrogen uptake depended on manure, recycled organic matter, biological fixation and indigenous/local supply through mineralization. With GR the use of N fertilizers has grown sevenfold and nowadays 30 – 80% of nitrogen applied to farmland is lost to surface and ground-waters, and to the atmosphere [13]. The environmental cost of N loss, in Europe, has been estimated at 70 – 320 billion Euros per year; a value outweighing the direct economic benefits of N in agriculture [14]. Moreover Green Revolution was enabled by the extraction of phosphates. Since the end of the last World War, global extraction of phosphate rock has tripled to meet industrial agriculture's requirements (90% of global use of phosphorus is for food production; ironically Africa is at the same time the world's largest exporter of phosphate rock and the continent with the largest food shortage, FAO 2006). This type of P use is not considered sustainable, since the fossil phosphate rock reserves are finite and located in few and controlled sites of the planet.

The model tested in the U.S. on corn had specific peculiarities: (a) the innovation was initially based on one single product; (b) the progress was driven by researchers with a very focused set of goals (*e.g.* development of a new variety with a higher potential yield); and (c) the process of innovation diffusion followed a top-down approach, from researchers towards farmer, which could adopts both the new variety the entire technology package. Since the 1960s this type of innovation model adopted inside free market conditions became **the model to be exported** all over the world, aiming at making agriculture develop everywhere. Kline and Rosenberg [15] defined this approach as a *linear model;* Chambers and Jiggins [16] as *transfer of technology model*, implying that: (a) it could be transposed with no significant changes in every other agricultural area of the world; (b) there was a unique development pattern fitting all areas; and (c) a causal mechanism was underlying the production, distribution and access to food processes, so that for facing food policy issues it was sufficient to increase yields, as occurred in the original model.

This innovation concept, generally speaking, has marked an evolution during the last five decades. Before considering how and when this occurred, it is useful to clarify some basic concepts.

The innovation can be applied at different levels: products, processes and systems, with another possible distinction of the singular *innovation* from the plural *innovations*. Innovation is something that integrates activities and processes linked with the creation, dissemination, adoption and management of new technological, institutional/social and managerial knowledge that, in turn, cause technical, socio-economic and environmental changes. Innovation is thus related to the development/diffusion of new products, technologies, markets, institutions or policies. The World Bank [17] defines *innovation system* as "(the) system that comprises the organizations, enterprises and individuals that demand and supply knowledge and technologies, and the policies, rules and mechanisms that affect the way in which different agents interact to share, access, exchange, and use knowledge." So, innovation can be described as the **emergent property of interactions** among the stakeholders/actors of a given system.

The modern concept of innovation implies technical and institutional innovation. The deep integration of the two dimensions requires inter-sectorial and interdisciplinary studies, reaching the **transdisciplinary level**.

In Table **1** a diagram with the stages of evolution of the concept of innovation in agriculture is shown. The passages from the transfer of technology to the so called innovation system are described, referring to the different phases from the 60s up to today.

Table 1. Evolution during the last 5 decades of the concept of innovation in agriculture. From production increase to promoting institutional change for sustainability and resilience of agro-food systems.

Approaches/ **Features**	*Technology Transfer*	*Farming System Research*	*Farmer participatory Research (FPR)*	*Innovation System*
Period	Until the 1960s	Beginning of 1970-80s	Beginning of 1990s	2000s
Mental models and activities	Provide technology through 'pipelines'	Understand the constraints of local farms through spatial surveys	Collaborate in the research and development	Develop joint research involving stakeholders in multiple partnerships and processes
Discipline-specific knowledge	Determined by a single discipline (genetics)	Multi-disciplinary (agronomy and agricultural economy)	Inter-disciplinary (with social scientists and expert farmers)	Trans-disciplinary analysis of complex system
Objectives	Production increase	Efficiency improvement	Provision of adequate livelihoods to the farm	Promote value chains and institutional change
Main elements	Technology packages	Technology packages updated to overcome constraints and obstacles	Joint production of knowledge and technologies	Promote value chains and institutional change
Drivers	Research offer	Assessment of boundaries and limits of farmers	Farmers' demand	Responsibility for change
Innovators	Researchers	Researchers and Extension agents	Farmers and researchers	Multiple stakeholders forming an innovation platform
Role of farmers	Adoption or Refusal	Information source	Active test agents	Partners, business agents, innovators
Role of scientists	Innovators	Experts	Cooperators	Partners as one of the parties answering to the needs
Perceived key changes	Behavior of farmers	Removal of the main constraints for farmers' work	Strengthening of farmers	Institutional change, innovation capacity
Expected Results	Adoption of technology	Adjustment of the farm system	Co-evolution of technologies with a more appropriate livelihoods system	Capacity to innovate, learn and change
Sustainability	Indefinite	Important	Explicit	Priority, regulated and multidimensional

From the above table it is easy to evolution of the concepts and the approach. The *innovation system approach* shifts the focus from top-down research and diffusion of knowledge and technology towards an interactive multi-stakeholder or multi-agents change of process. Technology dissemination and market development, of course, are some of the elements of the system, to be analyzed in the integrated frame.

It is possible to note that the historical trend was the constant updating of this notion through the decades: the meaning of innovation itself has changed from the push of new technologies, during the early stages, to the recent projecting/ planning of opportunities through **institutional development**. This deep change implies that the concept of innovation in agriculture requires system analysis to be framed in an integrated set of technical, agro-ecological, organizational, institutional and political components. That is to say that innovation in agriculture requires to move far beyond the old linear model, nonetheless still adopted by various international initiatives and/or proposed by some international agro-food industries.

Many Analysts working on the topic of innovation in agriculture pointed out that learning capacity of farmers is the critical point. Innovation processes at all scales of the system, must include all system stakeholders and actors, from farmers, to extensionists, consultants, researchers, civil society, private/institutional sector *etc.* – all together considered as innovation *students* [18, 19]. This approach, allowing creating links, relationships and alliances, is based on the principles of **mutual learning, resource and knowledge sharing** in ways for facilitating market and institutional change. This type of new partnership has been recently defined as *participation* and indicated as one of the five guiding principles of the activities (partnership, complementarity, subsidiarity, relevancy, participation) promoted by the European Initiative for Agricultural Research for Development (EIARD).

According to Pimbert [20] participation process should be described in different ways, with respects to its kind of relationships and research activities. From the so called passive participation (the target communities are only informed on what is happening) it is possible to implement functional participation (the target communities participate by composing groups to achieve some early goals) or interactive participation (the target communities participate in the joint analysis required for producing action/plans; interdisciplinary methodologies are employed for reaching different goals).

Pimbert [20] defines *auto-mobilization,* an additional form of participation, when communities participate by taking independent initiatives, even not directly related to the other actors, for changing (innovating) the existing conditions. This form of participation is similar to the Landcare experience [21] where Australian farmers and landowners drove an independent process for solving soil erosion, salinization and desertification, caused by the previous not appropriate adoption of European agricultural techniques in the Australian reality. This innovation process, born for developing new site-specific, appropriate land management techniques and for reactivating ecosystem services at landscape scale, required a great collective integrate effort, joint actions, new organizational structures and infrastructure, social learning and strong participation.

As mentioned before, innovation can also be described as the emergent property of interaction among stakeholders and actors in a natural resources or economic system service [22]. Roling [12] observed: "where the degradation of the resource or service is the collective outcome of each stakeholder's trying to satisfy his/her individual preferences, more sustainable management of the resource or service necessarily must emerge from collective processes – social learning, conflict and negotiation, agreement, reciprocal sacrifice or benefits and privileges, and leadership – that lead to concerted action" and "when innovation is the emergent property of interaction, promoting innovation becomes a matter of facilitating the interaction process, and the institutional support and favorable policies at higher level are essential ingredients for success at the local level". This last element is strictly connected to the epoch-making article that appeared in Science in 1968 "The tragedy of the commons" [23].

Given that innovation i) is a complex and adaptive process; ii) requires an appropriate period of time and resources for evolving and for producing significant assessment over its linkages to empowerment, environment and sustainability issues, it seems that **an appropriate logical and a rigorous conceptual framework** would be needed. With regards to the agricultural sector, **agro-ecology represents an ideal approach to filling this gap**.

An Agro-ecological Framework For Agricultural Development

The complexity of the above-described picture requires on one side a continuous updating of the basic concepts and the reference paradigm, on the other side a clear definition of the appropriate strategies to select conceptual and pragmatic tools. We can describe innovation strategy with the following scheme including three different levels of change(Table **2**) [24, 25].

1) The first level is called **substitution strategy**, when the existing farming systems are slightly adapted, not modified; the plot size (few m^2) is usually the scale of experimental activities and a single discipline drives research.
2) The second agro-ecological level is when the strategy aims at **building innovative technical scenarios** at least at farm level. The farming systems are innovated relying on biological processes and regulation in integrated multi crop production schemes; the research is carried out at farm or regional scale within the agro-ecological framework and with a multidisciplinary approach.
3) The third **level** analyzes and innovates agro-food systems with a global view; interdisciplinary, inter-sectorial, transdisciplinary research is carried out with the aim of tackling agro-food issues at the global scale. Linking local to global with

inter-scale studies, re-assessing the relationships of the agri-food sector to the society as a whole and/or focusing on specific issues, such as intensive agriculture and its links to the economic-industrial model or its failure **in sustainability terms**. New trends in agro-ecology and new time/space scales of agronomic research, ranging from food shed to large regional areas are possible.

Agro-ecology is gaining importance and it has been acknowledged as a strategic approach for pursuing sustainability agricultural system planning and management, particularly through the consolidation of stability and resilience of both the anthropic and the natural ecosystems. Indeed, the complex of scientific, conceptual and practical tools framed into the agro-ecology domain seems suitable for planning development initiatives with multiple purposes such as i) ensuring a sustainable management of all the resources involved in agricultural production processes (soil, air, water, biodiversity, human labor), ii) promoting food security and sovereignty, iii) protecting the landscape; iv) assuring equity. By analyzing a target agro-ecosystem, it is possible to identify and characterize the relationships between both the internal components, the system structure, the functions and the evolution at different scales of plot, farm and regional area, without neglecting the interactions among scientific, technological and socio-economic factors, and developing towards a transdisciplinary approach aimed at conflict resolution, inequality reduction for achieving the Millennium Development Goals (MDGs).

Table 2. Levels of innovation strategy (adapted from Evan and Fisher, 1999; Bocchi *et al.*, 2013).

Levels of Innovation Strategies	***Meaning***	***Example***	***Scale of research and disciplinarity***
Substitution strategy	Existing farming systems are only slightly adapted and fundamentally not altered (Rosset and Altieri, 1996)	Toxic chemicals and fertilizers should be replaced by less pollutant, less persistent in soil and less energy-consuming compounds.	- Plot - Monodisciplinary
Agro-ecological strategy	Build innovative technical scenario relying on biological regulation in integrated crop production scheme (Altieri, 1987, 1999)	Applying ecological concepts and principles, e.g. biodiversity, intercropping rotation, agroforestry.	- Farm or large territory; - Multidisciplinary (agronomy, landscape ecology, geography etc.).
Global strategy	Solve agricultural issues at the global scale, rethinking its relation to society. Failure of intensive agriculture are linked to its economic-industrial model. New trends in agro-ecology (Gliessman, 2007).	Studying the relationships between production and marketing; linkages farms and consumption; marketing networks.	- Foodshed, regional, - Inter-disciplinary, trans-disciplinary, cross-sectorial (agronomy, ecology, sociology, economics, politics)

Given the level of challenges that agro-food system is going to face in the next decades, it is indeed essential to support the planning and implementation of agro-ecological policies, starting from the detection of an appropriate set of advanced tools. Particularly, a key for empowering the actors involved in the activities of research, development, dissemination and application of new methods is the **identification of information and communication technologies** that farmers and policy makers need for coping with fast-changing conditions in a complex system.

CONCLUDING REMARKS

Croplands cover currently 1.5 billion hectares (12% of Earth's ice-free land), pasture cover around 3.4 billion hectares (26% of Earth's ice-free land). Agriculture altogether represents the main land use on the planet.

Land suitable for agricultural activities (production, ecosystem services *etc.*) is a finite and vulnerable resource globally speaking, but there are big differences between areas. Global average of arable land available per capita amounts to about 0.45 ha, but there are strong declines to 0.10 ha taking place in the most densely populated regions of the world. In these areas food security is declining.

Soil degradation has been estimated to affect 16 – 40% of the terrestrial surface [26], everywhere meaningful soil losses causing reduced yields are forecast for the next future .

Closing yield gap, through agro-ecological principles and practices, could increase food supplies. If yields of 16 important food and feed crops were brought up to 75% of their potential, global production would increase by 1.1 billion tons (2.8 x 10^{15} kcal), that is to say 28% increase [27].

The rise of meat consumption causes a sharp increase in the use of cereals for feed. Ironically, **current diets are both not appropriate** for human health (individually and socially) and are not compatible with sustainable resource use.

Moreover, globally more than a third of harvested food is thrown away. In industrialized countries, 40% wastes occur at retail and consumer level, whereas in the poorer countries 40% losses occur at post-harvest and processing levels.

Inefficient use of food stocks occurs also by feeding cereals and fodder starch to animals, very poor energy converters. Livestock requires about 7 kcal input (from 3 for broiler chickens to 16 for beef) (cereal grain feed) for every kcal generated. Cereals fed to livestock currently make up 30 – 50% of global cereal production (Corn contributes for 70%). Cereals and grain legumes convert energy into protein much more efficiently than animals. Shifting 16 major crops to 100% human food could add over 1 billion tonnes to global production (an additional 28% increase).

Shifting diets from beef to poultry or from grain-fed to pasture-fed beef would already increase significantly the food supply by closing a **Diet Gap** [28], improving people health, and reducing the impact on resources.

As stated by the International Assessment of Agricultural Knowledge, Science and Technology for Development (IASTD, 2008), "technologies such as high-yielding crop varieties, agrochemicals and mechanization have primarily benefited the better resourced groups in society and transnational corporations, rather than the most vulnerable ones. To ensure that technology supports development and sustainability goals, strong policy and institutional arrangements are needed…".

Like research and development in agriculture in general, **seed policies** must be driven, not by a preconceived view about the benefits technology can bring to farming, but by a careful and broad examination of their impacts on food security and sovereignty and, specifically, on the possibility of the vulnerable farmers to improve their livelihoods.

In the field of agricultural research for development the quantitative level of investments and number of organizations (research centers, universities, networks) does not seem the main limiting factor, which may rather be the set of strategic decisions taken by the international stakeholders and their coherency with the basic principles that should characterize this field. The sphere of agro-food research requires a clear, non-reductionist approach that would not limit its vision to the provision of products or processes *(silver bullets)* mainly originated by industries. Research **should not only be focused on yield, productivity and profitability** of agro-food value chains, but on agro-ecosystem functioning (stability, resilience, nutrient cycling) and agro-ecosystem services (biodiversity, carbon sequestration, water harvesting, landscape management), on agro-food systems management (sustainable healthy diets, food security, systems sustainability and resilience, food policy).The new approach is expected to review and replace the linear model by applying the agro-ecology principles and techniques at different scales, from farm to the whole agro-food system, addressing also landscape-related agronomy and bionomy issues [29].

As a result, it would be appropriate a revision of the funding system for agricultural research, specifically by conveying adequate resources to research programs and projects aimed at improving the whole agriculture and agri-food system (*e.g.* agro-ecology and agro-forestry, soil management techniques, composting, water management, agronomic practices, drought resistant varieties) or strengthening the institutional environment, particularly at community level (*e.g.* community seed banks, seed fairs and farmer field schools), rather than merely focusing on a single crop or variety.

Global food production is dependent on not only fertile soils, fresh water, biodiversity and informed and trained farmers, but also cheap energy, locally available. Alternatives to fossil-fuel-dependent agricultural systems will be required in the next future. Sustainable agriculture combines three main objectives: environmental health, ethical soundness, socio-economic profitability. **Education plays a central role**.

Strategies in the public agricultural research sector should be based on a set of shared values that can distinguish between the needs of society and environment *versus* those of big business firms, as well as the interest of farmers *versus* those of large international organizations that control the world markets.

There is a need to strengthen the inter-disciplinary attributes of innovation, as an integrated concept linking various scientific, policy and socio-economic fields. In this sense, innovation could lead to the improvement and conservation of local natural resources through participation of the agro-food systems' actors, as well as strengthening full food sovereignty and guaranteeing that the results of public research could be systematically accessible thanks to open-source policies. Transdisciplinary approach is required in addressing the basic issue of the sustainable diet at both individual and global scales.

CONFLICT OF INTEREST

The author confirms that author has no conflict of interest to declare for this publication.

ACKNOWLEDGEMENTS

Declared none.

REFERENCES

[1] Fischer RA. Definitions and determination of crop yield, yield gaps, and rate of change. Field Crops Res 2015; 182: 9-18.
[http://dx.doi.org/10.1016/j.fcr.2014.12.006]

[2] Evan LT, Fisher RA. Yield potential: its definition, measurement and significance. Crop Sci 1999; 34: 1544-51.
[http://dx.doi.org/10.2135/cropsci1999.3961544x]

[3] Loomis RS, Connor DJ. Crop Ecology, productivity and management in agricultural systems. Cambridge Univ. Press 2012; p. 537.

[4] Van Ittersum MK, Cassman KG, Grassini P, *et al.* Yield gap analysis with local to global relevance – A review. Field Crops Res 2013; 143: 4-17. [http://dx.doi.org/10.1016/j.fcr.2012.09.009]

[5] Lobell DB, Cassman KG, Field CB. Crop yield gaps: their importance, magnitudes and causes. Annu Rev Environ Resour 2009; 34: 179-204. [http://dx.doi.org/10.1146/annurev.environ.041008.093740]

[6] Grassini P, Eskridge KM, Cassman KG. Distinguishing between yield advances and yield plateaus in historical crop production trends. Nat Commun 2013; 4: 2918. [http://dx.doi.org/10.1038/ncomms3918] [PMID: 24346131]

[7] Halweil B. Can organic farming feed the world? World Watch 2006; 19: 18-24.

[8] Barret CB, Bellemare MF, Hou JY. Reconsidering conventional explanations of the inverse productivity-size relationship. World Dev 2009; 38: 88-97. [http://dx.doi.org/10.1016/j.worlddev.2009.06.002]

[9] De Shutter O. Report on the right to food. UN Human Rights Council 2011. http:://rs.resalliance.org./?p=4612.

[10] Horlings LG, Marsden TK. Toward the real green revolution? Exploring the conceptual dimensions of a new ecological modernization of agriculture than could feed the world. Glob Environ Change 2011; 441-52.

[11] Ryan B, Gross N. The diffusion of hybrid seed corn in two Iowa communities. Rural Sociol 1943; 8: 15-24.

[12] Roling N. Conceptual and methodological developments in innovation. In: Sanginga P, Ed. Innovation in Africa. 2009; pp. 10-34.

[13] Goulding K, Jarvis S, Whitmore A. Optimizing nutrient management for farm systems. Philos Trans R Soc Lond B Biol Sci 2008; 363(1491): 667-80. [http://dx.doi.org/10.1098/rstb.2007.2177] [PMID: 17652069]

[14] Sutton MA, van Grinsven H. The European Nitrogen Assessment. Cambridge: Cambridge Univ Press 2011. [http://dx.doi.org/10.1017/CBO9780511976988]

[15] An overview of innovation. In: Landau R, Rosenberg N, Eds. The positive sum of strategy: harnessing Technology for economic growth, national Academic Press. Washington, DC 1986; pp. 275-306.

[16] Chambers Jiggins. Agricultural research for resource-poor farmers. Part 1. Transfer-of-technology and farming system research part II: a parsimonious paradigm Agricultural Administration and Extention 1987; 27: 35-52. 109-28 [http://dx.doi.org/10.1016/0269-7475(87)90008-0]

[17] Enhancing agricultural innovation: how to go beyond the strengthening of research systems. Washington, DC: World bank 2006.

[18] Competence developmenting soft skill/personal mastery, report on design of a learning programme at Makerere University, Uganda . Nairobi, Kenya: Rockfeller Foundation 2002.

[19] Challenges to strengthening agricultural innovation systems: where do we go from here? Farmers first revisited: Innovation for agricultural research and development. IDS: Brighton 2007.

[20] Participatory Research and On-Farm management of Agricultural Biodiversity in Europe. IIED: London 2011.

[21] Community shaping the Land and the Future Allan and Unwin. St. Leonards: Australia 1994.

[22] Bawden RJ, Pacham R. System praxis in the education of the agricultural systems practitioner. Syst Pract 1993; 6: 7-19. [http://dx.doi.org/10.1007/BF01059677]

[23] Hardin G. The tragedy of the commons. The population problem has no technical solution; it requires a fundamental extension in morality. Science 1968; 162(3859): 1243-8. [PMID: 5699198]

[24] MacRae RJ, Hill SB, Henning J, Mehuys GR. Agricultural Science and Sustainable Agriculture: a Review of the Existing Scientific Barriers to Sustainable Food Production and Potential Solutions. Biol Agric Hortic 1989; 6: 173-219. [http://dx.doi.org/10.1080/01448765.1989.9754518]

[25] Bocchi S, Christiansen S, Oweis T, *et al.* Research for the Innovation of the Agri-food system in International Cooperation. It J Agron 2012; 7(3): 262-73.

[26] Chappel MJ, LaValle LA. Food security and biodiversity: can we have both? Agric Human Values 2011; 28: 3-26. [http://dx.doi.org/10.1007/s10460-009-9251-4]

[27] Foley JA, Ramankutty N, Brauman KA, *et al.* Solutions for a cultivated planet. Nature 2011; 478(7369): 337-42. [http://dx.doi.org/10.1038/nature10452] [PMID: 21993620]

[28] Tscharntke T, Clough Y, Wagner TC, *et al.* Global food security, biodiversity conservation and the future of agriculture intensification. Biol Conserv 2012; 151: 53-9. [http://dx.doi.org/10.1016/j.biocon.2012.01.068]

[29] Ingegnoli V. Landscape Bionomics Biological-Integrated Landscape Ecology. 2015. [http://dx.doi.org/10.1007/978-88-470-5226-0]

CHAPTER 2

Increasing Plant Breeding Efficiency through Evolutionary-Participatory Programs

Salvatore Ceccarelli[1,*]

Consultant, Rete Semi Rurali, Scandicci, Firenze, Italy

Abstract: One fundamental problem in plant breeding is the relationships between selection and target environments. Selection theory shows that response to selection (genetic gains) depends on this relationship because of genotype x environment interactions. Therefore, response to selection can be increased by making the selection environment as similar as possible to the target environment (decentralized breeding). However, this does not yet guarantee farmers' acceptance of the new variety, which we argue is a more correct way of measuring plant breeding efficiency than variety release as usually done by public breeding programs. Using selection theory, the chapter shows that the probability that a new variety is accepted by farmers, thus impacting their livelihood, increases by selecting in the target environment (decentralized selection) in collaboration with farmers. Decentralized-participatory plant breeding also increases agrobiodiversity and makes plant breeding more cost-effective. The proclaimed efficiency of private breeding program, which can claim a wide farmers' adoption, is actually driven by a seed market monopoly, which severely limits farmers' choice of which seed to buy. However, the weak point of decentralized-participatory plant breeding is the unreliability and unpredictability of Institutional participation. Evolutionary-participatory plant breeding may overcome the limitations of participatory plant breeding, because farmers can handle evolutionary populations independently from Institution, yet without excluding them from participating. Because in evolutionary-participatory plant breeding the unit of selection becomes the individual plant rather than a plot, a much higher selection intensity is possible, thus increasing even further the efficiency.

Keywords: Biodiversity, Climate change, Efficiency, Evolutionary plant breeding, Genetic gains, Genomic selection, Genotype x environment interaction, Human health, Participation, Response to selection, Seed.

INTRODUCTION

Biodiversity decline, climate change, hunger and malnutrition, poverty, water and the increased frequency of a number of diet related diseases such as diabetes as

* **Corresponding author Salvatore Ceccarelli:** Consultant, Rete Semi Rurali, Scandicci, Firenze, Italy; Tel: +39 3497707311; E-mail: ceccarelli.salvatore83@gmail.com

Roberto Pilu and Giuseppe Gavazzi (Eds.)

well as diseases associated with overweight and obesity are currently major global problems. All of these problems are related to seed.

Seed is related to water because, at the global level, agriculture uses 70% of the total water consumption, and the development of crops that can produce an economic yield with less water, will make more water available for human use. Seed is related to poverty through malnutrition: poor nutrition in the first 1000 days of life does affect the mental development of children [1]. Seed is related to climate change because farmers will need seed of crop varieties better adapted to the climate of the future. This is a particularly intriguing problem because of the uncertainty of the expected changes in temperature and rainfall [2, 3]. Therefore, plant breeding programs aiming at improving crop adaptation to climate change are actually addressing a moving target and probably a different target in different areas [3, 4]. Breeding for adaptation to climate change implies also breeding for resistance to new insect pest and diseases, which have been shown to have altered their latitudinal ranges in response to global warming [5]. An additional effect of climate change is on malnutrition, as the increase of CO_2 in the atmosphere is expected to decrease in C3 crops the content of iron and zinc whose deficiency is already causing the loss of 63 million lives annually [6].

Hunger and food security continue to be staggering challenges as about 800 million people are still undernourished and about half of the world population is lacking one or more essential nutrients [6, 7]. Whether the problem is insufficient agricultural production (but 30% of agricultural production is wasted annually) or non-equitable distribution of available food [8], seed, and the way in which it is produced, is central.

Therefore, talking about seed is not only talking about the major global problems but also about our health because most of our food comes from seeds, and food affects our health. A number of modern diseases are associated with food, such as the well-known case of celiac disease [9, 10], and the decrease of diversity is possibly related with the increased frequency of inflammatory diseases [11]. Overweight and obesity, largely associated with diet, have become a major global health challenge [12]; similarly mortality rates due to diabetes have increased [13] and non-alcoholic steatohepatitis (NASH) is becoming epidemic [14].

WHERE THE SEED COMES FROM?

Plant breeding, and the way in which it evolved from the way in which it was practiced by farmers for millennia to modern or "scientific" breeding, offers an understanding of how the problems discussed above developed and how they can be solved.

Over the years and before harvesting, farmers have selected the best plants to obtain the seed for the next cropping season, and this was done individually in each farmer's field: in other words they selected for millennia for specific adaptation producing what today we call ancient, old or heirloom varieties.

When plant breeding started to be done on scientific basis, there was a shift from selecting from specific adaptation to selecting for wide adaptation. This was done at the global level by the Green Revolution, which developed varieties able to make full use of fertilizers, pesticides, irrigation water and mechanization. If some elements of the package were missing, the varieties alone did not have any specific advantage over those that the farmers already had. On one hand, the Green Revolution averted the danger of extensive famine, but, on the other hand, it had a number of negative consequences [15 - 17]; eventually the poorest farmers could not benefit because they were not able to afford some of the components of the package [18]. GMOs, cannot be the solution to these problems because they ignore the Fundamental Theorem of Natural Selection (FTNS) by which the organisms to control evolve resistance [19]. The resistance to antibiotics, a phenomenon that is becoming widespread, and that can be very rapid [20], is based on the same fundamental biological principle. An agro ecological model of agriculture, such as different forms of organic agriculture, could be a solution, but is considered unable to produce enough food to feed a growing population, raising doubts on whether food security and food safety can be compatible objectives. The argument that organic conditions are associated with lower yields is biased by the fact that many of the meta-analysis used varieties not specifically selected for organic conditions.

In addition, the type of plant breeding, which has emerged with the Green Revolution and which is still largely followed today, particularly in public breeding, is not even the most efficient.

THE EFFICIENCY OF PLANT BREEDING

In public Institutions such as Ministries of Agriculture and the Centers of the CGIAR, the number of varieties released is the most common way of measuring plant breeding efficiency. A more scientific measure of a breeding programmes efficiency is the selection or genetic gain (or response to selection) obtained at the end of a breeding cycle [21]. Another measure of efficiency is the ratio between benefice and cost; this has been used by economists [22], but almost never by breeders. The number of varieties released is used as measure of plant breeding efficiency because is easy to measure; however, this measure ignores that a variety generated by plant breeders produces benefits only when it is accepted and grown by the farmers [22, 23]. The number of varieties released is also one of the criteria used for the professional recognition of the public breeders; in this case the lack of adoption is usually attributed to the non-availability of seed of the released varieties. The fallacy of using the number of varieties released as a measure of efficiency is also due to the weakly scientific basis on which varieties are released such as poorly designed and unrepresentative trials [24]. While variety release is necessary to legally commercializing the seed, in many developing countries the "legal" seed often represents well below 10% of what farmers sow. All the rest, particularly in marginal areas, is "illegally" produced and "illegally" exchanged by farmers [25].

Tripp *et al.* [24] extensively reviewed all the problems of the variety release system used in several countries (Table **1**). In addition to the organizational problems, the system of variety release is often based on obsolete experimental designs and statistical analysis, ignoring the most modern methods to adjust for spatial variability [26, 27]. Furthermore, the tendency of modern plant breeding is to test the breeding material in as many locations as possible by reducing the number of replication, which is possible, for example by using partially replicated design (p-rep) [28] using DiGGer (http://www.austatgen.org/software) to optimize randomization. In Syria, the variety release committee rejected three barley varieties because in the official trials the yield advantage was not sufficiently large: however, when they were included in the participatory trials, they were adopted on areas of between 10,000 and 50,000 ha [29, 30].

Table 1. Problems associated with the way trials for variety release are organized and conducted. (Modified from Tripp *et al.*, 1997).

Problems	Examples
Efficiency	low frequency of variety replacement
	uneven resource allocation to different trial stages
	prolonged variety testing
	inappropriate site selection
	inappropriate zoning
Standards	unrepresentative trials management
	trials' analysis biased against poor environments
	lack of attention to farmer-relevant variety traits
Participation	lack of participation by related organizations
	lack of farmers' participation
Transparency	lack of coordination between national and regional testing systems and lack of accountability and linkages

Consequently, in developing countries, it is common to find farmers growing varieties which were not released, and varieties which were released but not adopted, or initially adopted and then rejected. This happens more often in the case of crops grown in marginal environments [31] (Table **2**), but also in China [32], for sorghum in Nigeria [33] and in sub-Saharan Africa [34]. In developed countries, the mismatch is avoided by the monopoly of the seed market.

Table 2. Adoption (% of the area) of varieties by crops and the region, and as a ratio over the number of varieties released. † represents 40% of the pearl millet area.

Crop	Country	Region	Adoption (% area)	Adoption (ratio adopted/released)	Quoted from
Wheat	India		83%		[90]
Rice	India		60%		[90]
Sorghum	India		32%		[90]
Pearl millet	India		49%		[90]
Maize	India		36%		[90]
Pearl millet	India	Gujarat, Tamil Nadu, and Andhra Pradesh	90%		[91]
Pearl millet	India	dry areas of Haryana, Karnataka, and Maharashtra†	30%		[91]
Rice	India			2/525	[92]
Barley	Algeria			2/16	[45]
Barley	Syria		10%	2/8	[93]
Rice	Nepal		10–11% of household		[94]
Rice	India	Kalahandi, Orissa	57%		[95]

One of the most common assumption made by plant breeders is that developing high yielding varieties is a guarantee for adoption. They obviously ignore that large body of literature, which shows that adoption is not necessarily driven by yield [35, 36], as shown also by Aw-Hassan [31]. Lack of adoption of a variety may also be an obstacle to the adoption of other agricultural technologies [37].

There are several studies on the issue of variety adoption [35, 38 - 46] and they suggest that an ***apriori*** prediction about the adoption of a variety is very difficult. This is also because in a conventional breeding program, it takes 5–6 years after official release before appreciable adoption commences [47, 48]. During this period, the objectives set at the onset of the breeding program can become obsolete [35, 49].

Genetic Gains (or Response to Selection)

The genetic gain is equal to

$$R = S\,h^2$$

where S is the selection differential (the difference between the mean of the selected individuals and the mean of the whole population) and h^2 is the heritability of the target traits [50]. In the equation above (also known as the breeders' equation), the selection differential depends on the intensity of selection (i), which is equal to S/σ_p, where σ_p is the square root of the phenotypic variance. The genetic gain is also equal to

$$R = i\,h^2\,\sigma_p$$

Since breeding methods differ in the time required per cycle, the formula has been modified by Eberhart [51] in

$$R = (i\,h^2\,\sigma_p)/t \qquad \textbf{(1)}$$

where t is the time and h^2 is equal to

$$h^2 = \sigma_g^2/\sigma_p^2 = \sigma_g^2/\sqrt{}\sigma_e^2/(r_e) + \sigma_{ge}^2/r + \sigma_g^2,$$

where σ_p^2, σ_e^2, σ_{ge}^2, and σ_g^2 are the phenotypic, environmental, genotypic × environment interaction (GEI), and genotypic variances, respectively, and r and e are the number of replications and the number of environments (locations, years, or location–years combinations), respectively.

The breeding implications of formula (1) have been discussed by Falconer [50] whose major contribution to quantitative genetics as related to breeding, is that one character measured in two different environments has to be dealt with, from a genetic and selection viewpoint, as two different characters. This implies that, since selection takes place in one or more research stations to obtain a variety for one or more target environments (TE) (the farmers' fields), we should calculate the genetic gains as the correlated response to selection (CR) in the TE, which is equal to

$$CR_t = R_s\,h_t^2/h_s^2\,r_g \text{ or } i\,h_t\,h_s\,r_g\,\sigma_{pt} \qquad \textbf{(2)}$$

[50], where R_s is the genetic gain in the selection environment (SE), h_t^2 is the heritability in the TE, h_s^2 is the heritability in the SE, r_g is the genetic correlation coefficient between the trait measured in the two environments, and σ_{pt} is the phenotypic standard deviation of the trait in the TE.

In the case of a target population of environments (TPE), CR_t can be calculated separately for each TE or across the TPE, if the same variety is selected for the entire TPE. CR_t should also be expressed in terms of cycle time, as in the case of R.

Plant breeders have discussed for a very long time the relative efficiency of selecting in the target environment (R_t), *vs* the selection in the research station (R_s), where heritability is expected to be higher [52 - 56]. The ratio between CR_t and R_t, if t is the same, is equal to

$$CR_t/R_t = r_g\,(h_s/h_t) \quad \textbf{(3)}$$

The formula applies when selecting for one trait and the genetic gain is measured on another trait [50].

When $h_t = h_s$, the maximum value of CR_t/R_t is 1, when $r_g = 1$. Therefore, when heritabilities are the same, direct selection will always be more effective ($R_t > CR_t$) because r_g will always be less than one. With low r_g (0.1–0.2), as between high-yielding and low-yielding TE [56], h_s must be at least 5–10 times higher than h_t for CR_t to be greater than R_t.

Therefore, heritability alone is not sufficient to determine the optimum selection environment because when r_g, is negative, as in the case of genotype × environment interactions (GEI) of crossover type, the magnitudes of h_s and h_t become irrelevant [57].

Genomic-assisted breeding does not change the terms of the problem. The use of molecular techniques increases the precision of selection but not necessarily the efficiency of plant breeding; in fact, it would be more correct to speak of molecular selection rather than molecular breeding.

The use of the various genomic tools ends with the two critical steps indicated as "Identify & recombine superior genotypes" and "Multi-environment testing of best lines". In these two critical steps between the use of genomic tools and the final products, namely improved varieties for farmers, the issues are who identifies superior genotypes and where, superior where, for which characteristics and for whom, who is choosing the best lines for multi-environment testing, how these environments are chosen and managed [58].

The participation of the farmers, who are recognized as the ultimate beneficiaries of the entire process is therefore critical for the production of varieties useful to them.

Increasing Plant Breeding Efficiency

Based on what discussed in the previous sections, we can increase plant breeding efficiency by either increasing the selection gain, the adoption of varieties, or the benefit/cost ratio (Fig. **1**). The selection gain in the TE can be increased by reducing the cycle time (see formula for CR). Off-season nurseries, single seed descent in self-pollinated crops [59], double-haploids, and QTL introgression [60] are commonly used to achieve this. However, the usefulness of genomic tools is still limited, particularly for complex traits like yield [61]. An example of reduction of the breeding cycle time is the development of rice varieties tolerant to salinity, when the savings was at least 2–3 years [62].

Increasing the number of replications is one way to increase heritability, but is expensive and might negatively affect the cost/benefit ratio. Also, it is contrary to the current trend in modern plant breeding of reducing replications to increase the number of locations in METs. Decentralized selection is one way of increasing the magnitude of r_g [56]. Decentralized selection also reduces σ_{ge}^2, because it subdivides GEI into genotype × locations (GL) and genotype × years (GY) within locations, by testing the repeatability of GL. If GL is repeatable, the breeder can subdivided the TPE in subgroups in a way that within each subgroup GEI is lower than that across the entire TPE [63]. Decentralized selection, defined as above [50, 64], is paramount in increasing selection gains because research stations do rarely represent farmers' agronomic management practices, including soil preparation and tillage, use of chemical inputs, irrigation and crop rotations.

Despite this, and despite the Australian breeding programs decided long ago to decentralize because of the poor correlation between research stations and farmers' fields [65], most national and international public plant breeding programs still conduct several cycles of selection on station, leaving only the final testing of few advanced lines in farmers' fields. The reluctance of breeders to decentralize is often justified by the perception that on-station trials are more precise, but ignores that the results obtained from those breeding trials are often irrelevant [66] as shown by the low levels of adoption of the varieties obtained. The second reason given to justify on-station selection and testing is that in the early stages of breeding program there are several entries and little amount of seeds per entry. This reason denotes the unawareness of the range of new experimental designs and data analysis available today; this is also shown by the almost "religious" and exclusive use of the randomized block design with three replications, which limits the possibility of testing in a larger number of locations. The use of the partially replicated (p-rep) designs in rows and columns combined with the use of optimized randomization (http://www.austatgen.org/software) and of spatial analysis [28], makes it possible to optimize the use of seed, to test in a larger number of locations with a sufficient precision considering that in the early stages of a breeding program, ranking of genotype is more important than predicting their yields [67].

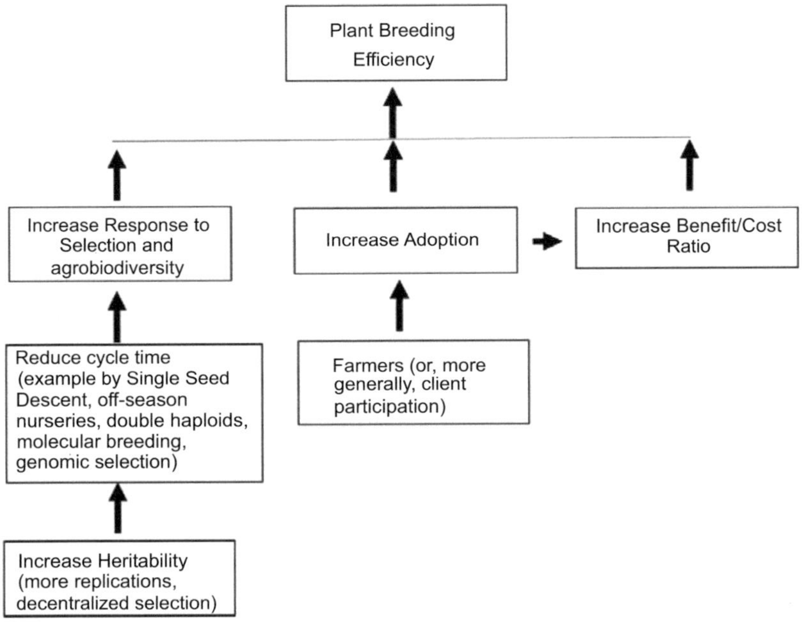

Fig. (1). Strategies to increase plant breeding efficiency defined as higher response to selection, higher agrobiodiversity, increased adoption and higher benefit/cost ratio.

However, even if the use of the best suite of methodologies available today increases response to selection, this does not necessarily increase variety adoption. The most obvious way to increase the efficiency of plant breeding, as defined earlier, is to combine decentralized selection with the participation of farmers [68] in a model of plant breeding known as participatory plant breeding (PPB) [69].

PARTICIPATORY PLANT BREEDING (PPB)

Participatory plant breeding is a plant breeding program organized in such a way as to shift the emphasis back to specific adaptation, thus contributing to both an increase in agricultural production at the farm level and to an increase in agro-biodiversity. With PPB, the breeder can exploit the advantages, shown earlier, of selection in the TE such as low input and organic agriculture with those of farmers'(men and women) participation in all the most important decisions [69 - 71] and with genomic selection.

Therefore, PPB puts farmers at the center of the process of developing new varieties, including seed production as recommended by the final report of the Special Rapporteur to the United Nations on the right to food ("Fund breeding projects on a large diversity of crops, including orphan crops, as well as on varieties for complex agro-environments such as dry regions, and encourage participatory plant breeding", pg 22) [72]. It also matches article 6c of the International Treaty on Plant Genetic Resources for Food and Agriculture, pg 7 [73] ".... promoting, as appropriate, plant breeding efforts, which, with the participation of farmers, particularly in developing countries, strengthen the capacity to develop varieties particularly adapted to social, economic and ecological conditions, including marginal areas".

The type of participation discussed here is very different from "farmer participation in technology verification and technology transfer", which is very popular in some international circles today. It rather corresponds to what has been discovered recently by CGIAR as "demand driven innovation" ignoring that the concept has been around for about 35 years.

PPB involves farmers in the "development" of a new technology (a new variety) and not merely in "verifying" if a technology developed by others, often in a different socio-economic and physical environment, is appropriate to them. The concepts applied in PPB in the case of plant breeding can easily be extended to other types of agricultural technologies.

One major characteristic of a PPB program is that it makes use of the same scientific principles, including advanced experimental designs and statistical analysis, as conventional plant breeding (CPB). The use of a sound scientific basis in a PPB program is the major responsibility of the participating scientist(s).

We initially implemented PPB in Syria in 1995 [71, 74], with the support of the German Government; later, with the support of IDRC, we extended PPB to Tunisia, Morocco and Jordan. Eventually, and with the support of a range of donors including IFAD, the Government of Italy, OPEC, the McKnight Foundation, the Government of Norway, FAO, PPB was introduced in Egypt, Eritrea, Algeria, Yemen, Iran, Ethiopia, Uganda, and in Italy with crops such as bread and durum wheat, barley, lentil, chickpea, faba bean, cowpea and tomato. PPB is a combination of modern science with the "local knowledge", and brings plant breeding back into farmers' hands – and not farmers back into breeding as suggested by Almekinders and Hardon [75]. PPB also increases agrobiodiversity.

The model is a typical plant breeding program based on four cycles of testing and selection usually called initial, preliminary, advanced and elite yield trials, or stage 1, stage 2, stage 3 and stage 4. To reduce the number of cycles of selection on station, we include in the initial trials (or stage 1) F_3 bulks each derived from a different cross (the F_3 bulks were used instead of the F_2 because of the greater availability seed to plant the trials (as p-rep) in as many locations as possible). Thus, the method relies on the selection *between* bulks from the F_3 to the F_6 generation, and then on the selection of pure lines *within* the superior bulks when genetic uniformity of the final variety is needed.

In this model, farmers (both men and women), but also consumers, traders, intermediaries, *etc.*, are involved in all the stages of the development of a variety – not just in testing the final few lines products of a scientific research as done in conventional (non-participatory) research. Therefore, PPB differ substantially from Participatory Variety Selection (PVS) which is often presented as an alternative to PPB but that, in practice, reduces significantly the "quality" of participation by delaying farmers' involvement to the last stages of a plant breeding program. In a PVS program, farmers are asked to choose among a number of varieties or of breeding lines which is considerably lower than in a PPB program: therefore, their choices are limited.

There are several differences between CPB and PPB: in CPB – and, with only few exceptions such as Australia, varieties are selected on research stations generally managed with large use of irrigation and inputs by breeders: only the lines candidates for release are tested on farm. In the research station, agronomic management including tillage, type and number of fertilizer applications, weed control, use of irrigation and rotations are different, and sometimes very different, from those used particularly by poor farmers in marginal areas. Adoption occurs at the end of the breeding process and there is usually a wide gap between varieties adopted and varieties released [76]. On the contrary, in a PPB program selection occurs in farmers' managed trials by both breeders and farmers. During the selection process, farmers start expressing their preferences, thus giving information on potential adoption. To be participatory, the program needs to be inclusive, encouraging the participation of women who, in low-income countries, play an important role in agriculture, and agriculture plays a critical role in their livelihoods [77]. Empowering women and focusing on their unique challenges will decrease poverty and will bring much wider gains in productivity [78].

CPB and PPB are based on the same scientific principles and they differ in key organizational aspects:

- The objectives are discussed together with farmers, who often decide the type of genetic material (landraces, modern varieties, populations, fixed lines) and the most important traits (seed color, plant height, fodder quality);
- The breeding material is tested in farmers' fields at a much earlier stage than in a

CPB program;

- Farmers are involved in all major decisions and particularly in deciding which material to select and which material to discard at the end of each cropping season. They also often suggest methodological innovations in the way selection is organized and conducted, or on how the trials are planted and with which seed rate. Details of who participates and how are given in Ceccarelli *et al.* [69, 79];
- Locations are chosen to sample as extensively as possible the TPE. Within each TE, the users are chosen to be as representative as possible of the target population of users. Each location is treated, during selection and data analysis, as an independent units by conducting independent selection within each location regardless of how the best breeding lines in that location perform in other locations. Similarly, data analysis is conducted within locations. Therefore PPB implement decentralization and selection for specific adaptation, while in CPB is strongly centralized and is for wide adaptation.
- The trials, particularly fertilizer applications and weed control, are managed by the farmers, who may decide to plant trials under organic or biodynamic conditions.
- The objectives of the programs are continuously monitored with the participating farmers.

PPB increases agrobiodiversity because, being based on decentralized selection, produces varieties which are different from country to country, from village to village within a country, and even within the same village responding to the need of different participating farmers. PPB increases agrobiodiversity both in space and in time because the process is continuous and highly dynamic producing a continues flow of new varieties, and an agricultural landscape is created, which is so variable that makes it difficult for insect pests and diseases to spread. The varieties selected by farmers through a PPB program are often not homogenous, – like the landraces – in contrast to the genetic uniformity of the varieties produced by CPB.

In addition to agrobiodiversity, PPB has had an impact on:

- *Adoption of varieties*: many new varieties have been adopted by farmers where PPB has been implemented; in Syria, between 2000 and 2011, farmers named and adopted more than 80 barley lines and/or populations from the PPB trials: these varieties were grown on about 80% of the barley dry areas. During the last 35 years the CPB program conducted by the Ministry of Agriculture only released 7 varieties of which only one had a modest level of adoption. In Jordan and Algeria, varieties obtained from PPB programs are under multiplication; in Eritrea, farmers selected and named three food barley, ten bread wheat and two durum wheat varieties; farmers in Yemen, adopted two barley and two lentil varieties; farmers in the North-West coast of Egypt adopted three barley varieties. In Tigray, a region often affect by drought and famine, two barley varieties were adopted and officially released. In Iran, the farmers selected and named four varieties and are currently testing various types of mixtures [80]. In Italy, three tomato lines were selected in organic conditions and out yielded significantly the commercial F_1 hybrids [81]. The best of these families out yielded significantly the commercial hybrid grown in the area. The third family out yielded the commercial hybrid by between 62 and 76%, in two of the four farms, but it was significantly lower yielding (-22%) on the research station. Had we conducted the breeding program only on station, we would have missed such a line. Yield advantages, as high as 50-70% were achieved in dry areas only by changing the variety – in these areas CPB failed at introducing any new variety. In several cases we obtained these yield advantages by using landraces strongly preferred by farmers in dry areas;

- *Institutional*: in a number of countries such as Yemen, Jordan, Algeria, Eritrea, Ethiopia, policy makers and scientists showed interest in PPB as an approach, which generates more relevant results faster and at a lower cost; participatory research and PPB have been and are included in recently announced and/or funded international research projects by the European Commission (www.h2020.net), IFAD (www.ifad.org/operations/projects/regions/pi/ir_all.htm) and the International Treaty on Plant Genetic Resources for Food and Agriculture (http://www.planttreaty.org/content/call-proposals-2010-2011);
- *Farmers' skills and empowerment* considerably improved because of the interactive nature of PPB. The participation into the process of plant breeding increased their ability to negotiate their dignity and their self-confidence. The skills they developed became very useful once we started evolutionary plant breeding (EP) that will be described later.

One of the advantages of PPB is that is a methodology able to increase agricultural production directly in the farmers' hands making therefore those increases readily available and accessible. PPB also can address the specific needs of family farms and to make them more productive, thus contributing to alleviate poverty and to meet local and global food demand, including food quality. PPB shifts research themes from those of large-scale farming to those of smallholdings, which are very different because they are based on crop rotation, crop-livestock integration, and the use of animal waste as fertilizer [82].

PPB is based on selecting for specific adaptation; hence is more efficient than conventional plant breeding [76], and is also able to generate varieties specifically adapted to an agro ecological agricultural model, which can reconcile food safety with food security.

EVOLUTIONARY-PARTICIPATORY PLANT BREEDING (EPB)

There are several examples of successful PPB in as many as 47 different countries with 23 different crops (http://www.solibam.eu/modules/wfdownloads/singlefile.php?cid=12&lid=33) However, despite its successes, PPB has a weakness because it requires a stable and continuous participation of a research institute to make available the breeding material and the scientific support such as experimental designs and statistical analysis. However, even when an institution supports PPB, there is no guarantee that the support will continue: examples of changes in the institutional commitment of an institution comes from Syria [83], where, after 13 years of institutional support, the new ICARDA management prevented the program from accessing external funds, and from Iran. Therefore, a PPB program depends on the long-term participation of a research institution, or, in other words, from its Institutionalization.

An alternative to PPB that makes farmers independent from Institutional support without excluding it, is Evolutionary (Participatory) Plant Breeding– participatory is in parenthesis because, though desirable, the participation of an Institution is not a limiting factor. The method was proposed back in 1956 [84] and consists in preparing mixtures (evolutionary populations, EP) of very many different genotypes of the same crop. These populations will be planted and harvested year after year following the agronomic practices commonly used for that particular crop in that particular area. Because of natural crosses (higher in cross-pollinated and less in self-pollinated crops), the seed which is harvested will be genetically different from the seed that was planted. Therefore, the population evolves and will become progressively better adapted to the environment (soil type, soil fertility, agronomic practices including organic systems, rainfall, temperature *etc.*) in which is grown. It will also become gradually adapted to climate change: although the weather conditions changes from year to year, the genetic composition of the population will fluctuate, but if the climate tends towards higher temperatures and lower rainfalls, the genotypes better adapted to those conditions will gradually become more frequent [19].

Farmers can make their own EPs by purchasing seed of several different varieties (including hybrids) of the same crop, mixing the seed and let the resulting mixture intercross freely: the seed so produced is the EP from which farmers, breeders or both can select.

When this is done, it is expected that, based on formula (1) given earlier, response to selection will increase because σ_p and i (which depends on population size) will be much higher than in most breeding programs because of the large population size of an EP.

This has been done in Italy [85] using a zucchini (summer squash) EP obtained by intercrossing 11 commercial hybrids. The farmer selected two varieties, differing in colour, that had a yield similar to the commercial hybrids. He also has already started selling the fruits of the EP. EPs of different crops (maize, barley, bread and durum wheat, common bean, tomato and summer squash are also grown in Jordan, Ethiopia, Iran, Italy (http://www.semirurali.net/), France, Portugal, and India. Farmers growing these populations report higher yields, lower weed infestation and disease presence, and lower insect damages [80]. The use of pesticides has consequently been reduced.

Iranian farmers growing an EP of wheat have marketed the bread obtained from the flour of the EP in local artisanal bakeries. The bread can be consumed also by customers allergic to gluten. The EP has a greater yield stability, and bread obtained from its flour has a better aroma and taste, which is now available in several artisanal bakeries in France and Italy.

EPB is a cheap and dynamic strategy to gradually improve adaption of crops to climatic changes, which include not only an increase in temperature and a decrease in rainfall but also to new insect diseases and weeds. Given that organic agriculture is highly location specific, EPB is precision breeding adapting crops to each specific type of organic agriculture. With EPB farmers can save and replant their seed, and exchange it with neighbours, thus maintaining a high level of agro biodiversity in the field without depending from external inputs. The diversity itself improves the resilience of the crop to short term whether fluctuations and to biotic stresses.

EPB can become participatory when scientists assist farmers in selecting from EP using genomic tools.

CONCLUDING REMARKS

In discussing about the global problems including the pandemic of obesity and diabetes, seldom it is recognized that the solution of these problems requires a change in the way seed is produced, because seed is related to all these problems. Conventional plant breeding conducted by large private seed companies needs to generate profit, and is difficult to change it from the current emphasis on wide adaptation supported by a consolidation of the seed industry [86, 87] to an emphasis on specific adaptation. This could be conveniently done, to some an extent, by small seed companies, but mostly by public breeding programs such as those conducted by CGIAR using both their large germplasm collections, amounting to about 700.000 seed samples, and their international network of collaborators in developing countries.

However, the very recent move of CGIAR towards public-private collaboration, is leading to the deployment of some of the public breeding programs to large seed corporations, the transfer of former top managers of some of the largest seed corporations into top-management positions in the CGIAR and the increasing role of private foundations [88] raise questions of weather this might lead in the future to a, at least partial, privatization of the CGIAR gene banks.

Whether this will happen or not, the EPs may play two important roles: firstly, in the hands of the farmers of developing countries may represent a continuous source of better adapted genetic material for their breeding program as an addition to or a replacement for the genetic material they usually receive from the CGIAR in the case events such as those in Syria prevent an International Centre to continue this service; secondly, in the hands of the farmers around the world, being non-patentable for their continuing evolving nature, they will remain as a unique publicly available genetic resources. "Once the farmers have the seed, they have the solution" [89].

CONFLICT OF INTEREST

The author confirms that he has no conflict of interest to declare for this publication.

ACKNOWLEDGEMENTS

The authors acknowledge several hundred farmers and many scientists, extension workers, seed specialists and non-governmental organizations for their contribution to the development, the refinement and the evolution of this program. We also thank the OPEC Fund for International Development, the Government of Italy, the Government of Denmark, Der Bundesminister für Wirtschaftliche Zusammenarbeit (BMZ, Germany), the International Development Research Centre (IDRC, Canada), the System Wide Program on Participatory Research and Gender Analysis (SWPPRGA), the International Fund for International Development (IFAD), and the Water and Food Challenge Program of the CGIAR for their financial support during the years.

REFERENCES

[1] Save the Children. Nutrition in the First 1,000 Days. State of the World's Mothers 2012.

[2] Nelson GC, Rosegrant MW, Koo J, *et al.* Climate Change: Impact on Agriculture and Costs of Adaptation. International Food Policy Research Institute 2009.

[3] Trenberth KE, Fasullo JT, Shepherd TG. Attribution of climate extreme events. Nat Clim Chang 2015; 5: 725-30.
[http://dx.doi.org/10.1038/nclimate2657]

[4] Ceccarelli S. Drought. In: Jackson M, Ford-Lloyd BV, Parry ML, Eds. Plant Genetic Resources and

Climate Change. CABI International 2014; pp. 221-35.

[5] Bebbe DP, Ramotowski MA, Gurr SJ. Crop pests and pathogens move polewards in a warming world. Nat Clim Chang 2013; 3: 985-8.
[http://dx.doi.org/10.1038/nclimate1990]

[6] Myers SS, Zanobetti A, Kloog I, *et al.* Increasing CO_2 threatens human nutrition. Nature 2014; 510(7503): 139-42.
[http://dx.doi.org/10.1038/nature13179] [PMID: 24805231]

[7] Tilman D, Clark M. Global diets link environmental sustainability and human health. Nature 2014; 515(7528): 518-22.
[http://dx.doi.org/10.1038/nature13959] [PMID: 25383533]

[8] Ehrlich PR, Harte J. Opinion: To feed the world in 2050 will require a global revolution. Proc Natl Acad Sci USA 2015; 112(48): 14743-4.
[http://dx.doi.org/10.1073/pnas.1519841112] [PMID: 26627228]

[9] van den Broeck HC, de Jong HC, Salentijn EM, *et al.* Presence of celiac disease epitopes in modern and old hexaploid wheat varieties: wheat breeding may have contributed to increased prevalence of celiac disease. Theor Appl Genet 2010; 121(8): 1527-39.
[http://dx.doi.org/10.1007/s00122-010-1408-4] [PMID: 20664999]

[10] Kissing Kucek L, Veenstra LD, Amnuaycheewa P, Sorrells ME. A Grounded Guide to Gluten: How Modern Genotypes and Processing Impact Wheat Sensitivity. Compr Rev Food Sci Food Saf 2015; 14: 285-302.
[http://dx.doi.org/10.1111/1541-4337.12129]

[11] von Hertzen L, Hanski I, Haahtela T. Natural immunity. Biodiversity loss and inflammatory diseases are two global megatrends that might be related. EMBO Rep 2011; 12(11): 1089-93.
[http://dx.doi.org/10.1038/embor.2011.195] [PMID: 21979814]

[12] Ng M, Fleming T, Robinson M, *et al.* Global, regional, and national prevalence of overweight and obesity in children and adults during 1980-2013: a systematic analysis for the Global Burden of Disease Study 2013. Lancet 2014; 384(9945): 766-81.
[http://dx.doi.org/10.1016/S0140-6736(14)60460-8] [PMID: 24880830]

[13] International Diabetes Federation. IDF Diabetes Atlas. 6th ed., Brussels: International Diabetes Federation 2013.

[14] Leslie M. The livers weighty problem. Science 2015; 349(6243): 18-20.
[http://dx.doi.org/10.1126/science.349.6243.18] [PMID: 26138960]

[15] Good AG, Beatty PH. Fertilizing nature: a tragedy of excess in the commons. PLoS Biol 2011; 9(8): e1001124.
[http://dx.doi.org/10.1371/journal.pbio.1001124] [PMID: 21857803]

[16] Gassmann AJ, Petzold-Maxwell JL, Keweshan RS, Dunbar MW. Field-evolved resistance to Bt maize by western corn rootworm. PLoS One 2011; 6(7): e22629.
[http://dx.doi.org/10.1371/journal.pone.0022629] [PMID: 21829470]

[17] Lu Y, Wu K, Jiang Y, *et al.* Mirid bug outbreaks in multiple crops correlated with wide-scale adoption of Bt cotton in China. Science 2010; 328(5982): 1151-4.
[http://dx.doi.org/10.1126/science.1187881] [PMID: 20466880]

[18] Baranski MR. Wide adaptation of Green Revolution wheat: international roots and the Indian context of a new plant breeding ideal, 1960-1970. Stud Hist Philos Biol Biomed Sci 2015; 50: 41-50.
[http://dx.doi.org/10.1016/j.shpsc.2015.01.004] [PMID: 25655670]

[19] Ceccarelli S. GMO, Organic Agriculture and Breeding for Sustainability. Sustainability 2014; 6: 4273-86.
[http://dx.doi.org/10.3390/su6074273]

[20] Baym M, Lieberman TD, Kelsic ED, *et al.* Spatiotemporal microbial evolution on antibiotic landscapes. Science 2016; 353(6304): 1147-51. [http://dx.doi.org/10.1126/science.aag0822] [PMID: 27609891]

[21] Ceccarelli S. Main stages of a plant breeding programme. In: Ceccarelli S, Guimaraes EP, Weltzien E, Eds. Plant Breeding and Farmer Participation. Rome: FAO 2009; pp. 63-74.

[22] Maredia MK, Raitzer DA. Estimating overall returns to international agricultural research in Africa through benefit–cost analysis: A "best-evidence" approach. Agric Econ 2010; 41: 81-100. [http://dx.doi.org/10.1111/j.1574-0862.2009.00427.x]

[23] Morris MA, Heisey PW. Estimating the benefits of plant breeding research: Methodological issues and practical challenges. Agric Econ 2003; 29: 241-52. [http://dx.doi.org/10.1111/j.1574-0862.2003.tb00161.x]

[24] Tripp R, Lowaars N, Van der Burg WJ, *et al.* Alternatives for seed regulatory reform. An analysis of variety testing, variety regulation and seed quality control. Agricultural Administration. London: Overseas Development Institute Network 1997; 69: pp. 1-25.

[25] Bishaw Z, Turner M. Linking participatory plant breeding to the seed supply system. Euphytica 2008; 163: 31-44. [http://dx.doi.org/10.1007/s10681-007-9572-6]

[26] Singh M, Malhotra RS, Ceccarelli S, *et al.* Spatial variability models to improve dryland field trials. Exp Agric 2003; 39: 151-60. [http://dx.doi.org/10.1017/S0014479702001175]

[27] Ceccarelli S. Plant breeding with farmers—A technical manual. Aleppo, Syria: ICARDA 2012.

[28] Cullis BR, Smith AB, Coombes NE. On the design of early generation variety trials with correlated data. J Agric Biol Environ Stat 2006; 11: 381-93. [http://dx.doi.org/10.1198/108571106X154443]

[29] ICARDA. ICARDA Annual Report. Aleppo, Syria: International Center for Agricultural Research in the Dry Areas 2006; p. vi+ pp 153. ISSN: 0254- 8313

[30] Ceccarelli S, Grando S, Winge T. Participatory barley breeding in Syria. In: Andersen R, Winge T, Eds. Realising farmers' rights to crop genetic resources: Success stories and best practices. Abingdon: Routledge 2013; pp. 97-116.

[31] Aw-Hassan A, Mazid A, Salahieh H. The role of informal farmer-to-farmer seed distribution in diffusion of new barley varieties in Syria. Exp Agric 2008; 44: 413-31. [http://dx.doi.org/10.1017/S001447970800642X]

[32] Li J, Lammerts van Bueren ET, Jiggins J, Leeuwis C. Farmers' adoption of maize (*Zea mays L.*) hybrids and the persistence of landraces in Southwest China: Implications for policy and breeding. Genet Resour Crop Evol 2012; 59: 1147-60. [http://dx.doi.org/10.1007/s10722-011-9750-1]

[33] Flower DJ. Physiological and morphological features determining the performance of the sorghum landraces of northern Nigeria. Exp Agric 1996; 32: 129-41. [http://dx.doi.org/10.1017/S0014479700026041]

[34] The African Center for Biodiversity. The expansion of the commercial seed sector in sub-Saharan Africa: Major players, key issues and trends 2015; 1-49.

[35] Sall S, Norman D, Featherstone AM. Quantitative assessment of improved rice variety adoption: The farmer's perspective. Agric Syst 2000; 66: 129-44. [http://dx.doi.org/10.1016/S0308-521X(00)00040-8]

[36] Jalleta T. Participatory evaluation of the performance of some improved bread wheat (*Triticum aestivum*) varieties in the Jijiga Plains of eastern Ethiopia. Exp Agric 2004; 40: 89-97. [http://dx.doi.org/10.1017/S0014479703001479]

[37] Byerlee D, de Polanco EH. Farmers' stepwise adoption of technological packages: Evidence from the Mexican Altiplano. Am J Agric Econ 1968; 68: 519-27. [http://dx.doi.org/10.2307/1241537]

[38] Weltzien E, Witcombe JR. Pearl millet improvement for the Thar Desert, Rajasthan, India. Proceedings of the XII Eucarpia Congress. 23-6.

[39] Ceccarelli S, Acevedo E, Grando S. Breeding for yield stability in unpredictable environments: Single traits, interaction between traits, and architecture of genotypes. Euphytica 1991; 56: 169-85. [http://dx.doi.org/10.1007/BF00042061]

[40] Saade M, Nassif F, Amri A, El Baghati H. Constraints to the adoption of barley varieties in Morocco Expert Report. Aleppo, Syria: ICARDA 1993.

[41] Brush SB. *In situ* conservation of landraces in centers of crop diversity. Crop Sci 1995; 35: 346-54. [http://dx.doi.org/10.2135/cropsci1995.0011183X003500020009x]

[42] Legg JP, Thresh JM. Cassava mosaic virus disease in East Africa: a dynamic disease in a changing environment. Virus Res 2000; 71(1-2): 135-49. [http://dx.doi.org/10.1016/S0168-1702(00)00194-5] [PMID: 11137168]

[43] Hossain M, Jaim WM, Paris TR, Hardy B, Eds. Adoption and diffusion of modern rice varieties in Bangladesh and eastern India. Los Baños, Philippines: International Rice Research Institute 2012.

[44] Li J, Lammerts van Bueren ET, Jiggins J, Leeuwis C. Farmers' adoption of maize (*Zea mays* L.) hybrids and the persistence of landraces in Southwest China: Implications for policy and breeding. Genet Resour Crop Evol 2012; 59: 1147-60. [http://dx.doi.org/10.1007/s10722-011-9750-1]

[45] Reguieg MM, Labdi M, Benbelkacem A, *et al.* First experience on participatory barley breeding in Algeria. J Crop Improv 2013; 27: 1-18. [http://dx.doi.org/10.1080/15427528.2013.794756]

[46] Yahiaoui S, Cuesta-Marcos A, Gracia MP, *et al.* Spanish barley landraces outperform modern cultivars at low-productivity sites. Plant Breed 2014; 133: 218-26. [http://dx.doi.org/10.1111/pbr.12148]

[47] Morris ML, Dubin JH, Pokhrel T. Returns to wheat research in Nepal. CIMMYT Economics Working Paper. Mexico: International Maize and Wheat Improvement Centre (CIMMYT) 1992; pp. 92-04.

[48] Witcombe JR, Packwood AJ, Raj AG, Virk DS. The extent and rate of adoption of modern cultivars in India. In: Witcombe JR, Virk DS, Farrington J, Eds. Seeds of choice: Making the most of new varieties for small farmers. London: Oxford and IBH Publishing Co., New Delhi, and Intermediate Technology Group 1998; pp. 53-68. [http://dx.doi.org/10.3362/9781780445854.005]

[49] Adesina AA, Baidu-Forson J. Farmers' perceptions and adoption of new agricultural technology: Evidence from analysis in Burkina Faso and Guinea, West Africa. Agric Econ 1955; 13: 1-9. [http://dx.doi.org/10.1016/0169-5150(95)01142-8]

[50] Falconer DS. Introduction to quantitative genetics. 2nd ed., London: Longmann Group Ltd. 1981.

[51] Eberhart SA. Factors affecting efficiencies of breeding methods. Afric Soils 1970; 15: 669-80.

[52] Allen FL, Comstock RE, Rasmusson DC. Optimal environments for yield testing. Crop Sci 1978; 18: 747-51. [http://dx.doi.org/10.2135/cropsci1978.0011183X001800050013x]

[53] Calhoun DS, Gebeyehum G, Miranda A, *et al.* Choosing evaluation environments to increase wheat grain yield under drought conditions. Crop Sci 1994; 34: 673-8. [http://dx.doi.org/10.2135/cropsci1994.0011183X003400030014x]

[54] Ceccarelli S. Specific adaptation and breeding for marginal conditions. Euphytica 1994; 77: 205-19.

[http://dx.doi.org/10.1007/BF02262633]

[55] Cooper M, Stucker RE, DeLacy IH, Harch BD. Wheat breeding nurseries, target environments, and indirect selection for grain yield. Crop Sci 1997; 37: 1168-76. [http://dx.doi.org/10.2135/cropsci1997.0011183X003700040024x]

[56] Atlin GN, Cooper M, Bjørnstad A. A comparison of formal and participatory breeding approaches using selection theory. Euphytica 2001; 122: 463-75. [http://dx.doi.org/10.1023/A:1017557307800]

[57] Ceccarelli S. Adaptation to low/high input cultivation. Euphytica 1996; 92: 203-14. [http://dx.doi.org/10.1007/BF00022846]

[58] Varshney RK, Terauchi R, McCouch SR. Harvesting the promising fruits of genomics: applying genome sequencing technologies to crop breeding. PLoS Biol 2014; 12(6): e1001883. [http://dx.doi.org/10.1371/journal.pbio.1001883] [PMID: 24914810]

[59] Goulden CH. Problems in plant selection. Proceedings of the Seventh International Genetics Congress. 132-3.

[60] Ribaut JM, Hoisington D. Marker-assisted selection: New tools and strategies. Trends Plant Sci 1998; 3: 236-9. [http://dx.doi.org/10.1016/S1360-1385(98)01240-0]

[61] Ribaut JM, de Vicente MC, Delannay X. Molecular breeding in developing countries: challenges and perspectives. Curr Opin Plant Biol 2010; 13(2): 213-8. [http://dx.doi.org/10.1016/j.pbi.2009.12.011] [PMID: 20106715]

[62] Alpuerto VE, Norton GW, Alwang J, Ismail AM. Economic impact analysis of marker-assisted breeding for tolerance to salinity and phosphorous deficiency in rice. Rev Agric Econ 2009; 31: 779-92. [http://dx.doi.org/10.1111/j.1467-9353.2009.01466.x]

[63] Windhausen VS, Wagener S, Magorokosho C, *et al.* Strategies to subdivide a target population of environments: Results from the CIMMYT-led maize hybrid testing programs in Africa. Crop Sci 2012; 52: 2143-52. [http://dx.doi.org/10.2135/cropsci2012.02.0125]

[64] Simmonds NW. Selection for local adaptation in a plant breeding programme. Theor Appl Genet 1991; 82(3): 363-7. [http://dx.doi.org/10.1007/BF02190624] [PMID: 24213182]

[65] Pederson DG, Rathjen AJ. Choosing trial sites to maximize selection response for grain yield in spring wheat. Aust J Agric Res 1981; 32: 411-24. [http://dx.doi.org/10.1071/AR9810411]

[66] Bänziger M, Cooper M. Breeding for low input conditions and consequences for participatory plant breeding: Examples from tropical maize and wheat. Euphytica 2001; 122: 503-19. [http://dx.doi.org/10.1023/A:1017510928038]

[67] Kempton RA, Gleeson AC. Unreplicated trials. In: Kempton RA, Fox PN, Eds. Statistical methods for plant variety evaluation. London: Chapman & Hall 1997; pp. 86-100. [http://dx.doi.org/10.1007/978-94-009-1503-9_6]

[68] Bellon MR. Crop research to benefit poor farmers in marginal areas of the developing world: A review of technical challenges and tools. CAB Rev Persp Agric Vet Sci Nutr Nat Res 2006; 1: 1-11.

[69] Ceccarelli S, Guimaraes EP, Weltzien E, Eds. Plant breeding and farmer participation. Rome: FAO 2009; p. 671.

[70] Halewood M, Deupmann P, Sthapit B, *et al.* Participatory plant breeding to promote Farmers' Rights. Biodiversity International, Rome, Italy 2007; 7: 1-7.

[71] Ceccarelli S, Grando S, Tutwiler R, *et al.* A Methodological Study on Participatory Barley Breeding. I.

Selection Phase. Euphytica 2000; 111: 91-104.
[http://dx.doi.org/10.1023/A:1003717303869]

[72] De Schutter O. Final report: The transformative potential of the right to food. Report of the special rapporteur on the right to food. United Nations General Assembly, A/HRC/25/57: Olivier De Schutter 2014; p. 28.

[73] FAO. The International Treaty on Plant Genetic Resources for Food and Agriculture, FAO, *via* delle Terme di Caracalla, 00153 Rome, Italy 2009.

[74] Ceccarelli S, Grando S. Decentralized-Participatory Plant Breeding: An Example of Demand Driven Research. Euphytica 2007; 155: 349-60.
[http://dx.doi.org/10.1007/s10681-006-9336-8]

[75] Almekinders C, Hardon J. Bringing Farmers Back into Breeding. Experiences with Participatory Plant Breeding and Challenges for Institutionalization. Agromisa Special 5, Agromisa, Wageningen. the Netherlands 2006; p. 140.

[76] Ceccarelli S. Efficiency of plant breeding. Crop Sci 2015; 55: 87-97.
[http://dx.doi.org/10.2135/cropsci2014.02.0158]

[77] Galiè A. Governance of seed and food security through participatory plant breeding: Empirical evidence and gender analysis from Syria. Nat Resour Forum 2013; 37: 31-42.
[http://dx.doi.org/10.1111/1477-8947.12008]

[78] Foresight The Future of Food and Farming Executive Summary. London: The Government Office for Science 2011; p. 211.

[79] Ceccarelli S, Galiè A, Grando S. Participatory breeding for climate change-related traits. In: Kole C, Ed. Genomics and breeding for climate-resilient crops. Berlin: Springer-Verlag 2013; Vol. 1: pp. 331-76.
[http://dx.doi.org/10.1007/978-3-642-37045-8_8]

[80] Rahmanian M, Salimi M, Razavi K, Haghparast R, Ceccarelli S. Living Gene Banks in Farmers Fields. Farming Matters 2014 March; 12-5.

[81] Campanelli G, Acciarri N, Campion B, *et al.* Participatory Tomato Breeding for Organic Conditions in Italy. Euphytica 2015; 204: 179-97.
[http://dx.doi.org/10.1007/s10681-015-1362-y]

[82] Godfray HC, Beddington JR, Crute IR, *et al.* Food security: the challenge of feeding 9 billion people. Science 2010; 327(5967): 812-8.
[http://dx.doi.org/10.1126/science.1185383] [PMID: 20110467]

[83] Ceccarelli S. Participatory barley breeding in Syria: policy bottlenecks and responses. Farmers' Crop Varieties and Farmers' Rights Challenges in Taxonomy and Law. Oxon, UK: Routledge, Abingdon 2016; pp. 84-96.

[84] Suneson CA. An Evolutionary Plant Breeding Method. Agron J 1956; 48: 188-91.
[http://dx.doi.org/10.2134/agronj1956.00021962004800040012x]

[85] Campanelli G, De Vita P, Fusari F, *et al.* Miglioramento genetico partecipativo ed evolutivo su zucchino e frumento duro in agricoltura biologica. Dal Seme 2014; 4: 63-5.

[86] Howard PH. Visualizing Consolidation in the Global Seed Industry: 1996–2008. Sustainability 2009; 4: 1266-87.
[http://dx.doi.org/10.3390/su1041266]

[87] Fuglie KO, Heisey PW, King JL, *et al.* Research Investments and Market Structure in the Food Processing, Agricultural Input, and Biofuel Industries Worldwide. ERR-130 US Dept of Agriculture, Econ Res Serv 2011 December; 147.

[88] Martens J, Seitz K. Philanthropic Power and Development. Who shapes the agenda? Brot für die Welt/Global Policy Forum/MISEREOR, Aachen/Berlin/Bonn/New York. 2015 November;

[89] Gilbert N. Frugal Farming. Nature 2016; 533: 308-10. [http://dx.doi.org/10.1038/533308a] [PMID: 27193660]

[90] Jansen HG, Walker TS, Barker R. Adoption ceilings and modern coarse cereal cultivars in India. Am J Agric Econ 1990; 72: 653-63. [http://dx.doi.org/10.2307/1243035]

[91] Kelley TG, Parthasarathy Rao P, Weltzien E, Purohit ML. Adoption of improved cultivars of pearl millet in an arid environment: Straw yield and quality considerations in western Rajasthan. Exp Agric 1996; 32: 161-71. [http://dx.doi.org/10.1017/S0014479700026077]

[92] Witcombe JR, Joshi A, Joshi KD, Sthapit BR. Farmer participatory crop improvement. I. Varietal selection and breeding methods and their impact on biodiversity. Exp Agric 1996; 32: 445-60. [http://dx.doi.org/10.1017/S0014479700001526]

[93] Ceccarelli S, Grando S, Capettini F. Barley breeding history, progress, objectives, and technology, Near East, North and East Africa and Latin America. Barley: Production, improvement and uses. Wiley-Blackwell: Ames (Iowa) 2011.

[94] Sthapit BR, Joshi KD, Witcombe JR. Farmer participatory crop improvement. III. Participatory plant breeding, a case study for rice in Nepal. Exp Agric 1996; 32: 479-96. [http://dx.doi.org/10.1017/S001447970000154X]

[95] Pandey S, Gauchan D, Malabayabas M, Eds. Patterns of adoption of improved rice varieties and farm-level impacts in stress-prone rainfed areas in South Asia. Los Baños, Philippines: International Rice Research Institute 2012. p. 318.

CHAPTER 3

Genetic Tools for Crop Improvement: Past, Present, and Future

Mario Motto[1,*]

Fondazione Istituto Tecnico Superiore per le nuove Tecnologie della Vita, Bergamo, Italy

Abstract: To respond to the contemporary increasing demand for food, feed, and feedstocks for bioenergy and bio-factory applications, there is an urgent need to improve agricultural plant production and quality-related features. Genetics and plant breeding are powerful strategies for increasing crop productivity. The objectives of this chapter are devoted to summarize i) historical developments of applied plant genetics or plant breeding, ii) fundamental principles affecting the current methods of molecular plant breeding, and iii) key factors that will affect the use of molecular breeding in crop improvement procedures. Additionally, the chapter takes a close look at the current exploitation of molecular plant breeding for the discovery of genes and their functions. These topics would disclose new perspectives for crucial plant biology research that will be beneficial to ensure food security to the rapidly growing world population and to sustainable agricultural systems. Moreover, they would open new doors to improve feedstocks to sustain non-food applications for the synthesis of high-added value products.

Keywords: Applied plant genetics, Biotech crops, Crop improvement, Genetic and genomic tools, Genetic engineering, Green revolution, Plant breeding.

INTRODUCTION

A recent report on the world population prospect, elaborated by the United Nations, indicates that people is projected to advance, on a global scale, from the current 7.2 billion to 9.6 billion by the middle of this century [1] (Fig. **1**). This trend will approach 11.0 billion by 2100.

To accommodate the additional demand for food and changing diets in developing nations, it is needed to supplement the global agricultural performance by approximately 100% by mid-century: namely doubling crop yields by 2050 [2]. The increased demand for crop and land derives not exclusively from the growing population and wealth, but also for ensuring non-food applications -*e.g.* energy

* **Corresponding author Mario Motto:** Fondazione Istituto Tecnico Superiore per le nuove Tecnologie della Vita, Bergamo, Italy; Tel: +39 0350789106; Fax: +39 0350789107; E-mail: mariomotto@libero.it

Roberto Pilu and Giuseppe Gavazzi (Eds.)

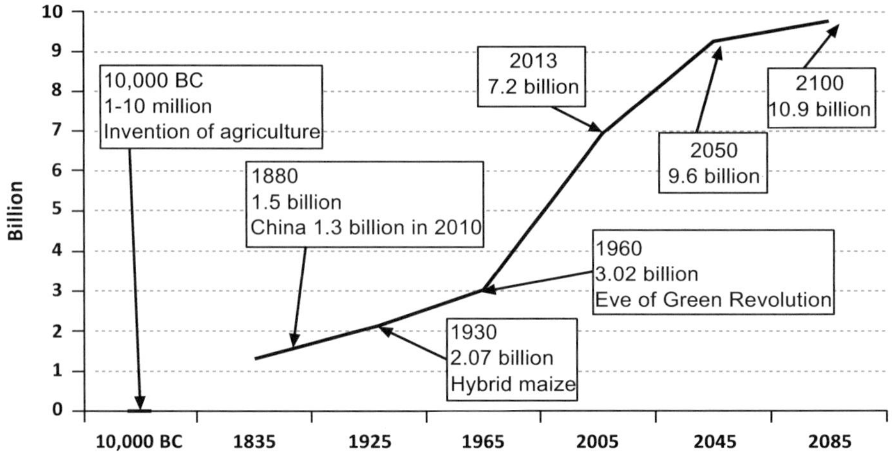

Fig. (1). U.N estimations of human population growth to 2100. Notably, from 1960 to presently, the human population has more than doubled to reach 7.2 billion people; by 2050, the population is forecasted to increase to approximately 9.6 billion and reach 10.9 billion by 2100. (Modified from Popp *et al.*).

access and chemicals- and climate changes. Moreover, it is worth noting that globally 805 million people are currently persistently starved and 2 billion are malnourished [3]. To double food yield by 2050, Ray and coworkers [4], have estimated that crop yield should increase at a rate of 2.4% per year. However the contemporary growth rates in global crop productivity of all major crops (*i.e.* maize, wheat, rice, soybean, roots and tubers) is declining from 2.5% and higher, since the mid-1990s, to less than 1.5% today [5]. This trend points out that the population growth rates have globally bypassed the linear rate of increases in food. In this scenario, raising crop yields to meet the expansion of food demands, rather than extend land for agriculture uses, has been highlighted as the most sustainable solution to meet this objective [4]. As a result, there is a pressing requirement to amplify agricultural productivity. It is also clear, at the same time, the necessity to diminish the contradictory influence of agriculture on the ecosystem stability: shortly a more sustainable and environmental friendly farming systems that will increase production [6, 7]. This may also be accomplished by considering, simultaneously, the tolerance to abiotic and biotic stress factors, the efficiency of nutrient and water use inputs, and the development of crops and cultivars having a broad range of end uses. The vast scientific progress in the last decades may help to accelerate these new challenges. In particular, genetic improvement of crops may assist to give a solution to this multifaceted target [8]. Accordingly, the production and diffusion of high-yielding varieties (HYVs) and the best achievements of biotechnology projects (*e.g.* pertinent Genetic Modified, GM, and/non-GM features), cultivated on the 1.5 billion hectares of arable land worldwide, will provide, in the near future, a fundamental contribution to sustainable agriculture. In addition, genetic crop interventions are an appealing goal. This approach is particularly ascribable to the potential of fast dissemination of the improved crops- that is seed distribution- and to their contribution in solving diverse features of contemporary crop biology. In this chapter I have briefly summarized i) historical developments of applied plant genetics or plant breeding, ii) fundamental principles affecting the current methods of molecular plant breeding, and iii) key factors that would sway the application of molecular breeding in crop improvement procedures. Additionally, in this chapter I have highlighted how the exploitation of molecular plant breeding is currently contributing to the discovery of genes and their functions. It is thereby predicted that these findings will open new perspectives in contemporary plant biological sciences.

CLASSICAL PLANT BREEDING

There are several approaches for increasing food production. These include: expanding the arable cropland, improving agronomic methods, extending the use of mechanization, as well as making perfect the supply of chemical fertilizers, and pesticides. However, the simple most agreeable factor in enhancing plant productivity is the genetic improvement of the crop plants themselves [9]. Specifically, plant breeding has had a dominant role to increase and safeguard potential, harvestable, and commercial productions. This scope was accomplished, *via* selecting for yield *per se*, addressing crop quality attributes, and developing plants exhibiting tolerance/resistance to disease and pest agents. It has also contributed timely to farmers by affording the best seeds of new developed HYVs. This has been implemented, in addition to higher yield, with other profitable features, able to upgrade farming incomes and sustainability. In the following sections, it is given a brief overview of genetic improvement from agricultural invention to present. Even though plant breeding emerged around 10-12,000 years ago, over the course of the human history, the general perception of genetic technology concepts is a relatively recent phenomenon (Fig. **2**).

Domestication and Empirical Plant Breeding

For 10-12,000 years, humankind has modified the genetic architecture of crop plants, initially *via* simple domestication (*i.e.* selection of plants having desirable traits to humans) and, in the last 2-300 years, employing more refined procedures.

For the domestication process, different plant species were selected for peculiar characteristics to satisfy human needs: cereals and pulses for their seeds, other plants for tubers, fruits or leaves. Evidence indicates that domestication or its syndrome incorporate i) combinations of various distinct attributes including: seed retention (non-shattering), enhanced fruit or seed size, alterations in branching and stature of plants, modifications in reproductive strategy, increased harvest index, and variations in secondary metabolite, ii) these features have permeated crop plants with uniformity, predictability, and greater performances in productivity [10]. These outcomes have also suggested that i) domestication and cultivation of plants are inextricably linked and ii) domestication attributes, which made some plants more attractive to the early plant farmers or domesticators, are governed by the expression of few genes. This is an event extremely vital to the early farmers to obtain a very quick selection and/or fixing those attributes within plant populations. Moreover, the history of plant domestication serves as an example to make clear how selection had altered the shape of living organisms

10,000BC-4,000BC -	Introduction of agriculture (plant selection)
100 BC -	First grafting application
1694-	Discovery of sexual reproduction in plants
1859-	Darwin outlines principles on origin of species *via* natural selection
1866 -	Mendel promotes a scientific revolution: birth of Genetics or basic principles of inheritance
1876 -	First intergeneric cross (wheat x rye) → triticale
1909 -	First protoplast fusion
1923 –	Sax performs studies on quantitative trait loci (QTLs) based on pigmentation and coloration traits of seeds in beans
1927 -	X-ray mutation breeding starts
1930 -	Beginning of hybrid maize production in the US
1950 -	McClintock outlines that genomes contain transposable genetic elements, able to restructure chromosomes and alter patterns of gene expression
1967 -	Plant regeneration from isolated cells
1968 -	Discovery of restriction enzymes able to digest specific DNA fragments
1973 -	First recombinant DNA molecules
1983 -	First GM plant (tabacco)
1994 -	First GM food approval
2000 -	Completion of whole genomes of plants (*e.g., Arabidopsis thaliana*)
2011 -	Large-scale second and third-generation sequencers permit low eukaryotic genomes to be analysed speedily (few hours).
2012 -	Technologies are used to alter temporal and spatial expression of genes in genome engineering and auxiliary gene removal
2013…	Extensive use of large-scale sequencing, macro- and micro-synteny, associative mapping, QTL cloning, molecular markers for genome-wide selection, wide-scale phenotyping, expansion of 'OMICS' use, multifold phenotypic features, broad use of bioinformatics, specialty GM crops, CRISPR-Cas9 technology applications

Fig. (2). Historical benchmarks related to key advancements in genetic tools useful for crop improvement.

Recent molecular studies, on the identification of genes affecting crop domestication and diversification, performed by analyzing diverse set of plant species, have advanced our knowledge of the genetic basis of the evolution of domesticated crops [11, 11a]. Notably, polyploidy is commonly believed to be associated with the domestication process. This is attributable to its concurrence with agriculturally favorable attributes and its widespread among the major plant crops. Evidence has suggested that the genetic consequences of polyploidy have provided genetic preconditions for fruitful domestication on several of these plants to emerge as crops. Other information showed that, out of nine domestication genes, eight are encoding transcriptional regulators. This revealed that transcription factors (TFs) play a central role in the domestication process. Examples in this field include: rice shattering genes *sh4* – a Myb TF- and *qSH1* –homeodomain motif- , and *Rc* – containing a βHLH motif- for seed color; maize *teosinte branched1 (tb1)*- that is a plant-specific TCP (TEOSINTE BRANCHED1/CYCLOIDEA/PROLIFERATING CELL NUCLEAR ANTIGEN FACTOR) member of a family of DNA binding proteins- involved in plant and inflorescence architecture, and *teosinte glume architecture*1(*tga1*- that is likely to encode a regulatory protein, belonging to the SBP (SQUAMOSA-pROMOTER BINDING PROTEIN)-domain family of TF- involved in seed casing; *AP2 (APETALA2)*-like wheat gene *Q* - a member of a large multi-gene family of TFs- conferring the free-threshing character [12]. Accordingly, these loci play a cardinal function during the plant life cycle, being key regulators of several developmental processes, such as inflorescence structure. Taken together from these investigations it was argued that, because domestication resulted relatively in a very quick life-time, on an evolutionary scale, mutations with great effects were easily detected and selected, and therefore easily fixed in plant genomes [13].

In agreement with the initially unconscious selection, farmers have finally initiated to systematically select plants consciously. This procedure was based on visible features - *e.g.* larger and more abundant kernels, types that did not shed from the parent plants, or those that were most responsive, for instance by germinating soon after planting [10, 14]. Furthermore, it has appeared gradually plain that not only observable plant attributes, such as grain size and quantity, are affected by selection, but also other characteristics such as tolerance against pests and diseases, and taste. This form of intentional selection, entirely based on observable evaluations and not on scientific knowledge, resembles what the contemporary plant breeders have largely applied. Therefore, this technology of plant breeding is frequently referred to as pre-scientific or “empirical breeding”: the gene concept was not yet defined at that time [14].

Modern Crop Improvement

Although, plant breeding as a technology derived from simple activities of people belonging to the past, it achieved the status of science, after Mendelian genetic laws were corroborated at the beginning of 1900s [15]. Additionally, Mendel's findings have stimulated, in the 20th century, the foundations for further research on inheritance. Furthermore, they have promoted a profound educational effect by encouraging further research in the field of genetics and biology.

Professional Cross Breeding: Generation of HYVs

The rediscovery of Mendel's principles on Genetics and the reliable development, on large-scale, of scientific plant breeding elaborated on those laws, were of central relevance in developing crops. Most commercial varieties that have been planted in modern agriculture were produced by applying classical selective breeding methods. These are programs and procedures outlined to change the phenotypes in economically important species of plants- to suitable plant materials. These methods are named classical or conventional to make a distinction from others, like those exploiting genetic engineering tools.

Briefly, conventional breeding procedures or selective systems (*i.e.,* an experimental technique that increases the recovery of specific - frequently rare-genotypes) can be summarized into three principal activities related to: i) detection of attribute(s) that support genetic yield capacity; ii) evaluation of genetic variability and its origin for suitable feature(s) to assist the detection of parental genetic resources –that is parental identification; and iii) incorporation of genes, controlling the specific trait(s), into varieties equipped with other favorable attributes [16]. Moreover, a breeding program designed to improve specific features in a cultivar cannot be considered as a confined research activity, but demands to be connected to other areas of research concerning that crop and related crops.

Typically plant breeding programs are expensive and time consuming. However, regardless of this, in the brief history of scientific plant breeding a body of evidence has been reported to corroborate that the classical or traditional selection methods have been effective in bringing about genetic progress in crop plants across years. Particularly, these methods have greatly upgraded crop yields of the dominant cereals (*e.g.* maize, rice, and wheat), which globally provide two-thirds of the world's food energy intake, and other crops.

A superlative example to emphasize Mendelian inheritance is related to the rediscovery of heterosis or hybrid vigor. This is a familiar phenomenon, noted in progenies by crossing distinct strains of diverse species, confined populations, or selected progenies within species or populations, whose exploitation into breeding gave rise to a spectacular agronomic performance in maize [17]. This plant was, and is, both a biological curiosity and a central cereal crop. With the discovery of the inheritance principles, maize gave a pivotal contribution in supporting the genetic laws and in bridging knowledge between theory and application. Specifically, maize hybrids inherit their economically favorable features - *i.e.* greater yield, resistance to disease and pest agents, and environmental stress tolerance- from two genetically different parental inbred lines. As a result, F_1 crosses manifested a considerable increase in yield performances. This outcome has promoted, around the 1930s, an increase in hybrid seed sales, with a remarkable escalation following the introduction of double cross hybrids. The cultivation of maize hybrids in the United States (US) was unprecedented in the history of agricultural innovations: by 1965, over 95% of the total area in US was cultivated with hybrid varieties. Therefore, the development of hybrid varieties is frequently considered one of the most outstanding genetic achievements in agriculture [18]. Nevertheless, when maize hybrids were first recommended in the U.S. cultivation, they experienced criticisms and resistances from farmers analogous to those raised at contemporary GM crops: the hybrids were laborious to put on a public performance and agriculture experimental stations were not interested to develop this type of activity. Furthermore, for this crop the substitution of double cross with single cross hybrids additionally improved yields in the 1960s to present days (Fig. **3**). Notably, the main findings emerging from studies [19 - 21] on the assistance of breeding to favor progresses in maize performance have indicated that i) grain yield is improved by approximately 100 kg ha^{-1} $year^{-1}$ or 2% $year^{-1}$ from the introduction of large-scale hybrids by growers in the late 1930s up to the first decade of the 21st century; ii) genetic improvements is accountable for about 50-60% of the on-farm progresses, while innovations in cultural managements are responsible for the rest; iii) increases in grain yield performances are centrally related with an enhanced kernel number, a persistently improved stay green, and a more protracted grain-filling duration; iv) stress tolerance is significantly enhanced: newer hybrids out-yielded the older hybrids not exclusively in high productive environments, but also when experimental trials are conducted under specific abiotic stress conditions (*e.g.* heat and drought) or under infestation of biotic stress agents (*e.g.* pathogens or pest insects); v) performance in actual heterosis for grain yield is improved modestly over the years, particularly when experimental trials are subjected to abiotic stress treatments: the increase is significantly inferior than total genetic progress.

Similarly to maize, yield in other crops, over the last sixty years, was increased steadily, as a consequence of improving their genetic ability. For example, total yields for rice, wheat, and soybeans, the primary resources, with maize, for human nutrition on a global scale, raised very quickly from 1961 to 2000: wheat grain yield performances increased by 2.6-fold, whereas rice and soybean grain productions grew, respectively, by 2.2- and 2.0-fold [22]. Modeling studies have also shown that crop productions of the main staples nourishing the globe would result 19.5-23.5% lesser at the beginning of the current millennium apart from intentional breeding efforts [23].

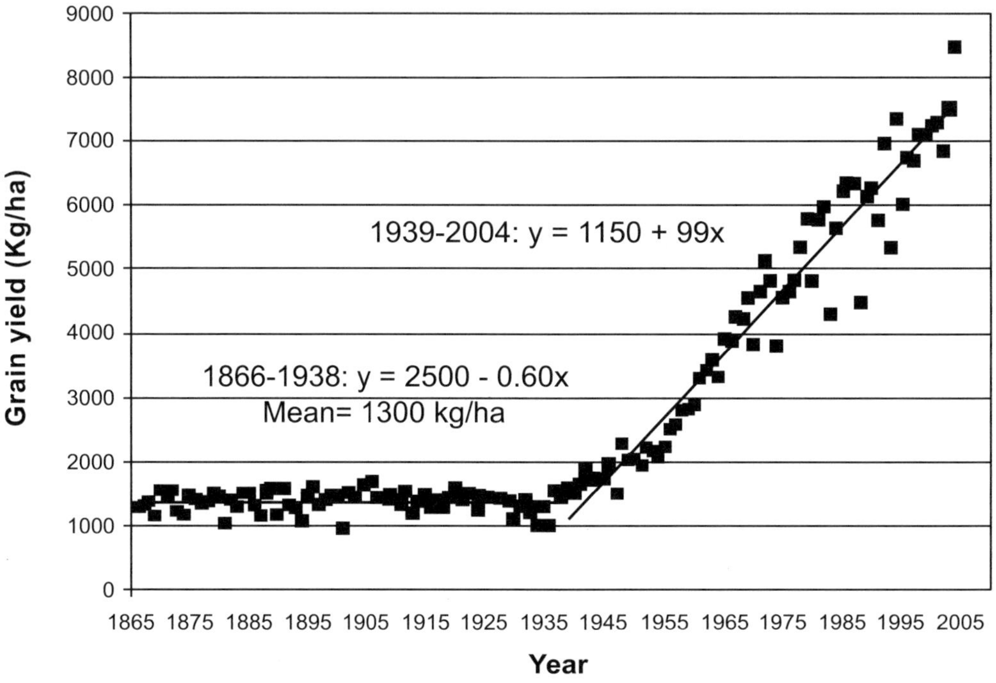

Fig. (3). Maize average grain yield in US from 1865 to 2004 in kg ha–1 (0% moisture) and linear regression of grain yield across years from 1866 to 1938 and from 1939 to 2004. Data compiled by the USDA. (From Tollenaar and Lee, 2006).

Although several other elements -*e.g.* availability of affordable mineral nitrogen (N) fertilizers, better pest and weed control, more widespread availability of irrigation, improved agronomic managements- have contributed to these gains in yield performance, experimental data have highlighted that genetic technology is responsible for at least 50% of the yield increases in some of the major crops currently cultivated [23, 24].

Green Revolution and Other Breeding Tools

The technological progresses that provided the outstanding goals realized in global food production, during the last 60 years, currently referred to as "Green Revolution", was related with the cultivation of HYVs of wheat and rice [23, 25]. These cultivars are reactive to fertilizer supplies, are tolerant to lodging, and their production potentiality is 2-3-fold higher that of cultivars planted before the Green Revolution. Moreover, the same cultivars brought about diversified resistances to disease and insect injuries and thereby exhibited a superior stability in production.

This technology was elaborated by Norman Borlaug, after the Rockefeller Foundation and Mexican government started in 1943 to collaborate to establish a research institution in Mexico that later became the International Maize and Wheat Improvement Center (CIMMYT) [25]. The target of this new institution was to improve local varieties of outstanding food crops to promote food security in Mexico and other developing countries. To Dr. Borlaug, generally credited to be the father of the Green Revolution, was assigned in 1970 the Nobel Peace Prize, for his contributions to human society. In this respect, it must be underlined that in his Nobel Peace Prize Acceptance Speech -December the 11th, 1970- he noted that: "*The green revolution has won a temporary success in man's war against hunger and deprivation; it has given man a breathing space. If fully implemented, the revolution can provide sufficient food for sustenance during the next three decades. But the frightening power of human reproduction must also be curbed; otherwise the success of the green revolution will be ephemeral only*" [26]. The concreteness of this forewarning was recognized at the beginning of 2008, when the commercial costs of wheat and maize doubled and that of rice tripled, promoting food turbulences in several developing nations.

Although, the exploitation of Green Revolution technology was helped by innovations in irrigation devices, accessibility of mineral fertilizers, and favorable government policies, its success was mainly due to the cultivation of HY semi-dwarf varieties of wheat and rice in association with the application of generous volumes of N fertilizers [27]. Notably, the varieties of wheat and rice cultivated before to the Green Revolution were high in stature, leafy with fragile stalks, and exhibited a harvest index of 0.3 - *i.e.* 30% grain and 70% straw. In fact, the stalks of toll wheat and rice varieties were not vigorous enough to withstand with the massive grain yield performance of the HYVs: consequently, plants were subjected to lodging which ultimately often result in a significant yield reduction. For instance, the presence of wheat mutant dwarfing alleles, at *Reduced height-1* (*Rht-B1* and *Rht-D1*) loci, have promoted the significant raises in global grain productions in the 1960s, attributable to progresses in both lodging tolerance and harvest index (~ 0.5) [28]. On account of this effect, *Rht-1* dwarfing alleles are currently largely employed in developing contemporary wheat varieties. Further research has shown that the wheat *Rht-1* alleles are encoding a mutant of DELLA proteins (a subfamily of the GRAS TF-family) that provides semi-dominant gibberellin (GA) insensitive dwarfism [29]. The GA hormones act throughout the life cycle of plants, influencing several aspects of plant growth and development, including stem elongation. Similarly several short straw varieties of rice which have a mutation in the *Semi-Dwarf-1(SD*1) locus, encoding for GA 20-oxidase-2, are impaired in GA biosynthesis. Evidence indicated that in barley this gene - termed *Denso* or *Semi-dwarf 1 (Sdw1*)- is also mutated in the majority of cultivars currently cultivated [30]. In Fig. (**4**) is given an example of different plant strains of rice introduced during the Green Revolution.

Fig. (4). Visual representation of distinct plant strains of rice. Left, tall conventional plant strain. Centre, improved high-yielding, high-tillering plant strains. Right, new plant strains ('Super Rice') having low tillering but robust stems, and possessing a superior number of grains per panicle. (Reproduced from Khush, 2001 [34]).

Owing to the Green Revolution, the HYVs of wheat and rice are currently cultivated on 84% and 74% of agriculture land, respectively. As a consequence, rice and wheat global yields were expanded from 127 million tons to above 760 million tons during the time period included from 1960s to 2000 [31]. On the other hand of the balance, the improved crop varieties have additionally expanded the quantities of fertilizers, pesticides, and water employed in agriculture, and generated a substantial reduction in the genetic diversity of crop plants [32]. It must be, however, emphasized that the specialization on the best performing species and varieties has profoundly increased the productivity of agricultural fields. This has reduced the necessity for agricultural expansion of land areas, protecting vulnerable zones from being converted to cropland and permitting biodiversity to persist in natural ecosystems [33].

Another technique, frequently used in breeding projects, for plant improvement was induced mutagenesis [35]. This task was achieved through chemical or radiation treatment. Although mutation breeding is no as familiar as the most recently refined method of transformation, it has been largely applied. Induced mutation is therefore an established crop improvement approach: there are more than 3, 200 mutant crop varieties and ornamental plants, catalogued in the FAO/ IAEA Mutant Varieties Database, being globally cultivated [36].

A recent innovation on mutagenesis, named Targeted Induced Local Lesions in Genomes (TILLING), simplify the detection and classification of gene variants underlying commercially relevant features [37]. This technique is an aggregation of typical chemical mutagenesis combined with high-throughput (HTP) molecular biology procedures which are devoted to detect valuable mutant variants. This strategy has been especially exploited for the improvement of under-investigated crop species. For instance, melon mutants that possess an increased shelf life or exhibit unisexual flowers, both of which are able to improve productivity, were isolated by TILLING approaches [38, 39]. It is predicted that this method will become an integral part of breeding programs, whereas more effective and distinct variant mutations may be identified by exploiting TILLING stocks.

A range of techniques - *e.g.* micro-propagation, haploid production, protoplast formation, haploid production, embryo and apical cultures, somatic embryogenesis- under the caption of tissue culture and interspecific hybridization, were established in the 1960s and aggregated to the collection of tools useful in crop improvement [40]. These devices and technologies have enabled to achieve considerable progresses in crop improvement during the last twenty years, thanks, in particular, to the development of biotechnology. Moreover, it has been reported that, although plant cell suspension cultures, producing phytochemicals, exhibit various benefits over plant cultivated in fields, have been not yet largely exploited. This event is likely attributable to scarce information concerning plant secondary metabolism and its *in vitro* control [41]. However, in this respect a positive example is represented in the case of *Lavandula vera* MM cell suspension [42]. For this plant a successful bioactive compound – that is rosmarinic acid- was obtained by the so-called "green cell factories" technology.

Transgenic Breeding

Biological scientists have established a range of technical protocols for *in vitro* manipulation of DNA, cells, and organisms. Recombinant DNA techniques and genetic engineering, including transgenic methodology for transferring extraneous genes into plant cells, have become available in the 1980s [43, 44]. Additionally, other devices and techniques were elaborated to introduce genes into plants *via* the adoption of the soil bacterium *Agrobacterium tumefaciens* - a natural vector to easily transports a fragment of DNA into a plant cell- or by a number of direct DNA introduction tools, such as particle bombardment *(i.e.* biolistic or gene gun), electroporation, and microinjection [45]. This combination of transfer methods and biological knowledge has permitted to transport a defined segment of DNA from the native or a related plant species or from a completely foreign entity inside practically any cultivated plant, generating what is commonly called as a "transgenic" plant.

Transgenic breeding is an attractive approach to plant breeders for generating completely new genetic variability that it is not possible to achieve *via* traditional methods. This tool has been the most rapid applied biotechnology in agricultural history: the global land area planted with genetically GM crops has raised more than 100-times, ranging from 1.7 million hectares in 1996 up to 180 million hectares in 2014 [46]. According to a recent report [46], in 2014 out of the 28 nations which cultivated biotech crops, 20 were developing and only 8 were industrial nations. The top 10 nations planting GM crops are situated in the Americas (USA, Canada, Argentina, Brazil, Uruguay, and Paraguay), Asia (China, India, and Pakistan), and South Africa. In the European Union (EU) countries, the area covered with GM crops is negligible, except Spain. This is mainly due to public acceptance issues and onerous regulatory legislations. From the previous report, it is also evident that the most popular GM crops are represented by the major commodity plant species, especially maize and soybean. The principal features that have been transferred into these crop plants are herbicide tolerance (HT) and insect resistance (IR). The last intervention is mainly attributable to distinct genes derived from the soil bacterium *Bacillus Thuringiensis* (*Bt*).

The leading technology is HT soybeans, which represent 50% of the total GM crop area and 82% of global soybean grain yield. GM maize is the second-most predominant crop and amounted for about 30% of the land grown worldwide with GM and 30% of maize yield on a global scale. GM maize comprises also HT and IR, partly as independent and partly as stacked technologies. In maize the *Bt* genes contrast efficiently the diffusion of different insect species (European corn borer, corn root-worm. and various stem-borers) [47, 48]. Interestingly, a GM maize with a 6-10% grain yield improvement, when grown under drought environments, it was recently provided by Monsanto [49]. This transgene exploited a cold shock gene (*cspB*) from *Bacillus subtilis*, and was commercialized since 2013. Other GM crop varieties, with sizeable area of cultivation shares, include cotton, canola, and sugar-beet. There are also a few other GM crops such as HT alfalfa and virus-resistant papaya, squash, and tomato that have been planted in specific countries, but they are currently covering only relatively modest areas.

A meta-analysis, recently reported by Klùmper and Qaim [50], has revealed that, on average, the current GM technology, mainly represented by HT and IR crops, has improved crop yields by approximately 21%. These yield improvements are not attributable to more extended genetic yield ability, but are due to a more powerful pest control that is reflected in a reduction of losses caused by those agents. Simultaneously, GM crops have diminished chemical pesticide applications by 37% and pesticide prices by 39%. In spite of this and of additional scientific efforts, to provide an objective view on this issue, conventional in comparison to biotech breeding is still an argument of discussion in the scientific and the civilian communities, especially in EU countries. Perhaps for GM crops disinformation or propaganda, motivated by political, commercial, and other interests for which scientific validity was at best of secondary relevance, seems to be more effective than scientific evidence.

CURRENT TOOLS FOR CROP IMPROVEMENT

From the beginning of agriculture, plant breeding programs have been mainly accomplished through methods of phenotypic selection. The assembling of HYVs *via* conventional breeding is a relatively laborious and time-consuming method. Notably, it takes 10-15 years from the first cross to the commercial release of a new variety. Therefore, technologies that are able to i) reduce the cost and time needed and ii) advance the probability of success in developing crop varieties with superior features, are central to achieve these goals. In this context, a number of molecular tools were provided and employed, over the last few decades, in plant breeding programs by using molecular biology technologies, particularly DNA markers, for plant improvement [8]. It is worth noting that these new tools have not modify the general concept of breeding new varieties, consisting of generating initial genetic variation and selecting the best performing candidates, for further testing in the field. However, the different methods used in biotechnology have contributed to broaden the efficiency of breeding programs.

A second sector for re-directing plant breeding in this century has been oriented to define objectives or traits to be selected for. These should include, in addition to the traditional objectives - *e.g.* crop yield and stability, tolerance to abiotic (nutrient and water uses) and biotic stresses - the following traits i) low-input eco-efficient farming systems, ii) improved nutritional quality attributes, and the multi-scope utilization of crop biomasses (*e.g.* bio-refining applications for fuels and chemicals). Plants may also be exploited as factories for the production of a wide range of proteins suitable for a multiplicity of applications -*e.g.* industrial enzymes for detergents, textile manufacturing, waste treatment, baking, proteins medically effective, including peptide hormones and therapeutic proteins like bioactive peptides and antibodies for disease therapies [51, 52].

New Breeding Technologies

Progresses in plant genetics and genomic tools are cooperating in speeding up gene discovery for the production of crop varieties. Notably, genomics - the study of an organism's complete DNA sequence- has been one of the most relevant instruments in biology and biosciences applications with a high power to change [53]. Sequencing of genome of organisms are central to i) gain insights into the roles of single genes and their connections, ii) define evolutionary similarities and processes, and iii) reveal earlier undefined mechanisms of control coordinating gene functions, with the inclusion of the most complex biological systems.

Several genomics technologies, such as high-throughput (HTP) sequencing strategies (*i.e.* next generation sequencing, NGS) [54, 55], have significantly reduced costs of sequencing, resulting in the sequencing of more than 100 plant genomes [56, 57]. More recently, Van Buren *et al.* [58], by running only single-molecule real-time (SMART) sequencing, have further improve NGS by obtaining, using the desiccation-tolerant plant *Oropetium thomaeum*, a 'near-complete' draft genome. This draft provides coverage without gaps over gene space in addition to intergenic sequences -*e.g.* telomeres, centromeres, transposons, and clusters of rRNA - that are usually not assembled in draft genomes up to now investigated.

Interestingly, the resulting DNA sequence information has also permitted to develop HTP marker genotyping that can be adopted not exclusively for constructing high density genetic and physical maps, but also for accelerating mapping of agronomic traits [59]. Notably molecular markers may be exploited routinely in crop breeding programs, for various scopes such as i) quick crop improvement and gene pyramiding or stacking, ii) genetic diversity analysis, iii) cultivar identification, iv) phylogenetic analysis, v) characterization of genetic resources, vi) and association with agronomic traits.

Typically, molecular genetic information about trait expression is used in marker-aided selection (MAS), sometimes also referred to as SMART (Selection with Markers and Advanced Reproductive Technologies) technology or precision breeding, transgenic breeding program, or in the more recent method, termed 'molecular strengthening' (MOST), as proposed by Hu and Luebbersted [60] in developing superior new crop varieties. According to the previous workers, the MOST technologies may be adopted to i) improve production stability, ii) monitor crop development and morphology, iii) decrease fertilizer supply or iv) temporally monitor attributes like time of flowering to make easier crosses for breeding purposes.

At the level of genome makeup, genomics and whole genome studies may also assist breeders to develop the most favorable recombination procedures and facilitate the application of optimal innovative breeding methodologies, like genomic selection (GS) or genomic-assisted breeding (GAB). These tools depend upon the phenotyping of broad breeding populations and are completed through the ranking of breeding values for any individual. The last operational stage is exclusively made *via* genotyping using markers associated with the trait to be selected for. Briefly GS or GAB allow integration and use of genomic devices in breeding protocols for developing new elite genotypes with superior biotic or abiotic stress tolerances and with an enhanced yield performance [61, 62]. It is evident that these tools are addressed to model genome-wide single nucleotide polymorphism (SNP) variation, without concern for HTP platforms (HTTPS). Furthermore, in this context image phenotyping has been suggested for thoroughly inspecting both growth and development of principal grain crop species by applying HTPPs [63]. According to the last authors, HTPPs are expected to provide quantitative plant data in large amount - hundreds or thousands of genotypes- employing either a minimally invasive or non-invasive integrated tools.

Information obtained by genotyping-by-sequencing (*i.e.* re-sequencing of a myriad of genomes of a plant species to identify ample amounts of SNPs) can also be directly employed for genome-wide association studies (GWAS) -a statistical association of a collection of SNPs with a trait of interest- [64, 65], or used in candidate gene-based association analyses [66, 67]. These approaches have unlocked a novel panorama for analyzing polygenic traits into genomic knowledge and for discovering genes/alleles and intragenic polymorphisms relating any given target trait [68, 69]. Additionally in this context, the discoveries of new genes have raised rapidly [70]. Similarly, functional [71] and comparative genomic studies [72] have been central from the advancements in sequencing technology, providing a fast identification and description of genes underlying any agronomic traits [73].

The approaches above mentioned, together with progress in transcriptomics [74], proteomics [75, 76], metabolomics [77], phenomics [63, 78], chemical genomics [79], and emerging hormone investigations [80], are highly powerful devices. This is particularly effective provided that HTP analytical methods are employed in association with computer-aided data processing and display [81]. Specifically, the later cited technologies may help in providing information on i) genome variability in crop species at the level nucleic acids and other chemical compounds, ii) gene and protein function, including gene networks and metabolic pathways, iii) protein and metabolic levels, and iv) genomic regions or genes involved in trait(s) expression. This information is central to i) improve crop yield and stability and ii) develop varieties with in-built novel characteristics, such as generating superior added value materials for industrial uses [82].

From the previous indications, it is clearly evident that genetic modification technologies have progressed largely over the past thirty years. However more recently, a series of stimulating developments provides important benefits, in addition to analyses of plant genomes, for its applications in crop improvement programs [83]. A replay to a request of data from the European Commission, regarding the state-of- the-art on the use and potential economic fallout of new plant breeding tools has been given by Lusser and collaborators [84]. Notably, they have provided a list of genetic approaches that appear central to the development of new genetic variability in plant species. Furthermore, they have suggested that the new crop varieties developed from those innovative tools may be disseminated inside 3 years. Briefly these tools and their characteristics are summarized below:

-*Zinc Finger Nuclease (ZFN)*: individual mutations or short indels are produced or novel genes are incorporated into pre-defined target regions of the genome [85]. Interestingly to generate site-specific breaks, in addition to ZNF, other different methodologies have been assembled. These include Transcription-Activator Like Effector Nucleases (TALENs), and Clustered Regularly Interspaced Short Palindromic Repeats (CRISPR)-Cas9 methods [85]. Specifically, CRISPR-Cas9, one of the most outstanding biotechnology progress since the polymerase chain reaction, has several advantages over the protein-based recruitment methods including i) ease and low cost of synthesizing new guide RNAs, ii) increased specificity of cleavage sites, and iii) ability to simultaneously edit multiple loci [86 - 88].

Briefly, CRISPR-Cas9 can target a specific genomic sequence, driven by an engineered 20-22-nt RNA sequence that binds to its DNA target by base-pairing rules, and possesses a vast potential for investigation of loss-of-function, gain-o- -function, and gene expression. More recent modification of the CRISPR system to use double nicking of single strands has significantly decreased off target cleavage events [89]. Moreover, it was found that this tool offers the ability to perform powerful 'genome editing', or genetic engineering, in several plant species including the creation of new traits [90, 91]. The accurate and powerful gene editing technologies permit to insert a gene that is devoted to generate fine, target mutations in an organism's own genome. These mutations may definitively modify the function of a host gene, alter its responsiveness to environmental cues or switch it off completely; the extraneous gene may then be eliminated in the course of specific breeding programs.

-*Oligonucleotide directed mutagenesis (ODM)*: specific mutations of one or some nucleotides are produced [92]. This tool provides a fast, accurate, and non-transgenic breeding variant for development of traits in crop plants.

-*Cisgenesis and intragenesis*: biotech crops are generated, in this case, by the introduction of DNA segments obtained from the plant species itself or from a sexually compatible species and are adjacent and unaltered (cisgenesis) or the introduced gene can be an original assemblage of DNA materials, but must still be derived from the species itself or from a crossable plant species (intragenesis) [93]. Cisgenic plants can accommodate one or more cisgenes, but they do not include any transgenes - *i.e.* cisgene contains all its innate components: promoter and terminator domains are in the typical sense orientation-, while an intragene includes natural functional members and components which might come from other crossable plant species -*i.e.* the gene component elements can derive from distinct genes and loci and from DNA sequences that are analogous to T-DNA borders from *Agrobacterium tumefaciens*. It is evident that cisgenic plants possess no extra hazards in comparison to plants developed *via* conventional or mutation breeding. As a result, it has been suggested that cisgenic plants should be excluded from the charges required in the current regulations prescribed for GM crop commercialization [94].

-*RNA-dependent DNA methylation (RdD*M): generates transcriptional gene silencing *via* methylation of promoter nucleotides and may drive epigenetic control of the genome in response to different cues [95 - 97]. In this context. small interfering RNAs (siRNAs), as central players in the epigenetic apparatus, have been endorsed for their leader function in gene silencing at a transcriptional as well as a post-transcriptional viewpoint. For example, in Arabidopsis, 24 nt siRNAs can direct DNA methylation and chromatin remodeling at their destination loci. Consequently, epigenome plasticity can permit to plant genomes to acquire up numerous epigenomes and furthermore to exhibit the appropriate epigenome at the correct time as a reaction to intracellular or extracellular signals. Variations in the methylation profiles generated by RdDM are heritable and – if arising in germline cells – are transmitted to the next generation. This may consent breeders to develop plants that do not include in their genomes foreign DNA sequences and in which no alterations or mutations are generated at the level of nucleotide sequence, although gene expression results epigenetically altered [98].

-*Grafting (on GM rootstoc*k): expected progresses are realized through grafting of non-transgenic scions inserted upon GM rootstocks [99]. In grafting events the variation created by the transgene lodged onto the GM rootstock may not at all or not carefully be distinguished in the resulting event. This technology is mainly used to produce disease resistance traits or induce a superior plant growth by employing a suitable GM rootstock and grafting a non-GM scion upon it. As a result the uppermost stalk of the resistant graft and obviously as predictable the fruits are not GM entities.

-*Reverse breeding*: is an arrangement of recombinant DNA methods and cell biology techniques used to develop appropriate transgene-free homozygous parental strains for quickly recreating elite heterozygous varieties [100]. The main goal of this technology is to develop homozygous parental lines (complementing parents) that can be crossed to regenerate a suitable heterozygous genotype that is represented by the initial hybrid genotype. An example of this approach has been provided by Wijnker and coworkers [101]. In this example, RNAi technology was adopted to knock down the function of the meiotic recombination protein DISRUPTED MEIOTIC cDNA1 (*DMC*1). In this respect, various fertility mutants were identified, including the semi-sterile phenotype *dmc*1 mutants of *A. thaliana*. Moreover, this example demonstrates the utility of RNAi in plants as a means of knocking down gene function and engineering desired traits [102].

-*Agro-infiltration*: is especially employed in research framework. For example, this technology is useful to i) investigate plant-pathogen interaction in living tissues, ii) choose parental lines, or iii) estimate the effectiveness of transgenes where a liquid suspension of *Agrobacterium sp.,* carrying the suitable gene(s), is employed to infiltrate plant tissues, mainly leaves, in a way that the genes are locally and transiently expressed in high amount [103, 104].

Besides the technologies previously indicated, to favor the use of directed genome engineering strategies in crop innovations, there are currently available a range of productive DNA delivery protocols and effective cell and tissue culture methods [45, 105]. These devices can permit, in a number of crops, including the three major cereals - *i.e.* maize, wheat, and rice-, an efficient manipulation of large numbers of cells. Moreover, it is worth noting that these procedures usually concern *Agrobacterium tumefaciens*-mediated transformation of cultured tissue or immature embryos, followed by callus cultures to regenerate plants. Unfortunately the ability of Agrobacterium to infect monocot plants is confined to a very narrow spectrum of genotypes. Moreover, the applicability of the method is additionally restricted by the recalcitrance of many genotypes to callus formation and regeneration. However, a breakthrough in this context has been recently reported by Lowe and colleagues [105a] *via* implementing a new approach devoted to increment monocot transformation rates in a large spectrum of genotypes. Specifically, these workers have taken advantage of overexpression of morphogenic regulatory genes, such as the maize *Baby boom* (*Bbm*) and *Wuschel2* (*Wus2*), to generate a biological context, promoting an outstanding level of transformation efficiency in monocots, after Agrobacterium-mediated transformation of immature embryos. For example, in maize the application of this strategy has shown a remarkable higher transformation rates, varying from 0 to 2% in the controls and from 25 to 51% in the presence of *Wus* and *Bbm* genes. Additionally, their results have also indicated that i) the level of recovery of transgenic plants is adequately high to be convenient for large-scale production both for commercial scopes and for research activities, ii) this strategy dramatically improves the range of genotypes suitable for transformation, iii) the method can be applied on a variety of initial tissues, including mature seed tissues or seedling leaf fragments, thus avoiding the typical explant culture step before transformation and simplifying large-scale transformation works, and iv) this morphogenic gene-based strategy is effective in other monocot crops by increasing transformation frequency rates: from 1.9% to 18% in sorghum and from 0–3% to 15–43% in rice, while for sugarcane callus they have found an almost shocking progress, varying from 2% to 274–885% (*i.e.*, manifold transgenic events recovered for initial starting tissue). Finally, it is worth noting that Lowe and collaborators, by expanding the range of species, genotypes, and tissues that can be exploited for efficient transformation, have given an astonishing blueprint of what may be accomplished by integrating basic research, technical knowledge, and understanding of practical problems facing ordinary requirements. Although , for many other crop species, these methods have still to be further developed or improved, it is also worth noting that research efforts have been mainly directed to develop i) more powerful and efficient synthetic promoters and TFs, for a precise regulation of targeted plant transgene expression [106] and ii) mini-chromosomes or synthetic chromosomes as a next generation vehicle for genetic engineering in plants [107]. More detailed information on plant biosynthetic pathways will also permit positive interventions concerning metabolic engineering, as well as to develop approaches for chemical syntheses that are exclusive to plants. It is predicted that genomic strategies will support plant biochemistry as scientists will explore more in deep metabolic pathways involved in the synthesis of high added-value compounds. Detecting rate-limiting steps in their production might contribute to identify suitable objectives for genetically engineering biochemical pathways. This will allow the synthesis of larger quantities of valuable or novel chemical products in plant cells.

Evidence indicates that plant-based production of original chemical compounds, including biopolymers (*e.g.* poly-hydroxy-alkanoates), in plant cell tissues has been achieved [108]. However, plant metabolic pathways are driven by complex mechanisms and often exhibited multiform levels of regulation, pointing out that will be demanding to target the most effective operational points and correctly forecast the results [109]. Nevertheless, progresses in knowledge-based metabolic engineering approaches (*i.e.* storing and mining of genomics and other omics information) and genetic tools, for overexpressing or silencing genes that may limit C fluxes to the biosynthesis of a target compound, should assist scientists in modeling metabolic pathways [110]. This information will be beneficial, in defining and refining the best interventional approaches, to reshape the metabolic networks involved in the synthesis of peculiar chemical compounds and materials that are more suitable to industrial purposes. Furthermore, the development of new biological disciplines, such as systems biology, synthetic biology, and evolutionary engineering, are revising the approaches in which metabolic engineering is carried out. Interestingly the synergistic integration of these disciplines has promoted development systems metabolic engineering technology. This last discipline is devoted to reach several objectives such as: i) development of new enzymes and new circuits involved in the synthesis of metabolites and their gene regulatory mechanisms; ii) detect fine regulation and extension of the metabolic fluxes favoring yield improvement and concentration of commercial products; iii) promote a greater tolerance of plant cells to the product and detrimental medium components; iv) generate limited amount or no byproducts; v) improve exploitation of appropriate C substrates; vi) optimize of cost-effective production and downstream procedures [111]. This information will allow to broad the range of metabolic engineering towards manufacturing efficiently various chemical building blocks and polymers for industrial applications.

New Breeding Objectives and Desired Traits

Plant breeders are invariably motivated by the necessity to enhance sustainable production, its stability, and quality attributes of cultivars by i) selecting for production *per se*, ii) addressing crop quality features, iii) developing crops and varieties having a broad range of end uses, iv) providing varieties with an increase tolerance to different abiotic stresses and with a better use of crops inputs, v) breeding for features conferring tolerance/resistance to disease and pest agents and thus less dependent on agrochemicals, whilst minimizing the impact of agriculture on the environment. Addressing further genetic gain in yield and quality-related traits, by developing new varieties that fit those scopes, will depend on detecting and using genetic variability in key morphological, structural, and physiological characters. Typically, these activities will deal with complex attributes, including those associated with plant growth and development, yield *per se* and its stability, water and nutrient use efficiency [112]. Yields may also be improved by manipulating other pathways, including those sustaining biotic and abiotic stress factors.

Yield Features

To realize yield improvement in developing new cultivars, it is central to: i) detect genes and genetic networks that are pivotal to biomass or grain productions, ii) identify the selective pressures *(i.e.* any reason for organisms with certain phenotypes to have either a survival benefit or disadvantage), acting within and among populations, that have forged the frequencies of these genes in developing contemporary varieties, and iii) ascertain which morphological and physiological features can ultimately drive to more productive crops. In this scenario, a large genetic variability has been often detected for yield traits (*i.e.* yield *per se*, yield stability, nutrient and water use efficiency). However, many of these characteristics, that are central in crop productivity, are represented by polygenic traits and are frequently laborious to select for.

QTLs mapping protocols are popular genomic tools to analyze the genetic make-up governing complex features and detect genomic regions underlying the expression of these traits [59]. Therefore, the genetic analysis of yield components – namely that influencing plant productivity- can assist to understand the physiological and biochemical programs, acting from gene to phenotype. An example of these possibilities for grain or biomass yields is given in Figs. (**5** and **6**), respectively. The approaches of genetic interventions therein reported, in combination with recent progresses in dissecting the basis in which plants function and develop, can assist to increase the potentiality and effectiveness of crop breeding programs. The information concerning gene and genome sequences, the control of gene expression, and the regulatory molecular and cellular mechanisms and networks underlying plant growth, development and function, may provide original choices for breeders. This information would allow them to readily modeling breeding projects for plant improvement. These approaches combine molecular tools, devoted to upgrading yields, and transgenic applications designed toward raising yield by genetic modification of various plant characteristics including hybrid vigor or heterosis.

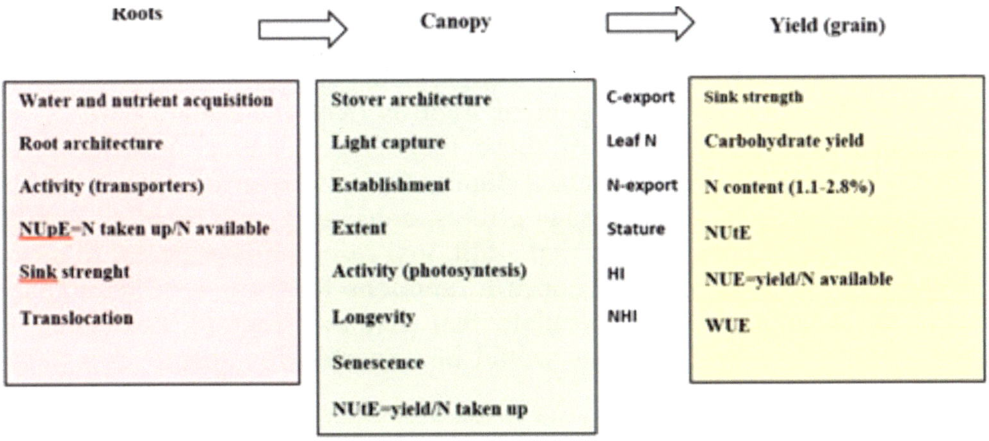

Fig. (5). Sketches of different strategies to enhance grain crop productivity. To simplify the identification of molecular targets in breeding programs, complex traits can be divided into sub-traits. Although these sub-traits are still complex, they depend upon current physiological and biochemical information (Modified from Parry and Hawkesford, 2012 [115]). Abbreviations: N, nitrogen; NUpE and NUtE, nitrogen use efficiency for whole plant or usable biomass, respectively; C, carbon; HI, harvest index; NHI, nitrogen harvest index; WUE, water use efficiency.

In this panorama, it is worth noting that heterosis has already shown a considerable economic importance in agriculture, particularly in maize as above described, and has exhibited substantial involvements in biology. However, identification of the genes and alleles that determine this phenomenon, as well as for extending allele-independent events, through genome-wide analyses, has not been yet successful in clarifying the genetic basis underling these mechanisms, both in maize or in other species. Typical arguments crosswise these investigations have pointed out that i) differences in sequence are essential, but not enough to drive the development of heterotic phenotypes, and ii) molecular routes that exhibit heterosis entangle different aspects such as: chromatin changes, transcriptional control, translation and protein processing, and interactions within and between developmental and biochemical metabolic circuits [113]. Collectively from these arguments, it has been speculated the existence of several and various molecular mechanisms translating DNAs into phenotypes. In this context, it has been guessed that the combination, very likely in a quantitative modality, of all those mechanisms across many genes is resulting in hybrid vigor or heterosis for complex traits [114]. A reinforcement of this hypothesis has been given from Huang and colleagues by studying in rice the genomic architecture of heterosis [114a]. Specifically, these workers have produced, sequenced, and recorded the phenotypes of a set of quantitative traits - primarily grain yield, flowering, and plant architecture- of more than 10,000 F2 lines derived from 17 elite representative of hybrid crosses [114a]. Accordingly, on the basis of hybrid-breeding strategies, the above cited workers were able to classify modern rice varieties into three principal groups. Although, they were unable to detect any heterosis-associated loci shared universally across all lines, within each group, a small number of genomic loci (QTLs) from female parents, which accounted for a sizable percentage of the yield gain of hybrids over their male parents, were highlighted. Notably, for some of these loci, Huang and collaborators, have provided support that additive, partial dominance, and overdominance can have been unveiled thank to allelic dosage effects, consistent with previous results in maize. Even if these data suggest that additional research is essential to clear up the new genetic associations and confirm functions for these candidate genes in heterosis, it is suggestive to speculate that genomic regions associated with heterotic effects behave largely as would be expected for quantitative traits.

Photosynthetic Process

Photosynthesis – *i.e.* the biological synthesis of chemical compounds in the presence of light- is the principal energy and C sources [117]. These essential sources are subsequently used to support plant growth and development and ultimately plant yields. Furthermore, it is projected that photosynthesis annually fixes approximately 10^{11} tons of C, which represents the storage of more 10^{18} kJ of energy (~7-times higher the world energy demand). Therefore, elevating the cumulative rate of photosynthesis over the growing season has been identified as a critical strategy for improving crop yields. In this context, molecular objectives were detected by investigating photosynthesis impediments and interventions to overtake these obstacles [117]. The data reported indicate that barriers are mainly attributable to mechanisms driving up-regulation or down-regulation of individual genes. In some examples, synthetic pathways were employed to counterbalance disadvantages in the endogenous photosynthetic pathways. In Table **1** is presented a list of potential objectives to enhance photosynthesis:

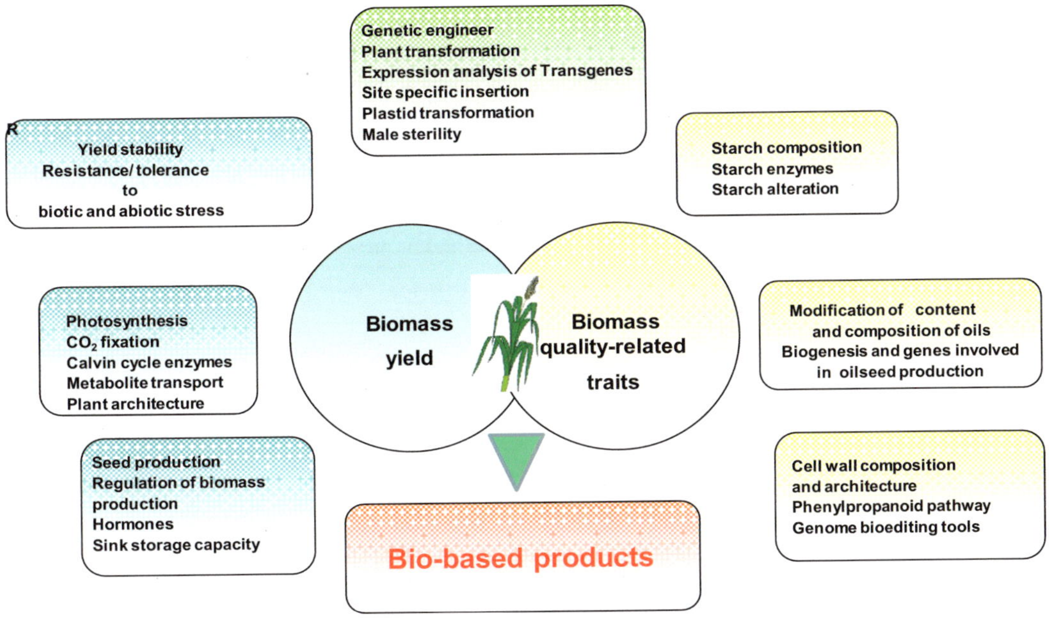

Fig. (6). Schemes of different strategies to enhance biomass crop productivity. The schemes indicate the two principal ways for improving plant biomass production and quality-related traits by genetic approaches. The first objective is directed to extend the biomass production per unit area of agricultural lands (*i.e.* the biomass production *per se* and its stability). The second objective is devoted to modify biomass features and composition to provide conversion process-friendly products for industrial uses, *i.e.* fuels, chemicals, and biomaterials (Modified from Lauria *et al.*, 2015 [116]).

A central aim for increasing plant source strength is the supply of photo-assimilates throughout photosynthesis. In this scenario, studies have been attempted to develop a more powerful C4-like photosynthetic system in C3 plants [118]. Likewise additional interventions have been devoted to upgrade the performance of photosynthetic C assimilation including i) altering the enzyme RUBISCO (Ribulose 1,5-bisphosphate carboxylase/oxygenase, a fundamental member of photosynthetic process in fixing CO_2 into organic compounds)

Table 1. Potential objectives for enhancing photosynthesis in plants (After Lauria *et al.*, 2015).

1. Enhancing Rubisco function
i. Enhancing Rubisco catalytic activity
ii. Modifying Rubisco quantity per leaf area
2. Rising thermo-stability of Rubisco activase to assist Rubisco activity at elevated temperature
3. Increasing CO_2 concentration surrounding Rubisco to magnify catalytic rate
And decrease photorespiration
i. Shifting C3 plants into C4 plants
ii. Introducing algal or cyanobacterial CCM into C3 plants
iii. Remodeling photo-respiratory metabolism
iv. Enhancing CO_2 transport pathways thorough stomata and/or mesophyll cells
4. Improving chloroplast rate for electron transport
i. Enhancing global chain for electron transport
ii. Altering light-harvesting processes
5. Improving enzyme activity of Calvin cycle (*i.e.* FBPase*, SBPase**)
6. Upgrading the ability of metabolite transport mechanisms and C employment
7. Intensifying levels of phyto-chrome expression

*. **, :FBPase,Fructose BisPhosphatase and SBPase, Sedoheptulose-BisPhosphatase, respectively.

ACTIVASE, a central regulator of RUBISCO activity [119] or improving the catalytic rate of RUBISCO itself [120], ii) upgrading flow of electrons *via* overexpressing CYTOCHROME C6 (CYTC6), a protein engaged in the photosynthetic chain for electron transfer [121], iii) assembling *via* metabolic engineering original networks into the chloroplast that are able to overpass photorespiration and release CO_2 strictly inside the stroma of chloroplasts [122], and iv) employing diverse 'add-ons' or CO_2-concentrating mechanisms (CCMs), to increase CO_2 amounts around RUBISCO [123] or stomatal CO_2 conductance by traits manipulation of these structures [124], and v) increasing *via* overexpression sucrose transporters or sucrose synthase activities in sink cells [125]. Collectively, these genetic interventions show a tendency to improve plant growth and yield. Notably, in this context, are already available appropriate tools for the nucleus genome or the chloroplast genome (*i.e.* trans-plastome) transformation [126, 127]. This would allow a more rigorous modification of the photosynthetic system. It is also anticipated that these plants could display a more effective photosynthetic process when subjected to controlled environments and manifest potential benefits related to greater productivity performances. It is also predicted that in the future, the coordinated exploitation of various approaches in this field would be central to extend photosynthetic potentiality and, consequently, plant growth and development, which, in turn, may increase crop yield.

Plant Structural Form

In recent years considerable advances has been fulfilled in clarifying the molecular processes that drive the makeup of plant structure. Analysis of plant structure, by the use of morphological mutants, has permitted to plant workers to gain insights into different features influencing plant architecture. This is exemplified by the study of the following features: shot apical meristem conservation and differentiation, axillary meristems initiation, axillary buds formation and outgrowth, stem elongation, and branches (*i.e.* tillers) number or angle [128]. Likewise, current researches have detected numerous genetic and environmental components affecting root growth apparatus that are playing a key role in the formation of root system architecture (RSA) and ultimately its nutrient acquisition [129, 130]. Interestingly, some of the genes detected have the capacity to reduce significantly crop production losses, brought about by environmental extremes, and are presently being exploited to trigger drought tolerance. It was speculated that altering the expression of these and additional genes that affect RSA might be pivotal for future improvement of crop on a global scale.

Despite our present knowledge of how a specific plant delineates its architectural makeup is still incomplete, the data currently achieved in dissecting plant architecture have already permitted to elaborate a molecular blueprint. This would unblock a door for developing better plant architecture in crop varieties and thus improving the productivity of biomass.

Regulation of Plant Biomass Production

The potentiality for increasing plant biomass productions has not until now been largely investigated. This is mostly due to the fact that conventional plant breeding efforts made on the major crops (*e.g.* maize, wheat, rice, and soybean) has been concentrated on selecting genotypes with elevated grain-yield features [131]. In spite of this, current molecular and genetic researches have detected several regulator factors that stimulate biomass yield. In Fig. (**7**) is presented a schematic representation, given by Demura and Ye [132], illustrating several regulatory components driving plant production.

From the panel, it comes out that yield improvements might be achieved through different routes, namely i) an expansion of the activity of vegetative apical meristem, mediated by KNOX (homeodomain TFs) proteins, and of the activity of the vascular cambium, triggered by KNOX along with cytokinins, ii) a repression of the transition into reproductive growth, by triggering the activity of a protein (FLC) - encoded by the *FLOWERING LOCUS C*- which promote flowering time, and iii) an inhibition of SOC1/ FUL/FT proteins - encoded by *FLOWERING LOCUS T (FT)* and *SUPPRESSOR OF OVEREXPRESSION OF CONSTANS 1(SOC1* loci, respectively. Notably both of these last compounds are implicated in several mechanisms regulating plant growth and development such as: meristems indeterminacy and plant longevity, cell elongation extension mediated by gibberellin, increasing in photosynthetic efficiency, and expansion of secondary cell wall deposition. For example, an increasing in the digestibility of cellulose, by varying properties affecting cell walls, might lead to improvements in biofuel yield performances. Thus, genes involved in determining set up of flowering process may be potentially controlled for developing grass crops with a more extended biomass production. Additionally, it has been reported that over-expression of some representative of NAC family -a plant-specific family of TFs- may as well as promote an upgrade in biomass production. For example, Arabidopsis transgenic strains, overexpressing *NAC1,* have greater plants with broader leaves, more compact stalks and a more profuse root apparatus in comparison to control plants [133]. Likewise, overexpression in Arabidopsis of another NAC-domain TF, named ATAF2, promotes improvement of biomass yield, mostly due to formation of wider leaves associated with bigger cells [134]. Moreover, recent discoveries about flowering time (FTi) regulators have indicated that this feature is controlled by a complex circuit associated with several (epi-) genetic regulator elements, reacting to environmental and endogenous signals that ultimately affect the correct onset of FTi [135]. These workers have also predicted that alterations in the expression of an individual FTi regulator are sufficient to modify dramatically FTi. Notably, this mechanism of control appears of crucial relevance for the genetic improvement of flowering in crop plants: flowering is a key developmental switch that changes plant growth stages from vegetative to reproductive, thus largely affecting plant productivity.

Plant Hormones

It is widely accepted that plant hormones accomplish central functions in growth and development of plants. This mechanism embraces various aspects related to meristematic activities and cell elongation control, processes that are essential for biomass production. For instance, the phyto-hormones auxins and brassino-steroids (BRs) are key regulators of plant growth, triggering both division and elongation of cells. Arabidopsis plants that are incapable to make up or recognize

BRs are typically dwarfs with rounded leaves and show a decrease in pollen fertility, which is reflected in a considerable prevention of flowering and leaf senescence [136]. Similarly, synthesis of gibberellins depends, to a large extend, on the activity of the enzyme GA20 OXIDASE, encoded by the *GA20OX* gene. Over-expression of GA20OX enhanced biomass production through increased height and girth of the main trunk in populous trees [137]. Other plant species that over-expressed GA20OX were taller and displayed longer thorns, a marker of juvenility. In contrast, suppression of GA20OX by gene silencing generated dwarf and bushy phenotypes with shorter thorns and larger, thicker leaves [138].

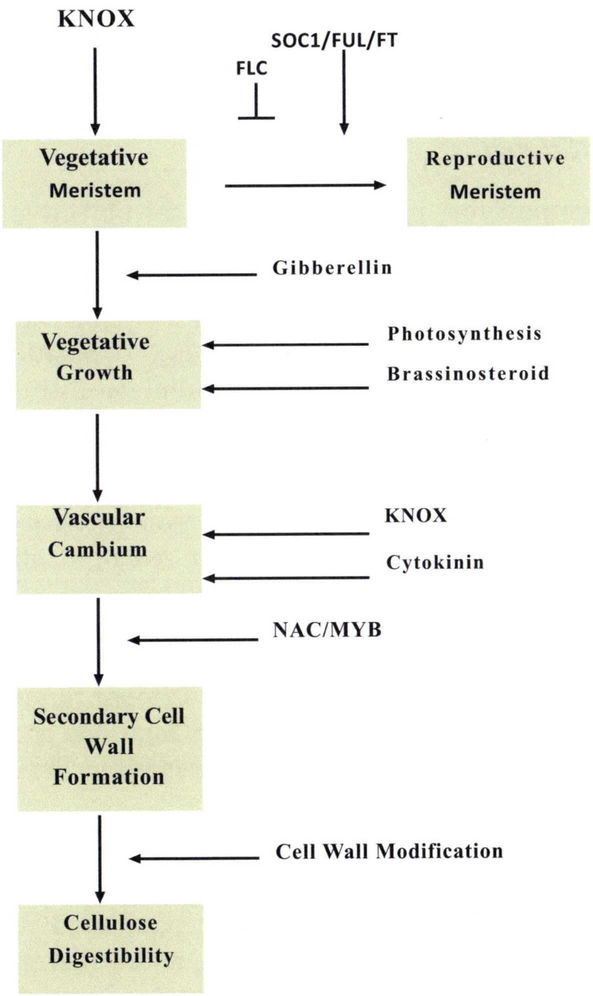

Fig. (7). Schematic representation showing several features influencing biomass production in plants (Redrawn from Demura and Ye, 2010).

A route by which auxin regulates final organ dimension is through the new TF ARGOS (AUXIN-REGULATED GENE INVOLVED IN ORGAN SIZE) [139]. This TF is acting upstream of AINTEGUMENTA (ANT) gene, a component of the AP2/ERF TF family. Mutations in either gene originate plants characterized by shorter lateral organs, whilst constitutive expression of these loci, driven by the 35S promoter, gives rise to plants featured with bigger lateral organs. Additionally, physiological investigations have provided information on a synergistic action of auxins and BRs in cell elongation processes. Similarly genome-wide microarray expression assays have detected an abundant class of genes that are activated by both auxins and BRs [140]. Combination of signal pathways for auxins and BRs appears to take place at the level of the auxin-response site (TGTCTC): this element is present in multiple copies in the promoters of BR-responsive genes [141].

Enhancing Sink Strength.

For this yield components, primary biochemical objectives are represented by enzymes. These biochemical compounds are directly or indirectly implicated in the processing of sucrose into starch. One of the targets to improve sink-strength has been devoted to modify in potato the adenylate pools: adenylate amounts have been found to be relevant factors affecting starch deposition in its tubers [142]. For example, down-regulation of the plastidial isoform of adenylate kinase produced a gain of starch up to 60% in potato and, more surprisingly, resulted in a 39% improvement in production of tubers [143].

Additional strategies in this field, devoted to yield increases, were in particular designed to advance starch biosynthesis. For example, in maize, Hannah and collaborators [144], by investigating the expression of a transgenic form (*HS33/Rev6 Sh2*) of ADP-glucose pyrophosphorylase - an enzyme central in the initiation of starch biosynthesis- characterized by an upgrade in heat stability and a decrease in P_i inhibition, recorded an improvement in grain yield above 60%. Notably, as earlier described in similar investigations in wheat and rice, this transgene strategy improve maize yield by upgrading the number of kernels.

Other Approaches

An additional intervention to modulate plant growth is targeted to i) alter the o cell cycle-related genes expression which is paralleled with improvements in general rates of growth and of leaf initiation, and a fast growth thought out all stages of plant development, ranging from seedling to maturity [145] or ii) promote rates of cell divisions by ectopic strategies resulting in plant characterized with leaves possessing more but smaller cells [146]. For example, a gene whose modulation improved biomass expansion is represented by the

CDC27a locus. This locus is encoding a protein that is a member of the ligase Anaphase-Promoting Complex (APC). The overexpression of the *CDC27a* gene of Arabidopsis in tobacco cells generated plants that are, at the flowering stage, up to 30% higher is size in comparison with controls [147]. Moreover, this feature in transgenic plants was associated with subtle modifications of apical meristems. The previous authors claimed that the improvement of the mechanism underlying growth, found in plants overexpressing *CDC27a,* is very likely attributable to the APC complex, rather than to the CDC27a protein itself. In fact, overexpression of at least two other subunits of the APC complex shows comparable improvement in plant growth.

A novel gene regulation mechanism, based on miRNAs, indicated that in Arabidopsis overexpression of miR396 exhibited an unfavorable reaction concerning cell proliferation during leaf development, because of the inhibition of GROWTH-REGULATING FACTOR (GRF) activity [148]. This event promotes a reduction of the expression of genes driving cell cycle. Similarly in Arabidopsis, over-expression of miR156 induces a delay in the transition of the vegetative-phase to reproductive one and an increase in the rate of leaf initiation [149], whereas over-expression of miR172 promotes adult leaf features including flowering [150]. The control of both miRNAs is tightly linked because their expression is altered by various factors, such as age, light, and temperature, acting in an antithetical modality. Furthermore, it was observed that at the molecular level, miR156 operates *via* the downregulation of genes encoding for the SQUAMOSA PROMOTER BINDING-LIKE (SPL) and the AP2-like TF, respectively [151]. These genetic factors are dominated by feedback control mechanisms by their target elements.

More recently, miRNAs have also gained *in planta* a growing attention as crucial regulators of the pathways of responses modulated by hormones as they influence their metabolism, distribution, and perception [152]. Yu *et al.* [153] have reported that GA in Arabidopsis induces flowering, by using a miR156-dependent pathway. Their findings suggest a contact between miRNA and hormone signaling pathways in shoot development regulation. Collectively, it appears from these results that miRNAs, as growth regulators, are intriguing powerful objectives to modify the expression of genes sustaining crop yield performances.

Quality-Related Features

Safeguarding and upgrading saleable yields are connected to improve crop quality-related traits to meet market requirements. Plant-based foods provide a wide range of nutritional elements that are central in human diet and benefit health. In spite of this, the principal staple crops on a global scale are frequently impaired in some of these nutritional building blocks [152, 153]. Moreover, functional foods and nutraceuticals have received, in the past decade, an ample attention mainly due to growing consumer awareness of the health benefits associated with food. Diet in human health is no preponderantly an issue of simple nutrition: consumers are more alerted and progressively concerned in the health benefits of functional foods and their function in the prevention of illness and chronic events. This aspect, associated with an aging population that pays attention not exclusively on longevity but also on life quality, has created a market for functional foods and nutraceuticals. Similarly plants can play a vital function in human life by providing multiple sources of metabolites, molded on a wide spectrum of skeletal systems and functional group compositions from a chemical perspective [154, 155]. Clearly these compounds can be exploited for the bulk production of value-added molecules, such as native and heterologous therapeutic enzymes or proteins.

Conventional agricultural strategies can slightly raise the nutritional value of specific foods, but progresses in molecular biology will be quickly used for genetically engineering crops with improved content of essential nutritional elements. These outcomes contain for example: high vitamin amount, better composition of fatty acids, production of very long chain polyunsaturated fatty acids, elevated mineral level, increased amino acid levels, reduction in the amount of anti-nutritional molecules, production of low calories sugars and up-grading of antioxidant levels [152, 153].

There is also evidence indicating that in many instances the benefits of bio-fortification -*i.e.* method of adding key micronutrients and other health-promoting substances to food- of crops for human nutrition and for feedstock crops exploitable for industrial conversion to biofuels, building blocks, or biomaterials, have not been achieved. This is mainly due to unreliable knowledge about different types of control (*i.e.* transcriptional, post-transcriptional, allosteric, and post-translational) that are split on finely detailed elements of regulation across metabolic pathways [156]. It is predicted that new genomics technologies will support plant biochemistry as scientists will attempt to clarify the metabolic pathways involved in the biosynthesis of suitable compounds. It is also anticipated that single steps in the metabolism can be modulated, or the expression of regulatory genes modified to provide cells with multiple controls throughout one or distinct biochemical networks. In fact, detecting rate-limiting steps in cell biosynthetic pathways might call attention to specific targets. These targets can be eventually exploited for genetically engineering biochemical pathways to achieve a profitable amount of endogenous or novel plant chemical compounds. Specific expression of a gene may be also exploited to direct metabolic C fluxes into original pathways, whereas gene-silencing techniques may be adopted to decrease or remove unwanted compounds or characteristics. Nevertheless, several attempts, using transgenic strategies, have been initiated to alter the composition of fatty acids for manufacturing bio-diesel, and nutrient oils to counteract heart diseases [157], micronutrients, resistant starch, and antioxidants [51]. For example, Golden rice, which synthesized β-carotene, a precursor of vitamin A, was bred to avoid insufficiencies in vitamins in human diets [158]. Other current attempts are devoted to the improvement of shelf life of fruits and vegetables (*i.e.* to decrease the post-harvest losses of crops liable to deteriorate), and to develop fruit crops for the yield of edible vaccines to contrast diffusion of principal human diseases, synthesis of antibodies, human or animal therapeutic proteins, antibiotics, industrial enzymes, and biomaterials [52, 155, 159, 160]. In this respect it was argued that modifications of plastid genome appears a favorable opportunity. This is attributable to high-level, simple, and safe expression of proteins, or other products, in cost-effective commercial applications [161].

CONCLUDING REMARKS

Plants are the source of organic C – that is the food- for the majority of non-photosynthetic organisms on this globe. In particular, they are indispensable for human life, as a central point of supply of carbohydrates, proteins, and fats useful for yield of food and feed, and fibers. Additionally, plants are a relevant source of a broad spectrum of secondary metabolites - *i.e.* pharmaceuticals, agrochemicals, flavors, fragrances, colors, and food additives- and are central for provision of O_2 and adsorption of CO_2.

To respond to the global increasing need of food and feed, as well as for meeting industrial demands for bio-factory plant applications devoted to boost bio-economy -the sector of economy dependent from biology and biosciences- there is an urgent need to increase crop productivity. This goal might be preferentially realized in a sustainable way, within social, political, economic, and environmental viewpoints. It is thus expected that genetic improvement of crops may help towards solving this multifaceted issue.

The information herein reported indicates that genetics and plant breeding are powerful strategies for increasing plant productivity through development of HYVs. Moreover, it is clearly evident that plant breeding has an extended history of incorporating the most recent technological progresses, provided by biology and biosciences, to advance crop improvement: the genetic revolutions of the 20th century have dramatically enhanced performance of crop productivity. It is therefore expected that plant genomics, in combination with genetics and new breeding technologies, would have again in this century a central role in crop improvement by i) ensuring food security to the rapidly growing world population and ii) providing feedstocks for upgrading bio-refining applications in the synthesis of high-added value products [162].

Interestingly, in recent years new genetic tools have been developed and are progressing very quickly for their efficient application in crop improvement. These achievements are disclosing new horizons in targeted breeding of specific traits and providing a powerful approach to upgrade yields and stability features, as well as quality-related attributes for a sustainable crop production [84, 163]. Additionally, these genetic innovations include genomic tools such as 'genome editing'. This technology will permit to target unique changes into plant genomes, including adding, removing or replacing DNA sequence at specific sites with a distinct one [86, 164]. Furthermore, it is also anticipated the potentiality to introduce epigenetic changes - histone modifications and/or DNA methylation- where the DNA sequence persists unmodified, although gene expression can be varied due to chromatin inheritable alterations [96].

Although, the challenges of engineering plants, exploiting genomic tools, are still complicated, plant genetics has already defined effective protocols for the routine genetic transformation of most of our dominant crops. Moreover, the applications of these devices, have allow to scientists to realize bigger gains in our knowledge regarding gene function and regulation in plant growth, development, and variation. This will also result greatly workable for the manipulation of plant primary and secondary metabolisms. Additional investigations concerning metabolic regulation, genetic information, and applications of appropriate technologies for genetic modifications may assist scientists to bypass the drawbacks to breed more productive crop varieties with high efficiency. These results will fulfill, in the near future, food and non-food demands and uses or eventually will help to breed plant varieties with in-built new features - *i.e.* to develop high-value metabolites for commercial exploitations. This should open markets completely news without effective competition and, thereby, avoiding issues related with commercial competitiveness.

CONFLICT OF INTEREST

The author confirms that author has no conflict of interest to declare for this publication.

ACKNOWLEDGEMENTS

I am sorry with all those whose contributions I did not include in this overview because of space constraints, personal preferences, or simple oversight. I wish also thank the anonymous reviewers for suggestions to improve the presentation of this chapter.

REFERENCES

[1] World Population Prospects: The 2012 Revision, Key Findings and Ad-vance Tables 2013. Working Paper No. ESA/P/WP.227.

[2] Popp J, Pető K, Nagy J. Pesticide productivity and food security. A review. Agron Sustain Dev 2013; 33: 243-55. [http://dx.doi.org/10.1007/s13593-012-0105-x]

[3] IFAD, WFP The state of food insecurity in the world 2014 Strengthening the enabling environment for food security and nutrition. Rome: FAO 2014.

[4] Ray DK, Mueller ND, West PC, Foley JA. Yield trends are insufficient to double global crop production by 2050. PLoS One 2013; 8(6): e66428. [http://dx.doi.org/10.1371/journal.pone.0066428] [PMID: 23840465]

[5] Qaim M, Klùmper W. Landwirtschaft für die Hungerbekämpfung Welternährung bis 2050. Chem Unserer Zeit 2013; 47: 318-26. [http://dx.doi.org/10.1002/ciuz.201300615]

[6] Spiertz H. Challenges for crop production research in improving land use, productivity and sustainability. Sustainability 2013; 5: 1632-44. [http://dx.doi.org/10.3390/su5041632]

[7] Venter O, Fuller RA, Segan DB, *et al.* Targeting global protected area expansion for imperiled biodiversity. PLoS Biol 2014; 12(6): e1001891. [http://dx.doi.org/10.1371/journal.pbio.1001891] [PMID: 24960185]

[8] Ronald PC. Lab to farm: applying research on plant genetics and genomics to crop improvement. PLoS Biol 2014; 12(6): e1001878. [http://dx.doi.org/10.1371/journal.pbio.1001878] [PMID: 24915201]

[9] Duvick DN. Plant breeding, an evolutionary concept. Crop Sci 1996; 36: 539-48. [http://dx.doi.org/10.2135/cropsci1996.0011183X003600030001x]

[10] Hancock JF. Plant evolution and the origin of crop species. 3rd ed. Wallingford, UK: CABI 2012; p. 256. [http://dx.doi.org/10.1079/9781845938017.0000]

[11] Purugganan MD, Fuller DQ. The nature of selection during plant domestication. Nature 2009; 457(7231): 843-8. [http://dx.doi.org/10.1038/nature07895] [PMID: 19212403]

[11a] Salman-Minkov A, Sabath N, Mayrose I. Whole-genome duplication as a key factor in crop domestication. Nature Plants 2016.

[12] Doebley JF, Gaut BS, Smith BD. The molecular genetics of crop domestication. Cell 2006; 127(7): 1309-21. [http://dx.doi.org/10.1016/j.cell.2006.12.006] [PMID: 17190597]

[13] Meyer RS, DuVal AE, Jensen HR. Patterns and processes in crop domestication: an historical review and quantitative analysis of 203 global food crops. New Phytol 2012; 196(1): 29-48. [http://dx.doi.org/10.1111/j.1469-8137.2012.04253.x] [PMID: 22889076]

[14] Murphy DJ. Plants, biotechnology and agriculture. Cambridge, UK: CABI Press 2011; p. 396. [http://dx.doi.org/10.1079/9781845939137.0000]

[15] Mayr E. The growth of biological though: diversity, evolution, and inheritance. Cambridge, MA: Harvard University Press 1982.

[16] Vasal SK, Singh NN, Singh BS, *et al.* Population improvement strategies for crop improvement. In: Jain HK, Kharkwal MC, Eds. Plant breeding: mendelian to molecular approaches. New Delhi, India: Narosa Publishing House 2004; pp. 391-406. [http://dx.doi.org/10.1007/978-94-007-1040-5_16]

[17] Crow JF. 90 years ago: the beginning of hybrid maize. Genetics 1998; 148(3): 923-8. [PMID: 9539413]

[18] Kloppenburg J. First the seed: The political economy of plant biotechnology. Madison, WI: The University of Wisconsin Press 2004; pp. 1492-2000.

[19] Duvick DN. The contribution of breeding to yield advances in maize (*Zea mays* L.). Adv Agron 2005; 86: 83-145. [http://dx.doi.org/10.1016/S0065-2113(05)86002-X]

[20] Tollenaar M, Lee EA. Dissection of physiological processes underlying grain yield in corn by examining genetic improvement and heterosis. Maydica 2006; 51: 399-408.

[21] Tollenaar M, Lee EA. Strategies for enhancing grain yield in maize. Plant Breed Rev 2011; 34: 37-83. [http://dx.doi.org/10.1002/9780470880579.ch2]

[22] Alston JM, Beddow JM, Pardey PG. Agriculture. Agricultural research, productivity, and food prices in the long run. Science 2009; 325(5945): 1209-10. [http://dx.doi.org/10.1126/science.1170451] [PMID: 19729642]

[23] Evenson RE, Gollin D. Assessing the impact of the green revolution, 1960 to 2000. Science 2003; 300(5620): 758-62. [http://dx.doi.org/10.1126/science.1078710] [PMID: 12730592]

[24] Fehr WR. Genetic contributions to yield gains of five major crop plants CSSA Special Publication No 7. Madison, Wisconsin: Crop Science Society of America and American Society of Agronomy 1984.

[25] Borlaug NE. The Green Revolution revisited and the road ahead special 30th anniversary lecture Sept 8. Oslo: The Norwegian Nobel Institute 2000.

[26] http://www.nobelprize.org/nobel_prizes/peace/laureates/1970/borlauglecture.html.

[27] Khush GS. Green Revolution: Challenges ahead. In: Tuberosa R, Ed. "In the wake of the double helix: from the green revolution to the gene revolution". In: Proc. Intern. Congress; 27-31 May 2003; Bologna, Italy, Avenue Media, Bologna, Italy. 2005; pp. 37-51.

[28] Hedden P. The genes of the Green Revolution. Trends Genet 2003; 19(1): 5-9. [http://dx.doi.org/10.1016/S0168-9525(02)00009-4] [PMID: 12493241]

[29] Hirano K, Asano K, Tsuji H, *et al.* Characterization of the molecular mechanism underlying gibberellin perception complex formation in rice. Plant Cell 2010; 22(8): 2680-96. [http://dx.doi.org/10.1105/tpc.110.075549] [PMID: 20716699]

[30] Pearce S, Saville R, Vaughan SP, *et al.* Molecular characterization of Rht-1 dwarfing genes in hexaploid wheat. Plant Physiol 2011; 157(4): 1820-31. [http://dx.doi.org/10.1104/pp.111.183657] [PMID: 22013218]

[31] FAO Statistical Pocketbook 2015.. World Food and Agriculture. FAO Rome, 2015, pp. 236. [http://www.fao.org/3/a-i4691e.pdf ISBN: 978-92-5-108802]

[32] Pingali PL. Significant environmental downsides to the Green Revolution. Proc Natl Acad Sci USA 2012; 109: 12302-8. [http://dx.doi.org/10.1073/pnas.0912953109] [PMID: 22826253]

[33] Hainzelin E. Cultivating biodiversity to transform agriculture. Springer Netherlands 2013; 261: 2-8. [http://dx.doi.org/10.1007/978-94-007-7984-6]

[34] Khush GS. Green revolution: the way forward. Nat Rev Genet 2001; 2(10): 815-22. [http://dx.doi.org/10.1038/35093585] [PMID: 11584298]

[35] Pathirana R. Plant mutation breeding in agriculture. Plant Sci Rev 2012; pp. 107-26.

[36] International Atomic Energy Agency (IAEA). Mutant varieties and genetic stocks. Database 2015. [https://mvd.iaea.org/]

[37] Tadele Z, Mba C, Till BJ. TILLING for mutations in model plants and crops. In: Jain SM, Brar DS, Eds. Molecular techniques in crop improvement, Springer Science+Business Media BV. 2010; pp. 307-42.
[http://dx.doi.org/10.1007/978-90-481-2967-6_13]

[38] Dahmani-Mardas F, Troadec C, Boualem A, *et al.* Engineering melon plants with improved fruit shelf life using the TILLING approach. PLoS One 2010; 5(12): e15776.
[http://dx.doi.org/10.1371/journal.pone.0015776] [PMID: 21209891]

[39] Foucart C, Boualem A, Lasseur B, Eleblu J, Fahraj I, Bendahmane A. [Sex determination in cucurbits]. Biol Aujourdhui 2012; 206(1): 57-62.
[http://dx.doi.org/10.1051/jbio/2012005] [PMID: 22463996]

[40] Trigiano RN, Gray DJ. Plant tissue culture, development, and biotechnology. Boca Raton: CRC Press 2010; p. 608.

[41] Wilson SA, Roberts SC. Recent advances towards development and commercialization of plant cell culture processes for the synthesis of biomolecules. Plant Biotechnol J 2012; 10(3): 249-68.
[http://dx.doi.org/10.1111/j.1467-7652.2011.00664.x] [PMID: 22059985]

[42] Khojasteh A, Mirjalili MH, Hidalgo D, Corchete P, Palazon J. New trends in biotechnological production of rosmarinic acid. Biotechnol Lett 2014; 36(12): 2393-406.
[http://dx.doi.org/10.1007/s10529-014-1640-0] [PMID: 25214214]

[43] Jain MS, Brar DS, Eds. Molecular techniques in crop improvement. Publisher Springer Netherlands. 2010, pp. IX, 772. [http://doi I10.1007/978-90-481-2967-6]

[44] Anami S, Njuguna E, Coussens G, Aesaert S, Van Lijsebettens M. Higher plant transformation: principles and molecular tools. Int J Dev Biol 2013; 57(6-8): 483-94.
[http://dx.doi.org/10.1387/ijdb.130232mv] [PMID: 24166431]

[45] DHalluin K, Ruiter R. Directed genome engineering for genome optimization. Int J Dev Biol 2013; 57(6-8): 621-7.
[http://dx.doi.org/10.1387/ijdb.130217kd] [PMID: 24166444]

[46] James C. Global status of commercialized biotech/ GM crops: 2014 ISAAA Brief No 49. Ithaca: ISAAA 2015.

[47] Batista R, Oliveira MM. Facts and fiction of genetically engineered food. Trends Biotechnol 2009; 27(5): 277-86.
[http://dx.doi.org/10.1016/j.tibtech.2009.01.005] [PMID: 19324440]

[48] Lu Y, Wu K, Jiang Y, *et al.* Mirid bug outbreaks in multiple crops correlated with wide-scale adoption of Bt cotton in China. Science 2010; 328(5982): 1151-4.
[http://dx.doi.org/10.1126/science.1187881] [PMID: 20466880]

[49] Nemali KS, Bonin C, Dohleman FG, Stephens M, Reeves WR, Nelson DE, *et al.* Physiological responses related to increased grain yield under drought in the first biotechnology-derived drought tolerant maize. Plant Cell Environ 2015; 38: 1866-80.

[50] Klümper W, Qaim M. A meta-analysis of the impacts of genetically modified crops. PLoS One 2014; 9(11): e111629.
[http://dx.doi.org/10.1371/journal.pone.0111629] [PMID: 25365303]

[51] Chen H, Lin Y. Promise and issues of genetically modified crops. Curr Opin Plant Biol 2013; 16(2): 255-60.
[http://dx.doi.org/10.1016/j.pbi.2013.03.007] [PMID: 23571013]

[52] Lau OS, Sun SS. Plant seeds as bioreactors for recombinant protein production. Biotechnol Adv 2009; 27(6): 1015-22.
[http://dx.doi.org/10.1016/j.biotechadv.2009.05.005] [PMID: 19460423]

[53] Koh HJ, Kwon SY, Thomson M. Current technologies in plant molecular breeding: a guide book of plant molecular breeding for researchers. 2015.
[http://dx.doi.org/10.1007/978-94-017-9996-6]

[54] Shendure J, Ji H. Next-generation DNA sequencing. Nat Biotechnol 2008; 26(10): 1135-45.
[http://dx.doi.org/10.1038/nbt1486] [PMID: 18846087]

[55] Chin CS, Alexander DH, Marks P, *et al.* Nonhybrid, finished microbial genome assemblies from long-read SMRT sequencing data. Nat Methods 2013; 10(6): 563-9.
[http://dx.doi.org/10.1038/nmeth.2474] [PMID: 23644548]

[56] Michael TP, VanBuren R. Progress, challenges and the future of crop genomes. Curr Opin Plant Biol 2015; 24: 71-81.
[http://dx.doi.org/10.1016/j.pbi.2015.02.002] [PMID: 25703261]

[57] List of sequenced plant genomes. URL: https://genomevolution.org/wiki/index.php/Sequenced plant_ genomes.

[58] VanBuren R, Bryant D, Edger PP, *et al.* Single-molecule sequencing of the desiccation-tolerant grass *Oropetium thomaeum.* Nature 2015; 527(7579): 508-11.
[http://dx.doi.org/10.1038/nature15714] [PMID: 26560029]

[59] Kujur A, Saxena MS, Bajaj D, Laxmi , Parida SK. Integrated genomics and molecular breeding approaches for dissecting the complex quantitative traits in crop plants. J Biosci 2013; 38(5): 971-87.
[http://dx.doi.org/10.1007/s12038-013-9388-6] [PMID: 24296899]

[60] Hu S, Lübberstedt T. Getting the MOST out of crop improvement. Trends Plant Sci 2015; 20(6): 372-9.
[http://dx.doi.org/10.1016/j.tplants.2015.03.002] [PMID: 25899781]

[61] Tester M, Langridge P. Breeding technologies to increase crop production in a changing world. Science 2010; 327(5967): 818-22.
[http://dx.doi.org/10.1126/science.1183700] [PMID: 20150489]

[62] Heslot N, Akdemir D, Sorrells ME, Jannink JL. Integrating environmental covariates and crop modeling into the genomic selection framework to predict genotype by environment interactions. Theor Appl Genet 2014; 127(2): 463-80.
[http://dx.doi.org/10.1007/s00122-013-2231-5] [PMID: 24264761]

[63] Ghanem ME, Marrou H, Sinclair TR. Physiological phenotyping of plants for crop improvement. Trends Plant Sci 2015; 20(3): 139-44.
[http://dx.doi.org/10.1016/j.tplants.2014.11.006] [PMID: 25524213]

[64] Ersoz ES, Yu J, Buckler ES. Applications of linkage disequilibrium and association mapping in crop plants. In: Varshney RK, Tuberosa R, Eds. Genomics-assisted crop improvement. Springer: Netherlands 2007; pp. 97-119.
[http://dx.doi.org/10.1007/978-1-4020-6295-7_5]

[65] Nordborg M, Weigel D. Next-generation genetics in plants. Nature 2008; 456(7223): 720-3.
[http://dx.doi.org/10.1038/nature07629] [PMID: 19079047]

[66] Rafalski JA. Association genetics in crop improvement. Curr Opin Plant Biol 2010; 13(2): 174-80.
[http://dx.doi.org/10.1016/j.pbi.2009.12.004] [PMID: 20089441]

[67] Thudi M, Upadhyaya HD, Rathore A, *et al.* Genetic dissection of drought and heat tolerance in chickpea through genome-wide and candidate gene-based association mapping approaches. PLoS One 2014; 9(5): e96758.
[http://dx.doi.org/10.1371/journal.pone.0096758] [PMID: 24801366]

[68] Liu S, Wang X, Wang H, *et al.* Genome-wide analysis of *ZmDREB* genes and their association with natural variation in drought tolerance at seedling stage of *Zea mays* L. PLoS Genet 2013; 9(9): e1003790.
[http://dx.doi.org/10.1371/journal.pgen.1003790] [PMID: 24086146]

[69] Varshney RK, Terauchi R, McCouch SR. Harvesting the promising fruits of genomics: applying genome sequencing technologies to crop breeding. PLoS Biol 2014; 12(6): e1001883.
[http://dx.doi.org/10.1371/journal.pbio.1001883] [PMID: 24914810]

[70] Bolger ME, Weisshaar B, Scholz U, Stein N, Usadel B, Mayer KF. Plant genome sequencing - applications for crop improvement. Curr Opin Biotechnol 2014; 26: 31-7.
[http://dx.doi.org/10.1016/j.copbio.2013.08.019] [PMID: 24679255]

[71] Morozova O, Marra MA. Applications of next-generation sequencing technologies in functional genomics. Genomics 2008; 92(5): 255-64.
[http://dx.doi.org/10.1016/j.ygeno.2008.07.001] [PMID: 18703132]

[72] Tello-Ruiz MK, Stein J, Wei S, *et al.* Gramene 2016: comparative plant genomics and pathway resources. Nucleic Acids Res 2016; 44(D1): D1133-40.
[http://dx.doi.org/10.1093/nar/gkv1179] [PMID: 26553803]

[73] Morrell PL, Buckler ES, Ross-Ibarra J. Crop genomics: advances and applications. Nat Rev Genet 2011; 13(2): 85-96.
[PMID: 22207165]

[74] Mudalkar S, Golla R, Ghatty S, Reddy AR. *De novo* transcriptome analysis of an imminent biofuel crop, *Camelina sativa* L. using Illumina GAIIX sequencing platform and identification of SSR markers. Plant Mol Biol 2014; 84(1-2): 159-71.
[http://dx.doi.org/10.1007/s11103-013-0125-1] [PMID: 24002439]

[75] Cramer R, Bindschedler L, Agrawal G. Plant proteomics in crop improvement. Proteomics 2013; 13(12-13): 1771.
[http://dx.doi.org/10.1002/pmic.201370104] [PMID: 23828769]

[76] Škultéty L, Danchenko M, Pretová A, Hajduch M. Techniques in plant proteomics. In: Hajduch M *et al.* (Eds), Molecular Techniques in Crop Improvement, Springer Netherlands 2010; pp. 469-91

[77] Sweetlove LJ, Obata T, Fernie AR. Systems analysis of metabolic phenotypes: what have we learnt? Trends Plant Sci 2014; 19(4): 222-30.
[http://dx.doi.org/10.1016/j.tplants.2013.09.005] [PMID: 24139444]

[78] Walter A, Liebisch F, Hund A. Plant phenotyping: from bean weighing to image analysis. Plant Methods 2015; 11: 14.
[http://dx.doi.org/10.1186/s13007-015-0056-8] [PMID: 25767559]

[79] Sadhukhan A, Sahoo L, Panda SK. Chemical genomics in plant biology. Indian J Biochem Biophys 2012; 49(3): 143-54.
[PMID: 22803329]

[80] Kudo T, Akiyama K, Kojima M, Makita N, Sakurai T, Sakakibara H. UniVIO: a multiple omics database with hormonome and transcriptome data from rice. Plant Cell Physiol 2013; 54(2): e9.
[http://dx.doi.org/10.1093/pcp/pct003] [PMID: 23314752]

[81] Jayashree B, Hoisington D. Bioinformatics tools for crop research and breeding. In: Jain S M, Brar DS, Eds. Molecular techniques in crop improvement:. 2nd Edition. Springer Science+Business Media B.V. 2010; pp. 97-116.
[http://dx.doi.org/10.1007/978-90-481-2967-6_4]

[82] Rhee SY, Mutwil M. Towards revealing the functions of all genes in plants. Trends Plant Sci 2014; 19(4): 212-21.
[http://dx.doi.org/10.1016/j.tplants.2013.10.006] [PMID: 24231067]

[83] Voytas DF, Gao C. Precision genome engineering and agriculture: opportunities and regulatory challenges. PLoS Biol 2014; 12(6): e1001877. [http://dx.doi.org/10.1371/journal.pbio.1001877] [PMID: 24915127]

[84] Lusser M, Parisi C, Plan D, Rodríguez-Cerezo E. Deployment of new biotechnologies in plant breeding. Nat Biotechnol 2012; 30(3): 231-9. [http://dx.doi.org/10.1038/nbt.2142] [PMID: 22398616]

[85] Gaj T, Gersbach CA, Barbas CF III. ZFN, TALEN, and CRISPR/Cas-based methods for genome engineering. Trends Biotechnol 2013; 31(7): 397-405. [http://dx.doi.org/10.1016/j.tibtech.2013.04.004] [PMID: 23664777]

[86] Doudna JA, Charpentier E. Genome editing. The new frontier of genome engineering with CRISPR-Cas9. Science 2014; 346(6213): 1258096. [http://dx.doi.org/10.1126/science.1258096] [PMID: 25430774]

[87] Belhaj K, Chaparro-Garcia A, Kamoun S, Patron NJ, Nekrasov V. Editing plant genomes with CRISPR/Cas9. Curr Opin Biotechnol 2015; 32: 76-84. [http://dx.doi.org/10.1016/j.copbio.2014.11.007] [PMID: 25437637]

[88] Liu D, Hu R, Palla KJ, Tuskan GA, Yang X. Advances and perspectives on the use of CRISPR/Cas9 systems in plant genomics research. Curr Opin Plant Biol 2016; 30: 70-7. [http://dx.doi.org/10.1016/j.pbi.2016.01.007] [PMID: 26896588]

[89] Ran FA, Hsu PD, Lin CY, *et al.* Double nicking by RNA-guided CRISPR Cas9 for enhanced genome editing specificity. Cell 2013; 154(6): 1380-9. [http://dx.doi.org/10.1016/j.cell.2013.08.021] [PMID: 23992846]

[90] Araki M, Ishii T. Towards social acceptance of plant breeding by genome editing. Trends Plant Sci 2015; 20(3): 145-9. [http://dx.doi.org/10.1016/j.tplants.2015.01.010] [PMID: 25726138]

[91] Schaeffer SM, Nakata PA. CRISPR/Cas9-mediated genome editing and gene replacement in plants: Transitioning from lab to field. Plant Sci 2015; 240: 130-42. [http://dx.doi.org/10.1016/j.plantsci.2015.09.011] [PMID: 26475194]

[92] Sauer NJ, Mozoruk J, Miller RB, *et al.* Oligonucleotide directed mutagenesis for precision gene editing. Plant Biotechnol J 2016; 14: 496-502. [http://dx.doi.org/10.1111/pbi.12496] [PMID: 26503400]

[93] Jacobsen E, Schouten HJ. Cisgenesis next step in classical plant breeding. In: S.M. Jain and D.S. Brar (eds.), Molecular Techniques in Crop Improvement, Springer Science+Business Media B.V. 2010; pp. 591-611.

[94] Shiva Prasad G, Raju BK, Kumar R. CISGENESIS – An alternate approach for development of genetically modified crops. Ann Biol Res 2013; 4: 109-15.

[95] Castel SE, Martienssen RA. RNA interference in the nucleus: roles for small RNAs in transcription, epigenetics and beyond. Nat Rev Genet 2013; 14(2): 100-12. [http://dx.doi.org/10.1038/nrg3355] [PMID: 23329111]

[96] Matzke MA, Mosher RA. RNA-directed DNA methylation: an epigenetic pathway of increasing complexity. Nat Rev Genet 2014; 15(6): 394-408. [http://dx.doi.org/10.1038/nrg3683] [PMID: 24805120]

[97] Saurabh S, Vidyarthi AS, Prasad D. RNA interference: concept to reality in crop improvement. Planta 2014; 239(3): 543-64. [http://dx.doi.org/10.1007/s00425-013-2019-5] [PMID: 24402564]

[98] Rajeevkumar S, Anunanthini P, Sathishkumar R. Epigenetic silencing in transgenic plants. Front Plant Sci 2015; 6: 693. [http://dx.doi.org/10.3389/fpls.2015.00693] [PMID: 26442010]

[99] Haroldsen VM, Chi-Ham CL, Bennett AB. Transgene mobilization and regulatory uncertainty for non-GE fruit products of transgenic rootstocks. J Biotechnol 2012; 161(3): 349-53. [http://dx.doi.org/10.1016/j.jbiotec.2012.06.017] [PMID: 22749907]

[100] Final Report. New techniques working group of the Commission, Brussels 2012.

[101] Wijnker E, van Dun K, de Snoo CB, *et al.* Reverse breeding in *Arabidopsis thaliana* generates homozygous parental lines from a heterozygous plant. Nat Genet 2012; 44(4): 467-70. [http://dx.doi.org/10.1038/ng.2203] [PMID: 22406643]

[102] Small I. RNAi for revealing and engineering plant gene functions. Curr Opin Biotechnol 2007; 18(2): 148-53. [http://dx.doi.org/10.1016/j.copbio.2007.01.012] [PMID: 17287115]

[103] Leckie BM, Neal Stewart C Jr. Agroinfiltration as a technique for rapid assays for evaluating candidate insect resistance transgenes in plants. Plant Cell Rep 2011; 30(3): 325-34. [http://dx.doi.org/10.1007/s00299-010-0961-2] [PMID: 21140154]

[104] Chen Q, Lai H, Hurtado J, *et al.* Efficient agroinfiltration of plants for high-level transient expression of recombinant proteins. Adv Tech Biol Med 2013; 1. [http://dx.doi.org/10.4172/atbm.1000103]

[105] Ji Q, Xu X, Wang K. Genetic transformation of major cereal crops. Int J Dev Biol 2013; 57(6-8): 495-508. [http://dx.doi.org/10.1387/ijdb.130244kw] [PMID: 24166432]

[105a] Lowe K, Wu E, Wang N, *et al.* Morphogenic regulators *Babyboom* and *Wuschel* improve monocot transformation. Plant Cell 2016; 28(1998–2015)

[106] Liu W, Stewart CN Jr. Plant synthetic promoters and transcription factors. Curr Opin Biotechnol 2016; 37: 36-44. [http://dx.doi.org/10.1016/j.copbio.2015.10.001] [PMID: 26524248]

[107] Birchler JA, Graham ND, Swyers NC, Cody JP, McCaw ME. Plant minichromosomes. Curr Opin Biotechnol 2016; 37: 135-42. [http://dx.doi.org/10.1016/j.copbio.2015.11.007] [PMID: 26723011]

[108] Snell KD, Singh V, Brumbley SM. Production of novel biopolymers in plants: recent technological advances and future prospects. Curr Opin Biotechnol 2015; 32: 68-75. [http://dx.doi.org/10.1016/j.copbio.2014.11.005] [PMID: 25437636]

[109] Yoon JM, Zhao L, Shanks JV. Metabolic engineering with plants for a sustainable biobased economy. Annu Rev Chem Biomol Eng 2013; 4: 211-37. [http://dx.doi.org/10.1146/annurev-chembioeng-061312-103320] [PMID: 23540288]

[110] Farré G, Twyman RM, Christou P, Capell T, Zhu C. Knowledge-driven approaches for engineering complex metabolic pathways in plants. Curr Opin Biotechnol 2015; 32: 54-60. [http://dx.doi.org/10.1016/j.copbio.2014.11.004] [PMID: 25448233]

[111] Lee JW, Kim HU, Choi S, Yi J, Lee SY. Microbial production of building block chemicals and polymers. Curr Opin Biotechnol 2011; 22(6): 758-67. [http://dx.doi.org/10.1016/j.copbio.2011.02.011] [PMID: 21420291]

[112] Vanhaeren H, Inzé D, Gonzalez N. Plant growth beyond limits. Trends Plant Sci 2016; 21(2): 102-9. [http://dx.doi.org/10.1016/j.tplants.2015.11.012] [PMID: 26739421]

[113] Schnable PS, Springer NM, Springer NM. Progress toward understanding heterosis in crop plants. Annu Rev Plant Biol 2013; 64: 71-88. [http://dx.doi.org/10.1146/annurev-arplant-042110-103827] [PMID: 23394499]

[114] Freeling M. Response to Birchler: Heterosis is partly a sub-problem of quantitative genetics, but its solution may depend on understanding the mysterious genetics of quantity. Maydica 2012; 57: 96-7.

[114a] Huang X, Yang S, Gong J, *et al.* Genomic architecture of heterosis for yield traits in rice. Nature 2016; 537: 629-33.

[115] Parry MA, Hawkesford MJ. An integrated approach to crop genetic improvement. J Integr Plant Biol 2012; 54(4): 250-9. [http://dx.doi.org/10.1111/j.1744-7909.2012.01109.x] [PMID: 22348899]

[116] Lauria M, Molinari F, Motto M. Genetic strategies to enhance plant biomass yield and quality- related traits for bio-renewable fuel and chemical productions. Plants for the Future, InTech 2015; pp. 99-133.

[117] Evans JR. Improving photosynthesis. Plant Physiol 2013; 162(4): 1780-93. [http://dx.doi.org/10.1104/pp.113.219006] [PMID: 23812345]

[118] Offermann S, Peterhansel C. Can we learn from heterosis and epigenetics to improve photosynthesis? Curr Opin Plant Biol 2014; 19: 105-10. [http://dx.doi.org/10.1016/j.pbi.2014.05.010] [PMID: 24912124]

[119] Kumar A, Li C, Portis AR Jr. Arabidopsis thaliana expressing a thermostable chimeric Rubisco activase exhibits enhanced growth and higher rates of photosynthesis at moderately high temperatures. Photosynth Res 2009; 100(3): 143-53. [http://dx.doi.org/10.1007/s11120-009-9438-y] [PMID: 19507049]

[120] Lin MT, Occhialini A, Andralojc PJ, Parry MA, Hanson MR. A faster Rubisco with potential to increase photosynthesis in crops. Nature 2014; 513(7519): 547-50. [http://dx.doi.org/10.1038/nature13776] [PMID: 25231869]

[121] Chida H, Nakazawa A, Akazaki H, *et al.* Expression of the algal cytochrome *c6* gene in Arabidopsis enhances photosynthesis and growth. Plant Cell Physiol 2007; 48(7): 948-57. [http://dx.doi.org/10.1093/pcp/pcm064] [PMID: 17548374]

[122] Kebeish R, Niessen M, Thiruveedhi K, *et al.* Chloroplastic photorespiratory bypass increases photosynthesis and biomass production in *Arabidopsis thaliana.* Nat Biotechnol 2007; 25(5): 593-9. [http://dx.doi.org/10.1038/nbt1299] [PMID: 17435746]

[123] Tsuchihira A, Hanba YT, Kato N, Doi T, Kawazu T, Maeshima M. Effect of overexpression of radish plasma membrane aquaporins on water-use efficiency, photosynthesis and growth of Eucalyptus trees. Tree Physiol 2010; 30(3): 417-30. [http://dx.doi.org/10.1093/treephys/tpp127] [PMID: 20124554]

[124] Tanaka Y, Sugano SS, Shimada T, Hara-Nishimura I. Enhancement of leaf photosynthetic capacity through increased stomatal density in Arabidopsis. New Phytol 2013; 198(3): 757-64. [http://dx.doi.org/10.1111/nph.12186] [PMID: 23432385]

[125] Baroja-Fernández E, Muñoz FJ, Li J, *et al.* Sucrose synthase activity in the *sus1/sus2/sus3/sus4* Arabidopsis mutant is sufficient to support normal cellulose and starch production. Proc Natl Acad Sci USA 2012; 109(1): 321-6. [http://dx.doi.org/10.1073/pnas.1117099109] [PMID: 22184213]

[126] Maliga P, Bock R. Plastid biotechnology: food, fuel, and medicine for the 21st century. Plant Physiol 2011; 155(4): 1501-10. [http://dx.doi.org/10.1104/pp.110.170969] [PMID: 21239622]

[127] Wani SH, Sah SK, Sági L, Solymosi K. Transplastomic plants for innovations in agriculture. A review. Agron Sustain Dev 2015; (35): 1391-430.

[128] Wang Y, Li J. Molecular basis of plant architecture. Annu Rev Plant Biol 2008; 59: 253-79. [http://dx.doi.org/10.1146/annurev.arplant.59.032607.092902] [PMID: 18444901]

[129] Kong X, Zhang M, De Smet I, Ding Z. Designer crops: optimal root system architecture for nutrient acquisition. Trends Biotechnol 2014; 32(12): 597-8. [http://dx.doi.org/10.1016/j.tibtech.2014.09.008] [PMID: 25450041]

[130] Rogers ED, Benfey PN. Regulation of plant root system architecture: implications for crop

advancement. Curr Opin Biotechnol 2015; 32: 93-8.
[http://dx.doi.org/10.1016/j.copbio.2014.11.015] [PMID: 25448235]

[131] Allwright MR, Taylor G. Molecular breeding for improved second generation bioenergy crops. Trends Plant Sci 2016; 21(1): 43-54.
[http://dx.doi.org/10.1016/j.tplants.2015.10.002] [PMID: 26541073]

[132] Demura T, Ye ZH. Regulation of plant biomass production. Curr Opin Plant Biol 2010; 13(3): 299-304.
[http://dx.doi.org/10.1016/j.pbi.2010.03.002] [PMID: 20381410]

[133] Xie Q, Frugis G, Colgan D, Chua NH. Arabidopsis NAC1 transduces auxin signal downstream of TIR1 to promote lateral root development. Genes Dev 2000; 14(23): 3024-36.
[http://dx.doi.org/10.1101/gad.852200] [PMID: 11114891]

[134] Huh SU, Lee SB, Kim HH, Paek KH. ATAF2, a NAC transcription factor, binds to the promoter and regulates NIT2 gene expression involved in auxin biosynthesis. Mol Cells 2012; 34(3): 305-13.
[http://dx.doi.org/10.1007/s10059-012-0122-2] [PMID: 22965747]

[135] Blümel M, Dally N, Jung C. Flowering time regulation in crops—what did we learn from Arabidopsis? Curr Opin Biotechnol 2015; 32: 121-9.
[http://dx.doi.org/10.1016/j.copbio.2014.11.023] [PMID: 25553537]

[136] Choe S, Fujioka S, Noguchi T, Takatsuto S, Yoshida S, Feldmann KA. Overexpression of DWARF4 in the brassinosteroid biosynthetic pathway results in increased vegetative growth and seed yield in Arabidopsis. Plant J 2001; 26(6): 573-82.
[http://dx.doi.org/10.1046/j.1365-313x.2001.01055.x] [PMID: 11489171]

[137] Eriksson ME, Israelsson M, Olsson O, Moritz T. Increased gibberellin biosynthesis in transgenic trees promotes growth, biomass production and xylem fiber length. Nat Biotechnol 2000; 18(7): 784-8.
[http://dx.doi.org/10.1038/77355] [PMID: 10888850]

[138] Fagoaga C, Tadeo FR, Iglesias DJ, *et al.* Engineering of gibberellin levels in citrus by sense and antisense overexpression of a *GA 20-oxidase* gene modifies plant architecture. J Exp Bot 2007; 58(6): 1407-20.
[http://dx.doi.org/10.1093/jxb/erm004] [PMID: 17317673]

[139] Krizek BA. Making bigger plants: key regulators of final organ size. Curr Opin Plant Biol 2009; 12(1): 17-22.
[http://dx.doi.org/10.1016/j.pbi.2008.09.006] [PMID: 18951836]

[140] Demura T, Fukuda H. Transcriptional regulation in wood formation. Trends Plant Sci 2007; 12(2): 64-70.
[http://dx.doi.org/10.1016/j.tplants.2006.12.006] [PMID: 17224301]

[141] Zhong R, Ye ZH. Regulation of cell wall biosynthesis. Curr Opin Plant Biol 2007; 10(6): 564-72.
[http://dx.doi.org/10.1016/j.pbi.2007.09.001] [PMID: 17950657]

[142] Loef I, Stitt M, Geigenberger P. Increased levels of adenine nucleotides modify the interaction between starch synthesis and respiration when adenine is supplied to discs from growing potato tubers. Planta 2001; 212(5-6): 782-91.
[http://dx.doi.org/10.1007/s004250000461] [PMID: 11346952]

[143] Regierer B, Fernie AR, Springer F, *et al.* Starch content and yield increase as a result of altering adenylate pools in transgenic plants. Nat Biotechnol 2002; 20(12): 1256-60.
[http://dx.doi.org/10.1038/nbt760] [PMID: 12426579]

[144] Hannah LC, Futch B, Bing J, *et al.* A *shrunken-2* transgene increases maize yield by acting in maternal tissues to increase the frequency of seed development. Plant Cell 2012; 24(6): 2352-63.
[http://dx.doi.org/10.1105/tpc.112.100602] [PMID: 22751213]

[145] Cockcroft CE, den Boer BG, Healy JM, Murray JA. Cyclin D control of growth rate in plants. Nature 2000; 405(6786): 575-9.

[http://dx.doi.org/10.1038/35014621] [PMID: 10850717]

[146] Dewitte W, Riou-Khamlichi C, Scofield S, *et al.* Altered cell cycle distribution, hyperplasia, and inhibited differentiation in Arabidopsis caused by the D-type cyclin CYCD3. Plant Cell 2003; 15(1): 79-92.
[http://dx.doi.org/10.1105/tpc.004838] [PMID: 12509523]

[147] Rojas CA, Eloy NB, Lima MdeF, *et al.* Overexpression of the Arabidopsis anaphase promoting complex subunit CDC27a increases growth rate and organ size. Plant Mol Biol 2009; 71(3): 307-18.
[http://dx.doi.org/10.1007/s11103-009-9525-7] [PMID: 19629716]

[148] Rodriguez RE, Mecchia MA, Debernardi JM, Schommer C, Weigel D, Palatnik JF. Control of cell proliferation in *Arabidopsis thaliana* by microRNA miR396. Development 2010; 137(1): 103-12.
[http://dx.doi.org/10.1242/dev.043067] [PMID: 20023165]

[149] Schwab R, Palatnik JF, Riester M, Schommer C, Schmid M, Weigel D. Specific effects of microRNAs on the plant transcriptome. Dev Cell 2005; 8(4): 517-27.
[http://dx.doi.org/10.1016/j.devcel.2005.01.018] [PMID: 15809034]

[150] Jung JH, Seo PJ, Kang SK, Park CM. miR172 signals are incorporated into the miR156 signaling pathway at the SPL3/4/5 genes in Arabidopsis developmental transitions. Plant Mol Biol 2011; 76(1-2): 35-45.
[http://dx.doi.org/10.1007/s11103-011-9759-z] [PMID: 21373962]

[151] Wu G, Park MY, Conway SR, Wang JW, Weigel D, Poethig RS. The sequential action of miR156 and miR172 regulates developmental timing in Arabidopsis. Cell 2009; 138(4): 750-9.
[http://dx.doi.org/10.1016/j.cell.2009.06.031] [PMID: 19703400]

[152] Curaba J, Singh MB, Bhalla PL. miRNAs in the crosstalk between phytohormone signalling pathways. J Exp Bot 2014; 65(6): 1425-38.
[http://dx.doi.org/10.1093/jxb/eru002] [PMID: 24523503]

[153] Yu S, Galvão VC, Zhang YC, *et al.* Gibberellin regulates the Arabidopsis floral transition through miR156-targeted SQUAMOSA promoter binding-like transcription factors. Plant Cell 2012; 24(8): 3320-32.
[http://dx.doi.org/10.1105/tpc.112.101014] [PMID: 22942378]

[154] Hirschi KD. Nutrient biofortification of food crops. Annu Rev Nutr 2009; 29: 401-21.
[http://dx.doi.org/10.1146/annurev-nutr-080508-141143] [PMID: 19400753]

[155] Motto M, Hartings H, Fracassetti M, Consonni G. Grain quality-related traits in maize: gene identification and exploitation. Maydica 2011; 56: 291-314.

[156] Campos-Vega R, Oomah BD. Chemistry and classification of phytochemicals. In: Tiwari BK, Ed. Handbook of plant food phytochemicals: sources, stability and extraction. Oxford: John Wiley & Sons Ltd 2013.
[http://dx.doi.org/10.1002/9781118464717.ch2]

[157] Osbourn AE, Lanzotti V. Plant-derived natural products – synthesis, function, and application. 1st ed., Heidelberg, London, New York: Springer 2009.
[http://dx.doi.org/10.1007/978-0-387-85498-4]

[158] Zhu C, Naqvi S, Gomez-Galera S, Pelacho AM, Capell T, Christou P. Transgenic strategies for the nutritional enhancement of plants. Trends Plant Sci 2007; 12(12): 548-55.
[http://dx.doi.org/10.1016/j.tplants.2007.09.007] [PMID: 18006362]

[159] Napier JA, Haslam RP, Beaudoin F, Cahoon EB. Understanding and manipulating plant lipid composition: Metabolic engineering leads the way. Curr Opin Plant Biol 2014; 19: 68-75.
[http://dx.doi.org/10.1016/j.pbi.2014.04.001] [PMID: 24809765]

[160] Mayo-Wilson E, Imdad A, Herzer K, Yakoob MY, Bhutta ZA. Vitamin A supplements for preventing mortality, illness, and blindness in children aged under 5: systematic review and meta-analysis. BMJ 2011; 343: d5094.

[http://dx.doi.org/10.1136/bmj.d5094] [PMID: 21868478]

[161] Liew PS, Hair-Bejo M. Farming of plant-based veterinary vaccines and their applications for disease prevention in animals. Adv Virol 2015.
[http://dx.doi.org/10.1155/2015/936940]

[162] Onrubia M, Cusidó RM, Ramirez K, *et al.* Bioprocessing of plant *in vitro* systems for the mass production of pharmaceutically important metabolites: paclitaxel and its derivatives. Curr Med Chem 2013; 20(7): 880-91.
[PMID: 23210777]

[163] Bock R. Genetic engineering of the chloroplast: novel tools and new applications. Curr Opin Biotechnol 2014; 26: 7-13.
[http://dx.doi.org/10.1016/j.copbio.2013.06.004] [PMID: 24679252]

[164] Carroll D. Genome engineering with targetable nucleases. Annu Rev Biochem 2014; 83: 409-39.
[http://dx.doi.org/10.1146/annurev-biochem-060713-035418] [PMID: 24606144]

CHAPTER 4

Role of Epigenetics in Crop Improvement

Michela Landoni[1] and **Roberto Pilu**[2,*]

[1] *Università degli Studi di Milano, Department of Biosciences, Milano, Italy*

[2] *Università degli Studi di Milano, Department of Agricultural and Environmental Sciences – Production, Landscapes, Agroenergy - Milano, Italy*

Abstract: It is time to start to take into consideration the role of epigenetics in breeding programs. So far, various authors have reported hereditable gene silencing phenomena generally affecting qualitative traits. Usually the genes involved in these phenomena determine characters which can easily be scored by visual inspection, such as modifications in the plant architecture and pigment accumulation, or characters like the loss of antibiotic resistance. But we have to take into account that the majority of genes are involved in the determination of quantitative characters, and the phenotypic modifications caused by QTL silencing will result in subtle variations which are difficult to detect. For this reason it would be very hard to find silencing phenomena involving quantitative traits. Therefore, assuming that epigenetics concerns not only the qualitative but also the quantitative traits, this phenomenon must be taken into account in breeding programs. In particular, the transcriptional state of the different epialleles should be considered. This chapter will start by defining what we mean by epigenetics, as numerous definitions are now used, starting from the original definition of Waddington and adapting the definition to the different fields of study. We will then describe the epigenetics marks, before going into more detail of the epigenomic studies on two model plants, arabidopsis and rice. Then we will present data concerning the interaction of epigenetics and the environment and the role of the epigenetic phenomena on crops and in particular, on yield improvement. A brief paragraph on the epigenetic phenomenon called paramutation will conclude the chapter.

Keywords: Crop improvement, Epiallele, Epigenetics, Gene silencing, Paramutation.

INTRODUCTION

The fine modulation of genome expression is essential to cell specialization in complex organisms. In fact, all the cells in a multicellular organism have the same genome, but, just as obviously, cells that are different at morphological, develop-

* **Corresponding author Roberto Pilu:** Department of Agricultural and Environmental Sciences -Production, Landscape, Agroenergy, Università degli Studi di Milano, Milano, Italia; Tel/Fax: +39 0250316549; E-mail: salvatore.pilu@unimi.it

Roberto Pilu and Giuseppe Gavazzi (Eds.)

mental and functional level, need different information and thus the activation of different subsets of genes [1].

Since the early studies on the inheritance of traits, it was clear that not all the changes in gene function segregated in accordance with Mendelian laws: among these exceptions are the epigenetic phenomena.

The term of epigenotype was first introduced by Waddington to define the "complex of developmental processes connecting genotype and phenotype" [2]. Over the years, new definitions have emphasized the meanings acquired by the term epigenetics with the increase of studies in this field [3]. In particular the definition of Russo, Martienssen and Riggs, currently one of the most used and accepted, revised the definition of epigenetics as "the study of mitotically and/or meiotically heritable changes in gene function that cannot be explained by changes in DNA sequence" [4], thus defining new levels of gene regulation.

An important breakthrough in the evolution of the initial definition of epigenetics was the work of Nanney who focused his studies on the concepts of stability and heritability of the expression states [5]. These concepts were also included in the definitions of Holliday ("study of the changes in gene expression, which occur in organisms with differentiated cells, and the mitotic inheritance of given patterns of gene expression" and "nuclear inheritance, which is not based on differences on DNA sequences") [6] and in the previously reported definition of Riggs and colleagues [4].

These more recent definitions fitted in particular with the genetic studies focused on the conservation of expression patterns through mitosis and meiosis, while studies of ecology and physiology, more concerned with gene-environment interactions, refer to the earlier definition of Waddington [3].

Epigenetics Marks

The studies of the molecular mechanisms responsible for the epigenetic variations disclosed the importance of two main type of DNA modification: DNA methylation and chromatin/histone modification. The covalent modifications of nucleotides, among which the most common is the methylation of cytosine, can be seen as a guide for a correct organization of the subsequent epigenetic modifications, *i.e.* the post translational modifications of the histone proteins [7].

These epigenetic marks, modifying the DNA packaging into nucleosomes, promote changes in chromatin structure, and thus DNA accessibility, regulate gene expression and gene function, and thus phenotypes, and furthermore because these modifications are enzyme-mediated, they are inherited even though they are not alterations of the nucleotide sequence [8].

Cytosine methylation can be in three possible contexts: CG, CHG and CHH (H: A, C, or T). The enzymes necessary for these modifications are maintenance DNA methyltransferases and *de novo* methyltransferase. In Arabidopsis the enzyme responsible for the *de novo* methylation, is the methyltransferase DNRM2 (DOMAINS REARRANGED METHYLTRANFERASE 2) [9].

With regard to the maintenance methyltransferases, MET1 (DNA METHYL-TRANSFERASE1), is responsible for maintaining the GC methylation [10, 11], the methyltransferase CMT3 (CHROMOMETHYLASE3) acts together with the histone methyltransferase KYP (KRYPTONITE) to maintain the CHG methylation [12, 13], and the CHH methylation is maintained by DRM2 (DOMAINS REARRANGED METHYLTRANSFERASE2) that is targeted on DNA by the 24-nucleotide small interfering RNAs bound by AGO4 (ARGONAUTE) [8]. Among the higher plants a high level of conservation of methyltransferases has been found, suggesting the evolutionary conservation of their function [13].

The mechanism used by the maintenance cytosine methyltranferases is based on the recognition of sequences containing hemimethylated cytosine, this is the signal allowing the methylation of the cytosine on the newly synthesized DNA sequence.

A similar mechanism is used to maintain histone modification during DNA replication, the existing modifications are the signal for the modification of the newly assembled nucleosomes, and this is possible also thanks to the association of the enzymes required for histone modification (histone methyltransferases and histone demethylases) with the DNA replication machinery [8].

The complex regulation of the propagation of the epigenetic marks has to satisfy two requirements: to allow the transmission of the epigenetic information *via* both mitosis (to daughter cells) or meiosis (to the next generation) but also to allow the meiotic resetting of the epigenetic information to enable the genome plasticity necessary to cope with environmental and developmental changes.

The propagation through mitotic divisions of the epigenetic marks is guaranteed by *de novo* methyl transferases and by the histone modification machinery, activated by small RNA molecules and sequence-specific binding factors [7].

With regard to the meiotic transmission of the epigenetic marks, there is much experimental evidence for the meiotic or developmental erasure of epigenetic variation, in plants as well as in animals [14 - 18]. However, different examples have been described in plants of epigenetic alleles, either natural (Table **1**) or obtained after transgenesis (Table **2**) and meiotically transmitted.

Table 1. Examples of epialleles in plants (modified from Richards, 2006; Pilu, 2015).

Species	**Gene**	**Phenotype**	**Stability**	**Ref.**
Maize	*a-m2-7991A1*	Pigmentation and transposition.	Metastable.	[79]
Maize	*r1*	Reduced pigmentation.	Metastable.	[80, 81]
Maize	*b'*	Reduced pigmentation.	Stable.	[82]
Maize	*p1*	Reduced pigmentation.	Stable.	[67]
Maize	*pl1*	Reduced pigmentation.	Metastable.	[83]
Maize	*lpa1*	Reduction of seed phytic acid.	Metastable.	[63]
Arabidopsis	*Fwa*	Delayed flowering.	Stable.	[84, 85]
Arabidopsis	pai2	Metabolic.	Metastable.	[86, 87]
Arabidopsis	Sup	Abnormal floral organ number.	Metastable.	[53]
Tomato	Sulf	Sectorial chlorophyll-deficient phenotype.	Metastable.	[88, 89]
Antirrhinum	Niv	Reduction of transposon frequency.	Metastable.	[73, 90]

Table 2. Examples of gene silencing in transgenic plants (modified from Pilu 2015).

Species	**Gene**	**Phenotype**	**Stability**	**Ref**
Nicotiana tabacum	*nptII (neomycin phospo-transferase II)* transgene	Loss of kanamycin resistance.	Stable.	[91]
Petunia hybrida	*a1(anthocyaninless1) transgene*	Reduction of the flower pigment.	Metastable.	[92]
Arabidopsis thaliana	*PAI* (phosphoribosyl-anthranilate-isomerase)	Gain of blue fluorescence.	Metastable.	[93]
Nicotiana tabacum	*hpt* (hygromycin phospho-transferase) transgene	Loss of hygromycin resistance.	Stable.	[94]
Nicotiana tabacum	*spt* (streptomycin phosphotransferase) transgene	Loss of streptomycin resistance.	Metastable.	[95]
Zea mays	*p1 (pericarp colour1) transgene*	Reduction of seed and cob pigment	Stable	[64]
Arabidopsis thaliana	*bal and cpr1-1*	Reversion of dwarf phenotype morphology	Metastable	[96]
Arabidopsis thaliana	*hpt(hygromycin phospho-transferase) transgene*	Loss of hygromycin resistance.	Stable.	[97]

In these cases the epigenetic variations are heritable in subsequent generations, allowing transgenerational epigenetic changes. In this way epigenetic mechanisms provide a molecular memory of the changed states that persists through cell division even in the absence of the inducing stimulus.

Soft inheritance [19] is defined as the genetic basis of characters modified by environment, use and disuse or intrinsic failure of constancy, transmitted to the progeny. It is the opposite of hard inheritance, based on the heritability of genetic material which is constant, except for random mutations. According to the definition of Mayr, therefore, the creation of epigenetic alleles by the effect of the environment, and their meiotic transmission, could be seen as a form of soft inheritance.

In plants, where reproductive lineages are derived from vegetative lineages late in development (while in animals this happens during embryogenesis) and epialleles erasure is less extensive than in animals, both the forms of soft inheritance, the extreme (involving transmission of adaptive acquired characters) and the less extreme (involving the transmission of environmentally induced or influenced alleles) are possible through epigenetic mechanisms [7]. In these terms, epigenetic variation may have the potential of being truly non-Mendelian inheritance.

Another important aspect of epigenetics is its possible involvement in the contribution of the so-called "missing heritability", the amount of heritable variation that cannot be completely explained by classical genetic variation [20]. The study of the contribution of epigenetic variants in natural phenotypic variation could shift attention onto epialles in crop improvement programs.

Epigenomic Studies On The Two Model Plants Arabidopsis And Rice

The two model plants *Arabidopsis thaliana* and rice (*Oryza sativa*) have been extensively studied at the epigenomic level. The analysis of the genome-wide chromatin modification profiles revealed that the methylations pattern of these two species show many similar characteristics. In both cases only GC methylation was present in genes, preferentially in moderately expressed genes, while non-GC methylation occurred in repeat elements and was associated with their silencing. The transcriptional start site (TSS) and the 3' end of genes showed a level of methylation inversely proportional with the level of gene transcription, while the methylation of the gene body was more frequent for moderately expressed genes than for genes with high or low expression [21, 22]. In both Arabidopsis and rice, genes with high levels of trimethylation of lysine 4 of histone H3 (H3K4me3), at the promoter and the 5' end, were actively transcribed, while the repressive mark H3K27me3 was specific for genes expressed at low level [23].

However compared to Arabidopsis the rice epigenome differs by having a higher level of overall methylation and a different landscape, probably due to a greater presence of transposable elements in non-centromeric regions in rice [23].

In Arabidopsis the epigenomic map obtained localizing the distribution of 12 histone modifications along the genome sequence revealed the presence of 4 main chromatin states, preferentially corresponding to active genes, repressed genes, silent repeat elements and intergenic regions, providing an important starting point for further studies of chromatin based regulatory mechanisms in plants [24]. Thus epigenomic studies can help in the functional annotation of the genome, in the identification of new genes, new gene functions and in predicting local genome activity [22]. Furthermore the study of the rice epigenome has shown that DNA methylation can contribute to phenotypic variation, either spontaneously or induced by *trans*-acting effects (siRNA), *cis*-acting effects (transposon or repeated sequences), or the environment [23]. The characterization of this epigenetic variation, based on transcriptional profiling, can be seen as a strategy to identify epialleles related to useful agronomic traits to be used in breeding programs to improve crop quality and production.

Epigenetics and Environments

The influence of the environment on the epigenetic state of a genome is a phenomenon which is well known.

As previously reported, the enzymes involved and the mechanisms needed to realize and maintain the pattern of DNA methylation are now fairly well known in detail, what remains is to clarify what initially identifies the region that will be methylated: it is at this level that the environment could act by influencing the epigenetic state of a genome.

There are different examples of natural variations in DNA methylation conferring specific phenotypes, but it is not always clear what is the effect of the environment on these phenomena. For example, while it is unclear what is exactly the role played by the environment on the variation of DNA methylation correlated with the non-ripening phenotype of tomato [25], the flower morphology of wild potato [26], and the transition from male to female flowers in melon [27], the environment seems to play important roles in the differential DNA methylation in natural populations growing in the different environments of perennial violet [28], mangrove tree [29] and certain orchids [30, 31].

In particular Guarino and colleagues investigated the epigenetic diversity in poplar populations sampled in different locations on the island of Sardinia (Italy) [32]. The authors showed that the low genetic variability detected is counterbalanced by the epigenetic variability between populations, strongly influenced by the environmental conditions. In facts the results presented demonstrated that the genome of individuals of the same clone showed different methylation patterns depending on their geographical position [32].

A similar conclusion was reported by the paper of Pilu and colleagues, which characterized the genetic structure of a collection of 87 Italian giant reed clones (*Arundo donax* L.) collected around Italy [33]. The authors found a high genetic uniformity among the populations analyzed by molecular markers, but also statistically significant differences between the clones for persistent inheritable phenotypic traits that could be exploited for clonal selection to improve this promising energy crop. Furthermore they suggested that this phenotypic variability might be due to epigenetic modifications induced by the different environmental conditions where the different clones were originally growing [33].

Another important example of environmentally induced epigenetic phenotypes in plants, is the effect of cold treatment (vernalization) on flowering time, which has been extensively studied in Arabidopsis.

Vernalization is the natural process allowing plants to flower in spring, *i.e.* immediately after exposure to the cold temperatures of winter.

In fact, since the plants cannot move their location, they need to coordinate flowering time to the period of most favorable climatic conditions for maximum seed production, and this is achieved by the process of vernalization which by monitoring the environmental conditions regulates the timing of flowering. As generally happens for environmentally induced epigenetic changes, the vernalized state is at rest during meiosis and thus each generation needs to be vernalized.

In Arabidopsis the vernalization process is accomplished by the epigenetic silencing of the floral repressor *FLC* (*FLOWERING LOCUS C*) [34, 35]. Prolonged exposure to cold progressively silenced the expression of this MADS box transcription factor, whose role is to repress the genes required to switch the vegetative meristem to floral meristem. The *FLC* repression is epigenetically maintained after the period of cold exposure and is reset during embryogenesis of the next generation. In the absence of the repression of *FLC* the plant is able to flower as soon as the conditions of light and temperature necessary for proper flower and fruit development are reached.

The expression of *FLC* is regulated very early during development and is subject to the effect of both natural variation and different regulators [36, 37]. The regulator *FRIGIDA* [38] increases *FLC* expression, allowing the trimethylation on histone H3 at lysine 4 and 36 [39, 40], while the autonomous pathway, requiring the demethylase *FLOWERING LOCUS D* [41, 42] and the activities of the different components of the Policomb repressing complex 2 (PCR2) [43, 44], reduces the methylation on H3K4 and H3K36, and increases H3K27 methylation: the balance of the activity of *FRIGIDA* and the autonomous pathway, determines the level of *FLC* transcript in seedlings and the need for vernalization for flowering.

When plants are exposed to prolonged periods of cold, *FLC* is epigenetically silenced, and the level of silencing depends on the duration of the exposure to cold. The PRC2 complex, also localized on the *FLC* gene before silencing [45], after cold exposure associates with the heterodimer formed by the association of two homeodomain proteins VIN3 [46], cold induced, and VNR5 [47], constitutively expressed, and this complex localizes on the first intron of the *FLC* gene, determining its silencing by increasing H3K27 methylation. The maintenance of *FLC* silencing, that will be reset only during the next embryogenesis, is guaranteed by the spreading of the PRC2 complex on the *FLC* locus determining high levels of H3K27 methylation along the entire gene, also after the winter, when temperatures rise [45].

The cold-induction of a set of the *FLC* antisense transcripts [48], alternatively spliced and polyadenylated, named COOLAIR, that precedes the recruitment of the different components of the Policomb complex, also plays a central role in *FLC* regulation in both warm and cold conditions [42, 49].

Analysis of *Arabidopsis thaliana* accessions revealed that variations in different steps of the vernalization process played important roles in the evolution of different reproductive strategies. In fact *FLC* variation is a major factor determining the need for different winter lengths in *A. thaliana* accessions, and the rapid cycling habit, typical of accessions not requiring vernalization that can reproduce multiple times per year, have evolved from independent mutations in *FRIGIDA*, an activator of *FLC* expression.

The study of the vernalization process has thus highlighted that epigenetic changes could produce adaptative variations in developmental phenotypes, useful starting points in breeding programs.

Epigenetics and Crop Improvement

Epigenetics phenomena are known to play crucial roles in natural variation, in regulation of gene expression during development and in response to the environment. These data strongly suggested a high potential for epigenetic approaches in crop improvement strategies, based on the selection of favorable epialleles, both natural and generated by mutagenesis or transgenic modification [50].

Traditional plant breeding programs, in fact, are based on the selection and crossing of the most favorable phenotypes to obtain improved offspring, generally regardless of whether the trait under selection is a genetic or an epigenetic variant, because natural variants can have both epigenetic and genetic components.

In the literature, various examples of naturally occurring epialleles affecting traits have been described, suggesting a role in natural phenotypic variation and a potential use for crop improvement. For example in Arabidopsis, seven independent epimutations named *clark kent* (*clk*) were identified showing phenotypes similar to the known *superman* (*sup*) mutants [51]. While wild type Arabidopsis flowers contain 6 stamens and two central carpels fused to form the pistil, the *sup5* allele, characterized by the nearly complete deletion of the SUP gene, produces an increased number of stamens (12.3$\pm$0.3) and carpels (2.9$\pm$0.1) [52], the *clk-3* allele has 7.8$\pm$0.3 stamens and 3.4$\pm$0.1 carpels, and the weaker *clk-1* allele has 6.4$\pm$0.1 stamens and 3.2 $\pm$ 0.1 carpels [53]. As expected, *clk* mutations fail to complement the *sup5* mutation although in F2 progenies about 1-3% of wild type plants were observed, indicating a partial instability of *clk* expression/phenotype. Analysis of the methylation patterns showed that *clk* epialleles, as with many other silenced alleles, are due to extensive methylation, covering the start of transcription and most of the transcribed region [53].

The characterization of the stability of epigenetic alleles and the contributions of epialleles in traits under selection may greatly help breeding programs.

A study on the stability of epialleles inheritance was conducted on maize, where the genome-wide profiling of DNA methylation was analyzed in two inbred genotypes, B73 and Mo17 [54]. The characterization of the distribution of cytosine methylation showed a high level of methylation in the pericentromeric regions and methylation patterns that are similar to the density of GC sites in gene bodies but not in the regions immediately upstream or downstream. Furthermore, while the distribution of GC is elevated throughout the gene body of short maize genes, in long genes the GC content is elevated only in the first 500-1000bp. The level of methylation was found to be lower in highly expressed genes, and in genes in synthenic positions relative to other grass species. In non-synthenic genes the high level of methylation correlated with a low level of expression, probably due to the presence of transposons that may have mediated their movement. The authors, by the comparison of the methylation profiles of the two inbred genotypes, B73 and Mo17, identified about 700 differentially methylated regions (DMRs). Some of these DMRs occurred in the absence of nearby genetic polymorphisms, suggesting that they could be considered examples of pure epigenetic variation [54]. In fact considering the dependence on the variation of the genome sequence, the epialleles can be defined as obligatory (the level of methylation exhibited by the epiallele is totally dependent on DNA changes, at linked or unlinked loci), facilitated (the epiallele exhibited a stochastic variation of methylation that can be dependent on genetic differences) or pure epialleles (the level of methylation is not dependent on DNA sequence changes) [7].

Furthermore the DMRs exhibited relatively stable trans-generational inheritance of the epigenetic variations, even in the absence of genetic differences, and thus may contribute to phenotypic variation which can be useful for natural and artificial selection [54].

Epigenetics and Yield Improvement

Yield is one of the most important but also complex quantitative traits, because of the high number of factors affecting the overall outcome of a crop. Indeed not only genetic and environmental but also epigenetic factors and their mutual interactions are involved in yield determination (Fig. **1**).

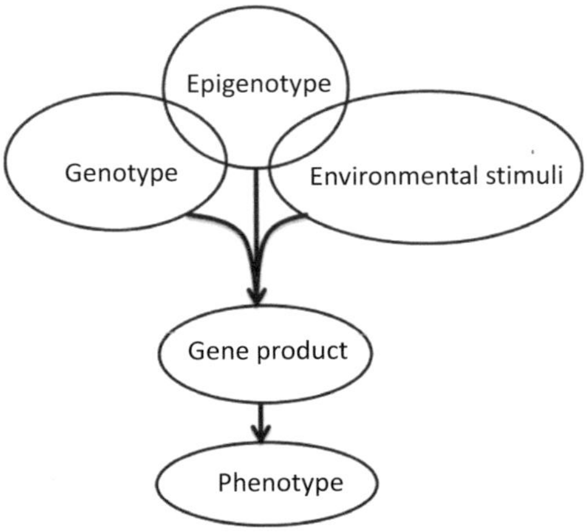

Fig. (1). The interaction among genotype, epigenotype and environment determine the phenotype of an organism.

Several papers suggest that recurrent selection can be used to improve quantitative traits by the selection of the epigenetic component. In fact, even if the epigenome is generally reset during meiosis, some epigenetic variations can be inherited through the generations. In the paper of Johannes and colleagues, two parents with different methylation profiles and few DNA sequence differences were used to generate a panel of epigenetic Recombinant Inbred Lines (epiRILs) in *Arabidopsis thaliana*. These epiRILs showed heritable variation for two quantitative traits analyzed, flowering time and plant height, over at least eight generations [55].

In canola (*Brassica napus*) Hauben and colleagues demonstrated that energy use efficiency (EUE), one of the major factors determining yield, can be improved through the selection of its epigenetic component [56].

These authors, through recurrent selection for respiration intensity, obtained different populations, for agronomic and physiological characteristics, starting from an isogenic canola population which was self-fertilized. In these populations, while the genome was identical, the epigenomes were different and these epigenomic states were heritable as well as the selected characteristics: in particular, the improved lines showed a 5% yield increase on top of heterosis. Starting from 200 seedlings from the doubled haploid population "Simon", the two plants with the highest and the lowest cellular respiration were selected for seed production by self-fertilization, allowing the production, after 4 rounds of selection, of 4 lines with lower and three lines with the higher respiration rate in comparison to the Simon population, used as a control. Fields trials showed that low-respiration lines, with a high EUE, showed a seed production increased up to 8% in comparison with the control, while seed production was reduced by up to 10% in high respiration lines with a low EUE. AFLP analysis showed no differences among the selected lines and the control, suggesting that selection may have been affecting the epigenetic state. A methylation-sensitive AFLP (ms-AFLP) analysis revealed 0.3% of differences in the methylation state between the selected lines and the Simon population. These differences, specific for the different lines, corresponded to coding sequences and were maintained for at least 8 cycles of self-fertilization in the greenhouse and three years of field trials. The epigenetic EUE component was also transmitted in reciprocal backcrosses with a control line.

Further analysis of the overall methylation and histone modification revealed that the genomic DNAs of the selected lines were hypomethylated and with a low level of histone methylation and acetylation. Thus the recurrent selection for high or low respiration had led to the selection for specific methylation states, the epigenetic EUE component. This selection started from an isogenic population with a variable epigenome and resulted in the selection of subpopulations with a partially fixed epigenome and the improved phenotype.

The authors are now testing the possibility of following this epigenetic stabilization by the analysis of the expression of specific miRNA. Moreover the characterization focused on identifying the genes which were differentially methylated and expressed and the different small RNAs in the selected sublines will shed light on the relationship between DNA methylation, histone modification, transcription and the resulting phenotypes. Preliminary results obtained in rice (*Oryza sativa*) and tomato (*Lycopersicon esculentum*), suggest that selection for EUE can potentially increase the yield of many cultivated species while also providing useful knowledge of epigenetic mechanisms [56].

An interesting example of how domestication acts even on epigenetic variability is shown in the paper of Li and colleagues [57]. The Chinese ginseng is a perennial plant, symbolic of traditional Chinese medicine, used to restore and enhance stamina. Ginseng plants, widely cultivated in northeastern China and Korea, derive from the progenitor *Panax ginseng* C.A. Meyer. Domestication did not involve dramatic morphological changes, except for the reduction of the growth cycle and the increase of root dimensions and branching. In populations of cultivated ginseng, molecular studies highlighted a high level of genetic and epigenetic diversity and the effect of domestication on this variation was analyzed in greater detail by Li and coauthors [57]. In this study the authors analyzed nucleotide diversity and cytosine methylation patterns in cultivated and wild ginseng using 49 nuclear genes and 720 methylation sensitive polymorphism (MSP) loci.

Although generally the domestication process is accompanied by a decrease in genetic variability in cultivated plants, this study demonstrated that cultivated and wild ginseng showed similar levels of genetic diversity. The epigenome analysis revealed that cultivated ginseng possessed a significantly lower level of cytosine methylation and a specific methylation pattern in comparison to wild ginseng, suggesting that domestication may have determined the morphological alterations observed at the level of root morphology and life cycle by acting on the methylation pattern, and thus on the epigenome.

These results suggest that in nature the adaptation and thus the genetic improvement is realized not only by selection of traits determined at the genetic level but also at the epigenetic level.

The Epigenetic Phenomenon Called Paramutation

Paramutation is an epigenetic event in which a silenced allele (named paramutagenic) is able to silence another allele (paramutable) in trans and this change is heritable, *i.e.* the silenced (paramutated) allele, even after segregation from the paramutagenic allele, acquires paramutagenic capacity in the next generations [58]. This phenomenon was described for the first time in the 1950s by Alexander Brink who studied the colored1 (*r1*) gene in maize, a complex locus (encoding myc-homologous transcription factors) that regulates the anthocyanin biosynthetic pathway [59]. Since then, different examples of paramutation and paramutation-like interactions have been described in plants and animals [58, 60 - 62]. As far as the stability of the epi-states is concerned, *r1* and *p1* are stable, while *b1*, *lpa1* and *pl1* (Fig. **2**), can change spontaneously to the silenced state with high frequency [63 - 65]. However, in every case of paramutation, the phenotypic modification is associated with a reduction of mRNA levels of the paramutated alleles [63 - 72].

Fig. (2). Gene silencing at *pl1* locus in maize.

The molecular bases of these phenomena are unknown but what is clear is that differently from a typical mutation, in which the change of the DNA sequence usually causes a switch off of the gene activity, paramutation generates different epialleles silenced with variable phenotypes due to modulation of gene expression [73 - 76]. Changes in DNA methylation, in chromatin structure and the activity of genes involved in RNA-directed transcriptional silencing are generally associated with the phenomenon of paramutation described in the literature. Considering the possible role of paramutation in the life cycle and in evolution we might conjecture that this phenomenon could be involved in a fast heritability of favorable epigenetic states across generations in a non-Mendelian way [77]. The fact that paramutation violates the Mendelian principle of segregation will result in difficult introgression and mapping of traits affected by this phenomenon, creating problems for breeding. It therefore seems important to reach a better understanding of paramutation phenomena, particularly regarding what makes certain alleles susceptible to paramutation, the prevalence of this phenomenon and how the paramutation events are inherited in natural or artificial populations [78].

CONCLUDING REMARKS

The increased interest in epigenetic phenomenon and in particular on the role of epigenetics in contributing to phenotypic variability will allow a better understanding of these phenomena and contribute to an effective use of epigenetic variability in crop breeding programs. In particular it will be important to clarify the relative roles of epigenetic and genetic variability on the determination of the phenotypic variation of the traits of interest and, furthermore, it will be also fundamental to determine the stability of the epigenetic inheritance to make possible the use of the selected epialleles in breeding programs. The analysis of the epigenome not only from a static but also from a dynamic point of view, during development in different environments, will further clarify the potential of epigenomics in crop improvement.

CONFLICT OF INTEREST

The authors confirm that they have no conflict of interest to declare for this publication.

ACKNOWLEDGEMENTS

Declared None.

REFERENCES

[1] Waddington CH. The epigenotype. Endeavour 1942; 1: 18-20. a

[2] Waddington CH. Canalization of development and the inheritance of acquired characters. Nature 1942; 150: 563-5. b
[http://dx.doi.org/10.1038/150563a0]

[3] Deans C, Maggert KA. What do you mean, epigenetic? Genetics 2015; 199(4): 887-96.
[http://dx.doi.org/10.1534/genetics.114.173492] [PMID: 25855649]

[4] Russo VE, Martienssen RA, Riggs AD. Epigenetic mechanisms of gene regulation. Woodbury, USA: Cold Spring Harbor Laboratory Press 1996.

[5] Nanney DL. Epigenetic control systems. Proc Natl Acad Sci USA 1958; 44(7): 712-7.
[http://dx.doi.org/10.1073/pnas.44.7.712] [PMID: 16590265]

[6] Holliday R. Epigenetics: an overview. Dev Genet 1994; 15(6): 453-7.
[http://dx.doi.org/10.1002/dvg.1020150602] [PMID: 7834903]

[7] Richards EJ. Inherited epigenetic variation revisiting soft inheritance. Nat Rev Genet 2006; 7(5): 395-401.
[http://dx.doi.org/10.1038/nrg1834] [PMID: 16534512]

[8] Feng S, Jacobsen SE, Reik W. Epigenetic reprogramming in plant and animal development. Science

2010; 330(6004): 622-7.
[http://dx.doi.org/10.1126/science.1190614] [PMID: 21030646]

[9] Cao X, Jacobsen SE. Role of the *arabidopsis* DRM methyltransferases in *de novo* DNA methylation and gene silencing. Curr Biol 2002; 12(13): 1138-44.
[http://dx.doi.org/10.1016/S0960-9822(02)00925-9] [PMID: 12121623]

[10] Bartee L, Bender J. Two *Arabidopsis* methylation-deficiency mutations confer only partial effects on a methylated endogenous gene family. Nucleic Acids Res 2001; 29(10): 2127-34.
[http://dx.doi.org/10.1093/nar/29.10.2127] [PMID: 11353082]

[11] Law JA, Jacobsen SE. Establishing, maintaining and modifying DNA methylation patterns in plants and animals. Nat Rev Genet 2010; 11(3): 204-20.
[http://dx.doi.org/10.1038/nrg2719] [PMID: 20142834]

[12] Lindroth AM, Cao X, Jackson JP, *et al.* Requirement of CHROMOMETHYLASE3 for maintenance of CpXpG methylation. Science 2001; 292(5524): 2077-80.
[http://dx.doi.org/10.1126/science.1059745] [PMID: 11349138]

[13] Pavlopoulou A, Kossida S. Plant cytosine-5 DNA methyltransferases: structure, function, and molecular evolution. Genomics 2007; 90(4): 530-41.
[http://dx.doi.org/10.1016/j.ygeno.2007.06.011] [PMID: 17689048]

[14] Morgan HD, Santos F, Green K, Dean W, Reik W. Epigenetic reprogramming in mammals. Hum Mol Genet 2005; 14(Spec No 1): R47-58.
[http://dx.doi.org/10.1093/hmg/ddi114] [PMID: 15809273]

[15] Zemach A, Kim MY, Silva P, *et al.* Local DNA hypomethylation activates genes in rice endosperm. Proc Natl Acad Sci USA 2010; 107(43): 18729-34. b
[http://dx.doi.org/10.1073/pnas.1009695107] [PMID: 20937895]

[16] Lauria M, Rupe M, Guo M, *et al.* Extensive maternal DNA hypomethylation in the endosperm of *Zea mays*. Plant Cell 2004; 16(2): 510-22.
[http://dx.doi.org/10.1105/tpc.017780] [PMID: 14729913]

[17] Gehring M, Bubb KL, Henikoff S. Extensive demethylation of repetitive elements during seed development underlies gene imprinting. Science 2009; 324(5933): 1447-51.
[http://dx.doi.org/10.1126/science.1171609] [PMID: 19520961]

[18] Hsieh TF, Ibarra CA, Silva P, *et al.* Genome-wide demethylation of Arabidopsis endosperm. Science 2009; 324(5933): 1451-4.
[http://dx.doi.org/10.1126/science.1172417] [PMID: 19520962]

[19] Mayr E. The Evolutionary Synthesis. Cambridge, Massachusetts; London, England: Harvard University Press 1980; pp. 1-48.
[http://dx.doi.org/10.4159/harvard.9780674865389.c2]

[20] Trerotola M, Relli V, Simeone P, Alberti S. Epigenetic inheritance and the missing heritability. Hum Genomics 2015; 9: 17.
[http://dx.doi.org/10.1186/s40246-015-0041-3] [PMID: 26216216]

[21] Zemach A, McDaniel IE, Silva P, Zilberman D. Genome-wide evolutionary analysis of eukaryotic DNA methylation. Science 2010; 328(5980): 916-9.
[http://dx.doi.org/10.1126/science.1186366] [PMID: 20395474]

[22] Roudier F, Teixeira FK, Colot V. Chromatin indexing in Arabidopsis: an epigenomic tale of tails and more. Trends Genet 2009; 25(11): 511-7. [Erratum in: Trends Genet 2010; 26: 102].
[http://dx.doi.org/10.1016/j.tig.2009.09.013] [PMID: 19850370]

[23] Chen X, Zhou DX. Rice epigenomics and epigenetics: challenges and opportunities. Curr Opin Plant Biol 2013; 16(2): 164-9.
[http://dx.doi.org/10.1016/j.pbi.2013.03.004] [PMID: 23562565]

[24] Roudier F, Ahmed I, Bérard C, *et al.* Integrative epigenomic mapping defines four main chromatin states in Arabidopsis. EMBO J 2011; 30(10): 1928-38. [http://dx.doi.org/10.1038/emboj.2011.103] [PMID: 21487388]

[25] Manning K, Tör M, Poole M, *et al.* A naturally occurring epigenetic mutation in a gene encoding an SBP-box transcription factor inhibits tomato fruit ripening. Nat Genet 2006; 38(8): 948-52. [http://dx.doi.org/10.1038/ng1841] [PMID: 16832354]

[26] Marfil CF, Camadro EL, Masuelli RW. Phenotypic instability and epigenetic variability in a diploid potato of hybrid origin, Solanum ruiz-lealii. BMC Plant Biol 2009; 9: 21. [http://dx.doi.org/10.1186/1471-2229-9-21] [PMID: 19232108]

[27] Martin A, Troadec C, Boualem A, *et al.* A transposon-induced epigenetic change leads to sex determination in melon. Nature 2009; 461(7267): 1135-8. [http://dx.doi.org/10.1038/nature08498] [PMID: 19847267]

[28] Herrera CM, Bazaga P. Epigenetic differentiation and relationship to adaptive genetic divergence in discrete populations of the violet Viola cazorlensis. New Phytol 2010; 187(3): 867-76. [http://dx.doi.org/10.1111/j.1469-8137.2010.03298.x] [PMID: 20497347]

[29] Lira-Medeiros CF, Parisod C, Fernandes RA, Mata CS, Cardoso MA, Ferreira PC. Epigenetic variation in mangrove plants occurring in contrasting natural environment. PLoS One 2010; 5(4): e10326. [http://dx.doi.org/10.1371/journal.pone.0010326] [PMID: 20436669]

[30] Paun O, Bateman RM, Fay MF, Hedrén M, Civeyrel L, Chase MW. Stable epigenetic effects impact adaptation in allopolyploid orchids (Dactylorhiza: Orchidaceae). Mol Biol Evol 2010; 27(11): 2465-73. [http://dx.doi.org/10.1093/molbev/msq150] [PMID: 20551043]

[31] Feil R, Fraga MF. Epigenetics and the environment: emerging patterns and implications. Nat Rev Genet 2012; 13(2): 97-109. [PMID: 22215131]

[32] Guarino F, Cicatelli A, Brundu G, Heinze B, Castiglione S. Epigenetic diversity of clonal white poplar (Populus alba L.) populations: could methylation support the success of vegetative reproduction strategy? PLoS One 2015; 10(7): e0131480. [http://dx.doi.org/10.1371/journal.pone.0131480] [PMID: 26147352]

[33] Pilu R, Cassani E, Landoni M, *et al.* Genetic characterization of an Italian Giant Reed (*Arundo donax* L.) clones collection: exploiting clonal selection. Euphytica 2014; 196: 169-81. [http://dx.doi.org/10.1007/s10681-013-1022-z]

[34] Michaels SD, Amasino RM. *FLOWERING LOCUS C* encodes a novel MADS domain protein that acts as a repressor of flowering. Plant Cell 1999; 11(5): 949-56. [http://dx.doi.org/10.1105/tpc.11.5.949] [PMID: 10330478]

[35] Sheldon CC, Burn JE, Perez PP, *et al.* The FLF MADS box gene: a repressor of flowering in Arabidopsis regulated by vernalization and methylation. Plant Cell 1999; 11(3): 445-58. [http://dx.doi.org/10.1105/tpc.11.3.445] [PMID: 10072403]

[36] Sheldon CC, Hills MJ, Lister C, Dean C, Dennis ES, Peacock WJ. Resetting of *FLOWERING LOCUS C* expression after epigenetic repression by vernalization. Proc Natl Acad Sci USA 2008; 105(6): 2214-9. [http://dx.doi.org/10.1073/pnas.0711453105] [PMID: 18250331]

[37] Choi J, Hyun Y, Kang MJ, *et al.* Resetting and regulation of *Flowering Locus C* expression during Arabidopsis reproductive development. Plant J 2009; 57(5): 918-31. [http://dx.doi.org/10.1111/j.1365-313X.2008.03776.x] [PMID: 19121105]

[38] Johanson U, West J, Lister C, Michaels S, Amasino R, Dean C. Molecular analysis of *FRIGIDA*, a major determinant of natural variation in Arabidopsis flowering time. Science 2000; 290(5490): 344-7.

[http://dx.doi.org/10.1126/science.290.5490.344] [PMID: 11030654]

[39] Xu L, Zhao Z, Dong A, *et al.* Di- and tri- but not monomethylation on histone H3 lysine 36 marks active transcription of genes involved in flowering time regulation and other processes in *Arabidopsis thaliana*. Mol Cell Biol 2008; 28(4): 1348-60.
[http://dx.doi.org/10.1128/MCB.01607-07] [PMID: 18070919]

[40] Jiang D, Gu X, He Y. Establishment of the winter-annual growth habit *viaFRIGIDA*-mediated histone methylation at *FLOWERING LOCUS C* in Arabidopsis. Plant Cell 2009; 21(6): 1733-46.
[http://dx.doi.org/10.1105/tpc.109.067967] [PMID: 19567704]

[41] Liu F, Quesada V, Crevillén P, Bäurle I, Swiezewski S, Dean C. The Arabidopsis RNA-binding protein FCA requires a lysine-specific demethylase 1 homolog to downregulate *FLC*. Mol Cell 2007; 28(3): 398-407.
[http://dx.doi.org/10.1016/j.molcel.2007.10.018] [PMID: 17996704]

[42] Liu F, Marquardt S, Lister C, Swiezewski S, Dean C. Targeted 3 processing of antisense transcripts triggers Arabidopsis *FLC* chromatin silencing. Science 2010; 327(5961): 94-7.
[http://dx.doi.org/10.1126/science.1180278] [PMID: 19965720]

[43] Wood CC, Robertson M, Tanner G, Peacock WJ, Dennis ES, Helliwell CA. The *Arabidopsis thaliana* vernalization response requires a polycomb-like protein complex that also includes VERNALIZATION INSENSITIVE 3. Proc Natl Acad Sci USA 2006; 103(39): 14631-6.
[http://dx.doi.org/10.1073/pnas.0606385103] [PMID: 16983073]

[44] Gendall AR, Levy YY, Wilson A, Dean C. The VERNALIZATION 2 gene mediates the epigenetic regulation of vernalization in Arabidopsis. Cell 2001; 107(4): 525-35.
[http://dx.doi.org/10.1016/S0092-8674(01)00573-6] [PMID: 11719192]

[45] De Lucia F, Crevillen P, Jones AM, Greb T, Dean C. A PHD-polycomb repressive complex 2 triggers the epigenetic silencing of *FLC* during vernalization. Proc Natl Acad Sci USA 2008; 105(44): 16831-6.
[http://dx.doi.org/10.1073/pnas.0808687105] [PMID: 18854416]

[46] Sung S, Amasino RM. Vernalization in Arabidopsis thaliana is mediated by the PHD finger protein VIN3. Nature 2004; 427(6970): 159-64.
[http://dx.doi.org/10.1038/nature02195] [PMID: 14712276]

[47] Greb T, Mylne JS, Crevillen P, *et al.* The PHD finger protein VRN5 functions in the epigenetic silencing of Arabidopsis *FLC*. Curr Biol 2007; 17(1): 73-8.
[http://dx.doi.org/10.1016/j.cub.2006.11.052] [PMID: 17174094]

[48] Swiezewski S, Liu F, Magusin A, Dean C. Cold-induced silencing by long antisense transcripts of an Arabidopsis Polycomb target. Nature 2009; 462(7274): 799-802.
[http://dx.doi.org/10.1038/nature08618] [PMID: 20010688]

[49] Hornyik C, Terzi LC, Simpson GG. The spen family protein FPA controls alternative cleavage and polyadenylation of RNA. Dev Cell 2010; 18(2): 203-13.
[http://dx.doi.org/10.1016/j.devcel.2009.12.009] [PMID: 20079695]

[50] Springer NM. Epigenetics and crop improvement. Trends Genet 2013; 29(4): 241-7.
[http://dx.doi.org/10.1016/j.tig.2012.10.009] [PMID: 23128009]

[51] Bowman JL, Sakai H, Jack T, Weigel D, Mayer U, Meyerowitz EM. SUPERMAN, a regulator of floral homeotic genes in Arabidopsis. Development 1992; 114(3): 599-615.
[PMID: 1352237]

[52] Gaiser JC, Robinson-Beers K, Gasser CS. The Arabidopsis SUPERMAN gene mediates asymmetric growth of the outer integument of ovules. Plant Cell 1995; 7(3): 333-45.
[http://dx.doi.org/10.1105/tpc.7.3.333] [PMID: 12242374]

[53] Jacobsen SE, Meyerowitz EM. Hypermethylated SUPERMAN epigenetic alleles in arabidopsis. Science 1997; 277(5329): 1100-3.

[http://dx.doi.org/10.1126/science.277.5329.1100] [PMID: 9262479]

[54] Eichten SR, Swanson-Wagner RA, Schnable JC. Heritable epigenetic variation among maize inbreds. Plos Genet 2011; 7: e1002372.
[http://dx.doi.org/10.1371/journal.pgen.1002372] [PMID: 22125494]

[55] Johannes F, Porcher E, Teixeira FK, *et al.* Assessing the impact of transgenerational epigenetic variation on complex traits. PLoS Genet 2009; 5(6): e1000530.
[http://dx.doi.org/10.1371/journal.pgen.1000530] [PMID: 19557164]

[56] Hauben M, Haesendonckx B, Standaert E, *et al.* Energy use efficiency is characterized by an epigenetic component that can be directed through artificial selection to increase yield. Proc Natl Acad Sci USA 2009; 106(47): 20109-14.
[http://dx.doi.org/10.1073/pnas.0908755106] [PMID: 19897729]

[57] Li MR, Shi FX, Zhou YX, *et al.* Genetic and epigenetic diversities shed light on domestication of cultivated ginseng (Panax ginseng). Mol Plant 2015; 8(11): 1612-22.
[http://dx.doi.org/10.1016/j.molp.2015.07.011] [PMID: 26278367]

[58] Pilu R. Paramutation phenomena in plants. Semin Cell Dev Biol 2015; 44: 2-10.
[http://dx.doi.org/10.1016/j.semcdb.2015.08.015] [PMID: 26335267]

[59] Brink RA. Change associated with the R locus in maize is directed and potentially reversible. Genetics 1956; 41(6): 872-89.
[PMID: 17247669]

[60] Romano N, Macino G. Quelling: transient inactivation of gene expression in Neurospora crassa by transformation with homologous sequences. Mol Microbiol 1992; 6(22): 3343-53.
[http://dx.doi.org/10.1111/j.1365-2958.1992.tb02202.x] [PMID: 1484489]

[61] Napoli C, Lemieux C, Jorgensen R. Introduction of a chimeric chalcone synthase gene into Petunia results in reversible co-suppression of homologous genes in trans. Plant Cell 1990; 2(4): 279-89.
[http://dx.doi.org/10.1105/tpc.2.4.279] [PMID: 12354959]

[62] Fire A, Xu S, Montgomery MK, Kostas SA, Driver SE, Mello CC. Potent and specific genetic interference by double-stranded RNA in *Caenorhabditis elegans*. Nature 1998; 391(6669): 806-11.
[http://dx.doi.org/10.1038/35888] [PMID: 9486653]

[63] Pilu R, Panzeri D, Cassani E, Cerino Badone F, Landoni M, Nielsen E. A paramutation phenomenon is involved in the genetics of maize low phytic acid1241 (lpa1241) trait. Heredity (Edinb) 2009; 102(3): 236-45.
[http://dx.doi.org/10.1038/hdy.2008.96] [PMID: 18781168]

[64] Sidorenko LV, Peterson T. Transgene-induced silencing identifies sequences involved in the establishment of paramutation of the maize p1 gene. Plant Cell 2001; 13(2): 319-35.
[http://dx.doi.org/10.1105/tpc.13.2.319] [PMID: 11226188]

[65] Chandler VL, Eggleston WB, Dorweiler JE. Paramutation in maize. Plant Mol Biol 2000; 43(2-3): 121-45.
[http://dx.doi.org/10.1023/A:1006499808317] [PMID: 10999400]

[66] Chandler VL, Stam M. Chromatin conversations: mechanisms and implications of paramutation. Nat Rev Genet 2004; 5(7): 532-44.
[http://dx.doi.org/10.1038/nrg1378] [PMID: 15211355]

[67] Das OP, Messing J. Variegated phenotype and developmental methylation changes of a maize allele originating from epimutation. Genetics 1994; 136(3): 1121-41.
[PMID: 8005419]

[68] Lund G, Das OP, Messing J. Tissue-specific DNAse-I-sensitive sites of the maizep-gene [??] and their changes upon epimutation. Plant J 1995; 7: 797-807. [Please check title].
[http://dx.doi.org/10.1046/j.1365-313X.1995.07050797.x]

[69] Della Vedova CB, Cone KC. Paramutation: the chromatin connection. Plant Cell 2004; 16(6): 1358-64.
[http://dx.doi.org/10.1105/tpc.160630] [PMID: 15178748]

[70] Stam M, Mittelsten Scheid O. Paramutation: an encounter leaving a lasting impression. Trends Plant Sci 2005; 10(6): 283-90.
[http://dx.doi.org/10.1016/j.tplants.2005.04.009] [PMID: 15949762]

[71] Alleman M, Sidorenko L, McGinnis K, *et al.* An RNA-dependent RNA polymerase is required for paramutation in maize. Nature 2006; 442(7100): 295-8.
[http://dx.doi.org/10.1038/nature04884] [PMID: 16855589]

[72] Chandler VL. Paramutation: from maize to mice. Cell 2007; 128(4): 641-5.
[http://dx.doi.org/10.1016/j.cell.2007.02.007] [PMID: 17320501]

[73] Harrison BJ, Carpenter RA. Comparison of the instabilities at the Nivea and Pallida loci in Antirrhinum majus. Heredity 1973; 31: 309-23.
[http://dx.doi.org/10.1038/hdy.1973.88]

[74] Sastry GR, Cooper HB Jr, Brink RA. Paramutation and somatic mosaicism in maize. Genetics 1965; 52(2): 407-24.
[PMID: 5861569]

[75] Mikula BC. Heritable changes in R-locus expression in maize in response to environment. Genetics 1967; 56(4): 733-42.
[PMID: 17248390]

[76] Meyer P, Linn F, Heidmann I, Meyer H, Niedenhof I, Saedler H. Endogenous and environmental factors influence 35S promoter methylation of a maize A1 gene construct in transgenic petunia and its colour phenotype. Mol Gen Genet 1992; 231(3): 345-52.
[http://dx.doi.org/10.1007/BF00292701] [PMID: 1538691]

[77] Pilu R. Paramutation: just a curiosity or fine tuning of gene expression in the next generation? Curr Genomics 2011; 12(4): 298-306.
[http://dx.doi.org/10.2174/138920211795860099] [PMID: 22131875]

[78] Springer NM, McGinnis KM. Paramutation in evolution, population genetics and breeding. Semin Cell Dev Biol 2015; 44: 33-8.
[http://dx.doi.org/10.1016/j.semcdb.2015.08.010] [PMID: 26325077]

[79] Banks JA, Masson P, Fedoroff N. Molecular mechanisms in the developmental regulation of the maize Suppressor-mutator transposable element. Genes Dev 1988; 2(11): 1364-80.
[http://dx.doi.org/10.1101/gad.2.11.1364] [PMID: 2463208]

[80] Brink RA. Change associated with the R locus in maize is directed and potentially reversible. Genetics 1956; 41(6): 872-89.
[PMID: 17247669]

[81] Brown DF, Brink RA. Paramutagenic action of paramutant Rr and Rg alleles in maize. Genetics 1960; 45(10): 1313-6.
[PMID: 17248001]

[82] Stam M, Belele C, Dorweiler JE, Chandler VL. Differential chromatin structure within a tandem array 100 kb upstream of the maize *b1* locus is associated with paramutation. Genes Dev 2002; 16(15): 1906-18.
[http://dx.doi.org/10.1101/gad.1006702] [PMID: 12154122]

[83] Hollick JB, Patterson GI, Coe EH Jr, Cone KC, Chandler VL. Allelic interactions heritably alter the activity of a metastable maize pl allele. Genetics 1995; 141(2): 709-19.
[PMID: 8647404]

[84] Lippman Z, Gendrel AV, Black M, *et al.* Role of transposable elements in heterochromatin and

epigenetic control. Nature 2004; 430(6998): 471-6.
[http://dx.doi.org/10.1038/nature02651] [PMID: 15269773]

[85] Soppe WJ, Jacobsen SE, Alonso-Blanco C, *et al.* The late flowering phenotype of fwa mutants is caused by gain-of-function epigenetic alleles of a homeodomain gene. Mol Cell 2000; 6(4): 791-802.
[http://dx.doi.org/10.1016/S1097-2765(05)00090-0] [PMID: 11090618]

[86] Bender J, Fink GR. Epigenetic control of an endogenous gene family is revealed by a novel blue fluorescent mutant of Arabidopsis. Cell 1995; 83(5): 725-34.
[http://dx.doi.org/10.1016/0092-8674(95)90185-X] [PMID: 8521489]

[87] Melquist S, Bender J. Transcription from an upstream promoter controls methylation signaling from an inverted repeat of endogenous genes in Arabidopsis. Genes Dev 2003; 17(16): 2036-47.
[http://dx.doi.org/10.1101/gad.1081603] [PMID: 12893775]

[88] Hagemann R. Somatic conversion (paramutation) at the sulfurea locus of Lycopersicon esculentum Mill. III. Studies with trisomics. Can J Genet Cytol 1969; 11: 346-58.
[http://dx.doi.org/10.1139/g69-043]

[89] Hagemann R, Berg W. Paramutation at the sulfurea locus of Lycopersicon esculentum Mill.: VII. Determination of the time of occurrence of paramutation by the quantitative evaluation of the variegation. Theor Appl Genet 1978; 53(3): 113-23.
[http://dx.doi.org/10.1007/BF00272688] [PMID: 24311343]

[90] Krebbers E, Hehl R, Piotrowiak R, Lönnig WE, Sommer H, Saedler H. Molecular analysis of paramutant plants of Antirrhinum majus and the involvement of transposable elements. Mol Gen Genet 1987; 209(3): 499-507.
[http://dx.doi.org/10.1007/BF00331156] [PMID: 17193710]

[91] Matzke MA, Primig M, Trnovsky J, Matzke AJ. Reversible methylation and inactivation of marker genes in sequentially transformed tobacco plants. EMBO J 1989; 8(3): 643-9.
[PMID: 16453872]

[92] Meyer P, Heidmann I, Niedenhof I. Differences in DNA-methylation are associated with a paramutation phenomenon in transgenic petunia. Plant J 1993; 4(1): 89-100.
[http://dx.doi.org/10.1046/j.1365-313X.1993.04010089.x] [PMID: 8220478]

[93] Mikula BC. Environmental programming of heritable epigenetic changes in paramutant r-gene expression using temperature and light at a specific stage of early development in maize seedlings. Genetics 1995; 140(4): 1379-87.
[PMID: 7498777]

[94] Park YD, Papp I, Moscone EA, *et al.* Gene silencing mediated by promoter homology occurs at the level of transcription and results in meiotically heritable alterations in methylation and gene activity. Plant J 1996; 9(2): 183-94.
[http://dx.doi.org/10.1046/j.1365-313X.1996.09020183.x] [PMID: 8820605]

[95] English JJ, Jones JD. Epigenetic instability and trans-silencing interactions associated with an SPT:Ac T-DNA locus in tobacco. Genetics 1998; 148(1): 457-69.
[PMID: 9475755]

[96] Stokes TL, Richards EJ. Induced instability of two Arabidopsis constitutive pathogen-response alleles. Proc Natl Acad Sci USA 2002; 99(11): 7792-6.
[http://dx.doi.org/10.1073/pnas.112040999] [PMID: 12032362]

[97] Mittelsten Scheid O, Afsar K, Paszkowski J. Formation of stable epialleles and their paramutation-like interaction in tetraploid *Arabidopsis thaliana*. Nat Genet 2003; 34(4): 450-4.
[http://dx.doi.org/10.1038/ng1210] [PMID: 12847525]

CHAPTER 5

Enhancing Photosynthesis: Different Strategies to Improve the Process at the Basis of Life on Earth

Monica Colombo[1] and **Paolo Pesaresi**[2,*]

[1] *Research and Innovation Centre, Fondazione Edmund Mach, San Michele all'Adige (Trento), Italy*

[2] *Department of Biosciences, Università degli Studi di Milano, Milano, Italy*

Abstract: Doubling agricultural production will be essential by 2050 to satisfy the demand of food of a constantly growing population, but climate change brings a lot of uncertainty and complexity to this challenge for agriculture. One of the most important changes that must be addressed is the increase in atmospheric [CO_2], which has increased from approximately 280 ppm in pre-industrial times to about 400 ppm nowadays and will further increase to values of 470–570 ppm by 2050 depending on the climate scenario (IPCC Synthesis report, Climate Change 2007). Although this increase in [CO_2] is expected to have a positive and significant effect on C_3 crops production, it is counteracted by the rise in temperature and the higher evaporative demand, with the increased risks for drought and heat likely to be progressive in all regions of our planet. As a matter of fact, the average stimulation of C_3 leaf photosynthesis under field conditions at elevated [CO_2] has been reported to be only 14% on average across FACE (550–600 ppm in Free Air CO_2 Enrichment) experiments, much lower than the expected increase of 38%.

Down-regulation of photosynthesis can be ascribe to multiple factors. These include the limited sink strength of the plants and the consequent accumulation of inhibitory photo-assimilates, the "hysterical" behavior of photosynthetic organisms to excess illumination, by either triggering EED (Excess Energy Dissipation) beyond the level effective for photo-protection or retaining a relevant fraction of quenching for extended periods after return to limiting light conditions, and the complex and multi-factorial network that controls CO_2 fixation and carbon allocation.

Here we describe the genetic constraints that limit yield potential and prevent it from being realized on the farm, in order to improve the understanding of plant responses under elevated [CO_2], and provide tentative biotechnological solutions to overcome the crop yield limitations.

It is worth noting that the huge improvements in agricultural production gained during the 'Green Revolution' were not directly related to manipulation of photosynthesis, therefore its modification remains an unexplored target for crop improvement.

* **Corresponding author Paolo Pesaresi**: Department of Biosciences, Università degli Studi di Milano, Milano, Italy; Tel/Fax: +39.02.50315057; E-mail: paolo.pesaresi@unimi.it

Keywords: Dark phase efficiency, Improvement strategies, Light phase efficiency, Photosynthesis, Yield potential.

INTRODUCTION

A steady and reliable supply of food is an essential factor in ensuring the survival of humankind and the stability of human societies. However, it is expected that the global population will increase from the current 7 billion to 9. 3 billion by mid-century before stabilizing at the end of the century [1, 2]. In addition to population growth, other factors are expected to put a strong pressure on agricultural production, such as the growing economic prosperity of the developing countries, which results in the formation of a middle class with increased demands for goods, and a strong growth in urban population. The final outcome will be an enlarged demand for processed foods and high-quality animal products, like dairy products and meat. Furthermore, taking in consideration the rapidly decreasing reserves of fossil fuels, such as oil, coal, and natural gas, the use of plants as a source of energy (biofuel) and as a commodity for industrial uses is expected to grow, making a substantial increase in agricultural output even more important and necessary. It is estimated that to meet these rising demands global agricultural production needs at least to be doubled by the year 2050, and this while adapting to climate change and adopting a global sustainable model of agricultural production, in order to take in consideration also the equally crucial needs for environmental protection [3].

Doubling yields will require considerable efforts. If in the coming years the rates of crop yield increase will simply be maintained at the current level, by the middle of this century we will be unable to meet the demand of food. For instance, it has been estimated that wheat, rice, maize and soybean, the most important global crops, have current yield improvements far below those deemed necessary to double the production by 2050 (0.9-1.6% per year, *versus* the required 2.4%) [4] (Fig. **1**).

Furthermore, the rate of yield improvement in some areas of the world is stagnating if not decreasing. Of the most important cereals, only maize maintained steady growth rates [2, 5]. The yields per hectare of rice in the most important producing countries (China, India and Indonesia) strongly increased between 1970 and 1990 but then they gradually downsized during the following years, despite continued genetic improvement [6] (Fig. **2**).

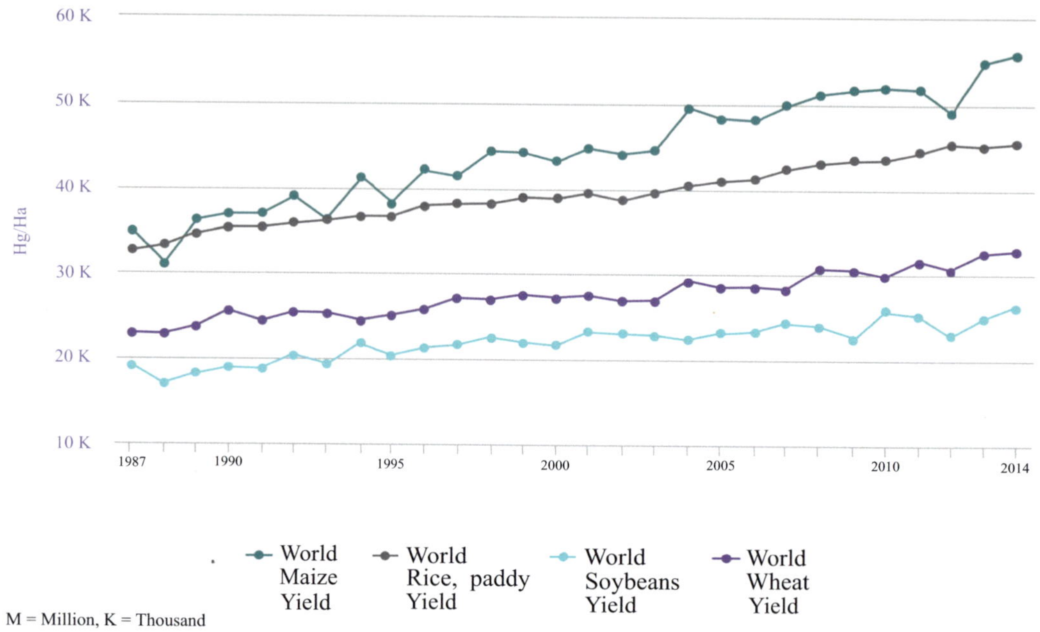

Fig. (1). Annual average global yields of the most important crops top (wheat, rice, maize, and soybean) from 1987 to 2014 (Source: Food and Agriculture Organization of the United Nations, FAOSTAT, 2015).

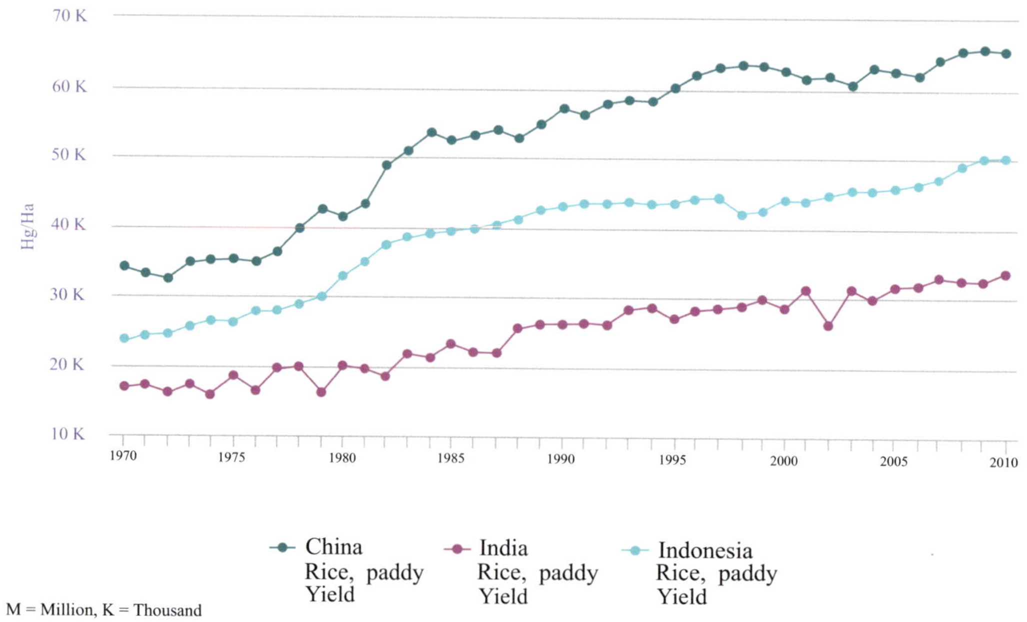

Fig. (2). Annual average yields of rice in China, India, and Indonesia from 1970 to 2010 (Source: Food and Agriculture Organization of the United Nations, FAOSTAT, 2015).

The consequences of these scarce rates of yield change, such as a rise in food prices, could be serious, especially in countries characterized by low-income and rapidly growing population. In addition, the situation could be even worse in the poorest tropical countries, since the increase in yield of the crops typical of these countries is slower. For instance, cassava, which is a major staple food in the developing countries, had a global average increase in yield per hectare equal to 63% between 1960 and 2010 whereas, over the same period, a 171% increase for wheat was achieved [6].

YIELD IMPROVEMENT AND PHOTOSYNTHETIC LIGHT CONVERSION EFFICIENCY

The significant gains in productivity attributable to the Green Revolution, occurred in the 20th century, were achieved mainly through the combination of high levels of investment in crop genetic improvement together with enhanced practices of crop management and protection and have led to the fulfilment of the crop genetic yield potential (see Table **1** for the complete definition). The yield potential of a given genotype, *i.e.* the yield that a crop can obtain under optimal management practices and in the absence of abiotic and biotic stresses, depends on the amount of solar radiation received and the efficiencies with which the crop intercepts the photosynthetically active light, converts it into biomass, and then partitions the biomass into the harvested part of the plant [3, 6].

Table 1. Glossary of terms.

Yield potential	yield (mass of harvested material per hectare of land) that a genotype of a crop can achieve in a given environment under optimal management practices and in the absence of abiotic and biotic stresses, therefore with good nutrition and in the absence of pests, diseases, and drought
Harvest index	proportion of the aboveground biomass of a plant that is invested into the harvested portion of the crop (*i.e.* the grain)
Interception efficiency	proportion of the photosynthetically active radiation (PAR) that is intercepted by the crop over the growing season
Conversion efficiency	ratio of biomass produced (in terms of mass or energy) over a given period to the solar energy or PAR intercepted or absorbed by the canopy over the same period

The increases in yield potential were obtained mainly through selection of genotypes able to allocate a substantial amount of their biomass into the harvested product (*i.e.* harvest index, see Table **1**). The classic example is represented by the selection in wheat of dwarf plants, which led to the accumulation of more biomass in the grain than in the stem. During the last 50 years this index has almost doubled in the most important grain crops, with the best germplasm that are currently able to accumulate over 50% of their total biomass into the grain, suggesting that the harvest index is now very close to its theoretical upper limit, with little room for further improvement [6].

Light **interception efficiency** (see Table **1**) of the photosynthetically active radiation (PAR), depends on several factors, such as leaf absorbance, the speed of canopy development and closure as well as canopy longevity, size, and architecture [7]. Similarly to the harvest index, also this parameter has been improved during the years, reaching values equal to 0.8–0.9 for modern crop genotypes, meaning that the crop leaf canopy is able to intercept over 90% of growing season radiation and suggesting that also this determinant is near to its biological limits [6].

The only determinant that until now has played a marginal role in the improvement of yield potential is the **conversion efficiency** (see Table **1**) of visible solar energy. Conversion efficiency depends on the efficiency of the photosynthetic process, net of respiratory losses by the crop. This parameter currently seems to be far away from its biological limits since it has been estimated to be at less than half of its theoretical maximum both in C_3 and C_4 plants, and therefore improvement of photosynthesis, an area where little work has been done so far in crops, can be considered a promising strategy to achieve a further increase in the yield potential of crops [6 - 11].

CARBONATING THE ATMOSPHERE: LESSONS FROM FACE

Atmospheric carbon dioxide concentration ([CO_2]) was stable at roughly 270 ppm (parts per million) for at least a millennium before the beginning of the Industrial Revolution. Since then, CO_2 started to accumulate in the atmosphere at an accelerating rate. In 2009 [CO_2] was 384 ppm [12], while recently it exceeded the threshold of 400 ppm [13], a level not seen in the last 20 million years [12]. If current trends in emissions continue it is expected that by the middle of this century [CO_2] will exceede 550 ppm, reaching 700 ppm by the end of the century [12]. It is interesting to note that in the last about 24 million years atmospheric CO_2 concentration appears to have remained below 500 ppm [14]. Therefore, the progenitors of our modern crops evolved under conditions of [CO_2], different from the ones they will start facing in the second half of this century.

The rise in atmospheric [CO_2] is the major factor that contributes to global warming and changes in precipitation regimes, which can in turn lead to more extreme weather events, such as heat waves and droughts [5, 9], that may constitute a serious threat to our agricultural and forestry production systems. On the other side CO_2 is a limiting substrate for photosynthesis in C_3 crops, so it has been demonstrated that CO_2 fertilization promotes higher levels of photosynthesis and increases the efficiency of light use, as well as resource use efficiency, leading to an increase in crop biomass and yield [15]. Understanding how plants and more generally ecosystems, are able to respond and adapt to this rapid change in [CO_2], is a key step in understanding the impact that global climate change will have on terrestrial ecosystems.

Within this frame, many studies of the effects of high concentrations of CO_2 on crop photosynthesis and production have been conducted over the past decades, in closed chambers or in open air, in order to understand how the relative stimulation of biomass yield can vary with rising temperatures and drought stress. Experiments performed in controlled laboratory chambers have generated a huge knowledge on how plants respond to elevated [CO_2] and which are the possible mechanisms of adaptation. Free-Air CO_2 Enrichment (FACE) experiments, on the other side, have provided an important reality check since they enable investigation of plant responses to elevated CO_2 levels under natural and fully open-air conditions. FACE technology, indeed, uses vent pipes, placed in the open, to release CO_2 at the periphery of vegetation plots. The released CO_2 is then dispersed across the experimental area by the combined action of natural wind and diffusion [16]. Results obtained from numerous FACE experiments, conducted in different parts of the world and in different environmental conditions, show that in open-air conditions the photosynthetic process does not reach the theoretical increase predicted from *in silico* calculations and from observations of the germplasm in protected environments [12, 16, 17]. Furthermore, the genotypic variability of grain yield in response to elevated [CO_2] among cultivars of C_3 crops has been shown to be substantial, implying that there is enough room for improving carbon gain. The response of rice has been examined in a number of studies demonstrating that considerable variation in the growth and yield response exists among different cultivars [18, 19]. In a screening performed on 16 genotypes, grown under normal or elevated concentrations of CO_2 in open top chambers, the percentage yield increases observed ranged from 3% to 175% [19] (Fig. **3a**). Moreover, it has been demonstrated that plant breeding in the past influenced CO_2 responsiveness. Studies performed on cultivars released in the last decades indicate that the yield response to [CO_2] has declined with the introduction of new varieties. A study performed on different cultivars of spring wheat (*Triticum aestivum* L.) introduced in Germany between 1890 and 1988 demonstrated that CO_2-associated increase in grain yield and biomass of the old cultivars (1890-1943) ranged between 40 and 50% whereas the average increase in the modern cultivars (1965-1988) amounted to circa 28% [20] (Fig. **3b**). Similarly, older cultivars of hard-red spring wheat showed greater [CO_2] yield sensitivity compared to the newer ones [21]. Taken together, these data demonstrate that genotypes with high CO_2 responsiveness are available for breeding programs aimed to the selection of new cultivars, able to be more productive in the forthcoming high CO_2 climate.

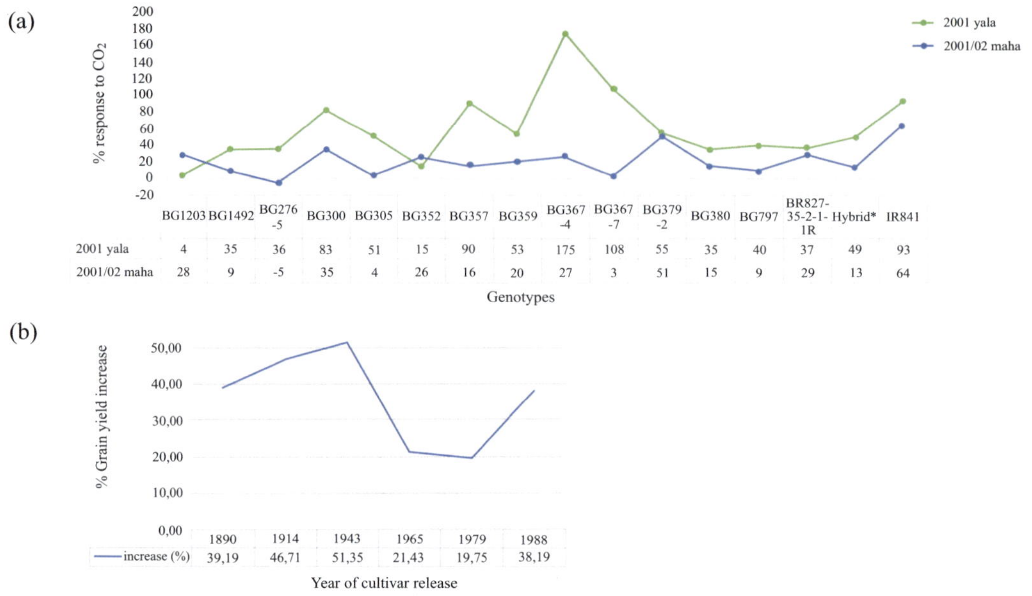

Fig. (3). (a) Yield response to increased atmospheric [CO_2] of different rice genotypes, as reported by De Costa *et al.*, 2007 [19]. Sixteen rice genotypes were evaluated in the study. These included cultivated varieties, inbred lines and one hybrid* (hybrid between IR58025A and BR827-35-2-1R). The plants were grown under ambient and elevated concentrations of CO_2 in open top chambers at the Rice Research and Development Institute in Sri Lanka, from May to August 2001 (locally known as yala season) and from November 2001 to March 2002 (maha season). All genotypes, with a few exceptions (*i.e.* BG276-5, for which a yield decrease in maha was observed), showed significant increases in grain yield at high concentrations of CO_2 compared to the ambient one, with significant variation between genotypes. The highest percentage increases in yield were observed in BG367-4 (175%) in yala and IR841 (64%) in maha, while the lowest performing genotypes were BG1203 (4%) in yala and BG367-7 (3%) in maha. (b) Effect of CO_2 enrichment on grain yield, as reported by Manderscheid & Weigel 1997 [20], in six different cultivars of spring wheat (*Triticum aestivum* L.) introduced in Germany between 1890 and 1988: cv. 'Heines Kolben' from 1890, cv. 'Janetzkis Friiher' from 1914, cv. 'Rimpaus Langensteiner' from 1943, cv. 'Adler' from 1965, cv. 'Turbo' from 1979, and cv. 'Nandu' from 1988. The different cultivars were cultivated in pots under normal and elevated CO_2 concentrations in open-top chambers. All cultivars showed an increase in grain yield per pot at elevated concentrations of CO_2 which ranged between circa 20 and 50%, with the greatest responses observed for the older cultivars.

Since in plants, carbon gain and yield are controlled by source and sink strength these results clearly indicate that nowadays source strength is still a limiting factor for productivity. On the other side, the observed increase was lower than predicted, suggesting that to fully exploit a stronger carbon source also more sink strength is required [11, 22, 23].

In general, these data demonstrate that improvement of photosynthesis would be clearly beneficial to crop yield [17].

TARGETS FOR INCREASING CROP PHOTOSYNTHETIC EFFICIENCY

As previously highlighted, the achieved net photosynthetic efficiency of crops is far below the theoretical one. The maximum efficiency of the photosynthetic process that could theoretically be achieved in C3 and C4 plants has been evaluated by Zhu *et al.*, 2008 [8]. They estimated the potential efficiency of each step, from interception of radiation to its conversion in plant biomass, identifying several steps where energy is lost (*i.e.* solar radiation outside the photosynthetically active spectrum, reflection and transmission of the light, photochemical inefficiency, thermodynamic limitations, carbohydrate biosynthesis, photorespiration in C3 plants and respiration). This analysis revealed that the theoretical maximum efficiency of conversion of the solar energy into biomass is 4.6% for C3 photosynthesis and 6% for C4 plants, when considering the total solar spectrum radiation at the surface of the earth. If this evaluation is limited only to the photosynthetically active radiation (PAR, 400–700 nm), these efficiencies are higher, becoming 9.4% for C3 and 12.3% for C4 [8]. However the highest conversion efficiency reported for crops across a full growing season is about 2.4% for C3 plants and about 3.7% for C4, with higher short-term conversion efficiencies observed only for brief periods [8]. Conversion efficiency of visible solar energy is, therefore, well below its theoretical maximum and several promising targets for its improvement have already been identified (for a summary of strategies aimed to improve photosynthesis efficiency, see Table **2** at the end of this chapter).

Improving Light Reactions

The first step in the photosynthetic process is the collection of the light energy, which is passed within the light-harvesting system and transferred to the reaction centre for the primary photochemical reaction. Light-harvesting complexes determine the quality and quantity of the sunlight available for photosynthesis.

Three approaches have been identified for improving the efficiency of light reactions (Fig. **4**): the first is the expansion of the solar spectrum available for the oxygenic photosynthesis, in order to augment the number of absorbed photons per unit of area irradiated and consequently the maximum amount of energy that can potentially be stored; the second is the fine tuning of the antenna size to increase the chances that an absorbed photon will be used to produce energy, therefore reducing the amount of excess light and favouring photosynthesis also in the lower leaves of the canopy; and the third one deals with faster relaxation of photoprotective mechanisms in order to maintain optimal photosynthesis efficiency even during the sudden changes of environmental light conditions (Table **2**).

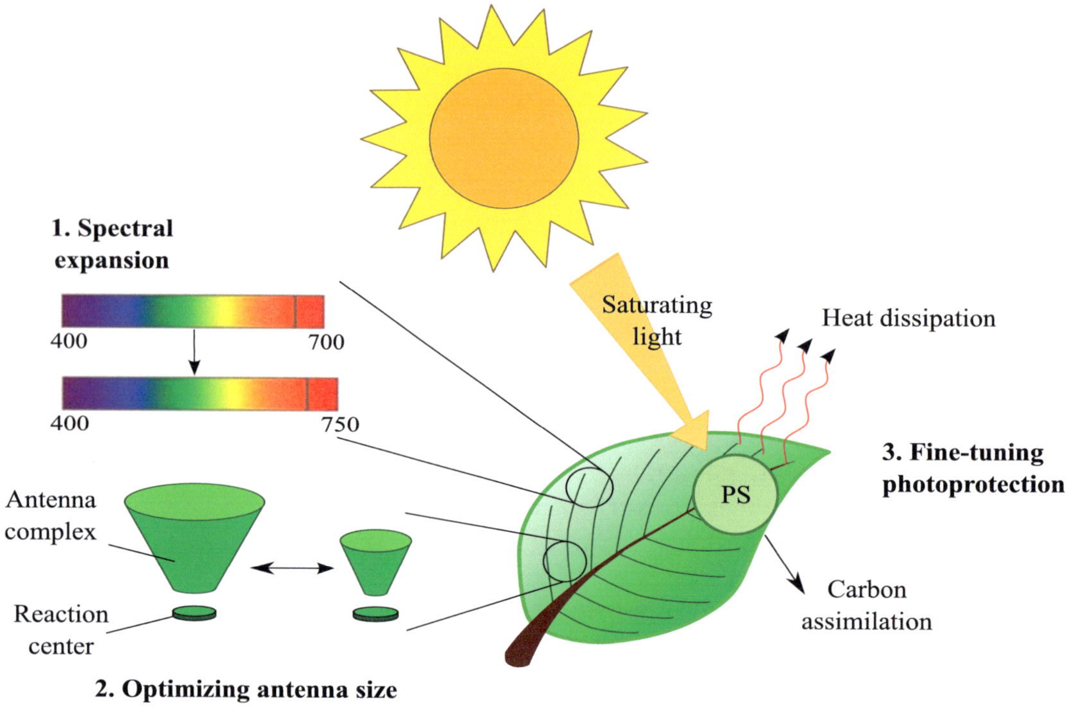

Fig. (4). Schematic representation of the three strategies that could improve the efficiency of the light reactions.

Spectral Expansion

Oxygenic photosynthetic organisms, such as plants and the green algae can only utilize the visible portion of the solar spectrum. This spectral range (400-700 nm) is called Photosynthetically Active Radiation (PAR) [24] and amounts to only 48.7% of the total solar energy available at the earth surface [8]. Expansion of the PAR range could therefore increase significantly the amount of energy available for the photosynthetic process.

For this purpose, the study of the photosynthetic organisms existing on Earth is essential. Photosynthetic organisms are indeed able to live nearly in every place where sunlight can penetrate since they have adapted their pigmentation to the specific spectral composition of light typical of their habitat. Photosynthetic bacteria, which rely on bacteriochlorophyll (Bchl) pigments for light interception, can use a broader range of wavelengths. For instance, many bacteria can capture and utilize wavelengths longer than 900 nm. Some Bchl *b*-containing purple bacteria (*i.e-Blastochloris viridis*), are even able to use wavelengths beyond 1000 nm. However, these organisms, unlike algae and plants, perform anoxygenic photosynthesis. In the case of oxygenic phototrophs, the higher energetic threshold required (which determines the wavelength limit of approximately 700 nm) was attributed to the elevated energy requirements of the water oxidation reaction. The threshold is indeed established by the longest wavelength photon able to drive the process, meaning that photons with longer wavelengths do not possess enough energy and are consequently not absorbed [25].

However, the discovery of first *Acaryochloris marina*, a Chl *d*-containing cyanobacterium [26], and more recently of *Halomicronema hongdechloris*, a Chl *f*-containing organism [27, 28] has compelled a revaluation of the minimum energetic threshold in oxygenic photosynthesis [25].

Acaryochloris marina is a Acaryochloris marina is a cyanobacterium containing, as primary photosynthetic pigment, chlorophyll *d*, a red-shifted chlorophyll with an absorbance maximum *in vivo* of 710 nm, meaning that it can use efficiently far-red light to drive the photosynthetic process [26]. Chlorophyll *f* produced by *Halomicronema hongdechloris*, is so far the most red-shifted chlorophyll observed in a cyanobacterium performing oxygenic photosynthesis [27, 28]. Re-engineering the photosystems of algae or higher plants with the introduction of these recently discovered new chlorophylls could allow the use of the portion of the solar spectrum ranging from 700 to 750 nm, otherwise worthless to the currently known eukaryotic photosynthetic organisms. This small increment in the exploitable solar spectral region can give access to a 19% additional photon flux, indicating that even a modest enlargement of the PAR can result in a significant increase in the amount of energy available for photosynthesis [24].

Although preliminary studies suggest that the capacity to store photosynthetic energy storage in *Acaryochloris* is the same of Chl *a*-containing cyanobacteria, it is still to prove whether this is true also for transgenic organisms engineered to produce Chl *d* [25]. Similarly, it is actually unknown whether it will be possible to further extend the limit for oxygenic photosynthesis into the near infrared. This is an important area for further research.

Optimizing Antenna Size of Crop Canopies

One of the main reasons for the lower observed photosynthetic efficiencies is light saturation of photosynthesis. It has been demonstrated that, in optimal conditions, the photosynthetic apparatus of leaves can already operate near the maximum efficiency under low light condition. For instance, the photosynthetic process of a C3 leaf is already saturated by approximately 25% of maximum full sunlight. As absorbed light increases, photosynthetic efficiency declines. This is due to limitation in either electron transport or the ability to utilize the NADPH and ATP produced by the primary photochemistry. The rate-limiting step is, generally, the carbon fixation reactions, in particular the reaction catalysed by the Ribulose-1, 5-

bisphosphate carboxylase/oxygenase (Rubisco) enzyme in the Calvin–Benson cycle [6, 24, 29]. The fact that the phenomenon of light saturation sets in at relatively low light intensities means that most of the photons normally absorbed during the day constitutes excess light, that cannot be used for photosynthesis and must be carefully dissipated to avoid photodamages. Photosynthetic organisms have evolved a plethora of mechanisms to prevent photodamage, principally as thermal dissipation, as well as repair mechanisms which take over when these first defences fail [30, 31], but the consequence is a significant loss of photosynthetic efficiency and a reduction in energy storage. In effect, it seems that the plant antenna systems are bigger than optimal, trapping more light energy than what can be efficiently used. The reason is that antenna size has been selected in order to maximize the reproductive success of the organism and indeed in the competitive natural conditions a bigger antenna system represents an important selective advantage [7]. First of all, it makes the organism able to efficiently absorb light energy also in condition of low light, such as cloudy days or at the beginning and the end of the day. At the same time it prevents competition from other photosynthetic organisms since an individual that can intercept more light deprives potential competing plants growing underneath.

From the farmer point of view, this behaviour is clearly disadvantageous. Reducing the size of the antenna systems in the leaves of the upper canopy can offer important advantages, such as saving the metabolic resources required for the production of the antenna complex while increasing the quantity of light able to reach the lower leaves and therefore reducing the loss of the excess photons in the form of heat and fluorescence by the upper canopy [24, 29]. There are now in literature a number of studies, which suggest that antenna truncation can be considered a valid approach to improve photosynthetic efficiency. Cell suspensions of *Synechocystis* PCC6714 and *Chlorella pyrenoidosa* with a reduced content of light-harvesting pigments showed a maximum photosynthetic activity 20–30% higher that the wild type [32]. An engineered *Chlamydomonas reinhardtii* strain with a small PSII antenna size exhibited a ~50% increase of photosynthetic efficiency in condition of saturating light [33]. In the green alga *Chlamydomonas reinhardtii* a partial reduction in chlorophyll *b* levels resulted in more than a two-fold increase in photosynthetic rate at high light intensities and a 30% increase of the growth rate at saturating light intensities [34]. However, in three mutant strains of *Synechocystis* sp. strain 6803 the progressive reduction of the peripheral antenna system resulted in a decreased productivity [35].

The hypothesis that a constitutive smaller antenna size in the upper leaves will improve canopy photosynthetic efficiency by minimizing the over-absorption of the incident sunlight and improving light penetration, is therefore a promising strategy to improve photosynthetic efficiency, and consequently yield and productivity, in the short term.

Fine-Tuning Photoprotection

Factors such as climate and geography, season, the time of the day, as well as the relative position of the leaf within the canopy and the position of the cell within the leaf, greatly influence the intensity and the spectral quality of the light which is available for the photosynthetic process. Plants evolved a plethora of protective mechanisms to counteract photodamage which are now tightly integrated within the photosynthetic process [31, 36]. Manipulating this photoprotection capacity can be another possible way to reduce losses in conversion efficiency. For instance, there are algal systems able to relax the photoprotective heat dissipation far more rapidly than plants. Engineering these systems in plants could lead to increases in potential assimilation. An additional possibility is to up-regulate the genes of the enzymes involved in the inter-conversion of intermediates of the xanthophyll cycle, this could result in a more rapid recovery of photosystem II efficiency [6].

Improving Dark Reactions

Working on the C3 Cycle

The C3 cycle represents the major pathway of carbon assimilation in most photosynthetic organisms. The intermediates here produced are essential for plant growth and development since they can exit the cycle and be used in the biosynthesis of several molecules, such as isoprenoids, sucrose and starch, amino acids, lignin or nucleotides. Understanding how the Calvin cycle responds to changes in the external environmental conditions, as well as to changes in demands for photosynthates within the plant, is an essential step for attempts aiming to increase crop yield [37].

The Calvin-Benson cycle can be divided in 3 phases (Fig. **5**). The first reaction in the C3 cycle is the carboxylation reaction catalysed by the enzyme Ribulose-1, 5- bisphosphate carboxylase/oxygenase, commonly known as Rubisco, which fixes CO_2 into the acceptor molecule ribulose 1,5-bisphosphate (RuBP) forming 3-phosphoglycerate (3-PGA). This step corresponds to phase 1, or carbon fixation, which is followed by a second phase (reductive phase) that leads to triose phosphate formation (Glyceraldehyde 3-phosphate) and finally by the third phase in which the acceptor molecule ribulose 1,5-bisphosphate is regenerated (regeneration phase).

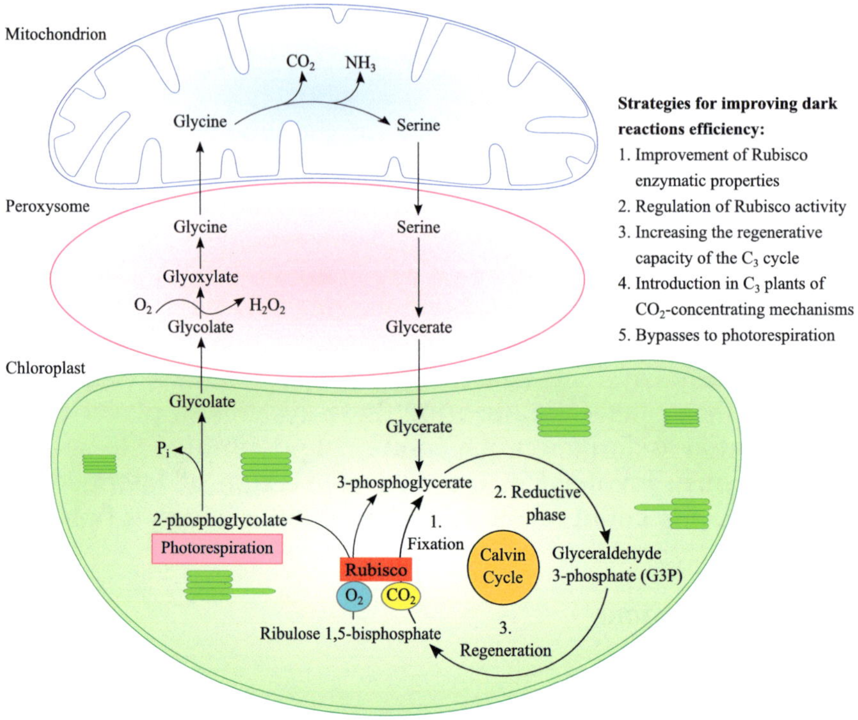

Fig. (5). Outline of the Calvin-Benson cycle (C3 cycle) and the pathway of photorespiration in higher plants. The major carbon intermediates and compartments are depicted. The stoichiometry of the reactions is not included. The main strategies proposed for the improvement of the dark phase of photosynthesis are emphasized (in bold font). During photorespiration, the 2-phosphoglycolate produced by the oxygenase reaction of Rubisco is hydrolysed in the chloroplast and the resulting glycolate is transported to the peroxisome. Here, it is oxidized to glyoxylate which is further transaminated to glycine. The glycine is transported to the mitochondria, where two molecules are converted to serine in a process which releases ammonia and CO_2. While the ammonia is reassimilated (not shown), the majority of the CO_2 liberated in the mitochondria escapes to the atmosphere and is not reassimilated in C3 plants. Back in the peroxisome, serine is converted to glycerate, which is then transported back to the chloroplast and recycled in the Calvin-Benson cycle.

The opposing oxygenation reaction of Rubisco fixes O_2 into the acceptor molecule RuBP, instead of CO_2, forming one molecule of 3-PGA and a two-carbon compound, 2-phosphoglycolate (2PG), a molecule of limited use to most organisms. Plants metabolize this product through a complex pathway, involving peroxisomes and mitochondria, to regenerate 3-PGA, which can re-enter the Calvin cycle, but this requires a significant amount of phosphorylating and reductive capacity, beyond the loss of CO_2. Furthermore organically-bound nitrogen is released as ammonia, which must be then re-assimilated [6, 38]. This process of oxygen consumption and CO_2 release is termed photorespiration and it is essential to limit the loss of fixed carbon due to the oxygenase activity of

Rubisco (see Fig. **5** for further details). Indeed it has been found in all the oxygenic photosynthetic organisms [39].

Application of modelling has highlighted that the photosynthetic process, at light saturation, is limited *in vivo* by capacity for carboxylation and capacity for regeneration of the acceptor molecule for carboxylation, ribulose 1,5-bisphosphate [6]. Several strategies have been proposed for overcoming these bottlenecks and improve photosynthetic efficiency, such as improving Rubisco activity, introducing a CO_2-concentrating mechanism into C3 plant or altering the photorespiratory pathway (Table **2**).

Improving Rubisco

Rubisco, the most abundant protein on earth, is an extremely slow catalyst, indeed large amounts of the enzyme are needed to support adequate photosynthesis rates [40]; moreover its carboxylase activity is subject to competitive inhibition by O_2, due to the higher O_2/CO_2 ratio in the atmosphere [41]. The specificity of Rubisco for CO_2 and the solubility of CO_2 relative to O_2 decline with increase in temperature, therefore loss due to photorespiration rises in hot climates [11, 42].

Engineering Rubisco, in order to obtain improved forms of the enzyme with increased activity and/or substrate specificity, is therefore a primary objective for enhancing photosynthesis efficiency. Being a very important rate-limiting step in carbon assimilation, even modest improvements in the overall performance of the enzyme could result in significant increases in plant yield, especially under stress conditions [43].

Rubisco is an ancient enzyme, arisen more than 3.5 billion years ago in an environment characterized by high atmospheric CO_2 and extremely low oxygen (O_2), since oxygen-producing photosynthesis was absent [44, 45]. The appearance of oxygenic photosynthesis, which resulted in the oxygenation of the atmosphere in a period of time ranging from 2.3 to 0.5 billion years ago, coupled with the carboxylase activity of Rubisco and other enzymes from different autotrophic assimilation pathways of inorganic carbon, in parallel with the geological long-term burial of organic carbon, changed the situation drastically [45]. The atmospheric levels of O_2 greatly increased while simultaneously CO_2 fell to very low levels, giving rise gradually to the current condition where CO_2 concentration in the atmosphere is limiting for Rubisco activity and O_2 can act as a competing substrate for the enzyme. Atmospheric conditions favouring significant levels of photorespiration, therefore, probably occurred only during the Carboniferous period (280–340 million years ago) and in the past 35 million years. Unfortunately, by that time Rubisco had already become the primary enzyme acting in the photosynthesis carboxylation step and was already well integrated into the plant primary metabolism the likelihood of its substitution with a different carboxylase was probably nil. The evolution of a Rubisco free of oxygenase activity seems to be equally unlikely due to the similarities in the oxygenase and carboxylase reactions which constrain the biochemistry of the active site [46].

To face these dramatic changes in the environmental conditions, Rubisco evolved better kinetic properties, acquiring an increased specificity for CO_2 at the expense of the velocity of carboxylation [44, 47].

It has been argued that through evolution, all Rubisco enzymes present in nature adapted to the natural conditions under which they operate, optimizing the compromise between CO_2/O_2 specificity and the maximum rate of catalytic turnover [47]. Indeed, when comparing Rubisco enzymes from a wide variety of organisms, green and non-green algae, cyanobacteria and higher plants, there is a significant variation in the catalytic properties of the enzyme as a consequence of land colonization which required the adaptation to the environmental conditions of many and different ecological niches [39, 48].

In general photosynthetic organisms living in condition of high CO_2 and low O_2 (*e.g.* proteobacteria) or organism that have evolved CO_2-concentrating mechanisms (*e.g.* many cyanobacteria and algae, as well as C4 plants), have Rubisco enzymes with lower CO_2 affinities, usually accompanied by lower specificities for CO_2 over O_2, but higher carboxylation rates.

In contrast, C3 plants (such as most crops) and algae lacking a CO_2-concentrating mechanism have higher CO_2 affinities, better CO_2/O_2 specificities, but slower carboxylation rates [48]. Galmés *et al.* [45] studied Rubisco kinetic parameters and chloroplast gene sequences in 28 plant species, representing different phylogenetic lineages, environmental adaptations and photosynthetic mechanisms. They observed high variability in Rubisco kinetic properties among groups, indicating that in nature deviations from the apparent trade-off between velocity and affinity are present. In particular, Rubisco from highly specialized groups, such as carnivorous plants, CAM species and C3 species from water-stress habitats, possessed higher carboxylase catalytic efficiency. The improved kinetics were attributed to the amino acids in the Large-subunit (RbcL), which were selectively conserved during evolution.

The study of natural variation in the catalytic properties of Rubisco among different species can therefore provide information, which can be valuable for Rubisco engineering. The substitution of Rubisco in C3 plants with more efficient forms, coming from other photosynthetic organisms or modified using genetic engineering, can be a valid approach to obtain better photosynthesis under a range of environmental conditions. It has been estimated that substitution of the existing

C3 crop Rubisco with the corresponding enzyme from *Amaranthus edulis* could increase carbon gain by 17% and the substitution with the enzyme of *Griffithsia monilis* (a non-green algae) could increase carbon gain by more than 25% [43].

Evolutionary analysis of kinetically distinct Rubisco enzymes from 15 species of *Flaveria*, which includes C3, C3–C4 intermediate, C4-like, and C4 species, demonstrated that changes in only two residues of RbcL distinguish C4 Rubiscos from the ancestral C3 species and these changes can explain most of the differences in catalytic properties found among the different enzymes [49]. These results were used to identify sequences in C4-Rubisco that can be transplanted into C3-Rubisco to improve carboxylation rate. Using transgenic tobacco lines expressing hybrid Rubisco enzymes, containing native and mutated *Flaveria* large subunits (RbCL) and tobacco S-subunits (RbcS), it was demonstrated that the simple substitution of amino acid 309 from Met to Ile in the L-subunit acts as a catalytic switch between C4 and C3 catalysis [50].

Furthermore, the introduction of subunits with high catalytic rates from a closely related species has been proposed as a valid approach to improve kinetic properties of Rubisco and consequently photosynthesis in C3 plants. For instance, the Rubisco small subunit from sorghum (*Sorghum bicolor*, a C4 plant), characterized by high carboxylation reaction rate, was overexpressed in rice (*Oryza sativa*). Partial replacement of the rice small subunit with the sorghum one resulted in higher catalytic rates of the chimeric enzyme in the transgenic plants [51]. However, their photosynthetic rate remained unchanged, possibly for the inadequate capacity of the electron transport chain to support the increase in Rubisco activity. These results indicate that approaches such as exchange of subunits as well as mutation of single amino acids can be used to manipulate Rubisco enzymatic properties.

However, two major challenges need to be addressed when bioengineering Rubisco in C3 plants, namely identifying the structural changes that can improve the performance of the enzyme and identifying the best ways to efficiently introduce the desired changes into the target plant [52]. Although enzymes with desirable characteristics can be currently identified or modified, improved knowledge and technical advances are required before their potential can be exploited in major crop plants. The recent significant advances both in understanding Rubisco biogenesis and in the available chloroplast transformation tools can be of great help in enhancing the efficiency of Rubisco catalysis.

Gene replacement is hindered by the complex arrangement of the genes encoding Rubisco subunits and its complex biogenesis in plant chloroplasts. The functional Rubisco enzyme in higher plants and green algae is composed of eight molecules of the large subunit (RbcL), encoded by the chloroplast *rbc*L gene, which includes the catalytic domain, and eight small subunits (RbcS) encoded by a family of *rbc*S genes in the nucleus. Several ancillary proteins are necessary to assemble correctly the different subunits into the multimeric enzyme. These complicated assembly needs make difficult, if not sometimes impossible, the biogenesis of Rubisco enzymes coming from other organisms, such as cyanobacteria, algae and higher plants [52]. For instance, the sunflower RbcL subunit and different *Flaveria* sp. ones showed really high differences in their capacity to form, in tobacco chloroplasts, hybrid Rubisco, containing tobacco small subunits, despite the fact that they shared the same genetic regulatory sequences and showed a very high amino acid identity. Therefore, it seems that the evolution of Rubisco catalytic function has been constrained to preserve compatibility with the chaperones necessary for its biogenesis [52].

The importance of this compatibility has been recently exploited by Whitney *et al.* [52]. Phylogenetic analysis revealed parallel evolutionary linkages between RbcL in *Arabidopsis* and its chaperone Rubisco accumulation factor 1 (RAF1). Tobacco plants expressing *Arabidopsis* Rubisco large subunits (AtRbcL) alone or co-expressed with its AtRAF1 chaperone were produced. Experimental results demonstrated that in presence of AtRAF1 more hybrid Rubisco enzymes, containing AtRbcL- and tobacco RbcS subunits, were assembled. The increase in Rubisco biogenesis resulted into improvements in CO_2-assimilation in leaf chloroplasts and growth of tobacco plants. These results therefore emphasize the importance of ancillary protein complementarity during Rubisco biogenesis in plastids, as well as the need to consider the specificity of this interaction when optimizing recombinant Rubisco in plants. For instance, since CbbX has been identified as an activase of red-type Rubisco [53], it appears that engineering of plants with this Rubisco will probably require the co-expression also of its activase, as well as of other accessory proteins (*e.g.* folding and assembly factors [54].

Furthermore, efforts toward redesigning Rubisco must be assisted by improvement in chloroplast transformation tools.

Nowadays plastid transformation of tobacco is a routine procedure, and several other Solanaceous species are transformable, including tomato, petunia, eggplant and potato [55]. Several other horticultural species can be efficiently transformed, *i.e.* lettuce, carrot, cauliflower and sugar beet [56]. Unfortunately, redesigning photosynthesis requires plastid transformation to be feasible in a greater number of species than the few listed above. Despite all efforts, little advances have been achieved in the development of plastid transformation protocols for additional species, and many important crops are still not transformable, among them all the monocotyledons, including cereals [3, 57]. The application of plastid technology to crops relies on the development of appropriate selectable markers and a highly efficient tissue culture and selection system [3, 55]. Major efforts therefore will be necessary to make plastid transformation a reality.

Regulating Rubisco Activity

The activation state of Rubisco, defined as the fraction of active sites that are catalytically competent, is dependent on the enzyme Rubisco activase (RCA), a member of the AAA+ family of ATPases. Rubisco is progressively inactivated by the formation of dead-end complexes with inhibitory sugar phosphates, produced either naturally as Rubisco regulators (such as CA1P) or originated from misfired reactions. The result of this inactivation is a significant decline in activity with time, necessitating the repeated intervention of Rubisco activase. RCA is a chloroplastic protein, encoded in the nucleus, which uses the energy of ATP to restore catalytic capacity by facilitating conformational changes in Rubisco in order to release of the inhibitory compounds from the active site. Since RCA regulates the portion of Rubisco that is catalytically active, its engineering seems to be a valuable approach to improve Rubisco performance in plant [40, 48].

In particular, while Rubisco is a thermostable enzyme, active even at high temperatures, RCA is unstable at elevated temperatures [58], therefore it has been proposed that the primary cause for reduction in photosynthesis during moderate heat stress is Rubisco deactivation caused by the inhibition of RCA. Strong evidence supporting this hypothesis has been obtained by the introduction into Arabidopsis *rca* mutants of more thermostable variants of RCA. These transgenic plants showed higher rates of photosynthesis, increased biomass and increased seed yield when grown at elevated temperatures, compared with control plants [59]. A recombinant chimeric RCA, where the Rubisco activation domain from the more thermostable tobacco RCA was fused to the Arabidopsis Rubisco recognition domain, showed both the greater heat stability of the tobacco enzyme and the high Rubisco activation capacity of the Arabidopsis one [59, 60].

However, it has also been suggested that Rubisco deactivation at moderately elevated temperatures could not be a direct effect of RCA instability, but rather a regulatory response to other limitations in the photosynthetic process (*i.e.* RuBP regeneration) or a protective measure adopted by the plant under conditions of stress, since naturally occurring inhibitors, such as CA1P (2-carboxy-D-arabinitol 1-phosphate), protect Rubisco from proteolytic degradation by inducing the closed conformation (reviewed in [54]).

RCA is probably involved also in regulation of Photosystem II (PSII) photochemistry. Its reduction in transgenic tobacco plants resulted in decreased levels of different PSII proteins (D1, D2, CP43, CP47, and PsbO), of the number of grana stacks in a chloroplast and of discs per grana stack [61], suggesting that this enzyme, in addition to the role in CO_2 assimilation regulation, it is important also in maintaining PSII function and chloroplast ultrastructure.

More physiological and biochemical work is therefore clearly needed to better understand RCA role in photosynthesis and to eventually engineer it to sustain photosynthetic rates and yield in preparation for global warming.

Increasing Regenerative Capacity of the C3 Cycle

The Rubisco carboxylation reaction, previously discussed, is not the only factor limiting the rate of photosynthesis. Another important limit is represented by RuBP regeneration capacity.

Indeed, under conditions of high CO_2 concentration or low irradiance, it is the RuBP regeneration capacity the key factor determining the photosynthetic rate [62].

Regeneration of the CO_2 acceptor molecule RuBP involves several enzymes. Using antisense plants it has been demonstrated that photosynthesis is sensitive to even small reductions in the levels of sedoheptulose-1,7- bisphosphatase (SBPase) and, to a lesser extent, to reductions in the activity of either transketolase (TK) and fructose 1,6-bisphosphate aldolase (aldolase), suggesting that increasing their activity can be a suitable approach to increase the photosynthetic capacity of plants [37].

SBPase has been identified as a major control point in the C3 cycle, since small reductions in SBPase activity resulted in a decrease in CO_2 fixation and growth. This enzyme is positioned at the branch point in the Calvin cycle between regeneration of the acceptor molecule RuBP and export to starch and sucrose biosynthesis. It therefore plays a critical role in maintaining the balance between the carbon needed for RuBP regeneration and the one leaving the cycle for biosynthesis. The level of control exerted varied with development and environmental conditions [37]. Overexpression of Arabidopsis SBPase in transgenic tobacco plants resulted in increased photosynthetic rates and improved biomass accumulation, results that however were not obtained if the plants were grown under short-day and low-light conditions [63]. Interestingly, the overexpressing plants showed an increased rate of growth especially during the early phase of development, which is important in agriculture since a rapid seedling establishment contributes to yield increases at maturity [37]. On the contrary, overexpression of SBPase in rice transgenic plants enhanced tolerance to salt or high temperature, but did not increased biomass accumulation [64]. The experiments with tobacco plants overexpressing Arabidopsis SBPase were repeated more recently by Rosenthal *et al.* [65]. Plants were grown under field conditions at ambient (ca. 385 ppm) and elevated [CO_2] (ca. 585 ppm). Transgenic tobacco plants improved yield compared to the wild type, confirming previous conclusions that upregulation of SBPase in tobacco can increase productivity. Moreover the increase in biomass was greater at elevated [CO_2] (22.7%), compared with plants grown at normal atmospheric conditions (12%). These results clearly demonstrated that transgenic tobacco plants with increased SBPase activity grown at elevated [CO_2], can enhance to a greater extent photosynthesis rate and biomass production compared to wild type plants. The simple increase in activity of a single native photosynthesis enzyme is therefore enough to enhance carbon fixation and growth in condition of high [CO_2], proving that it is possible to successfully and easily engineer C3 crops to face rapidly changing environmental conditions [65].

Plastid transketolase (TK) is the second enzyme in the Calvin cycle with a relevant effect on photosynthesis. Reduction in the activity of this enzyme resulted in decreased rates of carbon fixation and strongly reduced levels of aromatic amino acids and phenylpropanoid pathway products, given that the reactions catalyzed by this enzyme supply the precursor for both the shikimic acid pathway and phenylpropanoid metabolism (erythrose 4-phosphate) and may also influence the level of G3P and ribose 5-phosphate, substrates for thiamine biosynthesis [66]. To examine the effect of increased transketolase levels on photosynthetic capacity and plant growth, transgenic tobacco plants overexpressing Arabidopsis transketolase were produced, with unexpected results. Indeed, these plants exhibited a chlorotic and slow-growth phenotype, which was almost completely complemented by addition, at early stage of germination, of thiamine or thiamine pyrophosphate to the growth medium, demonstrating that increased plastid TK expression resulted in a thiamine auxotrophic phenotype [67]. Plants therefore can put up with only small changes in the level of transketolase enzyme, that is probably tightly controlled even if the mechanism remains to be elucidated. These findings highlight first of all the important role played by the enzymes of the C3 cycle in plant growth and development and consequently that engineering plants with enhanced desirable traits requires a deeper knowledge of the components and regulatory mechanisms involved. This understanding is essential for the successful design of dynamic kinetic models.

Aldolase is the last identified enzyme able to affect photosynthesis efficiency. Overexpression of Arabidopsis plastidial aldolase in Arabidopsis or tobacco plants resulted in increased photosynthetic capacity, better growth and increased biomass, particularly under high CO_2 concentration (700 ppm), highlighting that also manipulation of aldolase level in chloroplasts can be a novel and useful biotechnological tool to generate crops able to obtain better performance in the future conditions of high atmospheric CO_2 [62, 68].

CO_2-Concentrating Mechanism: Improving Carbon Uptake

The introduction in C3 plants of CO_2-concentrating mechanisms (CCM) from algae or cyanobacteria has to be considered as an alternative synthetic biology strategy to improve net photosynthetic rates.

Indeed, a limiting factor in carbon fixation is that C3 plants rely on passive diffusion to deliver CO_2 from the atmosphere to the site of carboxylation, located in the chloroplast stroma of the mesophyll cells. CO_2 faces a series of resistances while moving across the several compartments present in the leaf (stomata, intercellular airspaces, cell walls, cell membranes, cytoplasm and the chloroplast). The diffusive resistance to CO_2 passage creates a gradient between CO_2 levels in the substomatal cavity of the leaf and the level of dissolved CO_2 in chloroplasts, meaning that even in condition of high atmospheric CO_2 its effective concentration in the chloroplast is very low and therefore Rubisco works in condition of low (not saturating) chloroplast CO_2 concentration [44, 69].

Elevating CO_2 concentration around Rubisco can be an approach to improve photosynthetic performance and avoid photorespiration. Indeed photosynthetic organisms have evolved different ways to concentrate CO_2 nearby Rubisco, such as CCM in cyanobacteria, similar CCMs in eukaryotic algae and C4 photosynthesis in higher plants [39]. Since membrane permeability for CO_2 is relatively high, concentrating CO_2 in proximity of Rubisco is potentially difficult. This limitation is bypassed by accumulating carbon in a form other than CO_2, as bicarbonate (cyanobacteria and eukaryotic algae) or as metabolites like malate or aspartate (C4 plants) which are not able to cross biological membranes as rapidly as CO_2. These molecules are then transported where Rubisco is present and CO_2 is regenerated [39].

CCM in Cyanobacteria and Algae

Diffusion rate of CO_2 in water is far lower than in air and in absence of CCMs aquatic photoautotrophs, such as cyanobacteria and algae, would experience deficiency of CO_2. These organisms have therefore acquired during evolution mechanisms to increase both the inorganic carbon flux into the cells and the CO_2 concentration nearby Rubisco. These CCMs rely on inorganic carbon transporters (Ci), located in the plasma membrane and/or chloroplast envelope, and carbonic anhydrases (CA), which catalyse the reaction of the reversible hydration of carbon dioxide and its conversion in bicarbonate in aqueous solutions.

Cyanobacteria evolved an extremely efficient CCM (Fig. **6a**), able to concentrate CO_2 by a factor up to 1,000-fold. As a consequence, they were able to retain a faster form of Rubisco with a lower CO_2 affinity. Moreover, cyanobacterial Rubisco operates at near CO_2 saturation, meaning that photorespiration is almost completely eliminated [44].

Cyanobacterial CCMs rely on carbonic anhydrases, bicarbonate transporters and unique microcompartments known as carboxysomes where Rubisco is localized, together with carbonic anhydrase. Bicarbonate transporters actively transport and accumulate bicarbonate into the cell. The bicarbonate then enters into the carboxysome through pores present in the shell. Here, carbonic anhydrase activity generates CO_2 from the cellular bicarbonate pool, resulting in the formation of a microenvironment with a high CO_2 concentration around Rubisco, sufficient to minimize oxygenation reaction [44].

Even though plant chloroplasts originated from a cyanobacterial progenitor, they do not possess this CCM, making its introduction a promising approach for improvement of plant photosynthetic efficiency. A recent analysis evaluated that it should be possible to obtain a crop yield increase higher that 30% as a result of the introduction of carboxysomes into chloroplasts coupled with the addition of a bicarbonate ion transporter and the removal of stromal carbonic anhydrase [70].

A first step towards this goal was the production of functional carboxysomes in *Escherichia coli* [71]. Recently, the possibility of introducing carboxysomes into chloroplasts has been proved. Several components of the β-carboxysome from *Synechococcus elongatus* PCC7942 were transiently expressed in chloroplasts of *Nicotiana benthamiana*, demonstrating that the shell proteins of the β-carboxysome can assemble in plant chloroplasts giving rise to highly organized structures resembling empty microcompartments. Moreover a foreign protein was targeted to the shell proteins inside chloroplasts [72]. An important and recent step towards the introduction in the chloroplasts of a functional CCM is represented by the simultaneous introduction into a C3 plant of a carboxysome protein and a functional cyanobacterial Rubisco. The genes of the large and small subunits of the Rubisco from the cyanobacterium *Synechococcus elongates* PCC7942 (Se7942), in combination with an internal carboxysomal protein (CcmM35), were introduced in tobacco. In the chloroplast stroma of the transformed transplastomic tobacco line, Se7942 Rubisco and CcmM35 formed complexes highly similar to procarboxysomes, which represent an important step during biogenesis of cyanobacterial β-carboxysomes. The introduced cyanobacterial Rubisco assembled into an active enzyme, able to support autotrophic growth, with higher rates of CO_2 fixation per unit of enzyme compared to the control. This transplastomic tobacco line therefore represents a valuable host for the next attempts to produce functional β-carboxysomes in chloroplasts [73].

Mechanisms for concentrating inorganic carbon are also present in eukaryotic algae. The CCM in green algae is similar to the cyanobacterial model, including active bicarbonate accumulation by bicarbonate transporters in the cell and chloroplast membrane, production of CO_2 in the chloroplast and Rubisco packaging. Indeed in these algae Rubisco concentrates in a region, the pyrenoid, typically surrounded by starch [74].

The best-characterized algal CCM is that of *Chlamydomonas reinhardtii* (**Fig. 6b**). In this green unicellular alga, bicarbonate is accumulated within the stroma and is then converted to CO_2 by CAH3, a carbonic anhydrase located in the thylakoid lumen while Rubisco is localized to the pyrenoid. To date, several components have been implicated in this mechanism. In Chlamydomonas several CAs have been found in different cell compartments (*i.e.* periplasmic space, plasma membrane, chloroplast). Inhibition of the activity of these CAs resulted in reduced rate of photosynthesis or impossibility to grow under normal CO_2 concentrations [75].

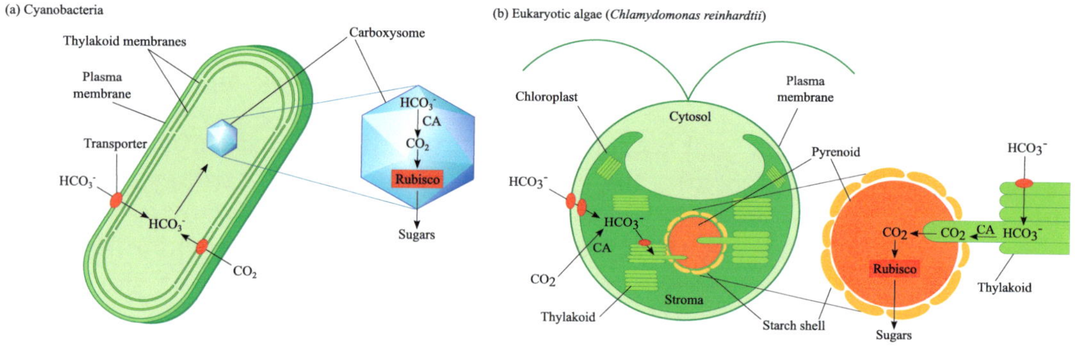

Fig. (6). The CO_2-concentrating mechanisms (CCM) in Cyanobacteria (a) and algae (b). (a) In a cyanobacterial cell specific bicarbonate transporters accumulate bicarbonate in the cytosol, and subsequently into the carboxysome. Inside the carboxysome a carbonic anhydrase converts HCO_3^- to CO_2 which is subsequently fixed into 3-phosphoglycerate (3-PGA) by Rubisco. (b) About half of the volume of *Chlamydomonas* cells is occupied by a single cup-shaped chloroplast. Bicarbonate transporters accumulate bicarbonate inside the cell. CO_2 can also diffuse into the cell, where is converted to bicarbonate by carbonic anhydrases. The bicarbonate, first accumulated in the chloroplast stroma, is transported into the thylakoid lumen, where is converted to CO_2 by a carbonic anhydrase. The thylakoids span through the Rubisco containing pyrenoid and here CO_2 fixation takes place. A starch shell surrounds the pyrenoid, minimizing CO_2 diffusion outside the site of fixation.

In order to test if the algal CCM can be useful in improving photosynthetic performance, the putative essential components of Chlamydomonas CCM were introduced in plant cells [69]. These included, for instance, putative Ci transporters probably located in the plasma membrane (LCI1 and HLA3) or in the chloroplast envelope (LCIA, CCP1 and CCP2), and carbonic anhydrases probably located in the periplasmic space (CAH1), the chloroplast stroma (CAH6) and thylakoid lumen (CAH3). First of all, the precise locations of these components were investigated in Chlamydomonas using fluorescent tags. Secondly, transient expression in tobacco demonstrated that quite all of these components can be successfully targeted to the appropriate subcellular locations in higher plants leaves, using modification of targeting sequences when necessary. Finally, Arabidopsis plants were transformed with LCIA-GFP or HLA3-GFP constructs and analysed. Transgenic Arabidopsis plants expressing LCIA or HLA3, grown under ambient CO_2 concentration, showed normal growth and photosynthetic characteristics, result that was not entirely unexpected. It has been therefore demonstrated that plant cells can express Chlamydomonas Ci transporters in the correct locations, without deleterious effect. Even if LCIA and HLA3 were active, their activity level was too low to adequately increase CO_2 levels around Rubisco and thus enhance CO_2 assimilation rate. It will be probably necessary the addition of further CCM components to achieve a significant enhancement in photosynthetic efficiency.

From C3 to C4 Photosynthesis

C4 photosynthesis, a process present in different higher plants, is another strategy to concentrate CO_2 around Rubisco [6, 39].

In most C4 plants, the leaf has a peculiar anatomy, known as Kranz anatomy: it contains two photosynthetic cell types, the mesophyll cells and the bundle sheath cells, which are structurally and biochemically specialized and able to function co-ordinately in carbon assimilation (Fig. **7**).

In the cytoplasm of mesophyll cells, atmospheric carbon dioxide is initially incorporated by carboxylation of phosphoenolpyruvate (PEP) to form oxaloacetate, which is converted to malate or aspartate. PEP carboxylase, the enzyme catalyzing this first reaction, can work efficiently even at low CO_2. Moreover, oxygen does not represent a competing substrate in this reaction. The C4 acids (malate or aspartate) are then transported from mesophyll to the inner bundle sheath cells, where Rubisco is present and PSII-mediated oxygen evolution is absent. Here, the C4 acids are decarboxylated, releasing CO_2 and pyruvate, which is then recycled back to the mesophyll cells for regeneration of PEP.

The amount of additional energy requested by the C4 cycle is, generally, less than what would be lost during photorespiration [6].

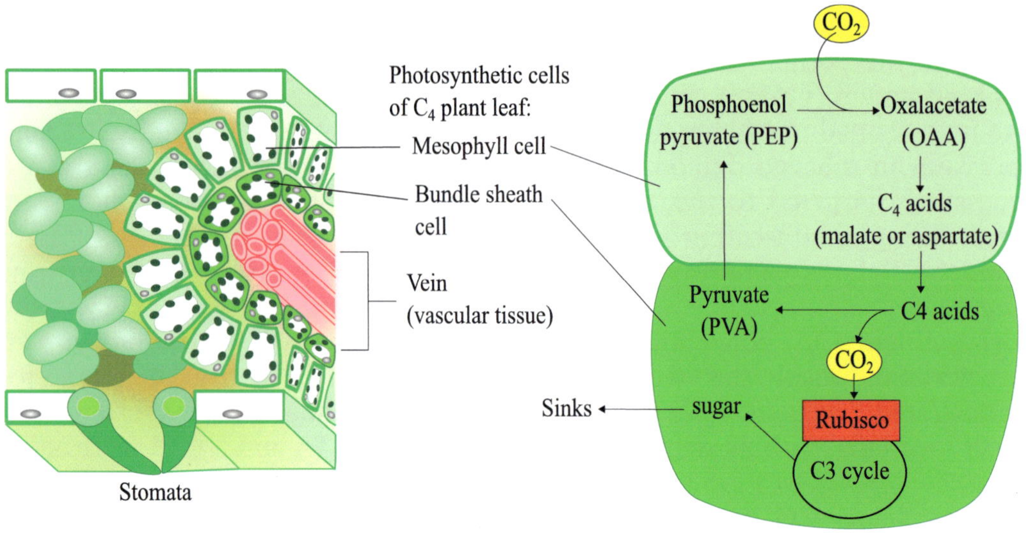

Fig. (7). Schematic representation of a C4 leaf structure (left) and the C4 metabolic pathway (right). Most C4 leaves possess a peculiar structure called Kranz anatomy. Their vascular tissue is surrounded by two ring of cells, the outer ring, formed by the mesophyll cells, and the inner ring, the bundle sheath cells, where Rubisco is present. The main function of this anatomy is to provide a site in which CO_2 can be concentrated around Rubisco, thereby suppressing the oxygenase activity of the enzyme and therefore avoiding photorespiration. Indeed in the bundle sheath cells, the atmospheric O_2 is absent and the Rubisco enzyme works in an environment saturated with the CO_2 released by the decarboxylation of the C4 acids (malate or aspartate) produced in the mesophyll cells. The arrows indicate enzymatic reactions or transport steps. The stoichiometry of the reactions is not included.

Similarly to what seen in cyanobacteria in presence of a CCM, C4 plants were also able to retain a faster Rubisco. Therefore, these plants generally have higher rates of photosynthesis, a better use of water and nitrogen and increased yield compared to C3 plants. The most productive plants and crops are C4 plants [6]. C4 photosynthesis is especially advantageous under conditions promoting photorespiration (*i.e.* saline, warm or, dry conditions, as well as some aquatic environments) [76]. In spite of its complexity, C4 photosynthesis has evolved independently over 60 times, with great variety of Kranz anatomy or biochemistry of the C4 cycle [77]. For all these reasons, the transfer of C4 traits to C3 plants has been proposed as one of the most promising ways for increasing photosynthetic efficiency of C3 crops, such as wheat and rice.

In the last 20 years several attempts were carried out to transfer C4 traits in C3 plants, for instance by the C4 Rice Consortium (http://irri.org/c4rice). Even though none of them has so far led to transgenic plants having significantly enhanced photosynthetic properties, our understanding of the C4 process has been greatly improved, including for instance understanding the genetic basis regulating Kranz anatomy or the development of C4-type chloroplasts and the biochemistry of the major C4 enzymes. This knowledge will be useful in future engineering attempts in C3 species. Furthermore, it might also suggest new approaches to improve the efficiency of existing C4 crops (*i.e.* to better resist to environmental stresses) [78].

Bypasses to Photorespiration

In parallel at reducing the oxygenase activity of Rubisco by CCM and C3-Ä transformation strategies, a further approach can be the engineering of a more efficient metabolism of the phosphoglycolate produced by the oxygenase reaction, able to reduce the losses of carbon, nitrogen, and energy and avoid the accumulation of photorespiratory intermediates.

This aim could be achieved through the, at least partial, replacement of the plant photorespiratory pathway with synthetic pathways characterized by higher efficiency (Fig. **8**). Whereas green algae and plants have a single, complex, resource-consuming pathway that metabolize phosphoglycolate, to date three simpler pathways have been identified in the prokaryotes, which can be used for this aim.

Kebeish *et al.* [79] introduced the *Escherichia coli* glycolate catabolic pathway into Arabidopsis. In such transgenic plants the glycolate, coming from dephosphorylation of 2-phosphoglycolate, was converted to glycerate in the chloroplast without ammonia release, which can result in an enhancement of the nitrogen use efficiency. Moreover, CO_2 release was shifted from mitochondria to chloroplasts, following the idea that CO_2 has a better chance to be refixed by Rubisco if it is released in the chloroplast rather than in the mitochondria. As a result of the increased concentration of CO_2 in the chloroplasts and the reduced energy demand of photorespiration, transgenic plants grew faster and produced more biomass indicating that the bypass effectively reduced photorespiration and enhanced photosynthesis [79, 80].

In the second bypass a pathway, similar to the one occurring in cyanobacteria, was introduced in plants [81]. Transgenic tobacco plants were generated containing two *E. coli* genes, targeted to the peroxisomes, in order to directly metabolize glyoxylate to hydroxypyruvate within these organelles. It was expected a competition between this pathway and the aminotransferases which convert glyoxylate to glycine, thus avoiding ammonia release. Unfortunately in no one of the transgenic lines analysed it was possible to detect the presence of the product of one of the two transgenes. Moreover, the transgenic plants exhibited stunted growth and chlorotic lesions when grown in ambient conditions whereas they seemed to grow normally in presence of high CO_2 concentration, a behaviour typical of the photorespiratory mutants isolated in Arabidopsis and barley, suggesting that the modifications were not able to completely bypass the photorespiratory cycle [81].

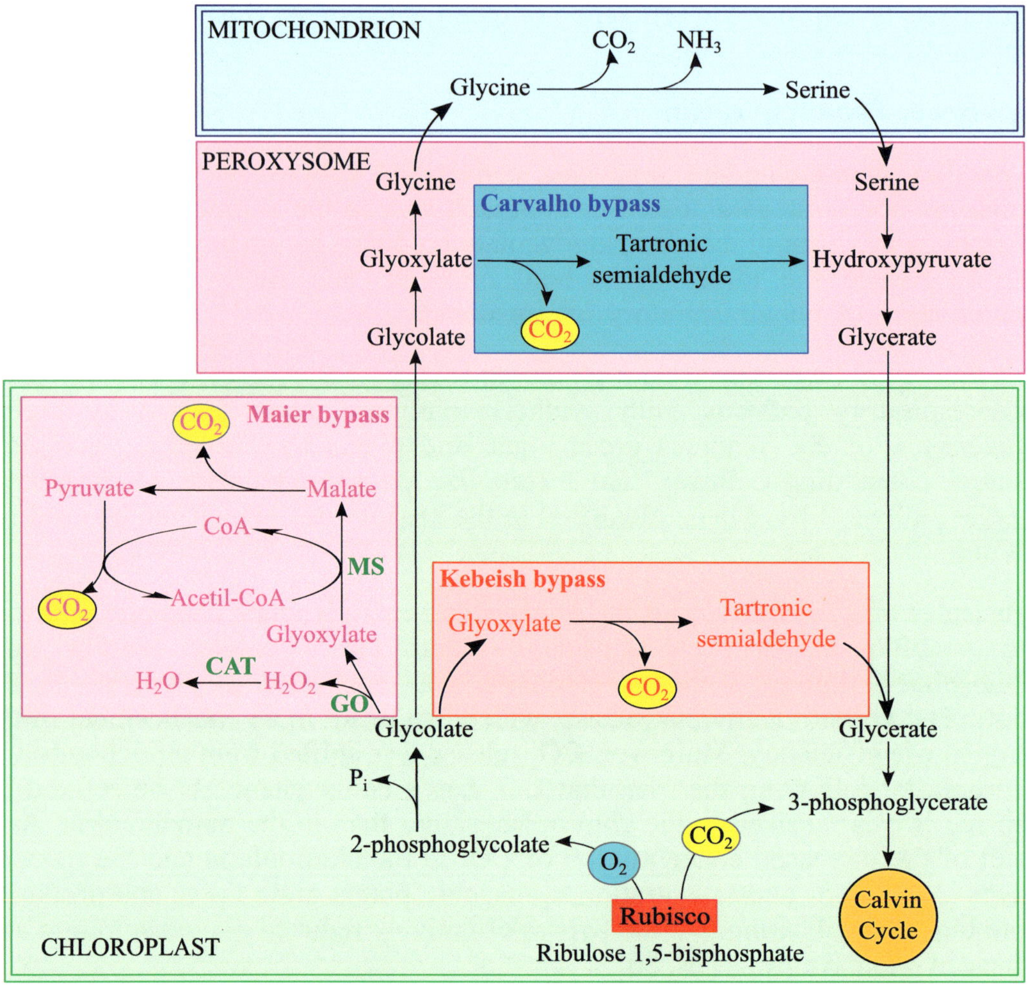

Fig. (8). Schematic representation of photorespiration and the three synthetic bypasses proposed for the reduction of photorespiratory losses. The photorespiratory pathway can be short-circuited in the chloroplast by the bacterial glycolate pathway (Kebeish bypass, red box), which converts glycolate to glycerate that can re-enter the Calvin cycle, or in the peroxisome (Carvalho bypass, blue box) where glyoxylate is converted in another photorespiratory intermediate, hydroxypyruvate. Alternatively, a complete glycolate oxidation pathway to CO_2 (Maier bypass, fuchsia box) can be engineered in the chloroplast. The stoichiometry of the reactions is not included. The transgenic enzymes engineered into Arabidopsis thaliana chloroplasts in the Maier bypass are highlighted in green. CAT, catalase; GO, glycolate oxidase; MS, malate synthase.

The third pathway cannot be considered a true bypass since there is no reconnection to the major pathway. Arabidopsis plants were engineered with the full glycolate catabolic cycle, that comprises the enzymes glycolate oxidase, catalase and malate synthase, which are able to completely oxidizes glycolate to CO_2 using a combination of endogenous and newly introduced enzymes [80, 82]. In this pathway, the glycolate is converted within the chloroplast into glyoxylate by the enzyme glycolate oxidase, relocated from the peroxisome to the chloroplast. Given that this enzyme produces hydrogen peroxide (H_2O_2) as a side product, a catalase is required in the chloroplast for detoxification. Malate synthase activity produces malate by condensing glyoxylate with acetyl-CoA. Malate is then decarboxylated to pyruvate by the chloroplastic NADP-malic enzyme, with the production of NADPH. Finally, pyruvate dehydrogenase converts pyruvate into acetyl- CoA, producing also NADH and another CO_2 molecule. This cycle is therefore able to convert one molecule of glycolate into two molecules of CO_2, which are released in the chloroplast where they can be directly refixed, with the concomitant production of reducing power and without release of ammonia. Some of the selected transgenic lines showed more leaves, more biomass, and higher photosynthetic rates.

These breakthroughs in synthetic photorespiratory bypasses set a precedent for future attempts. Moreover, pyruvate or intermediates in the synthetic bypass could be redirected for biotechnological applications, such as biofuel production or replacements for chemical feedstocks that are currently petroleum-derived [3, 83].

The biggest challenge that needs to be addressed is that Calvin cycle intermediates are totally integrated into the plant primary metabolism. The introduction of these synthetic pathways could therefore strongly affect it, possibly requiring an extensive re-engineering of the overall plant metabolism [6].

Sink Strength and Biomass Production

Increasing agronomical yield means basically increasing the amount of carbohydrates accumulated in the plants in different forms, such as starch in grains (*e.g.* wheat, barley and maize), tubers (potato) or roots (cassava), sugars in crops with high sucrose content (*i.e.* sugarcane, sugar beet and sweet sorghum), oils in oil-producing species (*e.g.* olive, sunflower, soybean, rapeseed *etc.*) or lignocellulose in biofuel crops.

Carbohydrates production is constrained by two factors tightly linked: the strength of the source and the strength of the sink. Whereas photosynthesis is the only source of carbon gain, its strength related mainly to the photosynthetic process efficiency and to the biomass of the photosynthetic organs, multiple sinks exist in different plant organs, either transitory (*i.e.* transitory starch) or permanent. Production of biomass and photosynthates conversion in starch in the leaf or export in heterotrophic sink organs, such as roots or seeds, are the most important sinks in the plant [23]. As already mentioned, FACE experiments demonstrated that the strength both of the source and the sink can be improved and that the sink strength is a limiting factor in conversion efficiency improvement. It has also been demonstrated that improving starch synthesis in sink organs is enough to enhance crop yield [19, 23], underpinning therefore the importance of the source/sink interaction. Hence, given the strong interrelationship between source and sink, improvement of photosynthetic efficiency must be coupled with a better use of photosynthates, otherwise the mechanisms of photosynthetic rate regulation by the sink could frustrate the attempts to directly manipulate photosynthesis.

Yield in cereals depends on the number of seeds and their weight, and both factors are determined by seed sink strength. ADP-glucose pyrophosphorylase (AGP) constitutes the rate-limiting step of the starch biosynthesis in the seed. Plant enzyme consists of two small and two large subunits. Expression in wheat (*Triticum aestivum* L.) of a modified maize AGP large subunit (Sh2r6hs) increased AGP activity in developing seed endosperm which resulted in an increase in seed yield and plant biomass [84], as well as in stimulation of photosynthesis. Sh2r6hs-expressing plants showed increased photosynthetic rates when grown under high light conditions, whereas no change of yield or photosynthetic rate was observed in plants grown in a suboptimal light environment, suggesting that the enhanced phenotype is probably due to an increase in photosynthetic rates established relatively early in seed development [84]. A similar correlation between the level of starch synthesis and the photosynthetic capacity was also demonstrated in the case of the starch present in the leaves which is used as a transient sink [10]. Transgenic Arabidopsis lines with increased capacity to produce leaf starch showed increased photosynthetic capacity, which resulted in enhanced growth and biomass. Rice lines engineered in a similar way exhibited comparable phenotypes, confirming that manipulation of leaf starch is a promising alternative approach to improve photosynthesis [85]. In a similar way, it has been demonstrated that the introduction of synthetic non-degradable carbon sinks is also able to enhance the efficiency of the photosynthetic process, *e.g.* the vacuolar production of the high-value sugar isomaltulose from sucrose in sugarcane leaves [23, 86].

Table 2. Brief summary of the strategies to be employed for photosynthesis improvement, as described in the text. (* As reported in Long *et al.*, 2015)

Strategy	Efficiency gain	Practical approaches	Organism	Outcome	Ref
Light phase					
1. Expansion of the usable solar spectrum	10%–30% *	Introduction of new chlorophylls (*e-f*), able to absorb in the 700–750 nm spectral region	working hypothesis	This modest expansion of PAR gives access to 19% additional photon flux	[24, 25]

(Table 2) contd.....

Strategy	Efficiency gain	Practical approaches	Organism	Outcome	Ref
2. Optimisation of the antenna size	20-50%	Reduction of light-harvesting pigments content in microalgal cells	*Synechocystis* PCC6714 and *Chlorella pyrenoidosa*	Maximum photosynthetic activity 20–30% higher in both organisms	[32]
		Reduction of PSII antenna size	*Chlamydomonas reinhardtii*	Circa 50% increase of photosynthetic efficiency at saturating light intensity	[33]
		Partial reduction in chlorophyll *b* levels	*Chlamydomonas reinhardtii*	More than a two-fold increase in photosynthetic rate at high light intensities, 30% increase in growth rate at saturating light intensities	[34]
		Progressive reduction of the peripheral antenna system (phycobiliproteins)	*Synechocystis* sp. strain PCC 6803	Decreased whole-culture productivity	[35]
3. Fine-Tuning Photoprotection	30% *	More rapid relaxation of heat dissipation at PSII			[6]
		up-regulation of the genes coding for the enzymes involved in interconversion of intermediates of the xanthophyll cycle			[6]
Dark phase					
1. Substitution of Rubisco with more efficient forms, possessing increased activity and/or substrate specificity	15%–30% *	Expression of hybrid Rubisco enzymes, containing tobacco S-subunits and L-subunits from *Flaveria pringlei* (C_3), *Flaveria floridana* (C_3-C_4), and *Flaveria bidentis* (C_4)	Tobacco	Identification of amino acid 309 in the L-subunit as a catalytic switch between C_4 (faster carboxylation rate, lower CO_2 affinity) and C_3 (slower carboxylation rate, higher CO_2 affinity) catalysis	[50]
		Overexpression of Rubisco S-subunit from the C_4 plant sorghum (*Sorghum bicolor*)	Rice (*Oryza sativa*)	Production of a chimeric enzyme with higher catalytic rates.	[51]
		Expression of Arabidopsis L-subunit, alone or co-expressed with its AtRAF1 chaperone	Tobacco	More hybrid Rubisco enzymes were assembled in presence of AtRAF1, resulting in increases in photosynthetic CO_2-assimilation rate and plant growth.	[52]

(Table 2) contd.....

Strategy	Efficiency gain	Practical approaches	Organism	Outcome	Ref
2. Rubisco Activase (RCA) engineering	> 20%	Expression of more thermo-stable RCA variants	Arabidopsis *rca* mutants	Plants with higher rates of photosynthesis, increased biomass and increased seed yield when grown at elevated temperatures	[59]
		Expression of a chimeric RCA (tobacco-arabidopsis)	Arabidopsis *rca* mutants	After a short exposure to higher temperatures, transgenic plants exhibited higher rates of photosynthesis and better recovery. Under extended exposure to moderately elevated temperature, transgenic lines had higher biomass and seed yield.	[60]
3. Optimize regeneration of RuBP. Working on different enzymatic steps of the C_3 cycle:	60% *				
sedoheptulose-1,7-bisphosphatase (SBPase)		Overexpression of Arabidopsis SBPase	Tobacco	Increased photosynthetic rates and improved biomass accumulation	[63]
		Overexpression of OsSBP from rice cultivar 9311 (*Oryza sativa* ssp. *indica*)	Rice cultivar zhonghua11 (*Oryza sativa* L. ssp. *japonica*)	Enhanced tolerance to salt or high temperature stress, but not increased biomass accumulation	[64, 87]
		Overexpression of Arabidopsis SBPase	Tobacco	Higher photosynthetic rates, greater carbon assimilation and improved yield	[65]
transketolase (TK)		Overexpression of Arabidopsis TK	Tobacco	Thiamine auxotrophic phenotype, slow-growth and chlorotic phenotype of the transgenic tobacco plants.	[67]
fructose 1,6-bisphosphate aldolase (aldolase)		Overexpression of Arabidopsis aldolase	in Arabidopsis or tobacco plants	Increased photosynthetic capacity, better growth and increased biomass, particularly under high CO_2 concentration	[62, 68]
4. Introduction of a complete and functional CCM into the chloroplasts of vascular plants					
Introduction of the cyanobacterial carboxysome system	60% *	Carboxysome operon from *Halothiobacillus neapolitanus*	*Escherichia coli*	Production of functional carboxysomes	[71]
		Expression of several protein components of the β-carboxysome from *Synechococcus elongatus* PCC7942	Tobacco (*Nicotiana benthamiana*)	Shell proteins of the β-carboxysome were able to assemble in plant chloroplasts.	[72]

(Table 2) contd.....

Strategy	Efficiency gain	Practical approaches	Organism	Outcome	Ref
		Simultaneous introduction of Rubisco (L- and S-subunits) from the cyanobacterium *Synechococcus elongates* PCC7942 (Se7942) and a carboxysomal protein (CcmM35)	Tobacco (*Nicotiana benthamiana*)	Formation of macromolecular complexes within the chloroplast stroma, highly similar in appearance to procarboxysomes. The introduced cyanobacterial Rubisco assembled into an active enzyme (higher rates of CO_2 fixation per unit of enzyme than the tobacco control).	[72]
Introduction of algal pyrenoid CO_2 concentrating system	60% *	Expression of putative essential components of Chlamydomonas CCM (several transporters and carbonic anhydrases)	Transient expression in tobacco	Quite all of these components successfully targeted to the appropriate subcellular locations in leaves of higher plants	[69]
		Expression of two inorganic carbon (Ci) transporters	Arabidopsis	Plants with normal growth. Stacking of further CCM components is probably required.	[69]
Engineering C_4 photosynthesis into C_3 crops	30% *	Various attempts		None of them have so far resulted in transgenic plants with significantly enhanced photosynthetic properties	
5. Synthetic bypasses to photorespiration	15%				
		Introduction of the glycolate catabolic pathway from *Escherichia coli* in the chloroplast	Arabidopsis	Plants grew faster, produced more biomass and contained more soluble sugars	[79]
		Introduction of two *E. coli* genes, targeted to the peroxisomes	Tobacco	Partial expression of the transgenes. Plants exhibited stunted growth and chlorotic lesions when grown in ambient air but normal growth in elevated CO_2	[81]
		Introduction in the chloroplast of a complete glycolate catabolic pathway	Arabidopsis	Some of the transgenic lines showed more leaves, higher fresh and dry weight, and displayed higher photosynthetic rates	[82]
6. Increasing sink strength		Expression of a modified maize ADP-glucose pyrophosphorylase (AGP) large subunit	Wheat (*Triticum aestivum* L.)	Increased AGP activity in developing seed endosperm resulting in an increase in seed yield and in total plant biomass as well as in stimulation of photosynthesis in high light	[84]

(Table 2) contd.....

Strategy	Efficiency gain	Practical approaches	Organism	Outcome	Ref
		Overexpression of starch synthase class IV (SSIV)	Arabidopsis	Increased levels (30 – 40%) of starch accumulated in the leaves, higher rate of growth	[10]
		Overexpression of Arabidopsis SSIV (AtSSIV)	potato (*S. tuberosum* L. cv. Désirée) plants	Increased tuber starch content and increased yield of starch production. No changes were observed in the morphology, size, biomass and photosynthetic parameters of the aerial part of the AtSSIV-expressing plants.	[10]
		Up-regulation of AGPase	Arabidopsis thaliana and rice (*Oryza sativa* L.) lines	Enhanced photosynthetic capacity, increased growth and biomass.	[85]
		Expression of a vacuole-targeted bacterial sucrose isomerase	Sugarcane	Accumulation in the storage tissues of the high-value sugar isomaltulose without any decrease in stored sucrose concentration. Lines with enhanced sugar accumulation also showed increased photosynthesis, sucrose transport and sink strength.	[86]

CONCLUDING REMARKS

Several factors make nowadays feasible the improvement of photosynthesis efficiency. First of all, in the last decades the knowledge regarding the photosynthetic process has exploded. This deeper knowledge, coupled with the progresses done in the *in silico* modelling and the emergence of high-performance computing, has allowed the realisation of kinetic models able to describe every single step of both the C3 and C4 photosynthesis as well as represents realistically the whole canopy of a crop. These models can therefore predict the possible results of different manipulations of the photosynthetic process, such as up- and downregulation of various photosynthetic genes, and hence may be useful in choosing the best targets for redesign of the photosynthetic process. In addition, the high level of conservation of the photosynthetic process among all plants can be advantageous since one single approach can be used in different species. Finally, genetic engineering techniques such as genome editing and synthetic biology, are now becoming available for an increasing number of plant species [3, 6].

Since several years, normally 10 to 15 if not more, are necessary to the breeders for the development and the commercialization of a new plant variety with improved quality, and taking in consideration the expected situation for the middle of this century, the achievement of higher yield in agriculture cannot be postponed further. It is estimated that manipulation of the different steps of the photosynthetic process, including the introduction of algal pyrenoid CO_2 concentrating system into C3 chloroplasts, may lead to a gain in photosynthesis efficiency up to 60% and simultaneously to an enhancement of both the efficiency of nitrogen use (NUE) and of water use (WUE) per unit of biomass, making the photosynthetic process ideal for manipulation in the second "Green Revolution".

CONFLICT OF INTEREST

The authors confirm that they have no conflict of interest to declare for this publication.

ACKNOWLEDGEMENTS

Declared None.

REFERENCES

[1] Lutz W, K C S. Dimensions of global population projections: what do we know about future population trends and structures? Philos Trans R Soc Lond B Biol Sci 2010; 365(1554): 2779-91. [http://dx.doi.org/10.1098/rstb.2010.0133] [PMID: 20713384]

[2] FAOSTAT. Food and Agriculture Organization of the United Nations, "FAOSTAT" [Online]. Available: http://faostat3.fao.org/home/E.

[3] Ort DR, Merchant SS, Alric J, *et al.* Redesigning photosynthesis to sustainably meet global food and bioenergy demand. Proc Natl Acad Sci USA 2015; 112(28): 8529-36. [http://dx.doi.org/10.1073/pnas.1424031112] [PMID: 26124102]

[4] Ray DK, Mueller ND, West PC, Foley JA. Yield Trends Are Insufficient to Double Global Crop Production by 2050. PLoS One 2013; 8(6): e66428. [http://dx.doi.org/10.1371/journal.pone.0066428] [PMID: 23840465]

[5] Long SP, Ort DR. More than taking the heat: crops and global change. Curr Opin Plant Biol 2010; 13(3): 241-8. [http://dx.doi.org/10.1016/j.pbi.2010.04.008] [PMID: 20494611]

[6] Long SP, Marshall-Colon A, Zhu XG. Meeting the global food demand of the future by engineering crop photosynthesis and yield potential. Cell 2015; 161(1): 56-66. [http://dx.doi.org/10.1016/j.cell.2015.03.019] [PMID: 25815985]

[7] Zhu XG, Long SP, Ort DR. Improving photosynthetic efficiency for greater yield. Annu Rev Plant Biol 2010; 61: 235-61. [http://dx.doi.org/10.1146/annurev-arplant-042809-112206] [PMID: 20192734]

[8] Zhu XG, Long SP, Ort DR. What is the maximum efficiency with which photosynthesis can convert solar energy into biomass? Curr Opin Biotechnol 2008; 19(2): 153-9. [http://dx.doi.org/10.1016/j.copbio.2008.02.004] [PMID: 18374559]

[9] Slattery RA, Ort DR. Photosynthetic energy conversion efficiency: setting a baseline for gauging future improvements in important food and biofuel crops. Plant Physiol 2015; 168(2): 383-92. [http://dx.doi.org/10.1104/pp.15.00066] [PMID: 25829463]

[10] Gámez-Arjona FM, Li J, Raynaud S, *et al.* Enhancing the expression of starch synthase class IV results in increased levels of both transitory and long-term storage starch. Plant Biotechnol J 2011; 9(9): 1049-60.
[http://dx.doi.org/10.1111/j.1467-7652.2011.00626.x] [PMID: 21645200]

[11] Long SP, Zhu XG, Naidu SL, Ort DR. Can improvement in photosynthesis increase crop yields? Plant Cell Environ 2006; 29(3): 315-30.
[http://dx.doi.org/10.1111/j.1365-3040.2005.01493.x] [PMID: 17080588]

[12] Leakey AD, Ainsworth EA, Bernacchi CJ, Rogers A, Long SP, Ort DR. Elevated CO_2 effects on plant carbon, nitrogen, and water relations: six important lessons from FACE. J Exp Bot 2009; 60(10): 2859-76.
[http://dx.doi.org/10.1093/jxb/erp096] [PMID: 19401412]

[13] "NOAA Earth System Research Laboratory Global Monitoring Division" [Online]. Available: http://www.esrl.noaa.gov/gmd/ccgg/trends/weekly.html.

[14] Pearson PN, Palmer MR. Atmospheric carbon dioxide concentrations over the past 60 million years. Nature 2000; 406(6797): 695-9.
[http://dx.doi.org/10.1038/35021000] [PMID: 10963587]

[15] Drake BG, Gonzalez-Meler MA, Long SP. MORE EFFICIENT PLANTS: A Consequence of Rising Atmospheric CO_2? Annu Rev Plant Physiol Plant Mol Biol 1997; 48: 609-39.
[http://dx.doi.org/10.1146/annurev.arplant.48.1.609] [PMID: 15012276]

[16] Ainsworth EA, Long SP. What have we learned from 15 years of free-air CO_2 enrichment (FACE)? A meta-analytic review of the responses of photosynthesis, canopy properties and plant production to rising CO_2. New Phytol 2005; 165(2): 351-71.
[http://dx.doi.org/10.1111/j.1469-8137.2004.01224.x] [PMID: 15720649]

[17] Long SP, Ainsworth EA, Leakey AD, Nösberger J, Ort DR. Food for thought: lower-than-expected crop yield stimulation with rising CO_2 concentrations. Science 2006; 312(5782): 1918-21.
[http://dx.doi.org/10.1126/science.1114722] [PMID: 16809532]

[18] Moya TB, Ziska LH, Namuco OS, Olszyk D. Growth dynamics and genotypic variation in tropical, field-grown paddy rice (*Oryza sativa* L.) in response to increasing carbon dioxide and temperature. Glob Change Biol 1998; 4: 645-56.
[http://dx.doi.org/10.1046/j.1365-2486.1998.00180.x]

[19] De Costa WA, Weerakoon WM, Chinthaka KG, *et al.* Genotypic Variation in the Response of Rice (*Oryza sativa* L.) to Increased Atmospheric Carbon Dioxide and its Physiological Basis. J Agron Crop Sci 2007; 193: 117-30.
[http://dx.doi.org/10.1111/j.1439-037X.2007.00255.x]

[20] Manderscheid R, Weigel HJ. Photosynthetic and growth responses of old and modern spring wheat cultivars to atmospheric CO_2 enrichment. Agric Ecosyst Environ 1997; 64: 65-73.
[http://dx.doi.org/10.1016/S0167-8809(97)00020-0]

[21] Ziska LH, Morris CF, Goins EW. Quantitative and qualitative evaluation of selected wheat varieties released since 1903 to increasing atmospheric carbon dioxide: can yield sensitivity to carbon dioxide be a factor in wheat performance? Glob Change Biol 2004; 10: 1810-9.
[http://dx.doi.org/10.1111/j.1365-2486.2004.00840.x]

[22] Ainsworth E, Rogers A, Nelson R, Long SP. Testing the 'source–sink' hypothesis of down-regulation of photosynthesis in elevated [CO_2] in the field with single gene substitutions in Glycine max. Agric Meteorol 2004; 122: 85-94.
[http://dx.doi.org/10.1016/j.agrformet.2003.09.002]

[23] Peterhansel C, Offermann S. Re-engineering of carbon fixation in plants - challenges for plant biotechnology to improve yields in a high-CO_2 world. Curr Opin Biotechnol 2012; 23(2): 204-8.
[http://dx.doi.org/10.1016/j.copbio.2011.12.013] [PMID: 22261558]

[24] Blankenship RE, Chen M. Spectral expansion and antenna reduction can enhance photosynthesis for energy production. Curr Opin Chem Biol 2013; 17(3): 457-61. [http://dx.doi.org/10.1016/j.cbpa.2013.03.031] [PMID: 23602382]

[25] Chen M, Blankenship RE. Expanding the solar spectrum used by photosynthesis. Trends Plant Sci 2011; 16(8): 427-31. [http://dx.doi.org/10.1016/j.tplants.2011.03.011] [PMID: 21493120]

[26] Swingley WD, Chen M, Cheung PC, *et al.* Niche adaptation and genome expansion in the chlorophyll d-producing cyanobacterium *Acaryochloris marina*. Proc Natl Acad Sci USA 2008; 105(6): 2005-10. [http://dx.doi.org/10.1073/pnas.0709772105] [PMID: 18252824]

[27] Chen M, Schliep M, Willows RD, Cai ZL, Neilan BA, Scheer H. A red-shifted chlorophyll. Science 2010; 329(5997): 1318-9. [http://dx.doi.org/10.1126/science.1191127] [PMID: 20724585]

[28] Chen M, Li Y, Birch D, Willows RD. A cyanobacterium that contains chlorophyll fa red-absorbing photopigment. FEBS Lett 2012; 586(19): 3249-54. [http://dx.doi.org/10.1016/j.febslet.2012.06.045] [PMID: 22796191]

[29] Ort DR, Melis A. Optimizing antenna size to maximize photosynthetic efficiency. Plant Physiol 2011; 155(1): 79-85. [http://dx.doi.org/10.1104/pp.110.165886] [PMID: 21078863]

[30] Li Z, Wakao S, Fischer BB, Niyogi KK. Sensing and responding to excess light. Annu Rev Plant Biol 2009; 60: 239-60. [http://dx.doi.org/10.1146/annurev.arplant.58.032806.103844] [PMID: 19575582]

[31] Murchie EH, Niyogi KK. Manipulation of photoprotection to improve plant photosynthesis. Plant Physiol 2011; 155(1): 86-92. [http://dx.doi.org/10.1104/pp.110.168831] [PMID: 21084435]

[32] Nakajima Y, Ueda R. Improvement of photosynthesis in dense microalgal suspension by reduction of light harvesting pigments. J Appl Phycol 1997; 9: 503-10.

[33] Beckmann J, Lehr F, Finazzi G, *et al.* Improvement of light to biomass conversion by de-regulation of light-harvesting protein translation in *Chlamydomonas reinhardtii*. J Biotechnol 2009; 142(1): 70-7. [http://dx.doi.org/10.1016/j.jbiotec.2009.02.015] [PMID: 19480949]

[34] Perrine Z, Negi S, Sayre RT. Optimization of photosynthetic light energy utilization by microalgae. Algal Res 2012; 1: 134-42. [http://dx.doi.org/10.1016/j.algal.2012.07.002]

[35] Page LE, Liberton M, Pakrasi HB. Reduction of photoautotrophic productivity in the *cyanobacterium Synechocystis* sp. strain PCC 6803 by phycobilisome antenna truncation. Appl Environ Microbiol 2012; 78(17): 6349-51. [http://dx.doi.org/10.1128/AEM.00499-12] [PMID: 22706065]

[36] Derks A, Schaven K, Bruce D. Diverse mechanisms for photoprotection in photosynthesis. Dynamic regulation of photosystem II excitation in response to rapid environmental change Biochim Biophys Acta 2015; 1847: 468-85.

[37] Raines CA. Increasing photosynthetic carbon assimilation in C3 plants to improve crop yield: current and future strategies. Plant Physiol 2011; 155(1): 36-42. [http://dx.doi.org/10.1104/pp.110.168559] [PMID: 21071599]

[38] Bauwe H, Hagemann M, Fernie AR. Photorespiration: players, partners and origin. Trends Plant Sci 2010; 15(6): 330-6. [http://dx.doi.org/10.1016/j.tplants.2010.03.006] [PMID: 20403720]

[39] Moroney JV, Jungnick N, Dimario RJ, Longstreth DJ. Photorespiration and carbon concentrating mechanisms: two adaptations to high O_2, low CO_2 conditions. Photosynth Res 2013; 117(1-3): 121-31.

[http://dx.doi.org/10.1007/s11120-013-9865-7] [PMID: 23771683]

[40] Parry MA, Andralojc PJ, Scales JC, *et al.* Rubisco activity and regulation as targets for crop improvement. J Exp Bot 2013; 64(3): 717-30. [http://dx.doi.org/10.1093/jxb/ers336] [PMID: 23162118]

[41] Spreitzer RJ, Salvucci ME. Rubisco: structure, regulatory interactions, and possibilities for a better enzyme. Annu Rev Plant Biol 2002; 53: 449-75. [http://dx.doi.org/10.1146/annurev.arplant.53.100301.135233] [PMID: 12221984]

[42] Sage RF. Variation in the k(cat) of Rubisco in C(3) and C(4) plants and some implications for photosynthetic performance at high and low temperature. J Exp Bot 2002; 53(369): 609-20. [http://dx.doi.org/10.1093/jexbot/53.369.609] [PMID: 11886880]

[43] Zhu X, Portis A, Long S. Would transformation of C3 crop plants with foreign Rubisco increase productivity? A computational analysis extrapolating from kinetic properties to canopy. Plant Cell Environ 2004; Spreitzer 1999: 155-65.

[44] Price GD, Badger MR, von Caemmerer S. The prospect of using cyanobacterial bicarbonate transporters to improve leaf photosynthesis in C3 crop plants. Plant Physiol 2011; 155(1): 20-6. [http://dx.doi.org/10.1104/pp.110.164681] [PMID: 20923885]

[45] Galmés J, Kapralov MV, Andralojc PJ, *et al.* Expanding knowledge of the Rubisco kinetics variability in plant species: environmental and evolutionary trends. Plant Cell Environ 2014; 37(9): 1989-2001. [http://dx.doi.org/10.1111/pce.12335] [PMID: 24689692]

[46] Sage RF. The evolution of C 4 photosynthesis. New Phytol 2004; 161: 341-70. [http://dx.doi.org/10.1111/j.1469-8137.2004.00974.x]

[47] Tcherkez GG, Farquhar GD, Andrews TJ. Despite slow catalysis and confused substrate specificity, all ribulose bisphosphate carboxylases may be nearly perfectly optimized. Proc Natl Acad Sci USA 2006; 103(19): 7246-51. [http://dx.doi.org/10.1073/pnas.0600605103] [PMID: 16641091]

[48] Whitney SM, Houtz RL, Alonso H. Advancing our understanding and capacity to engineer natures CO_2-sequestering enzyme, Rubisco. Plant Physiol 2011; 155(1): 27-35. [http://dx.doi.org/10.1104/pp.110.164814] [PMID: 20974895]

[49] Kapralov MV, Kubien DS, Andersson I, Filatov DA. Changes in Rubisco kinetics during the evolution of C4 photosynthesis in Flaveria (Asteraceae) are associated with positive selection on genes encoding the enzyme. Mol Biol Evol 2011; 28(4): 1491-503. [http://dx.doi.org/10.1093/molbev/msq335] [PMID: 21172830]

[50] Whitney SM, Sharwood RE, Orr D, White SJ, Alonso H, Galmés J. Isoleucine 309 acts as a C4 catalytic switch that increases ribulose-1,5-bisphosphate carboxylase/oxygenase (rubisco) carboxylation rate in Flaveria. Proc Natl Acad Sci USA 2011; 108(35): 14688-93. [http://dx.doi.org/10.1073/pnas.1109503108] [PMID: 21849620]

[51] Ishikawa C, Hatanaka T, Misoo S, Miyake C, Fukayama H. Functional incorporation of sorghum small subunit increases the catalytic turnover rate of Rubisco in transgenic rice. Plant Physiol 2011; 156(3): 1603-11. [http://dx.doi.org/10.1104/pp.111.177030] [PMID: 21562335]

[52] Whitney SM, Birch R, Kelso C, Beck JL, Kapralov MV. Improving recombinant Rubisco biogenesis, plant photosynthesis and growth by coexpressing its ancillary RAF1 chaperone. Proc Natl Acad Sci USA 2015; 112(11): 3564-9. [http://dx.doi.org/10.1073/pnas.1420536112] [PMID: 25733857]

[53] Mueller-Cajar O, Stotz M, Wendler P, Hartl FU, Bracher A, Hayer-Hartl M. Structure and function of the AAA+ protein CbbX, a red-type Rubisco activase. Nature 2011; 479(7372): 194-9. [http://dx.doi.org/10.1038/nature10568] [PMID: 22048315]

[54] Mueller-Cajar O, Stotz M, Bracher A. Maintaining photosynthetic CO_2 fixation *via* protein

remodelling: the Rubisco activases. Photosynth Res 2014; 119(1-2): 191-201. [http://dx.doi.org/10.1007/s11120-013-9819-0] [PMID: 23543331]

[55] Hanson MR, Gray BN, Ahner BA. Chloroplast transformation for engineering of photosynthesis. J Exp Bot 2013; 64(3): 731-42. [http://dx.doi.org/10.1093/jxb/ers325] [PMID: 23162121]

[56] Rogalski M, do Nascimento Vieira L, Fraga HP, Guerra MP. Plastid genomics in horticultural species: importance and applications for plant population genetics, evolution, and biotechnology. Front Plant Sci 2015; 6: 586. [http://dx.doi.org/10.3389/fpls.2015.00586] [PMID: 26284102]

[57] Scharff LB, Bock R. Synthetic biology in plastids. Plant J 2014; 78(5): 783-98. [http://dx.doi.org/10.1111/tpj.12356] [PMID: 24147738]

[58] Crafts-Brandner SJ, Salvucci ME. Rubisco activase constrains the photosynthetic potential of leaves at high temperature and CO_2. Proc Natl Acad Sci USA 2000; 97(24): 13430-5. [http://dx.doi.org/10.1073/pnas.230451497] [PMID: 11069297]

[59] Kurek I, Chang TK, Bertain SM, *et al.* Enhanced Thermostability of Arabidopsis Rubisco activase improves photosynthesis and growth rates under moderate heat stress. Plant Cell 2007; 19(10): 3230-41. [http://dx.doi.org/10.1105/tpc.107.054171] [PMID: 17933901]

[60] Kumar A, Li C, Portis AR Jr. *Arabidopsis thaliana* expressing a thermostable chimeric Rubisco activase exhibits enhanced growth and higher rates of photosynthesis at moderately high temperatures. Photosynth Res 2009; 100(3): 143-53. [http://dx.doi.org/10.1007/s11120-009-9438-y] [PMID: 19507049]

[61] Cai B, Zhang A, Yang Z, Lu Q, Wen X, Lu C. Characterization of photosystem II photochemistry in transgenic tobacco plants with lowered Rubisco activase content. J Plant Physiol 2010; 167(17): 1457-65. [http://dx.doi.org/10.1016/j.jplph.2010.05.004] [PMID: 20576317]

[62] Uematsu K, Suzuki N, Iwamae T, Inui M, Yukawa H. Increased fructose 1,6-bisphosphate aldolase in plastids enhances growth and photosynthesis of tobacco plants. J Exp Bot 2012; 63(8): 3001-9. [http://dx.doi.org/10.1093/jxb/ers004] [PMID: 22323273]

[63] Lefebvre S, Lawson T, Zakhleniuk OV, Lloyd JC, Raines CA, Fryer M. Increased sedoheptulose-1,--bisphosphatase activity in transgenic tobacco plants stimulates photosynthesis and growth from an early stage in development. Plant Physiol 2005; 138(1): 451-60. [http://dx.doi.org/10.1104/pp.104.055046] [PMID: 15863701]

[64] Feng L, Wang K, Li Y, *et al.* Overexpression of SBPase enhances photosynthesis against high temperature stress in transgenic rice plants. Plant Cell Rep 2007; 26(9): 1635-46. [http://dx.doi.org/10.1007/s00299-006-0299-y] [PMID: 17458549]

[65] Rosenthal DM, Locke AM, Khozaei M, Raines CA, Long SP, Ort DR. Over-expressing the C(3) photosynthesis cycle enzyme Sedoheptulose-17 Bisphosphatase improves photosynthetic carbon gain and yield under fully open air CO(2) fumigation (FACE). BMC Plant Biol 2011; 11: 123. [http://dx.doi.org/10.1186/1471-2229-11-123] [PMID: 21884586]

[66] Henkes S, Sonnewald U, Badur R, Flachmann R, Stitt M. A small decrease of plastid transketolase activity in antisense tobacco transformants has dramatic effects on photosynthesis and phenylpropanoid metabolism. Plant Cell 2001; 13(3): 535-51. [http://dx.doi.org/10.1105/tpc.13.3.535] [PMID: 11251095]

[67] Khozaei M, Fisk S, Lawson T, *et al.* Overexpression of plastid transketolase in tobacco results in a thiamine auxotrophic phenotype. Plant Cell 2015; 27(2): 432-47. [http://dx.doi.org/10.1105/tpc.114.131011] [PMID: 25670766]

[68] Singh J, Pandey P, James D, *et al.* Enhancing C3 photosynthesis: an outlook on feasible interventions

for crop improvement. Plant Biotechnol J 2014; 12(9): 1217-30.
[http://dx.doi.org/10.1111/pbi.12246] [PMID: 25196090]

[69] Atkinson N, Feike D, Mackinder LC, *et al.* Introducing an algal carbon-concentrating mechanism into higher plants: location and incorporation of key components. Plant Biotechnol J 2015. [PMID: 26538195]

[70] McGrath JM, Long SP. Can the cyanobacterial carbon-concentrating mechanism increase photosynthesis in crop species? A theoretical analysis. Plant Physiol 2014; 164(4): 2247-61. [http://dx.doi.org/10.1104/pp.113.232611] [PMID: 24550242]

[71] Bonacci W, Teng PK, Afonso B, *et al.* Modularity of a carbon-fixing protein organelle. Proc Natl Acad Sci USA 2012; 109(2): 478-83. [http://dx.doi.org/10.1073/pnas.1108557109] [PMID: 22184212]

[72] Lin MT, Occhialini A, Andralojc PJ, *et al.* β-Carboxysomal proteins assemble into highly organized structures in Nicotiana chloroplasts. Plant J 2014; 79(1): 1-12. [http://dx.doi.org/10.1111/tpj.12536] [PMID: 24810513]

[73] Lin MT, Occhialini A, Andralojc PJ, Parry MA, Hanson MR. A faster Rubisco with potential to increase photosynthesis in crops. Nature 2014; 513(7519): 547-50. [http://dx.doi.org/10.1038/nature13776] [PMID: 25231869]

[74] Rawat M, Henk M, Lavigne L, Moroney J. Chlamydomonas reinhardtii mutants without ribulose-1,--bisphosphate carboxylase-oxygenase lack a detectable pyrenoid. Planta 1996; •••: 198.

[75] Karlsson J, Clarke AK, Chen ZY, *et al.* A novel alpha-type carbonic anhydrase associated with the thylakoid membrane in Chlamydomonas reinhardtii is required for growth at ambient CO_2. EMBO J 1998; 17(5): 1208-16. [http://dx.doi.org/10.1093/emboj/17.5.1208] [PMID: 9482718]

[76] Studer RA, Christin PA, Williams MA, Orengo CA. Stability-activity tradeoffs constrain the adaptive evolution of RubisCO. Proc Natl Acad Sci USA 2014; 111(6): 2223-8. [http://dx.doi.org/10.1073/pnas.1310811111] [PMID: 24469821]

[77] Sage RF, Sage TL, Kocacinar F. Photorespiration and the Evolution of C4 Photosynthesis. Annual Review of Plant Biology 2012; 111(63): 19-47. [http://dx.doi.org/10.1146/annurev-arplant-042811-105511]

[78] Peterhansel C. Best practice procedures for the establishment of a C(4) cycle in transgenic C(3) plants. J Exp Bot 2011; 62(9): 3011-9. [http://dx.doi.org/10.1093/jxb/err027] [PMID: 21335437]

[79] Kebeish R, Niessen M, Thiruveedhi K, *et al.* Chloroplastic photorespiratory bypass increases photosynthesis and biomass production in *Arabidopsis thaliana*. Nat Biotechnol 2007; 25(5): 593-9. [http://dx.doi.org/10.1038/nbt1299] [PMID: 17435746]

[80] Peterhansel C, Blume C, Offermann S. Photorespiratory bypasses: how can they work? J Exp Bot 2013; 64(3): 709-15. [http://dx.doi.org/10.1093/jxb/ers247] [PMID: 22996676]

[81] Carvalho JdeF, Madgwick PJ, Powers SJ, Keys AJ, Lea PJ, Parry MA. An engineered pathway for glyoxylate metabolism in tobacco plants aimed to avoid the release of ammonia in photorespiration. BMC Biotechnol 2011; 11: 111. [http://dx.doi.org/10.1186/1472-6750-11-111] [PMID: 22104170]

[82] Maier A, Fahnenstich H, von Caemmerer S, *et al.* Transgenic Introduction of a Glycolate Oxidative Cycle into *A. thaliana* Chloroplasts Leads to Growth Improvement. Front Plant Sci 2012; 3: 38. [http://dx.doi.org/10.3389/fpls.2012.00038] [PMID: 22639647]

[83] Shih PM, Zarzycki J, Niyogi KK, Kerfeld CA. Introduction of a synthetic CO_2-fixing photorespiratory bypass into a cyanobacterium. J Biol Chem 2014; 289(14): 9493-500.

[http://dx.doi.org/10.1074/jbc.C113.543132] [PMID: 24558040]

[84] Smidansky ED, Clancy M, Meyer FD, *et al.* Enhanced ADP-glucose pyrophosphorylase activity in wheat endosperm increases seed yield. Proc Natl Acad Sci USA 2002; 99(3): 1724-9. [http://dx.doi.org/10.1073/pnas.022635299] [PMID: 11830676]

[85] Gibson K, Park JS, Nagai Y, *et al.* Exploiting leaf starch synthesis as a transient sink to elevate photosynthesis, plant productivity and yields. Plant Sci 2011; 181(3): 275-81. [http://dx.doi.org/10.1016/j.plantsci.2011.06.001] [PMID: 21763538]

[86] Wu L, Birch RG. Doubled sugar content in sugarcane plants modified to produce a sucrose isomer. Plant Biotechnol J 2007; 5(1): 109-17. [http://dx.doi.org/10.1111/j.1467-7652.2006.00224.x] [PMID: 17207261]

[87] Feng L, Han Y, Liu G, *et al.* Overexpression of sedoheptulose-1,7-bisphosphatase enhances photosynthesis and growth under salt stress in transgenic rice plants. Funct Plant Biol 2007; 34: 822-34. [http://dx.doi.org/10.1071/FP07074]

CHAPTER 6

Seed Size: an Important Yield Component

Giuseppe Gavazzi* and **Stefano Sangiorgio**

Department of Agricultural and Environmental Sciences- Production, Landscapes, Agroenergy, Università degli Studi di Milano, Milano, Italy

Abstract: Grain yield in cereals is considered one of the most important traits in the perspective of global food supply. It largely depends on several components including grain size and grain number. Several studies have been conducted to understand the genetic and molecular basis of these two traits. In this article we review the information so far obtained on the mechanisms governing seed size in crop plants particularly in cereals and discuss the potential uses of this information to improve the productivity of seeds in cereal crops.

Keywords: *Arabidopsis*, Cryptic variation, Grain weight, Hormones, Maize, Seed size, Tillers.

INTRODUCTION

Small seeds are a prerogative of wild species. Because of the small size they can be produced in high numbers and are easily dispersed. These traits have presumably increased the fitness of the plant and for this reason they have been a target of natural selection. On the other hand, starting from domestication, cultivated plants have been manipulated to improve seed size. The interest in studying seed size relies on the fact that this trait is an important yield component, particularly in cereals. Functional analysis of single genes affecting seed development and in particular seed size is fundamental for unraveling the complex genetic network underlying the processes of seed formation. It may also be instrumental for establishing classical as well as advanced breeding programs aimed at improving cereal productivity. Recently the advent of new technologies allowing the precise manipulation of single genes has renewed interest in understanding the role of genetic factors underlying these complex processes.

Genes that retain key functions will become the object of this approach since they will enable us to obtain gene variants with better performance with respect to

* **Corresponding author Giuseppe Gavazzi**: Department of Agricultural and Environmental Sciences- Production, Landscapes, Agroenergy, Università degli Studi di Milano, Milano, Italy; Tel: +39 3348919202; Fax: 0250316521; E-mail: giuseppe.gavazzi@unimi.it

Roberto Pilu and Giuseppe Gavazzi (Eds.)

grain yield, which is the main trait of agronomic importance in cereal improvement.

In this review we will concentrate on the analysis of factors, acting at the genetic, epigenetic and physiological level, involved in the control of seed size as outlined through several studies. We also discuss future perspectives in improving this character.

Yield Improvement By Seed-Size Selection

Farm-saved seed accounts for the greatest proportion of seeds used by farmers especially in low-income countries. Farmers have been able to improve the quality of the farm-saved seed by applying seed selection at harvest time. As a result of seed selection, yield increases have been reported in wheat, up to 50%, 24% in rice, 15.3% in okra and 13% in maize. In maize the selection is based on the choice of kernel size and ear dimension. In an experiment reported by Msuya and Stefano [1], selection of the larger seeds located in the basal portion of the ear and exclusion of the smaller ones, in the apical region, resulted in improved germination and seedling vigor, and yielded larger seeds in the progeny. Even though the authors did not analyze heritability of the improvement in seed size we assume that the observed gain in seed size is inherited since the selection applied is the one traditionally used by farmers to improve this crop. In an accurate study Odhiambo and Compton [2] by subjecting the Krug maize variety to 20 cycles of mass selection for seed size, obtained two populations with large and small seed sizes confirming the heritability of the trait selected.

Origin Of The Seed

The seed in flowering plants represents a crucial phase of the plant cycle, ensuring the reproduction of the plant and, through its dispersal, its spreading around over the surrounding soil surface. Its formation requires a double event of fertilization of the female gametophyte, the embryo sac, by the two sperms of the male gametophyte, the pollen grain. One of the two sperms unites with the egg cell, and the ensuing zygote, carrying one paternal and one maternal genome, undergoes several cell divisions resulting in the formation of the embryo, the progenitor of a plant which will form the next generation. The second sperm unites with the central cell of the embryo sac obtained by the fusion of two identical maternal genomes, giving rise to a triploid cell which is the precursor of the endosperm. Embryogenesis is characterized by different morphological stages, corresponding to the acquisition and development of new functions. After several cycles of cell divisions, the embryo acquires polarity followed by a morphogenetic program leading to the formation of root and shoot primordia at the two poles of the embryonic axis followed by a maturation phase, dehydration and quiescence. The endosperm shows four developmental phases: syncytium formation, cellularization, differentiation and cell death. The mature endosperm consists of four domains: the starchy endosperm, the single aleurone layer, the embryo surrounding region and the basal endosperm transfer layer [3, 4]. Since the embryo sac develops within the ovule, a sporophytic organ, the seed is surrounded by maternal sporophytic tissues (Fig. **1**). So the seed consists of two main compartments, firstly the embryo and the endosperm with 2n and 3n ploidy levels respectively, and secondly the seed coats with a 2n maternal genotype (Fig. **2**).

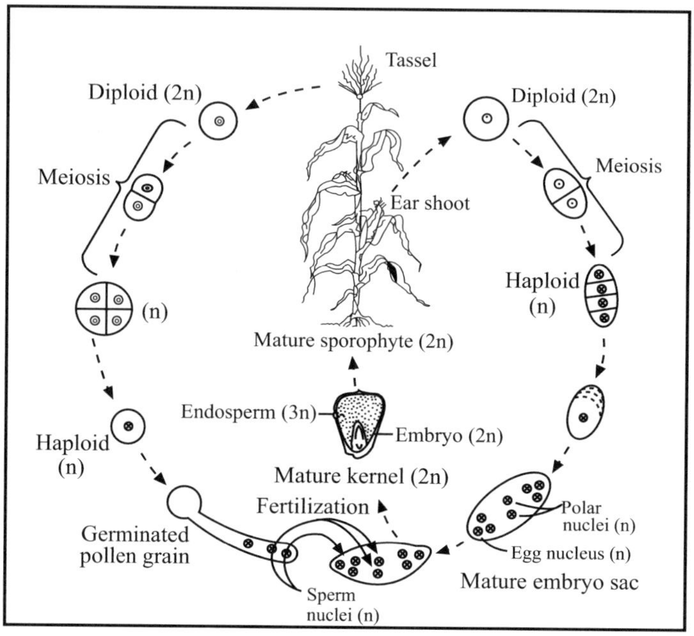

Fig. (1). Life cycle of *Zea mays* L. (adapted from Mc Clintock, 1983).

Seed development implies the formation of embryo and endosperm as well as interaction between the two and between the developing seed and the maternal tissues. Several factors influence seed size: ecogeographical factors such as photoperiod, edaphic conditions, precipitations, biotic and abiotic stresses [5]; the coordinated growth of the different seed components (embryo, endosperm and pericarp) [6] and hormones (auxin, cytokinin and brassinosteroids). In studying it we should be aware that seed size is the end result of the interplay of several factors. A comprehensive view of the mechanisms underlying seed size requires an analysis not only of the effect of one or a few genes at a time but also the application of modern-scale genomic approaches.

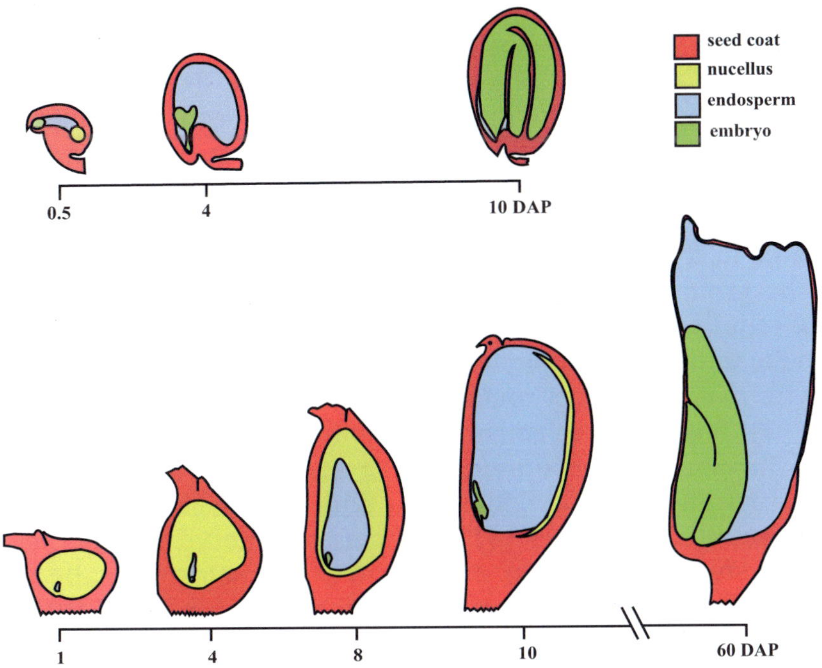

Fig. (2). Seed development in *Arabidopsis* and maize. Schematic drawing of key developmental stages of *Arabidopsis* (top) or maize seeds (bottom). The time scale is in days after pollination (DAP). Red color indicates the tissues of maternal origin. (From: Vernoud *et al.*, 2005).

Endosperm Failure Following Interploidy Crosses

In a range of species reciprocal crosses between diploid and tetraploid individuals result in abnormal endosperm development. In maize, abnormal endosperm development is observed following crosses of a normal male parent with a maternal parent carrying indeterminate gametophyte (*ig*), a mutant inducing a variable number of cells and nuclei in the embryo sac [7]. The genetic bases of endosperm failure are unclear and alternative hypotheses such as gene dosage effects [8] or imprinting of regulatory genes [9] have been proposed as an explanation. Even though the mechanisms leading to endosperm failure in intraspecific crosses with different ploidy levels are still poorly understood, the picture emerging from a large number of observations is that endosperm development relies on a range of epigenetic mechanisms referred to as imprinting.

Genes Controlling Seed Size

An approach to the study of the genetics of seed size is based on the detection of gene mutants impairing seed development and their analysis. These mutants can be ascribed to different categories.

Higher plants have different ways to exert a maternal control over growth and development of their offspring. During plant reproduction the gametophyte requires the expression of genes from both its own haploid genome and the diploid genome of the parent organism. In particular, during seed development, the embryo sac develops within the ovule so that the seed, following fertilization, develops within a maternal tissue. Consequently genes acting in this tissue are expected to have an effect on seed development. These genes have been identified and classified as maternal effect genes. They are required in the maternal sporophyte or in the female gametophyte for proper embryo sac development. They can be grouped into three distinct classes [10]: genes expressed in the sporophyte required for either embryo sac or embryo development, and genes required in the female gametophyte (gametophytic maternal effect genes). Several gametophytic maternal effect mutants have been identified in Arabidopsis and maize (Fig. **3**). Possible mechanisms accounting for the maternal effect include changes in functional gene dosage in the endosperm for mutations in genes required in two doses in the endosperm with two maternal genomes and one paternal genome (*e.g.*, floury3 [11];), abnormal embryo sac morphology (*e.g.*, baseless [12]; stunter [13];), loss of embryo sac proteins normally stored in the cytoplasm and acting after fertilization (perdurance) (*e.g.* PROLIFERA [14];) or imprinting (Polycomb group genes in maize and Arabidopsis [15]).

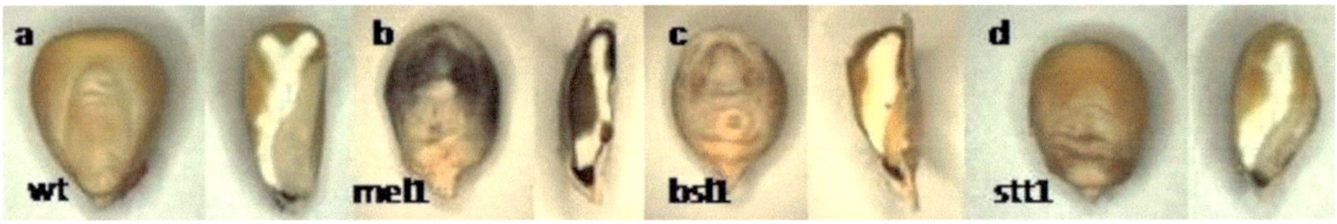

Fig. (3). Maternal effect mutants. All seeds with the germinal face on the left and a cross section on the right (a) Wild type (b) *mel1*. The embryo appears reduced in size and with an irregular profile. (c) *bsl1* loose pericarp and aborted embryo. (d) *stunter1*. Normal seed but smaller than wild type. From:https://dpb.carnegiescience.edu/labs/evanslab/projects/maternal-regulation-seed-development with the permission of Evans.

Maternal Effect Genes

An interesting maternal effect mutation, maternal effect lethal1 (*mel1*), has been described by Evans and Kermicle (2001) [10]. Ears produced by heterozygote *mel1* plants pollinated with a wild-type male yield defective kernels while homozygous wild-type females pollinated with a *mel1* heterozygous male show only normal kernels (Fig. **4**) suggesting either a maternal effect or a dominant mutation unable to be transmitted through the pollen. Paternal transmission of *mel1* occurs at a much reduced frequency if compared to wild-type and does not affect kernel development. Further experiments indicate that the low frequency of paternal *mel1* transmission reflects reduced transmission of the *mel1* carrying chromosome. Taken together these results are in accord with the interpretation of the *mel1* mutation as a representative of a gametophytic maternal effect.

Fig. (4). Pattern of transmission and phenotypes of *mel1* mutant. (A) Ear obtained by crossing *mel1* female with a wild type male. *mel1* seeds exhibit defective growth and appear etched and small. (B) Ear obtained by crossing a homozygous wild type female with a *mel1* heterozygous male. All seeds appear normal. (C) Normal seeds shown with germinal face up. (D) *mel1* seeds (germinal face) showing twin embryos (far left) embryos with reduced size (middle two) and devoid of embryo (far right) (From: Evans and Kermicle, 2001).

Mutants Impaired In The Grain Filling Process

Mutants impaired in seed size and development are formed with high frequency following induced mutagenesis. They generally have both embryo and endosperm defective or their defects are confined to the embryo or, less frequently, only to the endosperm. In maize, the analysis of a large collection of EMS-induced *dek* (defective kernel) mutants has led to their classification, after embryo culture of immature embryos, as nutritional-type and developmental-type mutants [16]. They may be ascribed to genes involved in the grain filling process. However, physiological characterization has been done in only a few cases of mutants with reduced seed size. Their analysis has revealed itself to be useful for elucidating the process of nutrient allocation in plant seeds. Developing seeds act as strong sink organs requiring the transfer of sugars from the leaves (source) through the phloem. To ensure successful seed development, there should be an uninterrupted flow of sugar reaching the seed through the pedicel. To promote transport into the seed, cereals have developed a transfer endosperm layer consisting of cells with ingrowths of the wall which amplify the surface area up to 20-fold, to facilitate maternal solute uptake. Transfer tissues are responsible for sugar translocation and partitioning by invertase enzymes and hence are seen as key determinants of sink strength and seed size. A functional endosperm transfer layer seems to be required for proper seed development as observed by analyzing the progeny of a cross between a diploid maize line as a female with its tetraploid male counterpart. The histological analysis of the seeds with an abnormal endosperm observed in the progeny of this cross reveals that the departure from the normal 2m:1p genomic ratio leads to the almost complete suppression of the formation of the transfer cell layer [17]. Because of the activity of cell wall INVs (invertases) and hydrolysis of sucrose to fructose and glucose, a sucrose concentration gradient between maternal symplast and endosperm apoplast is formed. Subsequently, an ETC-specific hexose transporter facilitates the import of hexoses into the endosperm [18 - 20]. One of the first mutants employed for the analysis of the transfer of photoassimilates and nutrients from maternal cells to developing seed was the miniature1(*mn1*) mutant in maize [19] (Fig. **5**).

Fig. (5). An ear segregating 1:1 for wt and *mn1* seeds derived from a backcross of +/*mn1* to homozygous *mn1* parent. (From MAIZE GDB).

In this species sucrose is hydrolyzed in the pedicel of the seed before entering into the endosperm. In the developing seed the majority of sucrose is hydrolyzed to glucose and fructose by invertases located in the basal endosperm cells [21].There are two forms of this enzyme in all organisms, a soluble one (in the cytoplasm) and a particulate one (bound to the cell wall). The *Mn1* gene codes for an endosperm-specific cell wall invertase designated CWI-2. The *mn1* mutant shows low levels of invertase activities. The enzyme deficiency gives rise to cellular destruction in the placento-chalazal region, causing a gap between the pedicel and the endosperm. Consequently, in the developing endosperm there is a reduced flux of nutrients from the mother plant slowing down its growth. Other mutants are known with reduced grain filling caused by effects on basal endosperm and pedicel development. One such mutation has been identified in maize and named *reduced grain filling1* (*rfg1*) (Fig. **6**) because of its significant reduction in weight and size compared to wild-type sibs. Characteristic of this mutant is its dosage-dependent expression in the endosperm, not observed in most defective endosperm mutants. The mutant shows defective pedicel development and reduced expression of endosperm transfer-layer specific markers. The reduction of starch accumulation observed in this mutant does not affect sugar pools or enzymes in starch biosynthesis. The reduction may be due to a reduced cell number, which determines kernel sink size early in maize development [22].

Therefore, the *Rgf1* gene not yet identified, may be involved in sugar sensing or transport in ETC. One possibility is that the *rfg1* mutation leads to an early reduction in sugar flux as well as in sink size.

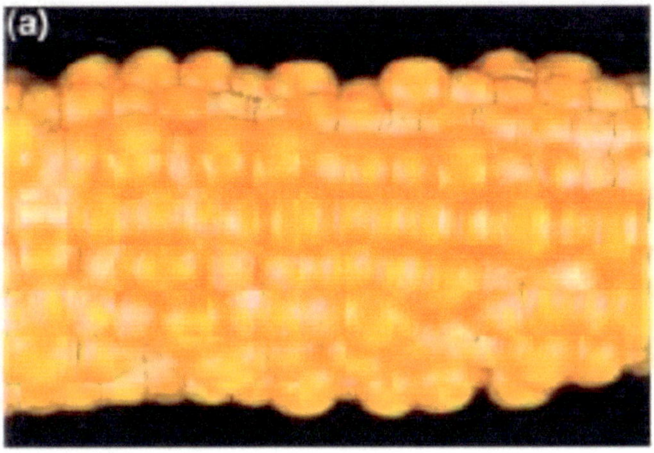

Fig. (6). An ear segregating for *rgf1* mutant kernels. (From: Maitz *et al.*, 2000).

Expression of some ETC-specific genes is under maternal control. One such gene, *Maternally Expressed Gene1* (*MEG1*), encodes an LTP with structural similarity to defensins. *MEG1* expression, confined to the BETL cells of maize endosperm shows a parent-of-origin pattern early in endosperm development and a biallelic one at later stages. Its product is glycosylated and localizes to the labyrinth of in-growths of the walls of transfer cells [23]. Stable *Meg1* RNAi transgenic maize lines have a small seed phenotype. The size reduction is due to defects in the endosperm transfer tissues, as shown by defects in the tissue morphology of endosperm transfer tissues as well as by changes in the pattern of expression of genes specifically expressed in these tissues. In addition, *Meg1* knock-down seeds exhibit a dramatic reduction in cell wall activity compared to wild-type seeds as well as in hexose sugar content, whereas sucrose content, similar to wild-type at 8 DAP (days after pollination), increased considerably thereafter. This suggests that *Meg1* is required for transfer cell function and sugar homeostasis during seed development. The formation of a functional BETL, the flux of sucrose from the maternal tissues to the endosperm, the role of the invertase in sucrose metabolism are all required for normal seed growth and development [24].

Gene Imprinting

Imprinting is an epigenetic change of the parental alleles leading to the differential expression of genes in a parent-of-origin manner. Many genes active during endosperm development are imprinted. The majority of those so far identified are active if contributed by the mother, while they are silenced if their contribution is paternal. Silencing is achieved by DNA methylation, histone methylation or acetylation or by chromatin remodeling. DNA methylation of cytosine causes silencing of gene transcription. In Arabidopsis DNA METHYL-TRANSFERASE (MET1) maintains cytosine DNA methylation. The analysis of the progeny of reciprocal crosses between antisense MET1 transgenic and wild-type plants shows that DNA hypomethylation causes a parent-of-origin effect on seed size. Similarly crosses between wild-type and homozygous *met1* parents indicate that hypomethylation of maternal and paternal genomes leads to larger and smaller F_1 seeds respectively [25]. However, imprinting is not necessarily the only mechanism accounting for parent-dependent seed impairment. Even though many parent-of-origin effects are viewed as the result of transcriptional imprinting, an alternative interpretation is the hypothesis that differential contributions of any dosage-sensitive regulator to the developing seed will show a parent-of-origin effect. Viewed in this way, the parent-of-origin events are interpreted as the result of a differential dosage .According to Birchler [26], genes exhibiting dosage effects are expected to code the subunits of macromolecular complexes that are impaired in their function and assembly by a decrease in one component. According to this view, the differential dosage hypothesis could be applied to interpret the parent-of-origin effect in interploid crosses, the aneuploidy effect as well as the result (see section on "The role of QTL on seed size determination") on seed size reduction observed in the progeny of B-A translocations [27].

PPR Genes Are Involved In The Control of Seed Size

RNA processing, maturation and editing of the transcripts of mitochondrial and plastidic genes are mediated by a class of RNA binding proteins, the pentatricopeptide repeat (PPR) proteins [28]. PPR proteins consist of a 35 amino-acid motif, loosely conserved and repeated in tandem up to 30 times. They are ubiquitous but particularly numerous in plants. They are RNA binding proteins predominantly targeted to plastids or mitochondria. Cytoplasmic male sterility and embryo lethality can be caused by mutations in genes encoding mitochondrial PPR proteins. Mutations of PPR genes coding for proteins that are targeted either to chloroplasts or mitochondria may lead to delay and arrest of embryo and endosperm development with concomitant reduction in seed size. Several PPR genes affecting seed size have been isolated. One extensively characterized at the molecular level is the *small kernel 1* (*smk1*) (Fig. **7**) gene in maize encoding a PPR-E type protein. The mutant, when homozygous, is lethal at the embryo stage and is recognizable, on a segregating ear, by the reduced size of mutant seeds. This gene, encoding an E-subclass pentatricopeptide repeat protein targeted to mitochondria, is needed for the C to U editing of a *nad7* transcript determining the incorporation of leucine rather than proline in the NAD protein at position 279. Loss of SMK1 function eliminates the C to U editing in the mitochondrial *nad7* transcript, causing retention of a proline in place of a leucine codon at position 279 [29].

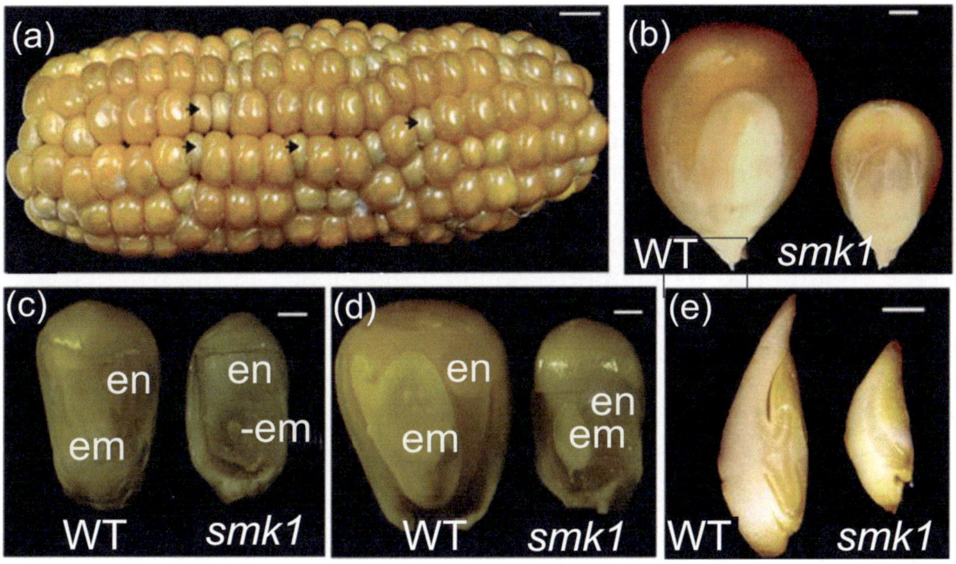

Fig. (7). The *smk1* mutant. (a) selfed +/*smk* ear segregating mutant seeds (arrows). (b) Wt and *smk1* seeds at 13 (c), 21 (d) and 35 (e) days after pollination (DAP). Scale bars: (a) 1 cm; (b,c,d,e) 1 mm. From Li *et al.*, 2014).

Other well characterized PPR genes essential for seed development in maize are *emp4* [30], *emp 5* [31], *emp7* [32] and *mppr6* [33, 34]. In *Arabidopsis,* among 250 embryo-defective mutants analyzed, 17 were found to be PPR genes essential to embryogenesis [35].

The Role of Quantitative Trait Loci (QTL) in Seed Size Determination

So far we have considered on one or a few genes at a time to explore the mechanisms governing seed development. However grain yield is a complex character determined by its three components: seed size and weight and number of seeds, that are quantitative traits. So the seed size could also be improved by manipulating the QTL governing this trait. The existence of multiple factors governing seed size is suggested by an analysis of aneuploids such as the ones originating in stocks of maize carrying B-A translocations. In the presence of the centromere of the B chromosome, there is nondisjunction of the B-A chromosome at the second mitotic division following meiosis in the pollen grain. As a consequence of this event, hypoploids, with zero copies and hyperploids, with two paternal copies of the B-A chromosome, endosperms or embryos are formed. The pioneer work in maize opens the way to the quantitative genetics of seed size in cereals. Birchler [36] first reported that the endosperm dimension is reduced if certain regions of the genome are missing in the paternal contribution of the sperm to the endosperm as the result of non-disjunction of a B-A translocation. This effect, originally observed with the loss of chromosome 1 long arm (1L) was later observed with six other B-A translocations causing loss of other chromosomal regions. The obvious explanation for this observation is that when only two, rather than three, maternal copies of the chromosomal segment are present in the endosperm, too few gene are expressed for normal seed development. However it was noticed that the additional copies of the missing segment introduced through the maternal side could not repair the reduced size phenotype conditioned by the corresponding deficiency in the paternal gamete. Other segments, when introduced maternally, produced an even smaller kernel when fertilized by a sperm missing the same A chromosome segment (Fig. **8**) or other ones that carry factors affecting endosperm size (Ef). These results suggest that the paternal Ef factors required for normal endosperm development, can enhance this effect of their own or of other Efs, when introduced as extra copies maternally.

The responsible genes appear to function when transmitted through the female. Since the maternal function is recognized only in combination with the small kernel effect, the maternal function appears to be involved in the same process. These results disclose a relationship between the sites that produce the small kernel effect and those that enhance it. The authors suggest that some threshold must be reached in order to recognize the effect of the maternally contributed doses. In fact when the egg carrying extra doses is fertilized by normal sperm, no visible effect on size is detected: only when the sperm is missing specific regions, does the effect become recognizable. For a more comprehensive view of this topic the reader is referred to the exhaustive review of Birchler [8]. In this regard the main point is the finding that several factors, identified with the help of B-A translocations and spread over the entire genome, are required in order to achieve a normal seed size. The recently developed molecular markers linkage maps have allowed the identification and mapping of hundreds of genes and quantitative trait loci (QTL) including those responsible for yield. The general picture emerging from a large number of studies, particularly in cereals, where large mapping populations or recombinant inbred lines (RIL) were analyzed is pointing to the existence of many QTL affecting grain yield scattered throughout the genome. A common observation based on the mapping studies is that some chromosomal regions with large effects on multiple traits form QTL clusters. In rice several QTLs with major effect on grain size and with pleiotropic effects have been cloned. Instrumental for the detection of QTL related to yield has been the isolation of F_2 populations from crosses between two *indica* varieties as well as recombinant inbred lines (RIL) [37]. Some of them play the role of negative regulators of grain shape or size while a few, like GS5, function as positive grain size regulators [38]. Thc map-based cloning of seed size related genes was successfully achieved in rice, leading to the identification of major QTL controlling grain length (qGL3) or grain size (GS3) or grain width and weight (GW2). Orthologs of the latter have been found in bread wheat (TaGW2) and two orthologs in maize. These two orthologs occupy duplicated chromosomal regions co-orthologous with the rice region containing GW2. So the results of QTL mapping represent useful information for uncovering the corresponding genetic loci by looking for their synthenic positions on molecular maps [39]. Accumulating these favorable genes could represent a promising strategy to improve the yield, and was successfully applied to the improvement of rice by QTL pyramiding beneficial genes with the help of MAS [40]. Some of these QTL genes have such a major effect on the trait under study as to make meaningless the classical separation between oligogenes and quantitative loci, each with a small additive effect on the trait analyzed. In rice, where QTLs for seed size have been extensively analyzed, grain yield depends on the number of panicles per plant, a trait determined by the ability of the plant to produce tillers, number of grains per panicle and grain weight.

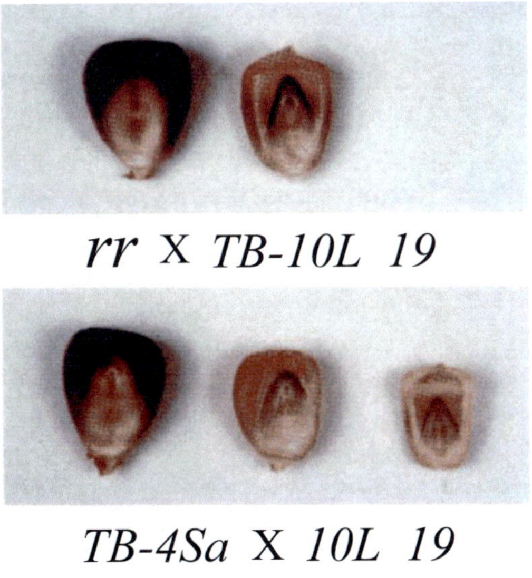

Fig. (8). Kernels from crosses of an *r* tester and hyperploid *TB-4Sa* by hyperploid *TB-10L19*. (*Top, left to right*) Normal kernel, hypoploid for *10L19* endosperm. (*Bottom, left to right*) Normal kernel, hypoploid for *10L19* endosperm, enhanced small kernel (also hypoploid endosperm). (From: Birchler and Hart 1987).

Number Of Tillers

Tillers are branches formed by the outgrowth of axillary buds, the activity of which is regulated by the interaction of different hormones. More specifically, auxin inhibits the growth of axillary buds while cytokinin removes this inhibition. Many mutants affecting tillering in rice have been detected. The first to be isolated was *moc1* with a monoculm phenotype [41]. MOC1 is a member of the plant-specific GRAS family of proteins, the function of which is to promote the initiation and growth of axillary buds. Several other genes have later been detected on the basis of the bushy phenotypes of their corresponding mutants, suggesting a role of these genes in suppressing bud activity. Their characterization disclosed their involvement in the biosynthesis of graft-transmissible signals of strictolactones. In maize the *teosinte branched 1* gene (*tb1*) functions in repressing the growth of axillary organs [42, 43] (Fig. **9**). Similarly, the role of its homolog in rice, *OsTB1,* encoding a basic helix-loop helix (bHLH) type of DNA-binding motif, is that of a negative regulator for lateral branches [44]. Understanding panicle development is important from an agronomic perspective, since the number of grains per panicle is proportional to the spikelet number. The spikelet is the basic unit of the inflorescence. Several genes affecting panicle development have been identified. The LAX1 gene determines the initiation and maintenance of axillary meristems in the panicle [45]. In addition genes are known for controlling cell proliferation important for panicle size and spikelet number (Fig. **10**). In the mutant *a*berrant *p*anicle *o*rganization (*apo1* [46]), the inflorescence meristem is converted to a spikelet meristem. APO1, a positive regulator of class C floral homeotic genes is an ortholog of Arabidopsis UNUSUAL FLORAL ORGAN (UFO) [47], encoding an F-box protein. Mutants with a high level of APO expression show an increase in the number of spikelets. A crucial role in panicle formation is played by cytokinins. By reducing the expression of APO, the level of this hormone in inflorescence meristems is enhanced, leading to increase in reproductive organs number and a higher grain yield. The opposite effect, *i.e.* a reduction in panicle size, is induced by the lonely guy (*log*) mutant with a defect in cytokinin biosynthesis [48]. Another hormone exerting an effect on seed size is auxin. The gene THOUSAND-GRAIN WEIGHT6 (TGW6 [49]), encoding indole 3 acetic acid glucose hydrolase, affects the timing of the passage from the syncytial to the cellular phase by controlling IAA supply, thus limiting cell number and grain length. In addition, TGW6 increases the accumulation of carbohydrates by causing higher translocation rates from source organs. So the pleiotropic effects of TGW6 on source ability as well as sink size are the bases of the higher yield promoted by it.

Fig. (9). (A) Teosinte plant (B) Maize plant (C) maize *tb1* mutant plant

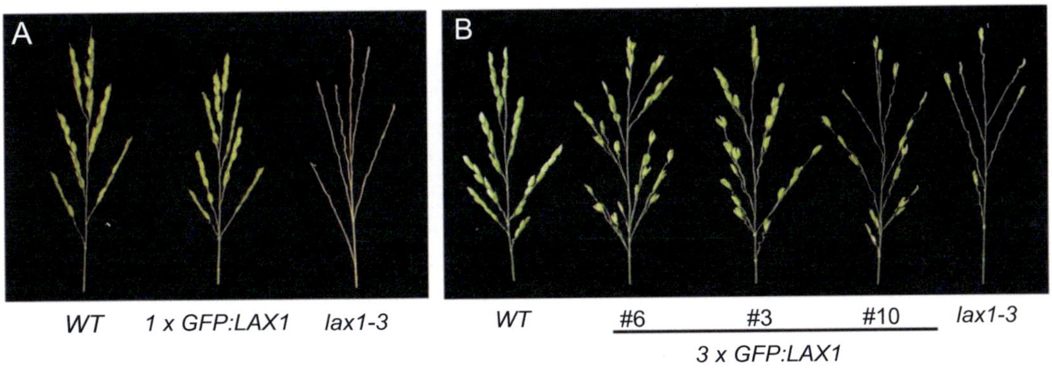

Fig. (10). (A) and (B) Panicle morphology of selected homozygous *lax1-3* transgenic plants harboring 1xGFP:LAX1 or 3xGFP:LAX1. (From: Oikawa and Kyozuka, 2009).

The first major QTL for grain length to be cloned was GS3 a negative regulator for seed size encoding a putative transmembrane. Later two other QTL genes (GW2 and GW5) were described, whose mutant alleles increased grain weight. They encode a RING type E3 ubiquitin ligase [50] and a nuclear polyubiquitin-binding protein [51, 52] respectively. Both of them are involved in the ubiquitin-proteasome pathway to regulate cell division during seed development. In contrast to these negative regulators of grain shape or size, GS5, recently identified and encoding a putative serine carboxypeptidase, appears to be a positive grain size regulator [38]. In this context the analysis of rare alleles of QTL has been very fruitful. For instance, OsmiR396c, a rare allele of GS2 affecting the binding site of a microRNA, induces high expression levels of GS2 leading to larger cells and higher numbers of cells, which thus enhances grain weight and yield [53].

Ubiquitin And The Control Of Seed Size In Plants

Ubiquitin is a protein that, through the sequential action of three enzymes, is attached to the target protein. At first, the ubiquitin activating enzyme 1 (E1) transfers the ubiquitin to a conjugating enzyme (E2). The E2 binds to the ubiquitin protein ligase E3 transferring ubiquitin to substrate proteins. The process is repeated several times forming a polyubiquitin chain leading the protein to the 26 proteasome for degradation [54]. Recently, a genetic screen for mutations affecting seed and organ size [55] by altering genes encoding ubiquitin receptors or E3 ubiquitin ligases, has uncovered ubiquitin-related-mutants. In rice such mutants include QTL loci such as GW2 or GW5, both of them involved in the ubiquitin-proteasome pathway by regulating cell division during development; or the *da1* (DA means "large" in Chinese) mutants in Arabidopsis or their homologs in other plants encoding a ubiquitin receptor and controlling seed and organ growth by limiting cell proliferation (see [29], for a review).

Hormones Involved In Seed Size

Hormones play an important role during plant development. For example, they promote the correct growth of roots, shoots, and apical meristems; they also promote the correct orientation of the leaves in order to achieve efficient photosynthesis. Two of them, ABA and ethylene, are involved in the senescence of the leaves, in the dormancy of seeds and buds and in the maturation of the fruits. Others such as gibberellins are involved in inter-node elongation, a response for "shade avoidance". In addition, a few hormones have an effect on seed size. The effect of promoting, regulating and controlling the seed size during the development of the seeds has been reported for auxin, cytokinins and brassinosteroids. A role of auxin in seed size is well documented in maize. The presence of IAA in the developing endosperm is explained by the needs of the developing embryo, as confirmed by the observation that only rare seed mutants have been described in maize associated to IAA deficiency, such as those with low IAA levels reported by Lur and Settler [56] and a viable defective endosperm mutant (*de18*) first described by Torti *et al.* [57]. The *de18* mutant phenotype was largely rescued by applying naphthalene acetic acid, a synthetic auxin, to developing seeds, suggesting impairment in IAA biosynthesis or metabolism as the basis of the mutant phenotype [57].

Cytokinins (CKs) are plant hormones with various effects on plant growth like bud formation, cell division and differentiation, leaf expansion, delayed leaf senescence and chloroplast development, morphogenesis in tissue culture (reviewed in [58]). Developing seeds are a rich source of CKs in different plants species. Zeatin, a major natural CK, was originally detected in developing seeds of maize [59].

CKs were first isolated on the basis of their promoting effect on cell division in plant meristems *in vitro*. Their high level of activity in the endosperm suggested that CKs exert an important role in the regulation of seed growth by controlling sink strength. This role is deduced from (a) the activity of CK in the promotion of cell division leading to an increase in the organ size of the sink tissues and (b) the correlation of CK accumulation with seed cell division profiles in several species. The plant cell cycle is modulated by cyclin-dependent kinases (CDKs) responsible for recognizing the target motif present in substrate proteins. Many plant cyclins have been identified. Those from class D seem to play a role as sensors integrating the cell cycle with the external signals. In maize, CKs stimulate the expression of CycD [60]. However, their precise role in the modulation of seed dimension still lacks precise definition.

To determine the role of CKs in maize a study was performed on an allelic series of the mutant *miniature1* (*mn1*). This mutant shows high levels of sucrose and low levels of hexoses in the basal region of the seed, while in its upper regions there are either similar or increased levels of sugars compared to those of the wild-type. These changes may be associated with the levels of CKs and auxin [61].

The results obtained confirm the linear gene-dose relationship of the *Incw2* transcripts with the number of *mn1-89* copies. They also show that the seed size differs significantly between the homozygous *Mn1* and the *mn1-89* series beginning at 12 to 16 DAP. Quantitative measurements of nine CK metabolites showed that the genotypes with reduced levels of invertase have lowered concentrations of zeatin-9-glucoside, a metabolically inactive CK and increased concentrations of active CK ribosides. In addition the *cytokinin-N-glucosyl-transferase* gene (*ZmCngt1*) was expressed at the highest level at 10 DAP and was greatly reduced in expression in the *mn1* mutant. This observation is in accord with a model [58] which proposes that high amounts of CKs and a higher ratio of their active forms may partially compensate the INCW2-deficiency in regulating CycD3 and CycD2 induced cell divisions by CKs and sucrose in the *mn1* endosperm. However, the precise molecular mechanism of CK in the modulation of seed dimensions is still lacking.

Probably the best picture of its action in this respect is provided by an analysis in the model species *Arabidopsis thaliana*. In this species endosperm size depends on both the genetic pathway HAIKU (IKU) and epigenetic mechanisms associated with parent-of-origin effects. To elucidate the effectors that integrate these pathways, Li [62] determined the gene expression profiles of wild type and homozygous *iku* mutants. This study led to the identification of a target of the IKU pathway, the gene *AT2g19550* encoding CYTOKININ OXIDASE 2 (CKX2), an enzyme involved in CK catabolism. They observed that the expression of CKX2 is strongly reduced in *iku* mutants, suggesting that the expression of CKX2 requires the IKU pathway. Apparently the increase in CK activity following the loss of IKU function is responsible for the decreased endosperm growth in *iku* mutants. By combining mutations that block CK signaling in a single plant, seeds larger than WT were obtained. Conversely, increased maternal genome size as obtained by crossing tetraploid (*4n*) with diploid (*2n*) parents produced seeds with reduced endosperm growth as observed in *iku* mutant seeds. A similar reduction is obtained by fertilization of WT ovules with pollen from mutants in DNA METHYLTRANSFERASE 1 (MET1) which maintains CK methylation. In both cases a clear reduction in CKX2 expression was observed. In summary, both an excess of maternal genome dosage and the paternal inheritance of *met1* suppress CKX2 transcription, causing seed size reduction. CKX2 expression also depends on H3K27me3 deposition, which fluctuates in response to maternal genome dosage imbalance and DNA demethylation of male gametes. Hence, the control of endosperm growth by CKX2 integrates genetic and epigenetic regulation.

Another class of hormones that have a role in seed size are the **Brassinosteroids (BRs)**.

They are steroidal plant hormones whose involvement in different developmental processes has been well analyzed in Arabidopsis, rice and tomato. They promote growth, cell division, elongation, and differentiation, reproductive development, defense against pathogens and stress tolerance. However only a few studies have considered their involvement in seed size. In one of them Wu [63] reported that increased BR levels affects plant development and seed weight in rice. When rice plants were transformed with the gene encoding the sterol C-22hydroxylase enzyme [64] more tillers and more seeds were produced than in non-transformed plants. Seeds, especially those at the bases of the spikes were heavier, whereas plants were taller, with increased leaf length and angle, height, and tiller diameter (Fig. **11**). Expression analysis of the C-22 hydroxylase in the embryo and endosperm did not affect seed weight, a conclusion reinforced by the results of microarray and photosynthesis analysis. These studies suggest that by manipulating the BR pathway it is possible to increase planting density, panicle number, light capture and seed filling. The more important determinants of seed filling and weight are photosynthetic efficiency and the levels of free sucrose and the control over the metabolic pathways leading to glucose, sucrose, and starch in the leaves and seed [65].

Fig. (11). Zm-CYP mutant phenotype. (A) WT *versus* mutant height. (B) Leaf sheath. (C) Leaf angle. (D) Leaf angle in the WT (E) and in Zm-CYP-1. (F) Tillering. (G) Panicle size (the panicles were chosen among the largest three for each plant). (H) and (J) Seed size in the WT (branch 9). (I) and (K) and in Zm-CYP-1 (branch 9). Arrows in (D) and (E) point to leaf joints. (From: Wu *et al.*, 2008).

In this context improvement can be made by channeling BR phenotypes in different parts of plants or by developing combinations of BR phenotypes and other phenotypes as well. This could be achieved in rice by increasing BR levels in the leaves to maximize photosynthetic efficiency and sugar conversions to glucose and sucrose, together with decreases in BR levels at the leaf joint to maximize planting density [64]. Similarly synergistic effects on grain productivity could be obtained by changing auxin levels to increase the numbers of leaves and buds on the stems [66] and cytokinin levels to delay senescence [67, 68].

Early Endosperm Development: Transition From The Syncytial To The Cellular Phase

To unravel the genetic basis of seed size in maize, Sekhon [69] examined transcriptomic profiles of developing seeds in two populations obtained by selecting large and small seed size. At the end of 30 cycles of selection the large seed population had a 4.7-fold greater weight than the small seed population. Measurement of seed weight during seed development shows that the process of grain filling starts earlier but at a slower rate in the small seed population. In accord with this observation, transcriptional analysis reveals that the larger seeds require a longer period of cell division and a higher rate of storage compounds accumulation compared to the small-seed population. It is a common observation in cereals that exposure to heat stress during seed development leads to a significant decrease in seed dimensions. To gain insight into its mechanism, rice seeds were exposed to controlled heat shock treatments and the small seeds so obtained were examined at both the cellular and molecular level and the results were compared to those of seeds not exposed to the heat shock. At the cell level it was noticed that a moderate heat stress (34°) caused a shorter syncytial stage and accelerated the transition to a cellularized state, while a 42° treatment caused lack of endosperm cellularization, leading to nonviable seeds. So precocious cellularization is the reason of the smaller seed size observed following moderate heat stress, an observation in accord with results reported for maize [69]. In Arabidopsis, early seed development and transition from the syncytial to the cellularized state is modulated by a group of genes encoding proteins that belong to the Polycomb group of general transcriptional repressors (Polycomb Repressor Complex, PRC). Loss of function of several of these PRC genes leads to abnormal endosperm development without fertilization. Of the six rice orthologues of the PRC2 complex tested, only *OsFie1* was found to be efficiently regulated in response to a moderate heat stress but repressed in presence of severe heat stress. Careful experiments on the expression of this gene in developing seeds grown at increasing temperature levels suggest that *OsFie1* induction during the course of seed development correlates with the transition from the syncytial to the cellularization stage. These data suggest that higher expression of this gene in rice seeds leads to precocious cellularization resulting in smaller seeds. In addition the DNA methylation state and a repressive histone modification of *OsFie1* are temperature sensitive. Taken together, these observations suggest that the thermal sensitivity of seed enlargement could be explained, at least in part, by altered epigenetic regulation of endosperm development during the transition from the syncytial to the cellularized state.

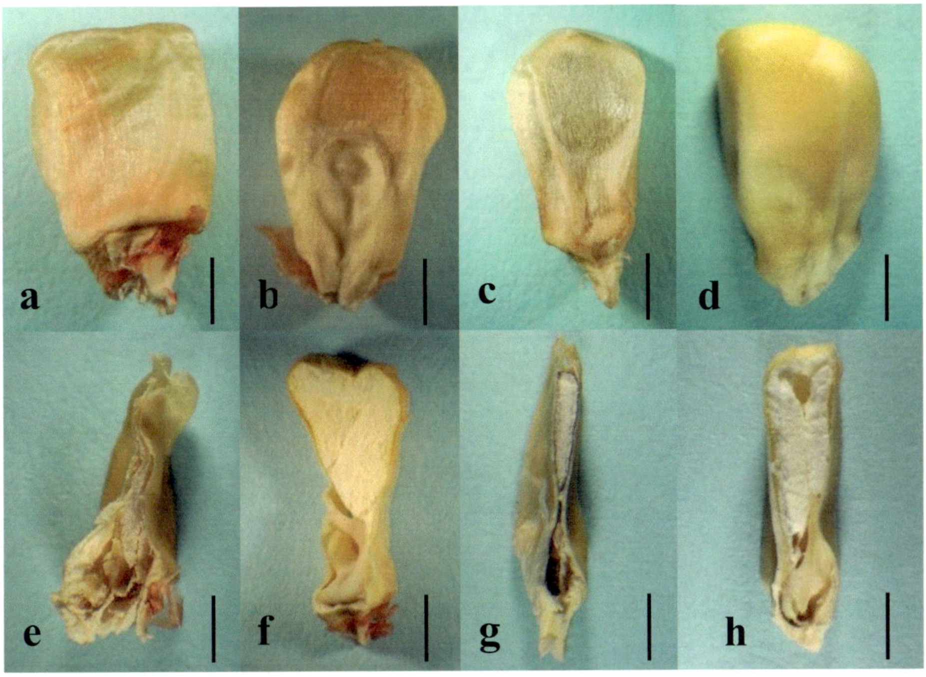

Fig. (12). Changes in the genetic backgrounds affect the morphology of mutant caryopses. From left to right: entire and longitudinal sections of homozygous *emp10-8376* mutant seeds in W64A (a, e), and in A636 (b, f), homozygous *emp4-9475* seeds in ACR (c, g) and in Mo17 (d, h). Scale bar = 1mm.

Cryptic Genetic Variation

Cryptic genetic variation (CGV) is defined as genetic variation that does not contribute to the normal range of phenotypes observed in a population, but that is unmasked after environmental changes or the introduction of novel alleles. Naturally occurring variation can be employed as a gene discovery tool, by detecting interactions between a defined mutation and natural alleles in other genes from various genetic backgrounds. This variation only becomes apparent when it is combined with a mutation of interest [70]. Ironically, CGV is often observed and deliberately discarded as the "background effect" that can confound the early analysis of a mutation. However, recent characterization of diverse germplasms for species like maize [71] and *Arabidopsis* [72] suggests that we should re-examine genetic background effects. Recently we made observations that seem to involve cryptic genetic variation affecting seed size [73]. The symbol *emp* (*empty pericarp*) refers to the phenotype of a group of defective kernel mutants with drastic reduction in endosperm tissue production and embryo lethals. They can be subdivided into two major subgroups: one including those with a flat appearance of the kernel and smooth pericarp and another one with a wrinkled pericarp. In all mutants, the histological analysis of the seed reveals the lack of morphogenesis in the embryo proper. Surprisingly, the introgression of these mutants in a different genetic background disclosed the existence of a cryptic genetic variation (CGV) detectable as a variable increase in the endosperm tissues (see Figs. **12** and **13**). The unmasking of CGV by introducing the mutants into different genetic backgrounds is the result of the interaction of the *emp* mutants with a modifier with no obvious phenotype of its own, present in the genetic background of the inbred lines where the *emp* mutants are transferred. Three cases of interaction between a specific *emp* and a modifier have been so far detected. The specificity of the modifier, *i.e.* its capacity to affect a given *emp* or several ones has not yet been elucidated. If these results are confirmed, *emp* mutants could be used as a tool for the detection of genetic factors enhancing the amount of endosperm tissue in the maize kernel to exploit in breeding programs.

Fig. (13). Two selfed ears of *emp4-9475* segregating for mutant seeds with the reference phenotype (left) in the ACR background and for mutant seeds with increased seed size (right) in the Mo17 background.

CONCLUDING REMARKS

Recently our understanding of the processes influencing grain size and grain number in crop plants has greatly advanced. Several genes and QTLs have been identified that are involved in specific pathways influencing these two traits. In future, cryptic variability will help in finding other genes useful in breeding programs aimed at an improvement in seed size.

CONFLICT OF INTEREST

The authors confirm that they have no conflict of interest to declare for this publication.

ACKNOWLEDGEMENTS

We thank Gabriella Consonni for critical reading of the manuscript and her suggestions.

REFERENCES

[1] Msuya DG, Stefano J. Responses of maize (*Zea mays*) seed germination capacity and vigor to seed selection based on size of cob and selective threshing. World J Agric Sci 2010; 6: 683-8.

[2] Odhiambo MO, Compton WA. Twenty cycles of divergent mass selection for seed size in corn. Crop Sci 1987; 27: 1113-6.

[http://dx.doi.org/10.2135/cropsci1987.0011183X002700060005x]

[3] Johnson DR, Tanner YW. Calculation of the rate and duration of grain filling in corn *Zea mays* L. Crop Sci 1972; 36: 301-6.

[4] Sabelli PA, Larkins BA. The development of endosperm in grasses. Plant Physiol 2009; 149(1): 14-26. [http://dx.doi.org/10.1104/pp.108.129437] [PMID: 19126691]

[5] Leishman MR, Wright IJ, Moles AT, Westoby M. The evolutionary ecology of seed size. Seed: the Ecology of Regeneration in Plant Communities. Wallingford, UK: Ed. by M. Fenner CAB International 2000; pp. 31-57. [http://dx.doi.org/10.1079/9780851994321.0031]

[6] Kesavan M, Song JT, Seo HS. Seed size: a priority trait in cereal crops. Physiol Plant 2013; 147(2): 113-20. [http://dx.doi.org/10.1111/j.1399-3054.2012.01664.x] [PMID: 22680622]

[7] Lin BY. Ploidy barrier to endosperm development in maize. Genetics 1984; 107(1): 103-15. [PMID: 17246209]

[8] Birchler JA. Dosage analysis of maize endosperm development. Annu Rev Genet 1993; 27: 181-204. [http://dx.doi.org/10.1146/annurev.ge.27.120193.001145] [PMID: 8122901]

[9] Haig D, Westoby M. Genomic imprinting in the endosperm: its effect on seed development in crosses between species, and between different ploidies of the same species, and its implications for the evolution of apomixis. Philos Trans R Soc Lond, B 1991; 333: 1-13. [http://dx.doi.org/10.1098/rstb.1991.0057]

[10] Evans MM, Kermicle JL. Interaction between maternal effect and zygotic effect mutations during maize seed development. Genetics 2001; 159(1): 303-15. [PMID: 11560906]

[11] Ma Y, Nelson OE. Amino acid composition and storage proteins in two high lysine mutants in maize. Cereal Chem 1975; 52: 412-9.

[12] Gutiérrez-Marcos JF, Costa LM, Evans MM. Maternal gametophytic baseless1 is required for development of the central cell and early endosperm patterning in maize (*Zea mays*). Genetics 2006; 174(1): 317-29. [http://dx.doi.org/10.1534/genetics.106.059709] [PMID: 16849604]

[13] Phillips AR, Evans MM. Analysis of *stunter1*, a maize mutant with reduced gametophyte size and maternal effects on seed development. Genetics 2011; 187(4): 1085-97. [http://dx.doi.org/10.1534/genetics.110.125286] [PMID: 21270392]

[14] Springer PS, Holding DR, Groover A, Yordan C, Martienssen RA. The essential Mcm7 protein PROLIFERA is localized to the nucleus of dividing cells during the G(1) phase and is required maternally for early Arabidopsis development. Development 2000; 127(9): 1815-22. [PMID: 10751170]

[15] Jullien PE, Katz A, Oliva M, Ohad N, Berger F. Polycomb group complexes self-regulate imprinting of the Polycomb group gene MEDEA in Arabidopsis. Curr Biol 2006; 16(5): 486-92. [http://dx.doi.org/10.1016/j.cub.2006.01.020] [PMID: 16527743]

[16] Neuffer MG, Sheridan WF. Defective kernel mutants of maize. I. Genetic and lethality studies. Genetics 1980; 95(4): 929-44. [PMID: 17249053]

[17] Charlton WL, Keen CL, Merriman C, *et al.* Endosperm development in *Zea mays*; implication of gametic imprinting and paternal excess in regulation of transfer layer development. Development 1995; 121: 3089-97.

[18] Miller ME, Chourey PS. The Maize Invertase-Deficient miniature-1 Seed Mutation Is Associated with Aberrant Pedicel and Endosperm Development. Plant Cell 1992; 4(3): 297-305.

[http://dx.doi.org/10.1105/tpc.4.3.297] [PMID: 12297647]

[19] Cheng WH, Taliercio EW, Chourey PS. The *Miniature1* seed locus of maize encodes a cell wall invertase required for normal development of endosperm and maternal cells in the pedicel. Plant Cell 1996; 8(6): 971-83.
[http://dx.doi.org/10.1105/tpc.8.6.971] [PMID: 12239408]

[20] Weschke W, Panitz R, Gubatz S, *et al.* The role of invertases and hexose transporters in controlling sugar ratios in maternal and filial tissues of barley caryopses during early development. Plant J 2003; 33(2): 395-411.
[http://dx.doi.org/10.1046/j.1365-313X.2003.01633.x] [PMID: 12535352]

[21] Ultrastructure of maize endosperm suspension cultures. Am J Bot 1987; 74: 1912-20.
[http://dx.doi.org/10.2307/2443975]

[22] Jones RJ, Schreiber EM, Roessler JA. Kernel sink capacity in maize: genotypic and maternal regulation. Crop Sci 1996; 36: 301-6.
[http://dx.doi.org/10.2135/cropsci1996.0011183X003600020015x]

[23] Gutiérrez-Marcos JF, Costa LM, Biderre-Petit C, *et al.* maternally expressed gene1 Is a novel maize endosperm transfer cell-specific gene with a maternal parent-of-origin pattern of expression. Plant Cell 2004; 16(5): 1288-301.
[http://dx.doi.org/10.1105/tpc.019778] [PMID: 15105441]

[24] Lopato S, Borisjuk N, Langridge P, Hrmova M. Endosperm transfer cell-specific genes and proteins: structure, function and applications in biotechnology. Front Plant Sci 2014; 5: 64.
[http://dx.doi.org/10.3389/fpls.2014.00064] [PMID: 24578704]

[25] Xiao W, Brown RC, Lemmon BE, Harada JJ, Goldberg RB, Fischer RL. Regulation of seed size by hypomethylation of maternal and paternal genomes. Plant Physiol 2006; 142(3): 1160-8.
[http://dx.doi.org/10.1104/pp.106.088849] [PMID: 17012404]

[26] Birchler JA, Bhadra U, Bhadra MP, Auger DL. Dosage-dependent gene regulation in multicellular eukaryotes: implications for dosage compensation, aneuploid syndromes, and quantitative traits. Dev Biol 2001; 234(2): 275-88.
[http://dx.doi.org/10.1006/dbio.2001.0262] [PMID: 11396999]

[27] Dilkes BP, Comai L. A differential dosage hypothesis for parental effects in seed development. Plant Cell 2004; 16(12): 3174-80.
[http://dx.doi.org/10.1105/tpc.104.161230] [PMID: 15579806]

[28] Barkan A, Small I. Pentatricopeptide repeat proteins in plants. Annu Rev Plant Biol 2014; 65: 415-42.
[http://dx.doi.org/10.1146/annurev-arplant-050213-040159] [PMID: 24471833]

[29] Li XJ, Zhang YF, Hou M, *et al.* Small kernel 1 encodes a pentatricopeptide repeat protein required for mitochondrial nad7 transcript editing and seed development in maize (*Zea mays*) and rice (*Oryza sativa*). Plant J 2014; 79(5): 797-809.
[http://dx.doi.org/10.1111/tpj.12584] [PMID: 24923534]

[30] Gabotti D, Caporali E, Manzotti P, Persico M, Vigani G, Consonni G. The maize pentatricopeptide repeat gene empty pericarp4 (emp4) is required for proper cellular development in vegetative tissues. Plant Sci 2014; 223: 25-35.
[http://dx.doi.org/10.1016/j.plantsci.2014.02.012] [PMID: 24767112]

[31] Liu YJ, Xiu ZH, Meeley R, Tan BC. Empty pericarp5 encodes a pentatricopeptide repeat protein that is required for mitochondrial RNA editing and seed development in maize. Plant Cell 2013; 25(3): 868-83.
[http://dx.doi.org/10.1105/tpc.112.106781] [PMID: 23463776]

[32] Sun F, Wang X, Bonnard G, *et al.* Empty pericarp7 encodes a mitochondrial E-subgroup pentatricopeptide repeat protein that is required for ccmFN editing, mitochondrial function and seed development in maize. Plant J 2015; 84(2): 283-95.

[http://dx.doi.org/10.1111/tpj.12993] [PMID: 26303363]

[33] Sosso D, Mbelo S, Vernoud V, *et al.* PPR2263, a DYW-Subgroup Pentatricopeptide repeat protein, is required for mitochondrial nad5 and cob transcript editing, mitochondrion biogenesis, and maize growth. Plant Cell 2012; 24(2): 676-91. [http://dx.doi.org/10.1105/tpc.111.091074] [PMID: 22319053]

[34] Manavski N, Guyon V, Meurer J, Wienand U, Brettschneider R. An essential pentatricopeptide repeat protein facilitates 5 maturation and translation initiation of rps3 mRNA in maize mitochondria. Plant Cell 2012; 24(7): 3087-105. [http://dx.doi.org/10.1105/tpc.112.099051] [PMID: 22773745]

[35] Cushing DA, Forsthoefel NR, Gestaut DR, Vernon DM. Arabidopsis emb175 and other ppr knockout mutants reveal essential roles for pentatricopeptide repeat (PPR) proteins in plant embryogenesis. Planta 2005; 221(3): 424-36. [http://dx.doi.org/10.1007/s00425-004-1452-x] [PMID: 15647901]

[36] Birchler JA, Hart JR. Interaction of endosperm size factors in maize. Genetics 1987; 117(2): 309-17. [PMID: 17246405]

[37] Fan C, Xing Y, Mao H, *et al.* GS3, a major QTL for grain length and weight and minor QTL for grain width and thickness in rice, encodes a putative transmembrane protein. Theor Appl Genet 2006; 112(6): 1164-71. [http://dx.doi.org/10.1007/s00122-006-0218-1] [PMID: 16453132]

[38] Li Y, Fan C, Xing Y, *et al.* Natural variation in GS5 plays an important role in regulating grain size and yield in rice. Nat Genet 2011; 43(12): 1266-9. [http://dx.doi.org/10.1038/ng.977] [PMID: 22019783]

[39] Gupta PK, Rustgi S, Kumar N. Genetic and molecular basis of grain size and grain number and its relevance to grain productivity in higher plants. Genome 2006; 49(6): 565-71. [http://dx.doi.org/10.1139/G06-063] [PMID: 16936836]

[40] Xue W, Xing Y, Weng X, *et al.* Natural variation in Ghd7 is an important regulator of heading date and yield potential in rice. Nat Genet 2008; 40(6): 761-7. [http://dx.doi.org/10.1038/ng.143] [PMID: 18454147]

[41] Li X, Qian Q, Fu Z, *et al.* Control of tillering in rice. Nature 2003; 422(6932): 618-21. [http://dx.doi.org/10.1038/nature01518] [PMID: 12687001]

[42] Doebley J, Stec A, Gustus C. *teosinte branched1* and the origin of maize: evidence for epistasis and the evolution of dominance. Genetics 1995; 141(1): 333-46. [PMID: 8536981]

[43] Studer A, Zhao Q, Ross-Ibarra J, Doebley J. Identification of a functional transposon insertion in the maize domestication gene *tb1*. Nat Genet 2011; 43(11): 1160-3. [http://dx.doi.org/10.1038/ng.942] [PMID: 21946354]

[44] Takeda T, Suwa Y, Suzuki M, *et al.* The OsTB1 gene negatively regulates lateral branching in rice. Plant J 2003; 33(3): 513-20. [http://dx.doi.org/10.1046/j.1365-313X.2003.01648.x] [PMID: 12581309]

[45] Komatsu M, Maekawa M, Shimamoto K, Kyozuka J. The LAX1 and FRIZZY PANICLE 2 genes determine the inflorescence architecture of rice by controlling rachis-branch and spikelet development. Dev Biol 2001; 231(2): 364-73. [http://dx.doi.org/10.1006/dbio.2000.9988] [PMID: 11237465]

[46] Ikeda K, Nagasawa N, Nagato Y. ABERRANT PANICLE ORGANIZATION 1 temporally regulates meristem identity in rice. Dev Biol 2005; 282(2): 349-60. [http://dx.doi.org/10.1016/j.ydbio.2005.03.016] [PMID: 15950602]

[47] Samach A, Klenz JE, Kohalmi SE, Risseeuw E, Haughn GW, Crosby WL. The UNUSUAL FLORAL ORGANS gene of *Arabidopsis thaliana* is an F-box protein required for normal patterning and growth

in the floral meristem. Plant J 1999; 20(4): 433-45.
[http://dx.doi.org/10.1046/j.1365-313x.1999.00617.x] [PMID: 10607296]

[48] Kuroha T, Tokunaga H, Kojima M, *et al.* Functional analyses of LONELY GUY cytokinin-activating enzymes reveal the importance of the direct activation pathway in Arabidopsis. Plant Cell 2009; 21(10): 3152-69.
[http://dx.doi.org/10.1105/tpc.109.068676] [PMID: 19837870]

[49] Ishimaru K, Hirotsu N, Madoka Y, *et al.* Loss of function of the IAA-glucose hydrolase gene TGW6 enhances rice grain weight and increases yield. Nat Genet 2013; 45(6): 707-11.
[http://dx.doi.org/10.1038/ng.2612] [PMID: 23583977]

[50] Song XJ, Huang W, Shi M, Zhu MZ, Lin HX. A QTL for rice grain width and weight encodes a previously unknown RING-type E3 ubiquitin ligase. Nat Genet 2007; 39(5): 623-30.
[http://dx.doi.org/10.1038/ng2014] [PMID: 17417637]

[51] Weng J, Gu S, Wan X, *et al.* Isolation and initial characterization of GW5, a major QTL associated with rice grain width and weight. Cell Res 2008; 18(12): 1199-209.
[http://dx.doi.org/10.1038/cr.2008.307] [PMID: 19015668]

[52] Shomura A, Izawa T, Ebana K, *et al.* Deletion in a gene associated with grain size increased yields during rice domestication. Nat Genet 2008; 40(8): 1023-8.
[http://dx.doi.org/10.1038/ng.169] [PMID: 18604208]

[53] Hu J, Wang Y, Fang Y, *et al.* A rare allele of GS2 enhances grain size and grain yield in rice. Mol Plant 2015; 8(10): 1455-65.
[http://dx.doi.org/10.1016/j.molp.2015.07.002] [PMID: 26187814]

[54] Pickart CM. Mechanisms underlying ubiquitination. Annu Rev Biochem 2001; 70: 503-33.
[http://dx.doi.org/10.1146/annurev.biochem.70.1.503] [PMID: 11395416]

[55] Li F, Vallabhaneni R, Yu J, Rocheford T, Wurtzel ET. The maize phytoene synthase gene family: overlapping roles for carotenogenesis in endosperm, photomorphogenesis, and thermal stress tolerance. Plant Physiol 2008; 147(3): 1334-46.
[http://dx.doi.org/10.1104/pp.108.122119] [PMID: 18508954]

[56] Lur HS, Setter TL. Endosperm development of maize defective kernel (*Dek*) mutants: auxin and cytokinin levels. Ann Bot (Lond) 1993; 72: 1-6.
[http://dx.doi.org/10.1006/anbo.1993.1074]

[57] Torti G, Manzocchi L, Salamini F. Free and bound indole-acetic acid is low in the endosperm of the maize mutantdefective endosperm-B18. Theor Appl Genet 1986; 72(5): 602-5.
[http://dx.doi.org/10.1007/BF00288997] [PMID: 24248069]

[58] Rijavec T, Kovač M, Kladnik A, Chourey PS, Dermastia M. A comparative study on the role of cytokinins in caryopsis development in the maize miniature1 seed mutant and its wild type. J Integr Plant Biol 2009; 51(9): 840-9.
[http://dx.doi.org/10.1111/j.1744-7909.2009.00863.x] [PMID: 19723243]

[59] Letham DS. Zeatin, a factor inducing cell division from *Zea mays*. Life Sci 1963; 8: 569-73.
[http://dx.doi.org/10.1016/0024-3205(63)90108-5] [PMID: 14058503]

[60] Gutiérrez R, Quiroz-Figueroa F, Vázquez-Ramos JM. Maize cyclin D2 expression, associated kinase activity and effect of phytohormones during germination. Plant Cell Physiol 2005; 46(1): 166-73.
[http://dx.doi.org/10.1093/pcp/pci007] [PMID: 15659448]

[61] Rolland F, Moore B, Sheen J. Sugar sensing and signaling in plants. Plant Cell 2002; 14 (Suppl.): S185-205.
[PMID: 12045277]

[62] Li J, Nie X, Tan JL, Berger F. Integration of epigenetic and genetic controls of seed size by cytokinin in Arabidopsis. Proc Natl Acad Sci USA 2013; 110(38): 15479-84.

[http://dx.doi.org/10.1073/pnas.1305175110] [PMID: 24003120]

[63] Wu CY, Trieu A, Radhakrishnan P, *et al.* Brassinosteroids regulate grain filling in rice. Plant Cell 2008; 20(8): 2130-45.
[http://dx.doi.org/10.1105/tpc.107.055087] [PMID: 18708477]

[64] Sakamoto T, Morinaka Y, Ohnishi T, *et al.* Erect leaves caused by brassinosteroid deficiency increase biomass production and grain yield in rice. Nat Biotechnol 2006; 24(1): 105-9.
[http://dx.doi.org/10.1038/nbt1173] [PMID: 16369540]

[65] Sakamoto T, Matsuoka M. Generating high-yielding varieties by genetic manipulation of plant architecture. Curr Opin Biotechnol 2004; 15(2): 144-7.
[http://dx.doi.org/10.1016/j.copbio.2004.02.003] [PMID: 15081053]

[66] Bainbridge K, Guyomarch S, Bayer E, *et al.* Auxin influx carriers stabilize phyllotactic patterning. Genes Dev 2008; 22(6): 810-23.
[http://dx.doi.org/10.1101/gad.462608] [PMID: 18347099]

[67] Smart CM, Scofield SR, Bevan MW, Dyer TA. Delayed leaf senescence in tobacco plants transformed with tmr, a gene for cytokinin production in Agrobacterium. Plant Cell 1991; 3(7): 647-56.
[http://dx.doi.org/10.1105/tpc.3.7.647] [PMID: 12324608]

[68] Ashikari M, Sakakibara H, Lin S, *et al.* Cytokinin oxidase regulates rice grain production. Science 2005; 309(5735): 741-5.
[http://dx.doi.org/10.1126/science.1113373] [PMID: 15976269]

[69] Sekhon RS, Hirsch CN, Childs KL, *et al.* Phenotypic and transcriptional analysis of divergently selected maize populations reveals the role of developmental timing in seed size determination. Plant Physiol 2014; 165(2): 658-69.
[http://dx.doi.org/10.1104/pp.114.235424] [PMID: 24710068]

[70] Sangster TA, Salathia N, Lee HN, *et al.* HSP90-buffered genetic variation is common in *Arabidopsis thaliana.* Proc Natl Acad Sci USA 2008; 105(8): 2969-74.
[http://dx.doi.org/10.1073/pnas.0712210105] [PMID: 18287064]

[71] Yu J, Holland JB, McMullen MD, Buckler ES. Genetic design and statistical power of nested association mapping in maize. Genetics 2008; 178(1): 539-51.
[http://dx.doi.org/10.1534/genetics.107.074245] [PMID: 18202393]

[72] Keurentjes JJ, Koornneef M, Vreugdenhil D. Quantitative genetics in the age of omics. Curr Opin Plant Biol 2008; 11(2): 123-8.
[http://dx.doi.org/10.1016/j.pbi.2008.01.006] [PMID: 18325828]

[73] Sangiorgio S, Carabelli L, Gabotti D, *et al.* A mutational approach for the detection of genetics factors affecting seed size in maize. Plant Reprod 2016; 29(4): 301-10.
[http://dx.doi.org/10.1007%2Fs00497-016-0294-6] [PMID: 27858171]

CHAPTER 7

Genetic Variability as a Means to Improve Seedling Emergence and Early Developmental Phases in Crop Plants

Martina Persico and **Gabriella Consonni**[*]

Department of Agricultural and Environmental Sciences – Production, Landscapes, Agroenergy, Università degli Studi di Milano, Milano, Italy

Abstract: Embryogenesis, germination and the early phases of seedling growth represent critical phases in the plant life cycle and are probably the most important events in determining the success of an annual plant. In the perspective of a more sustainable agriculture, we aim to achieve a robust seedling phase with improved resistance to abiotic as well as biotic environmental stress. Genetic improvement has forced the developmental pathway of crop plants toward the realization of highly productive species in which resource allocation processes are optimized at the expense of defense processes. In this context the discovery of key factors underlying the developmental process and at the same time playing an important role in the interplay between the young individuals and the environment is crucial for designing future gene manipulation approaches. Among the different aspects affecting seedling development, the two that will be analyzed in this chapter also play an important role in the interplay with the environment. Hormones are endogenous signals governing seedling growth and architecture establishment but at the same time are able to induce plant responses to environmental stress. Wax deposition is required for determining correct embryo and seedling development, and provides, besides that, a protective barrier that plants produce in their early developmental phases to defend themselves from pathogens as well as from variation in environmental abiotic components, such as temperature and water availability. We will explore the genetic, biochemical and physiological factors implicated and highlight the most significant aspects that might be taken into consideration in future breeding programs.

Keywords: Abiotic stress, Biotic stress, Cuticle, Cuticular wax, Embryogenesis, Germination, Heterosis, Phytohormones, Seedling growth, Transgenic plants.

INTRODUCTION

Embryogenesis, germination and early phases of seedling growth represent criti-

[*] **Corresponding author Gabriella Consonni**: Department of Agricultural and Environmental Sciences – Production, Landscapes, Agroenergy, Università degli Studi di Milano, Milano, Italy; Tel:+39.0250316634; E-mail: gabriella.consonni@unimi.it

Roberto Pilu and Giuseppe Gavazzi (Eds.)

cal phases in the plant life cycle and are probably the most important events in determining the success of an annual plant. Embryogenesis can be seen as the first phase of a continuous process, only temporarily interrupted by dormancy [1]. During the succeeding phase, germination, the embryo becomes the seedling. Thus, proper embryo development is a prerequisite for successful seedling emergence.

Germination and seedling emergence occur when environmental conditions for establishing a new plant generation are likely to be suitable. A rapid and robust emergence positively influences the capacity of the plant to take advantage of the growth-favouring environment and to compete with its neighbors. From an agronomic point of view, synchrony in germination and seedling emergence of a crop is a desirable character for a cultivated plant since it will allow the optimization of weed control practices. The possibility of predicting and synchronizing the time of emergence will reduce chemical use and will allow this strategy to be combined with more sustainable approaches, such as biological and physical weed control [2]. In the perspective of a more sustainable agriculture, certain specific characters are envisaged for a crop seedling, such as resistance to environmental critical abiotic as well as biotic factors. For these reasons, key factors controlling subtending plant developmental processes and contributing to the achievement of a productive and robust plant have to be searched for within the genetic network that controls embryo and seedling development.

There is evidence that heterosis is already determined at early developmental stages. Various studies on heterosis have shown that some traits related to this phenomenon are already evident during early phases of plant development, as clearly described for young maize roots [3, 4]. It has also been reported, in Arabidopsis and in crops, that F_1 seedlings are already larger than their parents [5, 6]. Greater cell number is the main determinant of the larger size of organs in heterotic plants [7]. It is conceivable that the rate of cell division, which is defined very early during embryogenesis, is higher in the progeny than in parental lines. The level of heterosis is most probably set very early and conditions the final organ size and numbers. On this basis the study of the mechanisms governing early phases of plant development, *i.e.* embryogenesis and seedling development, are appealing not only for investigating the molecular and genetic network underlying these processes, but also for the identification of genetic tools that might be of interest in breeding programs.

Maize embryogenesis leads to the formation of two main structures, a well differentiated embryo axis and a storage organ, the scutellum. The mature embryo axis comprises at the two poles the embryonic primary root and the embryonic shoot, separated by the scutellar node. They are both enclosed in protective structures, respectively the coleorhiza for the root and the coleoptile for the shoot. The shoot stem comprises a first internode, called the mesocotyl, that is located between the scutellar node and the coleoptilar node and five or six short internodes, depending on the genetic background, located above the coleoptilar node, with a leaf primordium attached to each node. Each leaf is rolled up inside those which enclose it, thus forming a cone shaped structure which encloses the shoot apical meristem (SAM). During embryogenesis, the coleoptile develops as a sheathing structure and envelops the stem tip and the embryonic seedling leaves [8, 9].

The scutellum, a massive organ, in which mainly lipids and proteins are accumulated, is attached to the scutellar node. For its functional equivalence, it is considered to be the single cotyledon in the embryo of monocotyledons. The epidermis of the scutellum differentiates in two regions. On the side facing the endosperm a scutellar epithelium is produced, while on the side adjacent to the coleoptile, scutellum cell walls develop a heavy cuticle [10].

Both coleoptile and the first set of seedling leaf primordia are initiated during maize embryogenesis at about 12-14 days after pollination [8] but, as shown by morphological analysis [9], they retain a different origin. The coleoptile arises as a ring of cells on the surface of the scutellum, whereas the first leaf is initiated at the basal face of the shoot apical meristem (SAM), where the coleoptilar ring closes, from the SAM cell population. This observation is also supported by the expression pattern analysis of different marker genes [11]. For instances the knotted gene is specifically expressed early in development on the anterior side of the embryo in two groups of cells that will give rise to shoot and root meristems, whereas its expression has not been found in the coleoptile founder cells that are visible in the scutellum [12]. Another example is the ZmWOX3A/B, the expression of which is specifically confined to a ring of peripheral cells, which marks the recruitment of cells to establish firm the P_0 primordium [13].

In maize, the coleoptile is the first organ that is produced when seed germination occurs. It appears as a cone shaped structure that elongates and pierces through the soil, enclosing and thus protecting the young leaves and the shoot apex, which comprises the shoot apical meristem (SAM), until they reach the above ground level. In this initial phase, first leaf elongation keeps pace with that of the coleoptile. Later on the coleoptile elongates and opens at its apex. The first leaf continues to grow and emerges from a coleoptile gap that is initiated at the top. The second and following leaves appear subsequently in a sequential manner, soon after the coleoptile dies.

Among the different aspects affecting seedling development, the two that will be analyzed in this chapter also play an important role also in the interplay with the environment. Hormones are endogenous signals governing seedling growth and architecture establishment but at the same time are able to induce plant responses to environmental stress. Wax deposition is required for determining correct embryo and seedling development, and provides, besides that, a protective barrier that plants produce in their early developmental phases to defend themselves from pathogens as well as from variation in environmental abiotic components, such as temperature and water availability. We will explore the genetic, biochemical and physiological factors implicated and highlight the most significant aspects that might be taken into consideration in future breeding programs.

Cuticular Wax Layer Protects The Plants Against Environmental Stresses

In all higher plants, the outer surfaces of the aerial parts are covered by a thin and continuous layer of predominantly lipid material, known as cuticle, which provides a primary waterproof barrier and a protection against different environmental stresses [14].

Cuticle synthesis starts in early stages of plant development and is co-regulated with plant growth to provide constant wax and cutin deposition, required during stem elongation [15]. Cuticle is involved in the regulation of epidermal permeability and nonstomatal water loss as well as protection against insects, pathogens, UV light and frost [16]. In addition, several studies have highlighted its functions in plant developmental processes, among them the prevention of postgenital organ fusion [16]. Many Arabidopsis transgenic plants with an impaired cutin load show organ fusion, suggesting that a proper cuticle is needed to define organ boundaries [16 - 19]. Moreover, changes in the cuticular wax component are often associated with morphological impairment of the epidermis, in particular of specialized cells such as trichomes and stomata. In Arabidopsis, the overexpression of the SHN1 gene, a transcription factor involved in wax biosynthesis, leads to structural defects in the cuticle and causes changes in epidermal differentiation, including trichome formation and stomata number [20].

The cuticle is a complex structure with two main components, the cutin, made of inter-esterified hydroxy fatty acid, and the waxes, which include a variety of different acyl lipids. On the basis of the different composition and reactivity to histochemical staining, the cuticle can be divided into two distinct layers. The first is the cuticular layer, which overlies the epidermis cell wall, and is composed of cutin, intracuticular waxes and polysaccharides. The second layer, the cuticle proper, which is less rich in polysaccharides, is made by cutin and intracuticular waxes. Waxes can also be deposited on the cutin surface as films or wax crystals, thus forming the outer layer of a plant tissue [21, 22].

Among different cuticle components, waxes and their biosynthesis have been extensively analysed, in particular in the model plant *Arabidopsis thaliana*. "Waxes are a heterogeneous mixture of very-long-chain fatty acids (VLCFAs) and their alkane, aldehyde, alcohol, ketone, and ester derivatives, which typically range from C24 to C32 in length" [23]. Their biosynthesis begins with the esterification of C16 and C18 fatty acids, which are *de novo* synthesized in the plastids of epidermal cells, to Co-enzyme A (CoA) and subsequent transfer to the endoplasmic reticulum (ER). In the ER the multi-enzymatic fatty acid (FAE) elongase complex catalyzes successive reactions in which the acyl-chain is extended by two carbons per cycle. "VLC-acyl-CoAs can be subsequently transformed into free VLCFAs and/or processed through two distinct pathways: an acyl reduction pathway, which gives rise to primary alcohols and wax esters, and a decarbonylation pathway, leading to the formation of aldehydes, alkanes, secondary alcohols and ketones" [21].

Different species have distinct epicuticular wax composition, and epicuticular wax composition can also differ among tissues within the same species as well as during development [24]. As to young organs and leaves, alkanes are major component in Arabidopsis, representing up to 70% of the total waxes in rosette leaves, followed by aldehydes (14%), and 44% in inflorescence stems, where the second components of importance are secondary alcohols, which accumulate to 23% [21]. In young leaves of barley, rice, and maize, the major cuticular wax components were shown to be primary alcohols, aldehydes, and fatty acids, whereas alkanes comprised less than 15% of the total wax amount [25 - 27]. In maize, where wax composition differs between juvenile and adult leaves [28] approximately 80% of the juvenile waxes are very-long-chain alcohols and aldehydes, whereas approximately 70% of the waxes produced throughout the life of a maize plant consist of esters [29].

Progress has recently been made to reveal the molecular and genetic mechanisms underlying wax production and deposition. In some species like Arabidopsis, barley and maize, genes involved in wax synthesis were first detected through forward genetic approaches based on the altered surface appearance of their mutant stem/leaf. In Arabidopsis 89 *eceriferum* (literally "not carrying wax") mutants were first detected for their shiny appearance that defined at least 21 CER genes [30]. In maize (*Zea mays*), at least 30 loci, namely the *glossy* loci, have been found to affect the quantity and/or the composition of cuticular waxes on the surface of seedling leaves [31] and in barley (*Hordeum vulgare*), about 80 independent *cer* loci had been defined [32].

Molecular analyses performed on these mutants and related genes have assigned functions to their gene products, some of which are enzymes catalyzing major steps in fatty acid elongation and wax biosynthesis, while others are transporter or regulatory factors. An updated list of the genes is provided in recent reviews [22, 33]. Gene functional analysis has indicated that, at least in Arabidopsis, total wax loads are mainly determined by regulatory mechanisms that act at the transcriptional level.

The total amount of cuticular waxes is modulated in response to changes in the environment; it is thus conceivable that waxes are actively involved in responding to environmental signals of a biotic as well as abiotic nature. It has been documented that drought, salt or osmotic stress conditions, or exogenous application of abscisic acid (ABA) cause an increase of two-fold in cuticular wax load in Arabidopsis [34]. Several studies have also revealed a positive correlation between wax amounts and resistance to water stress in crop species. Cuticular wax deposition was increased by approximately 1.5- to 2.5-fold by drought or water-deficiency stress in leaves of various plants including soybean (*Glycine max*), sesame (*Sesamum indicum*), sorghum, wheat (*Triticum aestivum*), maize, oats (*Avena sativa*) and cotton (*Gossypium hirsutum*) [35 - 38].

Regarding wax involvement in plant/pathogen interactions, there are two interesting examples in Arabidopsis, showing how wax composition can influence the capability of the pathogen to penetrate the cuticle barrier. One example is provided by the work of Weis and collaborators [39], which showed that the enhanced resistance of cyp83A1 plants to *Erysiphe cruciferarum* was based primarily on the deficiency of long chain aldehydes in the cuticular waxes. The author suggested that these molecules therefore seem to be critical to initiate fungal development. A second example shows that AtMYB30, which positively regulates the accumulation of alkanes in cuticular waxes [40], is also a crucial regulator of the hypersensitive response (HR). It was shown that its over-expression can stimulate the production of VLCFAs and waxes, and it was proposed that these cuticle compounds could have a role in the HR processes which lead to programmed cell-death limiting the propagation of pathogens [41].

Transcriptional regulators also mediate the interplay between wax biosynthesis and environmental factors. These regulatory genes contribute to wax regulation and modulate the interaction between waxes and abiotic stresses. WAX INDUCER1/SHINE1 (WIN1/ SHN1) was the first transcriptional regulator characterized. It is an AP2-EREBP-type transcription factor and activates cuticular wax biosynthesis by up-regulation of *KCS1*, *CER1*, and *CER2* genes. Overexpression of WIN1 confers drought tolerance in Arabidopsis [20, 42]. In this case, it was indeed demonstrated that the WIN1/ SHN1 gene product directly activated the promoters of several cutin biosynthetic genes, indicating that the gene has a primary role in cutin regulation with a downstream effect on wax biosynthesis [43].

Other studies have shown that response to environmental variations is mediated by the activity of regulatory genes with their targets specifically involved in wax biosynthesis, among them some members of the MYB family. MYB96, whose direct targets are *KCS1, KCS2*, *KCS6*, *KCR1*, and *CER3*, up-regulates wax biosynthesis under drought stress [33, 44]. Moreover, plants ectopically expressing MYB96 showed enhanced drought tolerance in transgenic Arabidopsis. In these plants, cuticular wax biosynthesis is also up-regulated [44, 45], while Arabidopsis plants overexpressing MYB96 also exhibited increased freezing tolerance. In this case a lipid transfer protein, LTP3, was identified as one of the targets of MYB96, which is activated during plant responses to both freezing and drought stress [46]. Another example is represented by MYB94, a transcription factor closely related to MYB96 that activates a set of distinct genes, *i.e. WSD1, CER2*, *FAR3*, and *ECR*, besides *KCS2/DAISY*, which is also regulated by MYB96 [47]. MYB94 contributes either independently or synergistically with MYB96 to the activation of wax biosynthesis under drought conditions [33].

Although regulatory mechanisms are less understood in crops, compared to the model species Arabidopsis, some regulatory factors involved in wax biosynthesis and deposition, have been characterized. Three genes have been isolated in maize. OCL1, a transcription factor of the homeodomain-leucine zipper IV (HD-ZIP IV) family, is an epidermis-specific positive regulator of wax biosynthesis [26]. 14 target genes up- or down-regulated in transgenic maize plants overexpressing OCL1 have been detected that encode proteins involved in lipid metabolism as well as in other processes, such as defense, or in biosynthesis of cuticle components. Moreover, OCL1 induced ZmWBC11a, an ortholog of AtWBC11 which is involved in the transport of wax and cutin molecules [26]. Another gene characterized in maize is GL3, encoding an MYB transcription factor. GL3 controls the level of aldehydes, which are the second most abundant wax compounds in maize leaves [48]. Both genes are expressed in seedling leaves. The *fused leaf1* gene, encoding a MYB protein, ZmMYB94, controls the accumulation of alcohols and aldehydes with molecules longer than 30 carbons during early phases of seedling growth [49]. "In rice WR1 (OsWR1), a homolog of Arabidopsis WIN1/SHN1 activates cuticular wax biosynthesis through the up-regulation of OsLACS1 and OsFAE1-L genes by direct binding to DRE and GCC cis-elements in their gene promoters, respectively" [50].

Despite these data, very little is known about the interactions between regulatory pathways and environmental stress in crop plants. Preliminary data have been obtained for the *fused leaves1* gene in maize showing that its transcript level is increased in drought stress conditions (manuscript in preparation). Better understanding of the molecular mechanisms underlying these processes in species of agronomic importance is urgently needed. Once their role is elucidated, transcription factors may be used as tools to enhance stress tolerance in crop plants [51].

Wax accumulation in young leaf tissues is an important physiological process because leaves are the primary photosynthetic organs and are often severely affected by environmental stress [52]. Further investigation of the molecular mechanisms underlying the regulation of cuticular wax biosynthesis in crop species under stress conditions may lead to the identification of additional genes involved in critical aspects of plant adaptation to drought stress. These will allow us to design new strategies to engineer crop cuticular lipids in plants in order to better withstand environmental conditions that severely limit plant growth, especially in the early phases of development, and that will have a positive impact on plant productivity.

Plant Growth And Response To Environmental Cues Are Largely Governed By Phytohormones

It is well established that plant growth, comprising cell division and expansion, is regulated by specific plant hormones *i.e.* auxin [indole-3-acetic acid (IAA)], cytokinins (CKs), gibberellins (GAs), brassinosteroids (BRs), and abscisic acid (ABA). It is also well know that cytokinins promote cell proliferation, while gibberellins stimulate cell elongation. Auxin and BRs are involved in both processes. After their discovery, the effects of these hormones on seedling growth have been demonstrated through different approaches, including *in vivo* assays and mutant phenotype analysis.

Auxin was the first hormone to be characterized in this context. The coleoptile unit, which includes the primary leaves and the coleoptilar node, is the main source of free IAA for the mesocotyl [53]. Dependence of the apical growing region of the mesocotyl on the coleoptile unit as a source of free IAA is almost total. One-half or more of the supply of IAA comes from the coleoptile tip, the rest mainly from the primary leaves. Removal of the coleoptile tip results in inhibition of mesocotyl elongation [54]. Elongations of maize coleoptile and pea internode were suppressed by application of auxin transport inhibitors [55]. On the other hand application of IAA stimulates the elongation of the hypocotyl of Arabidopsis [56]. The discovery of GAs, a large group of cyclic diterpene compounds soon revealed that, beside auxin, this class of hormones is also implicated in promoting the elongation of young stems during seedling growth [57]. It was soon evident that GAs promote cell elongation in the internodes, due to relaxation of the cell wall [58]. The action of BRs in promoting cell elongation brassinosteroids (BRs), commonly occurring steroid hormones that regulate multiple aspects of plant growth and development, was first observed in a 'bean second internode test' [59] and the structure of the stimulating molecule was years later determined and it was named brassinolide [60].

Dwarf mutants have been isolated in different species, such as maize, pea, tomato and Arabidopsis, and their identification allowed the isolation of structural as well of regulatory genes. In particular, mutants whose growth defects were rescued by exogenous hormone administration have been instrumental in defining genes encoding enzymes involved in specific steps of the different pathways. On the other hand mutants exhibiting insensitivity to hormone treatment were diagnostic for genes involved either in the perception or in the transduction of the signal.

For example, one important step in the comprehension of the role of GAs in promoting growth elongation derived from the cloning of the Arabidopsis GAI cDNA and its mutant allele *gai*, which produces GA-insensitivity [61]. The product of GAI is a member of the DELLA subgroup of the GRAS family of transcriptional regulators. DELLA proteins are located in the nucleus where they act as repressors of genes involved in promoting plant growth [62]. It was shown that an increase in GA level triggers DELLA degradation, *via* the ubiquitination–proteasome pathway, thus releasing the promoter of genes involved in growth from its inhibitory effect.

For these hormones the biosynthetic pathways as well as the genetic control underlying them along with the molecular mechanism related to their perception and responses have been extensively analysed. A comprehensive description of the data so far obtained and eventually of the models formulated on their bases is available in the literature [63, 64]. A general view/theme emerging from all these studies is that these hormones do not act in isolation but are interrelated by synergistic or antagonistic cross talks so that they modulate each other's biosynthesis or responses. This aspect of hormone action was shown in particular for cell elongation, probably because it is experimentally easy to work with. Some examples of this are provided below.

GA and IAA most probably interact in promoting stem elongation *via* different as well as antagonistic mechanisms. A study reported that auxin increases GA biosynthesis [65] while another has more recently shown that GA is required for auxin transport [66]. A synergistic relationship of brassinosteroids with auxin for hypocotyl elongation was demonstrated by simultaneous application of BL (the most biologically active BR) and IAA in Arabidopsis. BRs can induce, separately from auxin action, hypocotyl elongation in light-grown seedlings, but, at the same time, auxin can potentiate BR action on the same event.

Other physiological studies demonstrated additive effects of GAs and BRs on elongation growth of organs in pea, mung bean, and cucumber [67, 68]. More recently, it has been reported in Arabidopsis that BRs stimulate GA biosynthesis. A study conducted by Unterholzner and co-authors [69] has shown that BES1, a transcription factor whose presence is regulated by BR, binds to a regulatory element in promoters of key genes of GA biosynthesis, namely GA_{20} and GA_3, and controls their expression. They also propose that degradation of the DELLA repressor, due to the increase of GAs, releases the repressive action of BES1 in the transcription of targets that act further downstream and are required for plant growth and development.

In addition to a role in development, some phytohormones exert anti-stress effects on plants and are essential for the ability of plants to adapt to abiotic stresses by mediating a wide range of adaptive responses [70 - 73].

Several studies have demonstrated "anti-stress" effects of BRs, most of which have employed treatment of plants with exogenous application of BRs. They have proven the role of these hormones in a variety of environmental stresses, including high and low temperatures, drought, salinity and pathogen attack [74, 75]. "*Brassica napus* and tomato seedlings grown in the presence of 24-epibrassinolide (EBR) are significantly more tolerant to a lethal heat treatment than are control seedlings grown in the absence of the compound" [76]. The application of brassinosteroids to two susceptible varieties of sorghum resulted in alleviation of the negative impact of osmotic stress on seed germination and seedling growth. BR applications led to a decrease in the activities of the enzymes peroxidase and ascorbic acid oxidase. Moreover, they enhanced the levels of soluble proteins and free proline, which are important manifestations of stress tolerance in many plant species [77 - 79] and promoted seedling growth under osmotic stress conditions [80]. Enhanced levels of soluble proteins and free proline were found also in rice (*Oryza sativa* L.) cultivar Supcr-Basmati foliar sprayed with BRs at the five-leaf stage under drought stress conditions. Exogenous application of BRs also improved carbon assimilation and maintenance of tissue water status, improving seedling growth [81].

Evidence for the role of BRs in pathogen defence has been found in potato, barley and cucumber plants, *i.e.* potato plants sprayed with BRs had a lower incidence of infection by *Phytophthora infestans* [82]. Moreover, wild-type tobacco treated with BL showed enhanced resistance to the viral pathogen tobacco mosaic virus (TMV), the bacterial pathogen *Pseudomonas syringae* pv. *tabaci* (Pst), and to the fungal pathogen *Oidium* sp. In rice plants, BL enhances resistance to the fungal pathogen *Magnaporthe grisea* and the bacterial pathogen *Xanthomonas oryzae* [83]. Application of the epibrassinolide (epiBL) to heads of 'Lux' barley reduced the severity of Fusarium head blight (FHB) caused by *Fusarium culmorum* by 86%. Growth of plants in soil supplemented with epiBL resulted in a 28 and 35% reduction in Fusarium seedling blight (FSB) symptoms on the 'Lux' and 'Akashinriki' barley, respectively. Studies of gene expression also showed that pathogenesis-related genes are activated in plants as a result of growth in epiBL [84].

Some reports have shown that hormone-related mutant plants have activated defence responses in a constitutive manner. However, in these plants the character is combined with stunted growth and retarded development [85]. The induction of immune responses needs reprogramming of transcription: priority is given to genes involved in defence over growth-related functions. This transition is controlled by a limited number of transcription factors. One of them, *TL1*-binding transcription factor1, or TBF1, regulates the expression of endoplasmic-reticulum-resident (ER) genes required for antimicrobial protein secretion. TBF1 controls not only immune response genes but also genes involved in growth and development; in particular it turns on multiple defences and inhibits primary growth and development. Knockout mutants in this gene exhibit partial suppression of growth inhibition associated with defence activation, and transcriptional profiling of these mutants showed a general promotion of growth-related genes and repression of defence-related genes involved in immune responses induced by salicylic acid (SA) and by microbe-associated molecular profile, elf18 [86]. SA and IAA are hormones known to control stress and defence responses as well as plant growth. Transcriptome analysis of a number of Arabidopsis F_1 hybrids conducted by Groszmann and co-workers [87] showed altered expression of genes of the SA biosynthesis pathway and auxin (IAA) biosynthesis pathway. IAA-targeted gene activity increased, whereas SA concentration and target gene responses were reduced and these changes correlated with greater leaf cell numbers and increased leaf cell size respectively.

Further studies are required to unravel the mechanisms of phytohormones' action in alleviating the impact of biotic and abiotic stress in model plants as well as in crops. The information arising from these studies may be of help to design gene manipulation approaches as well for potential use of such compounds in agriculture and horticulture. However, a drawback in the application of these findings is that there is, in most cases, a competition between plant immunity, which is stimulated by hormones, and plant growth for resource allocation.

CONFLICT OF INTEREST

The authors confirm that they have no conflict of interest to declare for this publication.

ACKNOWLEDGEMENTS

Declared None.

REFERENCES

[1] Vernoud V, Hajduch M, Khaled AS, *et al.* Maize embryogenesis. Maydica 2005; 50: 469-83.

[2] Buhler DD, Liebman M, Obrycki JJ. Theoretical and practical challenges to an IPM approach to weed management. Weed Sci 2000; 48: 274-80. [http://dx.doi.org/10.1614/0043-1745(2000)048[0274:TAPCTA]2.0.CO;2]

[3] Hoecker N, Keller B, Piepho HP, Hochholdinger F. Manifestation of heterosis during early maize (*Zea mays* L.) root development. Theor Appl Genet 2006; 112(3): 421-9. [http://dx.doi.org/10.1007/s00122-005-0139-4] [PMID: 16362278]

[4] Hoecker N, Keller B, Muthreich N, *et al.* Comparison of maize (*Zea mays* L.) F1-hybrid and parental inbred line primary root transcriptomes suggests organ-specific patterns of nonadditive gene expression and conserved expression trends. Genetics 2008; 179(3): 1275-83. [http://dx.doi.org/10.1534/genetics.108.088278] [PMID: 18562640]

[5] Meyer RC, Törjék O, Becher M, Altmann T. Heterosis of biomass production in Arabidopsis. Establishment during early development. Plant Physiol 2004; 134(4): 1813-23. [http://dx.doi.org/10.1104/pp.103.033001] [PMID: 15064384]

[6] Meyer S, Pospisil H, Scholten S. Heterosis associated gene expression in maize embryos 6 days after fertilization exhibits additive, dominant and overdominant pattern. Plant Mol Biol 2007; 63(3): 381-91. [http://dx.doi.org/10.1007/s11103-006-9095-x] [PMID: 17082872]

[7] Birchler JA, Yao H, Chudalayandi S, Vaiman D, Veitia RA. Heterosis. Plant Cell 2010; 22(7): 2105-12. [http://dx.doi.org/10.1105/tpc.110.076133] [PMID: 20622146]

[8] Randolph LF. Developmental morphology of the caryopsis in maize. J Agric Res 1936; 53: 882-916.

[9] Abbe EC, Stein OL. The growth of the shoot apex in maize: embryogeny. Am J Bot 1954; 41: 173-284. [http://dx.doi.org/10.2307/2438600]

[10] Wolf MJ, Seckinger HL, Rosewall EC, *et al.* Studies on water-insoluble hemicelluloses of the endosperm cell walls in relation to milling quality of seven Pacific Northwest wheat varieties. Cereal Chem 1952; 29: 399-406.

[11] Nardmann J, Reisewitz P, Werr W. Discrete shoot and root stem cell-promoting WUS/WOX5 functions are an evolutionary innovation of angiosperms. Mol Biol Evol 2009; 26(8): 1745-55. [http://dx.doi.org/10.1093/molbev/msp084] [PMID: 19387013]

[12] Smith LG, Jackson D, Hake S. Expression of knotted1 marks shoot meristem formation during maize embryogenesis. Dev Genet 1995; 16: 344-8. [http://dx.doi.org/10.1002/dvg.1020160407]

[13] Nardmann J, Zimmermann R, Durantini D, Kranz E, Werr W. WOX gene phylogeny in Poaceae: a comparative approach addressing leaf and embryo development. Mol Biol Evol 2007; 24(11): 2474-84.

[http://dx.doi.org/10.1093/molbev/msm182] [PMID: 17768306]

[14] Post-Beittenmiller D. Biochemistry and molecular biology of wax production in plants. Annu Rev Plant Physiol Plant Mol Biol 1996; 47: 405-30. [http://dx.doi.org/10.1146/annurev.arplant.47.1.405] [PMID: 15012295]

[15] Suh MC, Samuels AL, Jetter R, *et al.* Cuticular lipid composition, surface structure, and gene expression in Arabidopsis stem epidermis. Plant Physiol 2005; 139(4): 1649-65. [http://dx.doi.org/10.1104/pp.105.070805] [PMID: 16299169]

[16] Sieber P, Schorderet M, Ryser U, *et al.* Transgenic Arabidopsis plants expressing a fungal cutinase show alterations in the structure and properties of the cuticle and postgenital organ fusions. Plant Cell 2000; 12(5): 721-38. [http://dx.doi.org/10.1105/tpc.12.5.721] [PMID: 10810146]

[17] Wellesen K, Durst F, Pinot F, *et al.* Functional analysis of the LACERATA gene of Arabidopsis provides evidence for different roles of fatty acid omega -hydroxylation in development. Proc Natl Acad Sci USA 2001; 98(17): 9694-9. [http://dx.doi.org/10.1073/pnas.171285998] [PMID: 11493698]

[18] Schnurr J, Shockey J, Browse J. The acyl-CoA synthetase encoded by LACS2 is essential for normal cuticle development in Arabidopsis. Plant Cell 2004; 16(3): 629-42. [http://dx.doi.org/10.1105/tpc.017608] [PMID: 14973169]

[19] Kurdyukov S, Faust A, Nawrath C, *et al.* The epidermis-specific extracellular BODYGUARD controls cuticle development and morphogenesis in Arabidopsis. Plant Cell 2006; 18(2): 321-39. [http://dx.doi.org/10.1105/tpc.105.036079] [PMID: 16415209]

[20] Aharoni A, Dixit S, Jetter R, Thoenes E, van Arkel G, Pereira A. The SHINE clade of AP2 domain transcription factors activates wax biosynthesis, alters cuticle properties, and confers drought tolerance when overexpressed in Arabidopsis. Plant Cell 2004; 16(9): 2463-80. [http://dx.doi.org/10.1105/tpc.104.022897] [PMID: 15319479]

[21] Bernard A, Joubès J. Arabidopsis cuticular waxes: advances in synthesis, export and regulation. Prog Lipid Res 2013; 52(1): 110-29. [http://dx.doi.org/10.1016/j.plipres.2012.10.002] [PMID: 23103356]

[22] Yeats TH, Rose JK. The biochemistry and biology of extracellular plant lipid-transfer proteins (LTPs). Protein Sci 2008; 17(2): 191-8. [http://dx.doi.org/10.1110/ps.073300108] [PMID: 18096636]

[23] Samuels L, Kunst L, Jetter R. Sealing plant surfaces: cuticular wax formation by epidermal cells. Annu Rev Plant Biol 2008; 59: 683-707. [http://dx.doi.org/10.1146/annurev.arplant.59.103006.093219] [PMID: 18251711]

[24] Mariani C, Wolters-Arts M. Complex waxes. Plant Cell 2000-98; 12(10): 1795-8. [http://dx.doi.org/10.1105/tpc.12.10.1795] [PMID: 11041876]

[25] von Wettstein-Knowles P. The molecular phenotypes of the eceriferum mutants. In: Nilan RA, Ed. 2nd Intl Barley Genetics Symp. 1969; Pullman, USA, Washington: State Universal Press 1971; pp. 146-93.

[26] Javelle M, Vernoud V, Rogowsky PM, Ingram GC. Epidermis: the formation and functions of a fundamental plant tissue. New Phytol 2011; 189(1): 17-39. [http://dx.doi.org/10.1111/j.1469-8137.2010.03514.x] [PMID: 21054411]

[27] Mao B, Cheng Z, Lei C, *et al.* Wax crystal-sparse leaf2, a rice homologue of WAX2/GL1, is involved in synthesis of leaf cuticular wax. Planta 2012; 235(1): 39-52. [http://dx.doi.org/10.1007/s00425-011-1481-1] [PMID: 21809091]

[28] Lawson EJ, Poethig RS. Shoot development in plants: time for a change. Trends Genet 1995; 11(7): 263-8. [http://dx.doi.org/10.1016/S0168-9525(00)89072-1] [PMID: 7482775]

[29] Bianchi G, Avato P, Salamini F. Biosynthetic pathways of epicuticular wax of maize as assessed by mutation, light, plant age and inhibitor studies. Maydica 1985; 30: 179-98.

[30] Koornneef M, Hanhart CJ, Hilhorst HW, Karssen CM. In vivo inhibition of seed development and reserve protein accumulation in recombinants of abscisic acid biosynthesis and responsiveness mutants in *Arabidopsis thaliana*. Plant Physiol 1989; 90(2): 463-9. [http://dx.doi.org/10.1104/pp.90.2.463] [PMID: 16666794]

[31] Schnable PS, Stinard PS, Wen TJ, *et al.* The genetics of cuticular wax biosynthesis. Maydica 1994; 39: 279-87.

[32] Lundqvist U, Lundqvist A. Mutagen specificity in barley for 1580 eceriferum mutants localized to 79 loci. Hereditas 1988; 1081: 1-12.

[33] Lee SB, Suh MC. Advances in the understanding of cuticular waxes in Arabidopsis thaliana and crop species. Plant Cell Rep 2015; 34(4): 557-72. [http://dx.doi.org/10.1007/s00299-015-1772-2] [PMID: 25693495]

[34] Kosma DK, Bourdenx B, Bernard A, *et al.* The impact of water deficiency on leaf cuticle lipids of Arabidopsis. Plant Physiol 2009; 151(4): 1918-29. [http://dx.doi.org/10.1104/pp.109.141911] [PMID: 19819982]

[35] Shepherd T, Wynne Griffiths D. The effects of stress on plant cuticular waxes. New Phytol 2006; 171(3): 469-99. [http://dx.doi.org/10.1111/j.1469-8137.2006.01826.x] [PMID: 16866954]

[36] Kosma DK, Jenks MA. Eco-physiological and molecular-genetic determinants of plant cuticle function in drought and salt stress tolerance Advances in Molecular Breeding toward Drought and Salt Tolerant Crops. Dordrecht, The Netherlands: Springer 2007; pp. 91-120. [http://dx.doi.org/10.1007/978-1-4020-5578-2_5]

[37] Kim K, Park S, Jenks M. Influence of water deficit on leaf cuticular waxes of soybean. Int J Plant Sci 2007; 168: 307-16. a [http://dx.doi.org/10.1086/510496]

[38] Kim KS, Park SH, Jenks MA. Changes in leaf cuticular waxes of sesame (*Sesamum indicum* L.) plants exposed to water deficit. J Plant Physiol 2007; 164(9): 1134-43. b [http://dx.doi.org/10.1016/j.jplph.2006.07.004] [PMID: 16904233]

[39] Weis C, Hildebrandt U, Hoffmann T, *et al.* CYP83A1 is required for metabolic compatibility of Arabidopsis with the adapted powdery mildew fungus Erysiphe cruciferarum. New Phytol 2014; 202(4): 1310-9. [http://dx.doi.org/10.1111/nph.12759] [PMID: 24602105]

[40] Raffaele S, Vailleau F, Léger A, *et al.* A MYB transcription factor regulates very-long-chain fatty acid biosynthesis for activation of the hypersensitive cell death response in Arabidopsis. Plant Cell 2008; 20(3): 752-67. [http://dx.doi.org/10.1105/tpc.107.054858] [PMID: 18326828]

[41] Mongrand S, Morel J, Laroche J, *et al.* Lipid rafts in higher plant cells: purification and characterization of Triton X-100-insoluble microdomains from tobacco plasma membrane. J Biol Chem 2004; 279(35): 36277-86. [http://dx.doi.org/10.1074/jbc.M403440200] [PMID: 15190066]

[42] Broun P, Poindexter P, Osborne E, Jiang CZ, Riechmann JL. WIN1, a transcriptional activator of epidermal wax accumulation in Arabidopsis. Proc Natl Acad Sci USA 2004; 101(13): 4706-11. [http://dx.doi.org/10.1073/pnas.0305574101] [PMID: 15070782]

[43] Shi JX, Malitsky S, De Oliveira S, *et al.* SHINE transcription factors act redundantly to pattern the archetypal surface of Arabidopsis flower organs. PLoS Genet 2011; 7(5): e1001388. [http://dx.doi.org/10.1371/journal.pgen.1001388] [PMID: 21637781]

[44] Seo PJ, Lee SB, Suh MC, Park MJ, Go YS, Park CM. The MYB96 transcription factor regulates cuticular wax biosynthesis under drought conditions in Arabidopsis. Plant Cell 2011; 23(3): 1138-52. [http://dx.doi.org/10.1105/tpc.111.083485] [PMID: 21398568]

[45] Lee SB, Kim J, Suh MC. Overexpression of myb94 transcription factor causes activation of Arabidopsis cuticular wax biosynthesis. Plant Cell Physiol 2014; pii: pcu142.

[46] Dubos C, Stracke R, Grotewold E, Weisshaar B, Martin C, Lepiniec L. MYB transcription factors in Arabidopsis. Trends Plant Sci 2010; 15(10): 573-81. [http://dx.doi.org/10.1016/j.tplants.2010.06.005] [PMID: 20674465]

[47] Guo L, Yang H, Zhang X, Yang S. Lipid transfer protein 3 as a target of MYB96 mediates freezing and drought stress in Arabidopsis. J Exp Bot 2013; 64(6): 1755-67. [http://dx.doi.org/10.1093/jxb/ert040] [PMID: 23404903]

[48] Liu S, Yeh CT, Tang HM, Nettleton D, Schnable PS. Gene mapping *via* bulked segregant RNA-Seq (BSR-Seq). PLoS One 2012; 7(5): e36406. [http://dx.doi.org/10.1371/journal.pone.0036406] [PMID: 22586469]

[49] La Rocca N, Manzotti PS, Cavaiuolo M, *et al.* The maize fused leaves1 (fdl1) gene controls organ separation in the embryo and seedling shoot and promotes coleoptile opening. J Exp Bot 2015; 66(19): 5753-67. [http://dx.doi.org/10.1093/jxb/erv278] [PMID: 26093144]

[50] Wang Y, Wan L, Zhang L, *et al.* An ethylene response factor OsWR1 responsive to drought stress transcriptionally activates wax synthesis related genes and increases wax production in rice. Plant Mol Biol 2012; 78(3): 275-88. [http://dx.doi.org/10.1007/s11103-011-9861-2] [PMID: 22130861]

[51] Hussain SS, Kayani MA, Amjad M. Transcription factors as tools to engineer enhanced drought stress tolerance in plants. Biotechnol Prog 2011; 27(2): 297-306. [http://dx.doi.org/10.1002/btpr.514] [PMID: 21302367]

[52] Jenks MA, Tuttle HA, Eigenbrode SD, Feldmann KA. Leaf epicuticular waxes of the eceriferum mutants in Arabidopsis. Plant Physiol 1995; 108(1): 369-77. [http://dx.doi.org/10.1104/pp.108.1.369] [PMID: 12228482]

[53] Iino M, Yu RS, Carr DJ. Improved procedure for the estimation of nanogram quantities of indole---acetic acid in plant extracts using the Indolo-alpha-pyrone fluorescence method. Plant Physiol 1980; 66(6): 1099-105. [http://dx.doi.org/10.1104/pp.66.6.1099] [PMID: 16661584]

[54] Iino M, Carr DJ. Sources of free IAA in the mesocotyl of etiolated maize seedlings. Plant Physiol 1982; 69(5): 1109-12. [http://dx.doi.org/10.1104/pp.69.5.1109] [PMID: 16662352]

[55] Haga K, Iino M. Auxin-growth relationships in maize coleoptiles and pea internodes and control by auxin of the tissue sensitivity to auxin. Plant Physiol 1998; 117(4): 1473-86. [http://dx.doi.org/10.1104/pp.117.4.1473] [PMID: 9701602]

[56] Tanaka K, Nakamura Y, Asami T, *et al.* Physiological roles of brassinosteroids in early growth of Arabidopsis: Brassinosteroids have a synergistic relationship with gibberellin as well as auxin in light grown hypocotyl elongation. J Plant Growth Regul 2003; 22: 259-71. [http://dx.doi.org/10.1007/s00344-003-0119-3]

[57] Phinney BO. The history of gibberellins The Chemistry and Physiology of Gibberellins. New York, NY: Praeger Press 1983; pp. 19-52.

[58] Cosgrove DJ, Sovonick-Dunford SA. Mechanism of gibberellin-dependent stem elongation in peas. Plant Physiol 1989; 89: 184-91. [http://dx.doi.org/10.1104/pp.89.1.184] [PMID: 11537446]

[59] Mitchell JW, Mandava N, Worley JF, Plimmer JR, Smith MV. Brassinsa new family of plant hormones from rape pollen. Nature 1970; 225(5237): 1065-6. [http://dx.doi.org/10.1038/2251065a0] [PMID: 16056912]

[60] Grove MD, Spencer GF, Rohwedder WK, *et al.* Brassinolide, a plant growth-promoting steroid isolated from Brassica napus pollen. Nature 1979; 281: 216-7. [http://dx.doi.org/10.1038/281216a0]

[61] Peng J, Carol P, Richards DE, *et al.* The Arabidopsis GAI gene defines a signaling pathway that negatively regulates gibberellin responses. Genes Dev 1997; 11(23): 3194-205. [http://dx.doi.org/10.1101/gad.11.23.3194] [PMID: 9389651]

[62] Bolle C. The role of GRAS proteins in plant signal transduction and development. Planta 2004; 218(5): 683-92. [http://dx.doi.org/10.1007/s00425-004-1203-z] [PMID: 14760535]

[63] Hedden P, Kamiya Y. Gibberellin biosynthesis: enzymes, genes and their regulation. Annu Rev Plant Physiol Plant Mol Biol 1997; 48: 431-60. [http://dx.doi.org/10.1146/annurev.arplant.48.1.431] [PMID: 15012270]

[64] Choi Y, Fujioka S, Harada A, *et al.* A new brassinolide biosynthetic pathway *via* 6-deoxocastasterone. Phytochemistry 1996; 43: 593-6. [http://dx.doi.org/10.1016/0031-9422(96)00342-1]

[65] Ross JJ, ONeill DP, Wolbang CM, Symons GM, Reid JB. Auxin-gibberellin interactions and their role in plant growth. J Plant Growth Regul 2001; 20(4): 336-53. [http://dx.doi.org/10.1007/s003440010034] [PMID: 11986760]

[66] Willige BC, Isono E, Richter R, Zourelidou M, Schwechheimer C. Gibberellin regulates PIN-FORMED abundance and is required for auxin transport-dependent growth and development in Arabidopsis thaliana. Plant Cell 2011; 23(6): 2184-95. [http://dx.doi.org/10.1105/tpc.111.086355] [PMID: 21642547]

[67] Mandava NB, Sasse JM, Yopp JH. Brassinolide, a growth promoting steroidal lactone. Activity in selected gibberellin and cytokinin bioassays. Physiol Plant 1981; 53: 453-61. [http://dx.doi.org/10.1111/j.1399-3054.1981.tb02730.x]

[68] Gregory LE, Madava NB. The activity and interaction of brassinolide and gibberellic acid in mung bean epicotyls. Plant Physiol 1982; 54: 239-43. [http://dx.doi.org/10.1111/j.1399-3054.1982.tb00253.x]

[69] Unterholzner SJ, Rozhon W, Papacek M, *et al.* Brassinosteroids are master regulators of gibberellin biosynthesis in Arabidopsis. Plant Cell 2015. tpc.15.00433 [http://dx.doi.org/10.1105/tpc.15.00433]

[70] Santner A, Estelle M. The ubiquitin-proteasome system regulates plant hormone signaling. Plant J 2010; 61(6): 1029-40. [http://dx.doi.org/10.1111/j.1365-313X.2010.04112.x] [PMID: 20409276]

[71] Argueso CT, Ferreira FJ, Kieber JJ. Environmental perception avenues: the interaction of cytokinin and environmental response pathways. Plant Cell Environ 2009; 32(9): 1147-60. [http://dx.doi.org/10.1111/j.1365-3040.2009.01940.x] [PMID: 19183294]

[72] Messing SA, Gabelli SB, Echeverria I, *et al.* Structural insights into maize viviparous14, a key enzyme in the biosynthesis of the phytohormone abscisic acid. Plant Cell 2010; 22(9): 2970-80. [http://dx.doi.org/10.1105/tpc.110.074815] [PMID: 20884803]

[73] Wang L, Wang Z, Xu Y, *et al.* OsGSR1 is involved in crosstalk between gibberellins and brassinosteroids in rice. Plant J 2009; 57(3): 498-510. [http://dx.doi.org/10.1111/j.1365-313X.2008.03707.x] [PMID: 18980660]

[74] Krishna P. Brassinosteroid-mediated stress responses. J Plant Growth Regul 2003; 22(4): 289-97.

[http://dx.doi.org/10.1007/s00344-003-0058-z] [PMID: 14676968]

[75] Divi UK, Krishna P. Brassinosteroid: a biotechnological target for enhancing crop yield and stress tolerance. N Biotechnol 2009; 26(3-4): 131-6. [http://dx.doi.org/10.1016/j.nbt.2009.07.006] [PMID: 19631770]

[76] Dhaubhadel S, Chaudhary S, Dobinson KF, Krishna P. Treatment with 24-epibrassinolide, a brassinosteroid, increases the basic thermotolerance of *Brassica napus* and tomato seedlings. Plant Mol Biol 1999; 40(2): 333-42. [http://dx.doi.org/10.1023/A:1006283015582] [PMID: 10412911]

[77] Wahid A, Ghazanfar A. Possible involvement of some secondary metabolites in salt tolerance of sugarcane. J Plant Physiol 2006; 163(7): 723-30. [http://dx.doi.org/10.1016/j.jplph.2005.07.007] [PMID: 16616583]

[78] Anuradha S, Rao SS. Effects of brassinosteroids on salinity stress induced inhibition of seed germination and seedling growth of rice (*Oryza sativa* L.). Plant Growth Regul 2001; 33: 151-3. [http://dx.doi.org/10.1023/A:1017590108484]

[79] Wahid A. Physiological implications of metabolite biosynthesis for net assimilation and heat-stress tolerance of sugarcane (*Saccharum officinarum*) sprouts. J Plant Res 2007; 120(2): 219-28. [http://dx.doi.org/10.1007/s10265-006-0040-5] [PMID: 17024517]

[80] Vardhini BV, Rao SS. Amelioration of water stress by brassinosteroids on seed germination and seedling growth of three varieties of sorghum. Plant Growth Regul 2003; 41: 25-31. [http://dx.doi.org/10.1023/A:1027303518467]

[81] Farooq M, Wahid A, Lee DJ, *et al.* Comparative time course action of the foliar applied glycinebetaine, salicylic acid, nitrous oxide, brassinosteroids and spermine in improving drought resistance of rice. J Agron Crop Sci 2010; 196: 336-45. [http://dx.doi.org/10.1111/j.1439-037X.2010.00422.x]

[82] Khripach VA, Zhabinskii VN, Litvinovskaya RP, *et al.* A method for protection of potato from phytophthorosis. Pat Appl 1996. BY 960346

[83] Nakashita H, Yasuda M, Nitta T, *et al.* Brassinosteroid functions in a broad range of disease resistance in tobacco and rice. Plant J 2003; 33(5): 887-98. [http://dx.doi.org/10.1046/j.1365-313X.2003.01675.x] [PMID: 12609030]

[84] Ali M, Sugimoto K, Ramadan A, Arimura G. Memory of plant communications for priming anti-herbivore responses. Sci Rep 2013; 3: 1872. [http://dx.doi.org/10.1038/srep01872] [PMID: 23695148]

[85] Heidel AJ, Clarke JD, Antonovics J, Dong X. Fitness costs of mutations affecting the systemic acquired resistance pathway in *Arabidopsis thaliana*. Genetics 2004; 168(4): 2197-206. [http://dx.doi.org/10.1534/genetics.104.032193] [PMID: 15611186]

[86] Pajerowska-Mukhtar KM, Wang W, Tada Y, *et al.* The HSF-like transcription factor TBF1 is a major molecular switch for plant growth-to-defense transition. Curr Biol 2012; 22(2): 103-12. [http://dx.doi.org/10.1016/j.cub.2011.12.015] [PMID: 22244999]

[87] Groszmann M, Gonzalez-Bayon R, Lyons RL, *et al.* Hormone-regulated defense and stress response networks contribute to heterosis in Arabidopsis F_1 hybrids. PNAS PLUS 2015; 112: pp. (46)E6397-406. [http://dx.doi.org/10.1073/pnas.1519926112]

CHAPTER 8

Natural Genetic Diversity and Crop Improvement

Elahe Tavakol[1,2,#], **Gianluca Bretani**[1,#] and **Laura Rossini**[1,3,*]

[1] *Department of Agricultural and Environmental Sciences – Production, Landscape, Agroenergy, Università degli Studi di Milano, Milano, Italy*

[2] *Department of Crop Production and Plant Breeding, College of Agriculture, Shiraz University, Shiraz, Iran*

[3] *Parco Tecnologico Padano - Lodi, Italy*

Abstract: With the human population expected to exceed 9 billion by 2050, food production will need to increase significantly in the coming years. In particular the forecast doubling in cereal demands requires improvement of yields of the top four cereal crops, maize, rice, wheat and barley. This goal is made more challenging by global environmental changes and the connected abiotic and biotic stresses. In this chapter we briefly discuss two different breeding strategies to increase cereal yield: the heterosis approach in maize and the ideotype approach based on knowledge of the genes controlling yield components in rice. We further discuss the importance of crop genetic diversity in connection to studies of the domestication history of maize, rice, wheat and barley. We present examples of how crop genetic resources including landraces and wild relatives have been used in genetic improvement of yield and adaptation to biotic and abiotic stresses. More extensive deployment of such resources to face future challenges is now empowered by new genomic tools enabling efficient exploration of genetic variability and innovative approaches to direct collection and conservation of genetic resources for food and agriculture.

Keywords: Adaptation, Barley, Breeding, Conservation, Crop genetic resources, Genetic diversity, Grain yield, Maize, Rice, Wheat.

INTRODUCTION

Out of the estimated 298.000 plant species present on earth [1], only 353 are domesticated food crops [2]. Cereals are the primary source of daily energy for humans and animals (FAO, 2013; http://faostat.fao.org/site/345/default.aspx). The first four cereals at the global level are grasses of the Poaceae family, maize

* **Corresponding author Laura Rossini**: Department of Agricultural and Environmental Sciences – Production, Landscape, Agroenergy, Università degli Studi di Milano, Milano, Italy; Tel: +39 02 50316581; Fax: +39 02 503 16521; E-mail: laura.rossini@unimi.it

# These authors contributed equally to this work.

Roberto Pilu and Giuseppe Gavazzi (Eds.)

(*Zea mays*), wheat (*Triticum aestivum*), rice (*Oryza sativa*) and barley (*Hordeum vulgare*) with a total annual production of 2.4 billion tons (FAOSTAT, Fig. **1**). Maize, wheat and rice together are estimated to provide nearly half of the world's food production. The consumption of these four major cereals varies based on region; maize is mainly used in central America, Mexico and Southern and Eastern Africa, wheat is preferred in south and north America, central and west Asia, and Europe, while rice is widely used in Asia [3].

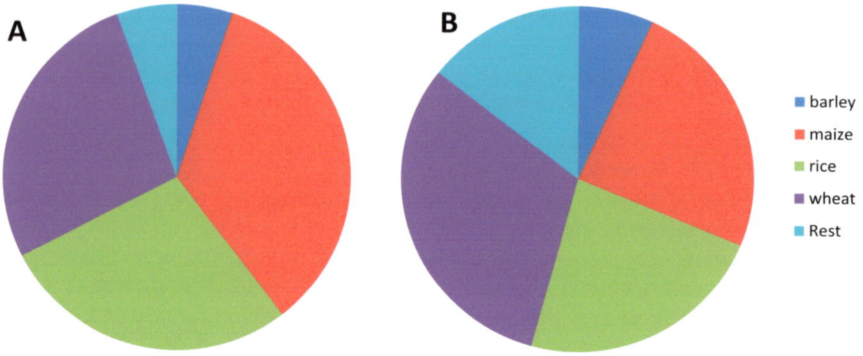

Fig. (1). Production (**A**) and areas harvested (**B**) of the four most important cereals in the world (FAOSTAT data http://faostat.fao.org/).

Maize

Maize is not cold tolerant, therefore it is cultivated only in temperate zones during spring or summer, it is also intolerant to water and nutrient deficiency. The top three maize-producing countries in the world are USA, China, and Brazil, producing ca. 563 of the 717 million tons/year [3]. Maize contains mainly starch that includes approximately 72% as well as 10% protein, and 4% fat, providing an energy density of 365 Kcal/100 g [4]. However, it has lower protein content compared to rice and wheat. In addition, maize provides essential minerals, many of the B vitamins, and fiber, while it is a modest source of calcium, iron and folate, and lacks vitamin B12 and vitamin C. Maize grains are mainly processed into flour and meal production for human uses and animal feed, and then for industrial products such as starch, oil, sweeteners, beverages, ethanol, and glue [3].

Genetically, maize is a diploid (2n=2x=20) and open-pollinated crop. Its genome was first completely sequenced in 2009 (inbred line B73). The maize genome is 2.3 Gb, 85% of which is composed of transposons and it harbors 32,540 genes [5].

Wheat

Wheat is an important staple food in both developing and developed countries worldwide. Bread wheat provides about 20% of total caloric requirement of humankind and is a primary source of protein, vitamins and minerals (World Agricultural Supply and Demand Estimates, WASDE, http://www.usda.gov/oce /commodity/wasde/). Wheat contains mainly carbohydrates, around 14% protein and 2.5% fat, supplying approximately an energy density of 327 kcal/100g. Besides being an important source of carbohydrates, wheat supplies ca. 20% of the protein for more than half of the world's population, and it also contains fiber, B vitamins, folic acid, magnesium, manganese, phosphorus and niacin, antioxidants and phytochemicals. Wheat can be used as human food in various ways such as bread, biscuits, pastry and cakes *etc*. Bran and straw are also used as animal feed. Moreover, it is sometimes used in preparation of paper, glue and some washing powders. The top three wheat-producing countries in the world are China, India and USA.

Genetically, bread wheat is an allopolyploid (2n=6x=42) and self-pollinated crop. Its genome was first sequenced with low coverage on a variety named Chinese Spring and published in 2012. The wheat genome is 17 Gb in size, with 94,000-96,000 predicted genes [6].

Rice

Rice is the major staple for about half of the human population [7]. Rice was traditionally grown in tropical and semitropical regions, and then it was rapidly dispersed in a wide range of environments.

The majority of production of rice is consumed as food for humans as a source of energy. The quality of rice protein is nutritionally ranked high among cereals, however its quantity is modest, the grain also contains vitamins, minerals, and fiber.

Genetically, rice is a diploid (2n=2x=12) and self-pollinated crop and it was the first crop whose genome was sequenced: draft genome sequences were first published in 2002 for both subspecies, japonica [8] and indica [9], and high quality complete genome was released in 2005 [10]. The rice genome is 389 Mb including 37,544 predicted genes [10].

Barley

Barley ranks in fourth position among cereals in terms of harvested area and production (http://faostat.fao.org). It is among the world's earliest domesticated crops [11], it is adapted to various harsh environments with better tolerance to abiotic stress in comparison to wheat. Barleys are grouped into winter and spring types depending on growth habit. The majority of barley production is used for animal feed, with around 20% usage for malting in (non)alcoholic beverages and minor usage as an ingredient in various food productions. Its grain is rich in high soluble dietary fiber [12].

Genetically, barley is a diploid (2n=2x=14) and self-pollinated crop and is recognized as a model for genetic and genomic research in the Triticeae tribe, which includes wheat and rye [13]. Its genome was first sequenced and published in 2012 when *de novo* assemblies of three barley varieties (Morex, Bowman and Barke) were reported. The barley genome is 5.1 Gb including 26,159 high-confidence genes [14].

GENETIC BASIS OF CEREAL YIELD

Grain yield is a complex polygenic trait strongly affected by environmental conditions. Increases in cereal yields have sustained a growing human population and genetics and breeding are estimated to have contributed about half of the yield gains achieved over the past century. With the human population forecast to exceed 9 billion by 2050, food production will have to increase significantly in the coming years. In particular to match doubling in cereal demand [15] an average yield increase of 2.4% per year is needed for the major cereals [16], but current trends are insufficient to reach this goal [17] and additional challenges come from global environmental changes, which threaten agricultural production at multiple levels [18].

Genetic analysis and improvement of yield has been a major research focus in maize [19], wheat [20, 21] rice [22], and barley [23]. In the next sections we present examples of two different approaches for yield improvement in cereals: the ideotype breeding approach as exemplified by the use of genes controlling yield components in rice, and the heterosis approach used in maize hybrid breeding.

Genetic Control Of Yield Components

In cereals, grain yield results from three main components: number of inflorescences, number of seeds per inflorescence and seed weight [22]. Indeed, these traits represent major defining features of the cereal ideotype, *i.e.* the model plant architecture pursued by breeders to improve crop production and harvest index (the proportion of harvested grain over total shoot dry matter) [24]. Notable success stories in ideotype breeding come from the Green Revolution, when introduction of semi-dwarfing genes produced high-yielding rice and wheat cultivars resistant to lodging, *i.e.* the bending of stems and resulting grain loss often caused by excessive plant growth under nitrogen fertilization.

The genetic control of yield in rice, as a model plant in monocots and cereals, has been investigated widely and consequently many genes controlling yield components are known and characterized at the molecular level [25]. In contrast, the genetic control of yield components in other cereals such as wheat and barley is less understood due to the complex nature of yield as a quantitative trait and more challenging genetics and genome complexity of these species. However, recent advances in genomic and bioinformatic tools and knowledge of chromosome colinearity among genomes allowing comparative mapping of traits and genes [26] can be used to transfer knowledge on genetic control of yield from rice to other cereals. Here we present some key genes related to yield components in rice.

Number of inflorescences is one of the important yield components and depends on the number of fertile tillers – *i.e.* lateral branches, which develop from vegetative axillary meristems (AXMs) located in the axils of basal leaves in rice, wheat and barley plants. Importantly, a trade-off between tiller number and vigor is needed as late developing tillers do not produce inflorescences and reduce the harvest index. While the regulatory signals and genes that regulate branching in dicots have been widely investigated [27, 28], they are not completely conserved and there are differences in details with tiller development in monocots [29]. In general rice tillering genes can be classified into those governing AXM initiation and those acting in the subsequent outgrowth of the axillary bud to develop a tiller [30]. While AXM initiation is largely genetically determined, tiller outgrowth depends on endogenous and external signals conveyed by multiple hormonal pathways including auxin (AUX), cytokinin (CK), gibberellins (GA), and strigolactones (SL). Since discovery of their role in repression of rice tillering [31], SLs have received particular attention and many genes involved in SL biosynthesis and signaling have been characterized in rice and other plant species – up-to-date information on this pathway can be found in excellent recent reviews [32, 33]. Despite this progress, knowledge in other major crops is still lacking. For example, in contrast to dozens of genes cloned and characterized in rice [30], among various mutants known to affect tiller number only few tillering genes have been identified and characterized in barley: *Uzu* encoding the barley ortholog of the rice brassinosteroid receptor BRI1 [34], *Uniculme4* (*Cul4*) encoding a *BLADE-ON-PETIOLE-like* gene [35], *low number of tillers* (*lnt1*) encoding the JuBel2 transcription factor [36] and *Many noded dwarf6* (*Mnd6*) encoding a member of the CYB78A subfamily of cytochrome P450 enzymes [37]. Interestingly, only the latter had been previously associated with tillering in rice, highlighting the importance of studies in different species to fully understand the mechanisms that preside to tiller development [30].

Beside number of tillers, the other two components of yield are related to the structure of inflorescences including the number of seeds and seed size. The maternal control of seed size in plants and in particular in Arabidopsis and rice was recently reviewed by [38]. Xing and Zhang [22] presented genes that may be used for designing yield-related traits in rice along with their chromosomal positions that can be used also in comparative studies for yield related traits in other cereals. The genes associated with grain weight and shape that have been identified in rice can be classified as three main groups based on their mutant phenotypes: i) the dwarfing genes with pleiotropic effects that reduce seed size such as *dwarf1 (D1), D2, D11*, and *D61* [39, 40 - 42]; ii) genes that specifically affect grain traits including *GS3* that controls grain length and weight [43]; *GW2, GW5, GS5*, and *GW8/SPL16* that play important roles in determination of grain width and weight [44 - 47], and *GIF1* known to be involved in grain filling [48]. iii) the small and round seed (SRS) genes that are involved in determination of cell length such as *SRS1, SRS3*, and *SRS5* [49, 50].

While possible negative correlations and compensation among different traits should be carefully considered in breeding for yield-related traits [51], some genes have pleiotropic effects that may be exploited for simultaneous improvement of multiple yield-related traits. For example semidominant alleles of the rice gene *OsSPL14* (encoding the SQUAMOSA PROMOTER BINDING PROTEIN-LIKE 14 transcription factor) have been associated with the *WEALTHY FARMER'S PANICLE* (*WFP*)/*Ideal Plant Architecture 1* (*IPA1*) locus and shown to increase grain yield by ca. 10% as a result of reduced number of unproductive tillers, improved lodging resistance and increased number of grains per inflorescence and grain weight [52, 53]

Heterosis

Heterosis was recently defined as "the emergent property wherein the progeny of a cross perform better than either parent" [19] and has long been investigated in both animals and plants. Heterosis in maize was initially studied by Darwin [54] and later by Shull [55] who coined this term in relation to hybrid vigour. In fact, the ease of outcrossing due to the separation of male and female inflorescences made maize an ideal model system for the study of this phenomenon. Its importance in modern maize breeding is clearly indicated by the 15% gain in grain yield and the use of hybrids for 65% of global maize production (reviewed in [56]).

Despite intensive research, the precise mechanisms subtending heterosis are not clear and may differ among species also in relation to different modes of reproduction (*e.g.* outcrossing species such as maize *vs.* inbreeding species like rice) [19, 56]. Three complementary models have been proposed to explain the superior performance of hybrids compared to their parents [19]. The dominance model proposes that complementation occurs between favourable dominant alleles and inferior recessive alleles. The epistasis model postulates that hybrids benefit from positive epistatic interactions among alleles contributed by the two parents. In the overdominance model, individuals that are heterozygous for two alleles outperform the respective homozygote genotypes. A variation on this theme is the pseudo-overdominance model, where hybrid vigour derives from favourable combinations of haplotypes of dominant and recessive alleles of tightly linked genes: such blocks of linked genes are essentially inherited as single loci [57] and occur especially in low-recombination regions, for example around centromeres in maize. Indeed evidence for the action of pseudo-overdominance has been reported in maize [19]. Beside these classical models, new insights into the genes involved in heterosis may come from QTL analyses [58] and trascriptomic studies [59].

Regardless of the specific mechanisms, the extent of hybrid vigour is correlated with genetic distance between the parents and to make high-yielding hybrids breeders cross inbred lines from different heterotic groups, highlighting the importance of genetic diversity for maximizing heterosis [19].

While hybrid production is more complicated in self-pollinating species, rice hybrids are grown in China offering a yield increase of about 10–15% over the best known inbred cultivars and cover 50% of rice production [60].

NATURAL GENETIC DIVERSITY

Origin And Diversity Centers

Genetic variation provides the foundation for any breeding program and natural genetic diversity represented historically the major source of variability for crop domestication, improvement and adaptation to a variety of local conditions, abiotic and biotic challenges. Indeed, in 1997 the use of wild relatives in breeding was estimated to afford crop yield improvements worth 115 billion USD per year globally [61] with this trend increasing in recent years [62, 63]. This justifies the growing emphasis on conserving and characterizing a wide range of crop genetic resources (cultivars, landraces and wild relatives) and understanding the domestication history of crops, as a basis to preserve biodiversity and support breeding for future food security and sustainability in changing environments [64, 65].

A milestone in the understanding of plant genetic diversity was the hypothesis of a plant's "center of origin" initially proposed by Alphonse Louis Pierre Pyrame de

Candolle (1806–1893) [66, 67]. In his book *Origine des Plantes Cultiveés* he pointed out that identification of a "center of origin" requires a multidisciplinary approach [66], connecting classical botany with plant geography and knowledge of spontaneous and ancestral species including plant history and phylogenesis [67, 68]. Another key figure was the Russian scientist Nikolai Ivanovich Vavilov (1887-1943), a pioneer in the study and collection of crop genetic resources who developed an inclusive approach from plant exploration to study of plant diversity and breeding [68]. Starting from the three primary areas of plant domestication identified by De Candolle (China, south west Asia/Egypt and tropical America), Vavilov developed his own theories concerning the evolution and domestication of wild and cultivated plants [69]. Integrating ideas from Darwin's work, Vavilov defined the "center of origin" of crop plants as a geographical area where a group of plants were first domesticated and identified it with the "center of diversity" [68]. Indeed the major hypothesis put forward by Vavilov was that the level of genetic diversity for a given species in a defined area reflects the duration of presence of the species in that location. According to this hypothesis, in the case of crops, areas with maximum genetic diversity should be considered as the "centers of origin" where the crop has existed for the longest time and was possibly domesticated [70, 71].

Vavilov conducted many scientific expeditions all around the world, with the aims of identifying those centers, through the study and classification of crop plants and their wild ancestors [68]. Finally Vavilov proposed that domestication of a wide range of crops can be traced back to seven centers of origin [72, 73]:

I. **East Asian**: western China, Korea, Japan, and Taiwan - native regions of soybeans, millet and many vegetables.
II. **Hindustani:** tropical India, Indochina, southern China, and the islands of southeastern Asia - rice, sugarcane, tropical fruit originated in this area.
III. **Southwest Asia**: Syria, Palestine, Trans-Jordania, Afghanistan, northwestern India, Transcaucasia, Iran and Turkmenistan. The native plants of this area are wheat, rye, as well as species of temperate fruit trees.
IV. **Mediterranean**: countries bordering on the Mediterranean Sea. While considering this area less important than the other eastern centers, Vavilov proposed it as native to some wheats, barleys, forage plants.
V. **Abyssinian:** of lesser importance, this was a refuge area for crops from other regions, especially wheats and barleys, spices and local grains.
VI. **The South Mexican and Central American**: southern North America, Mexico, and the west Indian Islands. Important plants such as maize, cotton, beans, pumpkins, cocoa, avocados, and subtropical fruits originated in this area.
VII. **Andean regions**: the Andean mountain range, Bolivia, Peru, Ecuador. This

area produced many tuber-bearing crops such as potatoes, quinine tree, and the coca bush.

According to this model and considering the spectrum of plant diversity in tropical and temperate regions, the most important food crops come mainly from mountains and valleys displaying an ample range of habitats [69, 71, 74].

While Vavilov's scientific legacy remains invaluable and the importance of the germplasm bank he founded in St. Petersburg is widely recognized [75], more recent studies led to distinguish more clearly the terms "center of origin" and "center of diversity" as two distinct concepts [70]. The "center of diversity" identifies a geographic area where the genetic diversity of a particular crop is especially high [76, 77]. This variability refers both to number of cultivated varieties, landraces and wild relatives and to the allelic diversity they harbor [74]. On the other hand the "center of origin" is the region where for the first time an ancestor of a plant showed specific morphological (and genetic) characteristics discriminating it from other related plants: in the case of crops, this is a region where domestication traits first appeared [71, 73]. A related question arising from these concepts is whether for a given crop a single *vs.* multiple center(s) of origin/diversity can be identified. Indeed for a given crop species the center of origin and diversity often overlap, but divergence between them is also possible [70]. For example, a local increase in a crop's genetic diversity may occur in response to environmental conditions and human activities in a distinct location compared to its center of origin. Thus, a plant species may have one center of origin but multiple diversity centers [70]. Vavilov himself admitted his original theory of a single center of origin was not perfect establishing the concept of "secondary center of origin"; indeed valuable forms of crop plants may appear far away from their primary area of origin, probably due to direct (breeding) or indirect (*e.g.* trade, construction of artificial barriers) humans activities: these areas were defined by Vavilov as secondary centers of origin and generally host more limited variability compared to the primary center of origin since their origin is more recent in time. Lesser genetic diversity, abundance of recessive alleles and a high levels of crossing over are typical of these areas usually related with human selection and breeding activities; in fact. those secondary centers often corresponded to ancient cultivated sites [71].

Beyond the theoretical developments of Vavilov's center of origin concept, later studies by Jack R. Harlan (1917-1998) and others noted some discrepancies in the location of some centers of diversity proposed by the Russian scientist. Harlan for example pointed out: "The idea of a center, an area in which things originate and out if which things are dispersed, is reasonable, logical and intellectually satisfying, but it does not always agree with the evidence" [70] and discussed the difficulty in precise delimitation of centers that cover very broad geographical areas [70].

Today, novel DNA analysis and genomics tools are being used to characterize genetic variation in crops and their wild relatives (see below). Such studies are offering unprecedented insights into the origin and history of cultivated plants and directing the exploitation of genetic resources for crop improvement [62]. Indeed several studies comparing the gene pool of crop wild relatives, traditional local varieties (landraces) and modern cultivars, showed that a progressive loss of genetic diversity accompanied domestication and breeding as a result of genetic bottlenecks and human selection [78]. At the DNA level diversity reduction was more pronounced for selected loci under domestication (and for linked neutral polymorphisms due to hitchhiking) compared to loci not affecting the selected phenotypes (neutral loci) [78]. The narrow genetic basis of modern cultivars is now recognized as a major limiting factor for the improvement of production and resilience in view of changing climatic conditions and sustainable agriculture [79]. Having continued to evolve under local variable conditions, crop wild relatives and landraces represent valuable sources of allelic diversity for a number of adaptive traits, such as resistance and tolerance to biotic and abiotic stresses, that will be key for future yield increase and stability [80]. Thus collection, maintenance and characterization of these genetic resources is important not only to conserve plant biodiversity, but also to explore a wider range of genetic diversity and identify new alleles for breeding [64].

In the following sections, we focus on each of the four most important cereals, summarizing current knowledge of their domestication history and presenting selected examples of the use of exotic germplasm for genetic improvement [62, 81].

Maize

Maize was domesticated from teosinte (*Zea mays* ssp. *parviglumis).* It is cultivated in a greater number of countries than any other crop and its adaptability allows it to grow in very different climates, from the southwest of USA to the high peaks of south America. The oldest archaeological remains were discovered in Mexico and date back to 6250 years [82]. Phylogenetic and archaeological analyses identified Mexico, specifically near the Balsa River, as the region where the nearest living maize relatives grow [83] and where likely maize domestication occurred, around 9000 years before present. The plant more closely related to modern cultivated maize is a variety of teosinte *Zea mays* ssp. *parviglumis*. Another teosinte (*Zea mays* ssp. *mexicana*) prefers to live in high areas and it can be found as a weed in maize fields [84, 85]. A scenario for maize spreading across

America was based on different genetic studies [83, 86]; likely the first maize cultivation moved rapidly at high and low latitudes in Mexico, then to the tropics and all around South America and Andes [87]. Maize arrived in North America later (3200 years ago) and findings in south-eastern USA support this hypothesis [88]. After the discovery of the Americas by Europeans, maize, like other American crops plants, spread to the old world and consequently worldwide [89]. One of the consequences of the domestication activities is the genome-wide demographic effect: selecting accessions that are morphologically suited for human needs, farmers founded new populations originated from very few genotypes. The resulting genetic drift had a major impact on genomic diversity of cultivated populations [79, 90]. Indeed, evidence of selection during or following domestication has been uncovered for 2 up to 5% of maize genes [79, 90]. Analysis at the DNA sequence level showed that traditional varieties exhibit 57% of the diversity present in teosinte [79, 89].

Tallury and Goodman [91] argued that due to the advent of hybrid maize agriculture all around the world, especially in the USA, in the 1930s, open-pollinated cultivars have been substituted by commercial hybrids, whereas those non-hybrid varieties are still cultivated by a large number of farmers in developing countries. Nowadays the elite inbred lines, which dominate hybrid production, are the results of breeding activities on a few open-pollinated cultivars. Hallauer and Carena [92] pointed out that in the period 1920–1950, as a consequence of the large use of hybrid lines in the USA, there was little development of maize germplasm resources, excluding more than 98% of the total maize germplasm offered by exotic/unadapted maize varieties; 87% of the genetic basis of hybrid maize cultivation came from five traditional corn varieties [89].

Inclusion of exotic (sub)tropical germplasm in breeding programs for temperate areas involves careful selection in order to minimize the effects of photoperiodism and fit modern hybrid standards for yield, pest tolerance, root and stalk robustness [92]. The introduction of new favorable traits in cultivated maize is needed in order to face the next generation food demand [64] and some positive examples indicate the feasibility of using exotic germplasm.

In a work of Feng *et al.* [93] an Introgression Lines (IL) population including 124 lines developed from a cross between an elite inbred extensively used in China (Huangzao 4) and a drought resistance variety (CML343) was used to successfully identify loci related to leaf-drying degree and root traits at the seedling stage. In a recent work, flooding tolerance was investigated by Mano and Omori [94] through the creation of a small IL population between Nicaraguan teosinte (*Z. nicaraguensis*) and the elite inbred line Mi29, created by the National Agriculture and Food Research Organization (NARO). Results show that *Z.*

nicaraguensis expresses different traits related to flooding tolerance, whereas Mi29 lacks these traits. Gray leaf spot (GLS; causal agent *Cercospora zeae-maydis* and *Cercospora zeina*) is a terrible corn pathology in the US. Trying to counter this threat, Lennon *et al.* [95] studied and identified alleles from teosinte that, when introduced into temperate maize germplasm, conferred significant levels of GLS resistance. A population of 693 near isogenic lines (NIL) was developed by crossing nine different teosinte accessions into the background of the maize inbred B73: in most cases, single-marker analysis validated predicted allelic substitution effects from the original NIL populations. The North Dakota State University maize breeding program through germplasm adaptation and cultivar improvement is developing drought and cold tolerant short-season crops with improved grain quality: in the last years, they developed 890 inbred lines combining good yield and earliness with drought and cold tolerance, green snap, fast dry down, and desirable genetic diversity [96].

Wheat

Wheat is one of the most important food crops all around the world, and two species are mainly cultivated: hexaploid bread wheat counting for 95% of world wheat production, and tetraploid durum for the remaining 5% [97]. Wheat is cultivated mainly in temperate areas including southern Russia, USA central plains, southern Canada, the Mediterranean Basin, north-central China, India, south America (Argentina) and south-west Australia [98]. Triticum species constitute a three level ploidy series: diploid (2n= 2x= 14, like *T. urartu*/einkorn), tetraploid (2n=4x= 28 like *T. turgidum*/wild emmer) and hexaploid (2n= 6x= 42 like *T. aestivum*) [99 - 101]. This is the result of two allopolyploidization events: an initial allotetraploidization between *T. urartu* (genome A^uA^u) and a diploid *Aegilops* species (*Aegilops speltoides*) from which originated allotetraploid wheat (*Triticum turgidum*, genome BBA^uA^u) about 0.36–0.7 million years ago [101 - 103]. Next, an allohexaploidization occurred between a domesticated form of *T. turgidum* (*T. turgidum* ssp. *dicoccon*) and the wild diploid species *A. tauschii* (genome DD), that gave rise to hexaploid wheat less than 10,000 years ago in the Caspian Sea region [97, 104, 105]. Several other diploid and tetraploid wheat species have been cultivated in the past and are now confined to niche productions [104]: during the last 5,000 years tetraploid and hexaploid wheat took the place of einkorn as staple food; now einkorn is a relict crop and is mainly grown on small scale in a few Mediterranean countries [106].

Since hexaploid bread wheat did not originate from a wild grass by domestication and selection, but from *T. turgidum* spp. *dicoccon* [76], here we discuss domestication research on einkorn wheat and emmer wheat. *T. urartu* occurs in specific regions of the Fertile Crescent and is a diploid wheat [107], it occupies a prominent place in wheat evolution, even if it was never domesticated, providing the A^u genome to modern wheat [108].

T. dicoccoides is the wild form of emmer, *T. turgidum* spp. *Dicoccon*, and its domestication probably took place in southeast Turkey [76, 109 - 112]. Phylogenetic analysis revealed two different races of *T. dicoccoides* exist: i) the western one can be found in Israel, Syria, Lebanon and Jordan and ii) the central-eastern was identified in Turkey. The latter is the more likely ancestor for domesticated wheat germplasm [109 - 112]. Archaeological emmer findings do not support the theory of domestication inside a small region, but are instead consistent with multiple cultivation origins of wheat in the Near East [104, 113]. Ozkan *et al.* [114] showed that wild emmer probably underwent two independent domestication events, one in the southern Levant and other in the northern Levant.

Several introgressions from landraces and wild relatives have been reported for increasing wheat pest resistance. For example, resistance to Hessian fly, a pest causing damage for millions of dollars in the USA alone, is present in spring wheat lines originated from *Aegilops tauschii* (https://www.ars.usda.gov/is/pr/2005/050211.htm).

Furthermore, synthetic hexaploid (SH) wheats are good examples of high yielding hybrid derivatives produced from wild relatives. Created by the International Maize and Wheat Improvement Center (CIMMYT, http://www.cimmyt.org), SH are made by crossing tetraploid wheat and *Aegilops tauschii* and subsequent artificial chromosome doubling to produce hexaploid individuals [115]. SH lines are then back-crossed with elite hexaploid wheat to obtain crops with improved characteristics. In 2003 a cross between a SH and a local Chinese cultivar named 'Chuanmai 42', was released in China, rising the production up to 20–35% (CIMMYT 2004) [62, 115, 116].

Rice

Rice belongs to genus *Oryza* which includes twenty wild species and two cultivated subspecies: *Oryza sativa indica* and *Oryza sativa japonica*. Cultivated rice originated from the wild species *Oryza rufipogon*; recently attention and efforts were spent in the identification of the geographic origin and domestication of rice. Different regions were proposed but at present many research groups accept the Yangtze River area in China as the center of rice domestication [117 - 120]. While research continues, many questions remain open about the timing and the series of events (agricultural, historical and genetic) leading to rice domestication [121]. In particular a major question is whether indica and japonica have a monophyletic (single) or bi/polyphyletic (multiple) origin(s). A recent work by Fuller *et al.* [122] recognizes south India as a hypothetic and independent center of agricultural origins emphasizing the Ganges area and the Deccan Plateau; proto-indica rice appear to have originated independently around Ganges plains and Deccan Plateau, but domestication occurred just when domesticated japonica was imported from China, starting a hybridization process 4,000 years before present [122, 123].

Important but not definitive results came from studies of neutral loci, not directly related with the domestication process [124 - 127]. Londo *et al.* showed phylogeographic evidences consistent with diverse origins of indica and japonica [126]: two likely origins were suggested for domesticated rice - indica was bred in eastern India and japonica in south China. In contrast, an evaluation of bottlenecks at corresponding loci in the two *Oryza* subspecies genomes was consistent with the single origin hypothesis [127]. Generally neutral loci support a polyphyletic relationship, and studies including *O. rufipogon* showed that japonica and indica were more genetically near to different populations of wild species than to each other, in agreement with polyphyletic origins [125, 126].

Beside neutral loci, additional insights come from characterization of domestication genes controlling shattering and pericarp color (*Sh4, qSH1, Rc*) [128 - 130]. Grain shattering (seed dispersal through abscission) is essential for reproduction of wild species, whereas reduction of this trait is considered a key step in cereal domestication, enabling harvest of seeds by humans. A point mutation in the *Sh4* gene causes reduced shattering and is shared across *japonica* and *indica* accessions: although this has been interpreted as evidence in favor of a single origin, the possibility that the reduced shattering allele was initially selected in one subspecies and later introgressed into the other cannot be excluded [128, 131]. During rice domestication, loss of red color of the pericarp (external cell layer of the seed coat) resulted in the white grains widespread in cultivated forms. Pericarp pigmentation is controlled by a transcription factor encoded by the *Rc* gene and a unique deletion in the gene (*rc* allele) is present in 98% of white pericarp rice accessions (n=337), and absent in 103 red pericarp landraces. Resequencing of a genomic region spanning the *Rc* gene across both *Oryza sativa* subpopulations demonstrated the origin of the white pericarp allele within the *japonica* subgroup followed by a spread through *indica* [128, 129]. Overall, in studies on domestication genes three genetic patterns emerged clearly. The domesticated allele can be: (a) unique to *japonica* [128, 132], (b) common to both *japonica* and *indica* [133] or (c) common to a large majority of *japonica* and *indica* varieties [121, 134]. These analyses revealed the complexity of the rice domestication process, generally pointing to a monophyletic origin for japonica and indica groups [121, 129].

To resolve discrepancies between results on domestication and neutral genes, Molina *et al.* analyzed 630 sequenced gene fragments from 20 *O. rufipogon*, 20 *O. sativa indica* and 16 *O. sativa japonica* accessions [135]. Results indicated a single origin as more probable for rice [135 - 137]. Another study by Huang *et al.* [123] used 42 gene fragments and 180 accessions, including 108 *O. rufipogon* genotypes [138]. Principal Component Analysis points to the existence of two groups within *O. rufipogon*. Surprisingly, results showed that *indica* rice was more related to the *O. rufipogn* subgroup from south China, previously believed to be closer to the japonica subspecies, which in this study showed no strong relationship with either *O. rufipogon* subgroup [123]. Based on these results, authors propose a single origin of domesticated rice in southern China, followed by shattering and hybridization with local *O. rufipogon* in Southeast Asia.

Beside their interest for a better understanding of the domestication history of rice, studies of wild and cultivated genetic diversity have provided useful materials and tools for yield improvement in modern cultivars. Progenies from a backcross between a Malaysian *O. rufipogon* accession and cultivated rice overcome the yield outcome of the recurrent parent by as much as 18% [139]. Two QTLs were identified from the wild parent, which are possibly responsible for yield rise. Similarly defined introgressions from a low-yielding wild variety, like *Oryza rufipogon,* significantly improved performance of an elite cultivated background after the introgression by more than 25% [140].

NERICA rice lines, developed at WARDA (Africa Rice Center) through hybridization between rice variety 'Morobrekan' and an accession of African *O. glaberrima*, exhibit higher yield and wide adaptation to sub-Saharan regions, performing better than the Green Revolution rice varieties in African environment [141]. Breeders at the International Rice Research Institute (www.irri.org) bred many high yield potential lines by crossing elite materials and exotic varieties of *Oryza* [24]. Another interesting example of genes from wild species that were transferred to commercial varieties is offered by the work of Ishimaru *et al.* who moved early-morning flowering (EMF) traits from *Oryza officinalis* into cultivar Koshihikari [142, 143].

Barley

Barley is one of the world's most ancient cereals. Archaeological studies show that barley domestication occurred for the first time in the Fertile Crescent 10,000 years before present [98, 144]. At least two domestication events were proposed for cultivated barley because of the existence of two distinct tough rachis alleles in modern landraces [145]. Evidence supporting multiple domestication events came from analysis of chloroplast microsatellite [146] and from molecular characterization of a region related to the tough rachis locus [147]; both works state that barley landraces can be grouped into at least two genetically separate groups, characterized by distinct geographic origins. A resequencing study on 18 loci clearly demonstrated the polyphyletic origin of cultivated barley, proposing that beside the Israel-Jordan area domestication also occurred in a region to the east of the Fertile Crescent near the Zagros mountains [148, 149]. Recently Poets *et al.* [150] demonstrated a pattern of genetic diversity in barley that does not fit with a mono/polycentric origin theory: studying 6152 SNPs in barley landraces, they found that barley is a collage from a number of wild populations that spread through the Fertile Crescent and far beyond [150, 151].

The picture of barley crop genomes evolved from a polycentric view in which distinct domestication regions were recognized by studying several genes [144], to a wide group of loci with alleles heralding from a wide range of localities [151].

In barley, the use of diverse germplasm in breeding focused on the gene pool of adapted landraces as a source of resistance against pests, pathogens and abiotic stress [98]. Ethiopian barley landraces provided resistance against *Blumeria graminis f. sp. hordeii* (mildew) due to the *mlo* locus. Indeed the *mlo-11* allele was introduced into modern barley cultivars conferring robust and durable resistance for more than 30 years [152]. In 2004, six drought tolerant barley cultivars originated from wild barley *Hordeum spontaneum* K. Koch were used in Syria [62]. Genetic stocks include lines harboring chromosomal fragments of the wild barley genome [153 - 155]. Back-cross breeding was used to introgress desirable alleles from exotic varieties into cultivated ones: powdery mildew and leaf rust (*Puccinia hordei*) resistances were improved by the work of Schmalenbach *et al.* [156] while Kalladan *et al.* [157] enhanced post-anthesis drought tolerance. Tantangara an Australian barley variety carries favorable resistance loci against leaf scald (caused by *Rhynchosporium commune*) introgressed from wild barley [158], and wild barley varieties are largely used in the International Center for Agricultural Research in the Dry Areas (ICARDA, http://www.icarda.org) for crossing programs aimed at improving abiotic stress tolerance [159, 160].

GENOMIC TOOLS FOR THE EFFICIENT USE OF CROP GENETIC RESOURCES

A major obstacle in harnessing the potential benefits of crop genetic resources has been the limited information on the useful traits carried by individual accessions due to the workload required by field-based evaluation in multiple environments. Since sequencing of the rice genome was completed, genomics opened new avenues to mine crop genetic diversity and accelerate breeding programs. Genome sequences [26], efficient genomic tools and statistical approaches provide unprecedented opportunities to identify genes and alleles to improve grain production and quality and resilience to changing environmental conditions and local adaptation [161]. Indeed systematic surveys of genetic diversity through genomic approaches have been proposed as the first of three key steps to mine biodiversity for food security [64].

Classically, Sanger sequencing is used to re-sequence the isolated known genes in various genotypes followed by bioinformatics analysis including alignment of sequences, identification of variations such as SNPs or insertion/deletions and finally prediction of possible functional variations based on non-synonymous variations in conserved functional regions of the gene. Consequently, new functional alleles are predicted and can be experimentally examined for phenotypic effects. However, despite its accuracy Sanger sequencing is limited to evaluation of only one sequence at a time that makes it a time consuming and costly approach for allele mining. This also restricts the number of genes that can be evaluated. Next Generation Sequencing (NGS) technologies are now enabling whole genome or exome resequencing approaches to obtain a comprehensive view of allelic diversity, explore and exploit the association between functional variants and desired traits. With NGS approaches vast quantities of data can be efficiently produced in a much shorter time in comparison to Sanger sequencing [162]. Thanks to a varied range of NGS approaches, the genomes of various crops including rice, maize, barley and wheat have been/are being sequenced and re-sequenced [26]. NGS has opened new avenues for molecular breeding also for allele mining, allowing investigation of allelic variation of all related genes of the same pathway involved in the interested trait(s) [163]. Indeed, several projects and consortia have been established to re-sequence the whole genome/exome of a number of species, following the example of the 1000 human genome project.

In the 3,000 rice genomes project, a collection of 3,000 rice accessions from 89 different countries were sequenced with an average sequencing depth of 14x, resulting in identification of approximately 18.9 million SNPs through alignment to the Nipponbare japonica reference genome. Such data provide fundamental information for high throughput discovery of new alleles for a range of traits using different bioinformatics and genetic approaches [164].

While this approach is feasible for the relatively compact rice genome, it is not convenient for barley and wheat due to the large size and abundance of repetitive elements (80%) of their genomes. An attractive solution to overcome these limitations are targeted resequencing approaches where sequence capture is performed to focus sequencing on coding regions (exome sequencing) or specific genomic regions associated with a particular trait [165]. Targeted resequencing starts with the identification of genes (whole exome, genes controlling a specific phenotype or class of genes) followed by a design step (designing of primers or hybridization-based capture tool) allowing to capture regions of interest and create short sequence fragments during a sequencing reaction, which can be used in identification of allelic variations of specific genes. For example, targeted resequencing of 43 genes for a total sequence of 2.2 Mb (in comparison to 2.3 Gb maize genome), resulted in identification of 2,500 SNPs. A broader view can be obtained by resequencing of the exome, which includes all protein-coding sequences [162]. Exome capture was used in allotetraploid wheat to target a total of 3.5 Mb exon regions to survey 3,497 genes that resulted in discovery of 4,386 SNPs [166]. Winfield *et al.* [167] used a targeted capture re-sequencing approach to capture and characterize 56.5 Mb of allohexaploid wheat genomic DNA. Using this method on eight wheat varieties they were able to identify over 500,000 SNPs; although 80% of these were variants distinguishing between the A, B and D homoeologous genomes, a significant number (20%) were putative varietal SNPs between the eight studied genotypes and KASPar genotyping of a subset of variants resulted in 94% validation efficiency. Using exome capture on eight varieties of hexaploid wheat, another study led to identification of 95,266 SNPs [168]. Mascher *et al.* [169] have developed an in-solution hybridization-based tool to capture a 61.6 Mb coding sequence target covering predicted genes in barley. This exome capture platform is being used in the context of ongoing international projects, *e.g.* WHEALBI [170], to analyze the natural genetic variation residing in the protein-coding portion of the barley genome and link it to natural variation for climate-related adaptive traits in landrace and wild accessions.

CONSERVATION OF GENETIC DIVERSITY

In view of the paramount importance of crop genetic resources for future food security, major efforts are dedicated at the national and international level to the collection and conservation of crop varieties, landraces and wild relatives. Landraces and wild relatives especially are important for adaptation as they display high genetic variability, phenotypic plasticity, and locally adaptive traits [65]. While up to now crop wild relatives have been mostly used to improve resistance to biotic stresses, they are attracting increasing attention as sources of traits for adaptation to abiotic stresses, *e.g.* drought and heat-tolerance for future yield stability under climate change [65].

Two complementary approaches are used for conservation of crop genetic resources: maintenance *ex situ* is possible in genebanks (infrastructures that store accessions usually as seed or vegetative material) beside conservation *in situ* in farmers' fields in the case of landraces or in protected areas or natural reserves in the case of wild plants [171]. These two approaches have different advantages and limitations.

Ex situ conservation is cheaper and convenient especially for seed-propagated crops, which can be efficiently shared among scientists and breeders, and is important to preserve accessions that have disappeared from fields and natural habitats. However, in the case of landraces and crop wild relatives genebank accessions represent only a fraction of the genetic variation harbored by the original population, no longer evolve in their natural environment and often loose genetic diversity due to inadequate storage conditions and other factors [171]. Vavilov was among the first to systematically collect plant genetic resources running over 100 expeditions around the globe: during his tenure, 250,000 accessions from more than 10,000 species were accumulated at the germplasm bank he founded, currently called the N.I. Vavilov Research Institute of Plant Industry [172, 173] It has been estimated that there are now more than 1,750 individual genebanks maintaining approximately 7.4 million plant accessions worldwide [174]. Based on surveys run by FAO, 25-30% of the total samples (ca. 1.9 - 2.2 million accessions) are distinct, while the rest represent duplicated entries maintained either in the same or, more often, a separate collection. In the case of cereal seeds, different conditions are used for short-medium *vs.* long-term storage [173]. An important long-term storage facility is the Svalbard Global Seed Vault: five years after its establishment in 2008, this collection comprised 774,601 seed samples, coming from 95% of the 193 United Nation member countries [175]. Today, collection campaigns benefit from new technologies such as GIS, so that each accession can be geo-referenced and associated to local environmental data that can be used for linking genetic diversity in specific areas and adaptive traits. For example, the Focused Identification of Germplasm Strategy (FIGS) combines geographical and environmental information with phenotypic data to locate areas likely to harbor genetic diversity for particular traits [170]. Other modeling methods have been developed to identify conservation gaps in gene pools of landraces and wild relatives, based on factors such as environmental variables, geography, taxonomy *etc* [176]. For example, Vincent *et al.* [177] focusing on wild Hordeum species proposed that the Mendoza Province in Argentina should be prioritized for *in situ* conservation. A recent global scale analysis identified gaps and priorities for further collection of wild relatives of a number of crop species, where specific taxa and areas were clearly defined [63]. The resulting inventory can be found at www.cwrdiversity.org/checklist/ and has been adopted for establishing a global network for *in situ* crop wild relatives conservation (FAO, 2013). Indeed, *in situ* conservation is also essential to preserve wider genetic diversity. Legislative measures and incentives are used for the protection of the areas, habitats and ecosystems in which the relevant plants have developed their distinctive traits, so that they continue to evolve in their environment. Under such conditions for example co-evolution between plants and their pathogens continues so that plant defense mechanisms adapt to the possible appearance of new pathogen strains, providing valuable sources of disease resistance genes for breeding. The major limitations of this approach are the costs and the potential loss of genetic diversity due to uncontrollable natural events and possible social and political difficulties for on-farm management. A growing number of *in situ* conservation areas are protected for example through on-farm conservation in traditional agricultural systems [171].

While conservation of crop genetic resources is the foundation for future agriculture, full exploitation of such resources requires appropriate strategies for characterizing, evaluating, cataloguing and policies for access and sharing of genetic materials to allow their effective use. In this respect, a central role is played by the International Treaty on Plant Genetic Resources for Food and Agriculture (ITPGRFA 2009 http://www.fao.org/docrep/011/i0510e/i0510e00.htm), which was devised to guarantee fair access and use of crop diversity. This is especially relevant considering the effects of climate change in increasing interdependency among countries for food supplies, highlighting the need for effective international cooperation programs [65]. Beside technical aspects, socio-economic, political and ethical aspects are of utmost importance in this scenario, where much of the genetic variation needed for future food security is likely to come from traditional varieties preserved by small-hold farmers in subsistence agricultural systems [171]. Indeed, participatory breeding approaches have been proposed to directly involve farmers in defining targets and identifying and evaluating materials in their fields, highlighting the value of local knowledge and resources in crop improvement [170].

CONFLICT OF INTEREST

The authors confirm that they have no conflict of interest to declare for this publication.

ACKNOWLEDGEMENTS

Declared None.

REFERENCES

[1] Mora C, Tittensor DP, Adl S, Simpson AG, Worm B. How many species are there on Earth and in the ocean? PLoS Biol 2011; 9(8): e1001127. [http://dx.doi.org/10.1371/journal.pbio.1001127] [PMID: 21886479]

[2] Meyer RS, Purugganan MD. Evolution of crop species: genetics of domestication and diversification.

Nat Rev Genet 2013; 14(12): 840-52.
[http://dx.doi.org/10.1038/nrg3605] [PMID: 24240513]

[3] Ranum P, Peña-Rosas JP, Garcia-Casal MN. Global maize production, utilization, and consumption. Ann N Y Acad Sci 2014; 1312: 105-12.
[http://dx.doi.org/10.1111/nyas.12396] [PMID: 24650320]

[4] Nuss ET, Tanumihardjo SA. Maize: A Paramount Staple Crop in the Context of Global Nutrition. Compr Rev Food Sci Food Saf 2010; 9: 417-36.
[http://dx.doi.org/10.1111/j.1541-4337.2010.00117.x]

[5] Schnable PS, Ware D, Fulton RS, *et al.* The B73 maize genome: complexity, diversity, and dynamics. Science 2009; 326(5956): 1112-5.
[http://dx.doi.org/10.1126/science.1178534] [PMID: 19965430]

[6] Brenchley R, Spannagl M, Pfeifer M, *et al.* Analysis of the bread wheat genome using whole-genome shotgun sequencing. Nature 2012; 491(7426): 705-10.
[http://dx.doi.org/10.1038/nature11650] [PMID: 23192148]

[7] Sasaki T, Burr B. International Rice Genome Sequencing Project: the effort to completely sequence the rice genome. Curr Opin Plant Biol 2000; 3(2): 138-41.
[http://dx.doi.org/10.1016/S1369-5266(99)00047-3] [PMID: 10712951]

[8] Goff SA, Ricke D, Lan TH, *et al.* A draft sequence of the rice genome (*Oryza sativa L. ssp. japonica*). Science 2002; 296(5565): 92-100.
[http://dx.doi.org/10.1126/science.1068275] [PMID: 11935018]

[9] Yu J, Hu S, Wang J, *et al.* A draft sequence of the rice genome (*Oryza sativa L. ssp. indica*). Science 2002; 296(5565): 79-92.
[http://dx.doi.org/10.1126/science.1068037] [PMID: 11935017]

[10] International Rice Genome Sequencing Project. The map-based sequence of the rice genome. Nature 2005; 436(7052): 793-800.
[http://dx.doi.org/10.1038/nature03895] [PMID: 16100779]

[11] Purugganan MD, Fuller DQ. The nature of selection during plant domestication. Nature 2009; 457(7231): 843-8.
[http://dx.doi.org/10.1038/nature07895] [PMID: 19212403]

[12] Blake T, *et al.* Barley Feed Uses and Quality Improvement, in Barley. Wiley-Blackwell 2010; pp. 522-31.

[13] Schulte D, Close TJ, Graner A, *et al.* The International Barley Sequencing Consortium—At the Threshold of Efficient Access to the Barley Genome. Plant Physiol 2009; 149(1): 142-7.
[http://dx.doi.org/10.1104/pp.108.128967] [PMID: 19126706]

[14] Mayer KF, Waugh R, Brown JW, *et al.* A physical, genetic and functional sequence assembly of the barley genome. Nature 2012; 491(7426): 711-6.
[PMID: 23075845]

[15] Tilman D, Balzer C, Hill J, Befort BL. Global food demand and the sustainable intensification of agriculture. Proc Natl Acad Sci USA 2011; 108(50): 20260-4.
[http://dx.doi.org/10.1073/pnas.1116437108] [PMID: 22106295]

[16] Ray DK, Ramankutty N, Mueller ND, West PC, Foley JA. Recent patterns of crop yield growth and stagnation. Nat Commun 2012; 3: 1293.
[http://dx.doi.org/10.1038/ncomms2296] [PMID: 23250423]

[17] Ray DK, Mueller ND, West PC, Foley JA. Yield Trends Are Insufficient to Double Global Crop Production by 2050. PLoS One 2013; 8(6): e66428.
[http://dx.doi.org/10.1371/journal.pone.0066428] [PMID: 23840465]

[18] Tester M, Langridge P. Breeding technologies to increase crop production in a changing world.

Science 2010; 327(5967): 818-22.
[http://dx.doi.org/10.1126/science.1183700] [PMID: 20150489]

[19] Wallace JG, Larsson SJ, Buckler ES. Entering the second century of maize quantitative genetics. Heredity (Edinb) 2014; 112(1): 30-8.
[http://dx.doi.org/10.1038/hdy.2013.6] [PMID: 23462502]

[20] Calderini DF, Dreccer MF, Slafer GA. Genetic improvement in wheat yield and associated traits. A re-examination of previous results and the latest trends. Plant Breed 1995; 114: 108-12.
[http://dx.doi.org/10.1111/j.1439-0523.1995.tb00772.x]

[21] Valluru R, Reynolds MP, Salse J. Genetic and molecular bases of yield-associated traits: a translational biology approach between rice and wheat. Theor Appl Genet 2014; 127(7): 1463-89.
[http://dx.doi.org/10.1007/s00122-014-2332-9] [PMID: 24913362]

[22] Xing Y, Zhang Q. Genetic and molecular bases of rice yield. Annu Rev Plant Biol 2010; 61: 421-42.
[http://dx.doi.org/10.1146/annurev-arplant-042809-112209] [PMID: 20192739]

[23] Rae SJ, Macaulay M, Ramsay L, *et al.* Molecular barley breeding. Euphytica 2006; 158: 295-303.
[http://dx.doi.org/10.1007/s10681-006-9166-8]

[24] Khush GS. Strategies for increasing the yield potential of cereals: case of rice as an example. Plant Breed 2013; 132: 433-6.

[25] Ikeda M, Miura K, Aya K, Kitano H, Matsuoka M. Genes offering the potential for designing yield-related traits in rice. Curr Opin Plant Biol 2013; 16(2): 213-20.
[http://dx.doi.org/10.1016/j.pbi.2013.02.002] [PMID: 23466256]

[26] Haberer G, Mayer KF, Spannagl M. The big five of the monocot genomes. Curr Opin Plant Biol 2016; 30: 33-40.
[http://dx.doi.org/10.1016/j.pbi.2016.01.004] [PMID: 26866569]

[27] Wang Y, Li J. Molecular basis of plant architecture. Annu Rev Plant Biol 2008; 59: 253-79.
[http://dx.doi.org/10.1146/annurev.arplant.59.032607.092902] [PMID: 18444901]

[28] Domagalska MA, Leyser O. Signal integration in the control of shoot branching. Nat Rev Mol Cell Biol 2011; 12(4): 211-21.
[http://dx.doi.org/10.1038/nrm3088] [PMID: 21427763]

[29] Kebrom TH, Spielmeyer W, Finnegan EJ. Grasses provide new insights into regulation of shoot branching. Trends Plant Sci 2013; 18(1): 41-8.
[http://dx.doi.org/10.1016/j.tplants.2012.07.001] [PMID: 22858267]

[30] Hussien A, Tavakol E, Horner DS, Muñoz-Amatriaín M, Muehlbauer GJ, Rossini L. Genetics of tillering in rice and barley. The plant genome 2014; 7(1): 1-20.

[31] Umehara M, Hanada A, Yoshida S, *et al.* Inhibition of shoot branching by new terpenoid plant hormones. Nature 2008; 455(7210): 195-200.
[http://dx.doi.org/10.1038/nature07272] [PMID: 18690207]

[32] Al-Babili S, Bouwmeester HJ. Strigolactones, a novel carotenoid-derived plant hormone. Annu Rev Plant Biol 2015; 66: 161-86.
[http://dx.doi.org/10.1146/annurev-arplant-043014-114759] [PMID: 25621512]

[33] Lopez-Obando M, Ligerot Y, Bonhomme S, Boyer FD, Rameau C. Strigolactone biosynthesis and signaling in plant development. Development 2015; 142(21): 3615-9.
[http://dx.doi.org/10.1242/dev.120006] [PMID: 26534982]

[34] Chono M, Honda I, Zeniya H, *et al.* A semidwarf phenotype of barley uzu results from a nucleotide substitution in the gene encoding a putative brassinosteroid receptor. Plant Physiol 2003; 133(3): 1209-19.
[http://dx.doi.org/10.1104/pp.103.026195] [PMID: 14551335]

[35] Tavakol E, Okagaki R, Verderio G, *et al.* The barley Uniculme4 gene encodes a BLADE-O-

-PETIOLE-like protein that controls tillering and leaf patterning. Plant Physiol 2015; 168(1): 164-74. [http://dx.doi.org/10.1104/pp.114.252882] [PMID: 25818702]

[36] Dabbert T, Okagaki RJ, Cho S, Heinen S, Boddu J, Muehlbauer GJ. The genetics of barley low-tillering mutants: low number of tillers-1 (lnt1). Theor Appl Genet 2010; 121(4): 705-15. [http://dx.doi.org/10.1007/s00122-010-1342-5] [PMID: 20407739]

[37] Mascher M, Jost M, Kuon JE, *et al.* Mapping-by-sequencing accelerates forward genetics in barley. Genome Biol 2014; 15(6): R78. [http://dx.doi.org/10.1186/gb-2014-15-6-r78] [PMID: 24917130]

[38] Li N, Li Y. Maternal control of seed size in plants. J Exp Bot 2015; 66(4): 1087-97. [http://dx.doi.org/10.1093/jxb/eru549] [PMID: 25609830]

[39] Ashikari M, Wu J, Yano M, Sasaki T, Yoshimura A. Rice gibberellin-insensitive dwarf mutant gene Dwarf 1 encodes the α-subunit of GTP-binding protein. Proc Natl Acad Sci USA 1999; 96(18): 10284-9. [http://dx.doi.org/10.1073/pnas.96.18.10284] [PMID: 10468600]

[40] Yamamuro C, Ihara Y, Wu X, *et al.* Loss of function of a rice brassinosteroid insensitive1 homolog prevents internode elongation and bending of the lamina joint. Plant Cell 2000; 12(9): 1591-606. [http://dx.doi.org/10.1105/tpc.12.9.1591] [PMID: 11006334]

[41] Hong Z, Ueguchi-Tanaka M, Umemura K, *et al.* A rice brassinosteroid-deficient mutant, ebisu dwarf (d2), is caused by a loss of function of a new member of cytochrome P450. Plant Cell 2003; 15(12): 2900-10. [http://dx.doi.org/10.1105/tpc.014712] [PMID: 14615594]

[42] Tanabe S, Ashikari M, Fujioka S, *et al.* A novel cytochrome P450 is implicated in brassinosteroid biosynthesis *via* the characterization of a rice dwarf mutant, dwarf11, with reduced seed length. Plant Cell 2005; 17(3): 776-90. [http://dx.doi.org/10.1105/tpc.104.024950] [PMID: 15705958]

[43] Mao H, Sun S, Yao J, *et al.* Linking differential domain functions of the GS3 protein to natural variation of grain size in rice. Proc Natl Acad Sci USA 2010; 107(45): 19579-84. [http://dx.doi.org/10.1073/pnas.1014419107] [PMID: 20974950]

[44] Song XJ, Huang W, Shi M, Zhu MZ, Lin HX. A QTL for rice grain width and weight encodes a previously unknown RING-type E3 ubiquitin ligase. Nat Genet 2007; 39(5): 623-30. [http://dx.doi.org/10.1038/ng2014] [PMID: 17417637]

[45] Weng J, Gu S, Wan X, *et al.* Isolation and initial characterization of GW5, a major QTL associated with rice grain width and weight. Cell Res 2008; 18(12): 1199-209. [http://dx.doi.org/10.1038/cr.2008.307] [PMID: 19015668]

[46] Li Y, Fan C, Xing Y, *et al.* Natural variation in GS5 plays an important role in regulating grain size and yield in rice. Nat Genet 2011; 43(12): 1266-9. [http://dx.doi.org/10.1038/ng.977] [PMID: 22019783]

[47] Wang S, Wu K, Yuan Q, *et al.* Control of grain size, shape and quality by OsSPL16 in rice. Nat Genet 2012; 44(8): 950-4. [http://dx.doi.org/10.1038/ng.2327] [PMID: 22729225]

[48] Wang E, Wang J, Zhu X, *et al.* Control of rice grain-filling and yield by a gene with a potential signature of domestication. Nat Genet 2008; 40(11): 1370-4. [http://dx.doi.org/10.1038/ng.220] [PMID: 18820698]

[49] Kitagawa K, Kurinami S, Oki K, *et al.* A novel kinesin 13 protein regulating rice seed length. Plant Cell Physiol 2010; 51(8): 1315-29. [http://dx.doi.org/10.1093/pcp/pcq092] [PMID: 20587735]

[50] Segami S, Kono I, Ando T, *et al.* Small and round seed 5 gene encodes alpha-tubulin regulating seed cell elongation in rice. Rice (N Y) 2012; 5(1): 4.

[http://dx.doi.org/10.1186/1939-8433-5-4] [PMID: 24764504]

[51] Peng S, Khush GS, Virk P, *et al.* Progress in ideotype breeding to increase rice yield potential. Field Crops Res 2008; 108: 32-8.
[http://dx.doi.org/10.1016/j.fcr.2008.04.001]

[52] Jiao Y, Wang Y, Xue D, *et al.* Regulation of OsSPL14 by OsmiR156 defines ideal plant architecture in rice. Nat Genet 2010; 42(6): 541-4.
[http://dx.doi.org/10.1038/ng.591] [PMID: 20495565]

[53] Miura K, Ikeda M, Matsubara A, *et al.* OsSPL14 promotes panicle branching and higher grain productivity in rice. Nat Genet 2010; 42(6): 545-9.
[http://dx.doi.org/10.1038/ng.592] [PMID: 20495564]

[54] Darwin CR. The effect of cross and self fertilization in the vegetable kingdom. London: John Murray 1876.
[http://dx.doi.org/10.5962/bhl.title.110800]

[55] Shull GH. The Composition of a Field of Maize. J Heredity 1908; os-4: 296-301.

[56] Lippman ZB, Zamir D. Heterosis: revisiting the magic. Trends Genet 2007; 23(2): 60-6.
[http://dx.doi.org/10.1016/j.tig.2006.12.006] [PMID: 17188398]

[57] Larièpe A, Mangin B, Jasson S, *et al.* The genetic basis of heterosis: multiparental quantitative trait loci mapping reveals contrasted levels of apparent overdominance among traits of agronomical interest in maize (*Zea mays* L.). Genetics 2012; 190(2): 795-811.
[http://dx.doi.org/10.1534/genetics.111.133447] [PMID: 22135356]

[58] Frascaroli E, Canè MA, Landi P, *et al.* Classical genetic and quantitative trait loci analyses of heterosis in a maize hybrid between two elite inbred lines. Genetics 2007; 176(1): 625-44.
[http://dx.doi.org/10.1534/genetics.106.064493] [PMID: 17339211]

[59] Birchler JA, Yao H, Chudalayandi S, Vaiman D, Veitia RA. Heterosis. Plant Cell 2010; 22(7): 2105-12.
[http://dx.doi.org/10.1105/tpc.110.076133] [PMID: 20622146]

[60] Jeon JS, Jung KH, Kim HB, *et al.* Genetic and Molecular Insights into the Enhancement of Rice Yield Potential. J Plant Biol 2011; 54: 1-9.
[http://dx.doi.org/10.1007/s12374-011-9144-0]

[61] Pimentel D, Wilson C, McCullum C, *et al.* Economic and Environmental Benefits of Biodiversity. Bioscience 1997; 47: 747-57.
[http://dx.doi.org/10.2307/1313097]

[62] Hajjar R, Hodgkin T. The use of wild relatives in crop improvement: a survey of developments over the last 20 years. Euphytica 2007; 156: 1-13.
[http://dx.doi.org/10.1007/s10681-007-9363-0]

[63] Vincent H, Wiersema J, Kell S, *et al.* A prioritized crop wild relative inventory to help underpin global food security. Biol Conserv 2013; 167: 265-75.
[http://dx.doi.org/10.1016/j.biocon.2013.08.011]

[64] McCouch S, Baute GJ, Bradeen J, *et al.* Agriculture: Feeding the future. Nature 2013; 499(7456): 23-4.
[http://dx.doi.org/10.1038/499023a] [PMID: 23823779]

[65] Winge T. Linking access and benefit-sharing for crop genetic resources to climate change adaptation. Plant Genetic Resources 2016; 14: 11-27.
[http://dx.doi.org/10.1017/S1479262114001038]

[66] Alfonse de Candolle. Origine des plantes cultivées, Librairie Germer Baillère et Cie 1883.

[67] Smith CE. The New World Centers of Origin of Cultivated Plants and the Archaeological Evidence. Econ Bot 1968; 22: 253-66.

[http://dx.doi.org/10.1007/BF02861958]

[68] Hummer KE, Hancock JF. Vavilovian Centers of Plant Diversity: Implications and Impacts. HortScience 2015; 50(6): 780-83.

[69] Vavilov NI. The origin, variation, immunity and breeding of cultivated plants, ed. C botanica. 1949.

[70] Harlan JR. Agricultural origins: centers and noncenters. Science 1971; 174(4008): 468-74. [http://dx.doi.org/10.1126/science.174.4008.468] [PMID: 17745730]

[71] Harlan JR. Anatomy of Gene Centers. Am Nat 1951; 85: 97-103. [http://dx.doi.org/10.1086/281657]

[72] Vavilov NI. The Phytogeographical Basis for Plant Breeding (1935). In: University of Caifornia, Division of Agriculture and Natural resources. http://pba.ucdavis.edu/files/198376.pdf

[73] Vavilov NI, Dorofeev FV. Origin and geography of cultivated plants. Cambridge University Press, 1992. ISBN0521404274, 9780521404273.

[74] Vavilov NI. The process of evolution in cultivated plant (1932). In: Proceedings of the sixth international congress of genetics, vol I. Transactions and general addresses. Electronic Scholarly Publishing. 1968, pp 331–342. http://www.esp.org/books/6th-congress/facsimile/

[75] Janick J, Vavilov NI. *Nikolai Ivanovich Vavilov* Plant Geographer, Geneticist, martyr of science. HortScience 2015; 50(6)

[76] Dvorak J, Luo MC, Akhunov ED. Vavilov's Theory of Centres of Diversity in the Light of Current Understanding of Wheat Diversity, Domestication and Evolution. Czech J Genet Plant Breed 2011; p. 47.

[77] Simmonds NW. Genetic Resources in Plants—Their Exploration and Conservation In: Frankel O H, Bennett E, Eds. Experimental Agriculture. Oxford: Blackwell Scientific Publications 1970; p. 554. 1972; 8: 87

[78] Doebley JF, Gaut BS, Smith BD. The molecular genetics of crop domestication. Cell 2006; 127(7): 1309-21. [http://dx.doi.org/10.1016/j.cell.2006.12.006] [PMID: 17190597]

[79] Wright SI, Bi IV, Schroeder SG, *et al.* The effects of artificial selection on the maize genome. Science 2005; 308(5726): 1310-4. [http://dx.doi.org/10.1126/science.1107891] [PMID: 15919994]

[80] Carvalho MA, Bebeli PJ, Bettencourt E, *et al.* Cereal landraces genetic resources in worldwide GeneBanks. A review. Agron Sustain Dev 2012; 33: 177-203. [http://dx.doi.org/10.1007/s13593-012-0090-0]

[81] Zamir D. Improving plant breeding with exotic genetic libraries. Nat Rev Genet 2001; 2(12): 983-9. [http://dx.doi.org/10.1038/35103590] [PMID: 11733751]

[82] Piperno DR, Flannery KV. The earliest archaeological maize (Zea mays L.) from highland Mexico: new accelerator mass spectrometry dates and their implications. Proc Natl Acad Sci USA 2001; 98(4): 2101-3. [http://dx.doi.org/10.1073/pnas.98.4.2101] [PMID: 11172082]

[83] Matsuoka Y, Vigouroux Y, Goodman MM, Sanchez G J, Buckler E, Doebley J. A single domestication for maize shown by multilocus microsatellite genotyping. Proc Natl Acad Sci USA 2002; 99(9): 6080-4. [http://dx.doi.org/10.1073/pnas.052125199] [PMID: 11983901]

[84] Fukunaga K, Hill J, Vigouroux Y, *et al.* Genetic diversity and population structure of teosinte. Genetics 2005; 169(4): 2241-54. [http://dx.doi.org/10.1534/genetics.104.031393] [PMID: 15687282]

[85] Piperno DR, Ranere AJ, Holst I, *et al.* Starch grain and phytolith evidence for early ninth millennium

BP maize from the Central Balsas River Valley. Mexico: PNAS 2009.

[86] Vigouroux Y, Glaubitz JC, Matsuoka Y, Goodman MM, Sánchez G J, Doebley J. Population structure and genetic diversity of New World maize races assessed by DNA microsatellites. Am J Bot 2008; 95(10): 1240-53.
[http://dx.doi.org/10.3732/ajb.0800097] [PMID: 21632329]

[87] Zarrillo S, Pearsall DM, Raymond JS, Tisdale MA, Quon DJ. Directly dated starch residues document early formative maize (*Zea mays* L.) in tropical Ecuador. Proc Natl Acad Sci USA 2008; 105(13): 5006-11.
[http://dx.doi.org/10.1073/pnas.0800894105] [PMID: 18362336]

[88] Fritz GJ. New Dates and Data on Early Agriculture: The Legacy of Complex Hunter- Gatherers. Ann Mo Bot Gard 1995; 82: 3-15.
[http://dx.doi.org/10.2307/2399975]

[89] Vigouroux Y, Barnaud A, Scarcelli N, Thuillet AC. Biodiversity, evolution and adaptation of cultivated crops. C R Biol 2011; 334(5-6): 450-7.
[http://dx.doi.org/10.1016/j.crvi.2011.03.003] [PMID: 21640954]

[90] Vigouroux Y, Mitchell S, Matsuoka Y, *et al.* An analysis of genetic diversity across the maize genome using microsatellites. Genetics 2005; 169(3): 1617-30.
[http://dx.doi.org/10.1534/genetics.104.032086] [PMID: 15654118]

[91] Tallury S, Goodman M. The state of the use of maize genetic diversity in the USA and sub-Saharan Africa. Brodening the Genetic Base of Crop Production. Rome: IPGRI/FAO 2001.
[http://dx.doi.org/10.1079/9780851994116.0159]

[92] Hallauer AR, Carena MJ. Maize. New York, NY: Springer US 2009; pp. 3-98.

[93] Feng X, Li W, Feng Y, *et al.* Identification of significant loci for drought resistance and root traits at seedling stage with a set of maize introgression lines. Maydica 2013; 58: 231-7.

[94] Mano Y, Omori F. Flooding tolerance in interspecific introgression lines containing chromosome segments from teosinte (*Zea nicaraguensis*) in maize (*Zea mays* subsp. mays). Ann Bot (Lond) 2013; 112(6): 1125-39.
[http://dx.doi.org/10.1093/aob/mct160] [PMID: 23877074]

[95] Lennon JR, Krakowsky M, Goodman M, *et al.* Identification of Alleles Conferring Resistance to GrayLeaf Spot in Maize Derived from its Wild Progenitor Species Teosinte. Crop Sci 2016; 56

[96] Carena MJ. Development of cold and drought tolerant short-season maize germplasm for fuel and feed utilization. Crop Breed Appl Biotechnol 2013; 13: 1-8.
[http://dx.doi.org/10.1590/S1984-70332013000100001]

[97] Peng JH, Sun D, Nevo E. Domestication evolution, genetics and genomics in wheat. Mol Breed 2011; 28: 281-301.
[http://dx.doi.org/10.1007/s11032-011-9608-4]

[98] Feuillet C, Langridge P, Waugh R. Cereal breeding takes a walk on the wild side. Trends Genet 2008; 24(1): 24-32.
[http://dx.doi.org/10.1016/j.tig.2007.11.001] [PMID: 18054117]

[99] Dubcovsky J, Dvorak J. Genome plasticity a key factor in the success of polyploid wheat under domestication. Science 2007; 316(5833): 1862-6.
[http://dx.doi.org/10.1126/science.1143986] [PMID: 17600208]

[100] Feldman M, Levy AA. Genome evolution due to allopolyploidization in wheat. Genetics 2012; 192(3): 763-74.
[http://dx.doi.org/10.1534/genetics.112.146316] [PMID: 23135324]

[101] Dvorak J, Deal KR, Luo MC, You FM, von Borstel K, Dehghani H. The origin of spelt and free-threshing hexaploid wheat. J Hered 2012; 103(3): 426-41.

[http://dx.doi.org/10.1093/jhered/esr152] [PMID: 22378960]

[102] Dvorak J, Akhunov ED. Tempos of gene locus deletions and duplications and their relationship to recombination rate during diploid and polyploid evolution in the Aegilops-Triticum alliance. Genetics 2005; 171(1): 323-32.
[http://dx.doi.org/10.1534/genetics.105.041632] [PMID: 15996988]

[103] Gornicki P, Zhu H, Wang J, *et al.* The chloroplast view of the evolution of polyploid wheat. New Phytol 2014; 204(3): 704-14.
[http://dx.doi.org/10.1111/nph.12931] [PMID: 25059383]

[104] Feldman M, Kislev ME. Domestication of emmer wheat and evolution of free-threshing tetraploid wheat. Isr J Plant Sci 2007; 55: 207-21.
[http://dx.doi.org/10.1560/IJPS.55.3-4.207]

[105] Liu C, Yang X, Zhang H, *et al.* Genetic and epigenetic modifications to the BBAA component of common wheat during its evolutionary history at the hexaploid level. Plant Mol Biol 2015; 88(1-2): 53-64.
[http://dx.doi.org/10.1007/s11103-015-0307-0] [PMID: 25809554]

[106] Nesbitt M, Samuel D. From staple crop to extinction? The archaeology and history of the hulled wheats. In: Padulosi S, Hammer K, Heller J, Eds. Proceedings of the 1st International Workshop on Hulled Wheats, held at Castelvecchio Pascoli. 21-22 July 1995; Tuscany, Italy. Rome, Italy: International Plant Genetic Resources Institute 1996; pp. 41-100.

[107] Zohary D, Hopf M, Weiss E. Domestication of plants in the Old World: The origin and spread of domesticated plants in south-west Asia, Europe, and the Mediterranean Basin. Oxford: Oxford University Press 2012.
[http://dx.doi.org/10.1093/acprof:osobl/9780199549061.001.0001]

[108] Dvorák J, Terlizzi P, Zhang HB, Resta P. The evolution of polyploid wheats: identification of the A genome donor species. Genome 1993; 36(1): 21-31.
[http://dx.doi.org/10.1139/g93-004] [PMID: 18469969]

[109] Özkan H, Brandolini A, Schäfer-Pregl R, Salamini F. AFLP analysis of a collection of tetraploid wheats indicates the origin of emmer and hard wheat domestication in southeast Turkey. Mol Biol Evol 2002; 19(10): 1797-801.
[http://dx.doi.org/10.1093/oxfordjournals.molbev.a004002] [PMID: 12270906]

[110] Ozkan H, Brandolini A, Pozzi C, Effgen S, Wunder J, Salamini F. A reconsideration of the domestication geography of tetraploid wheats. Theor Appl Genet 2005; 110(6): 1052-60.
[http://dx.doi.org/10.1007/s00122-005-1925-8] [PMID: 15714326]

[111] Mori N. Wheat domestication: when, where and how? Plant Cell Physiol 2003; 44: S2.

[112] Luo MC, Yang ZL, You FM, Kawahara T, Waines JG, Dvorak J. The structure of wild and domesticated emmer wheat populations, gene flow between them, and the site of emmer domestication. Theor Appl Genet 2007; 114(6): 947-59.
[http://dx.doi.org/10.1007/s00122-006-0474-0] [PMID: 17318496]

[113] Kislev ME, Nadel D, Carmi I. Festschrift For Professor Van Zeist Epipalaeolithic (19,000 BP) cereal and fruit diet at Ohalo II, Sea of Galilee, Israel. Rev Palaeobot Palynol 1992; 73: 161-6.
[http://dx.doi.org/10.1016/0034-6667(92)90054-K]

[114] Özkan H, Willcox G, Graner A, *et al.* Geographic distribution and domestication of wild emmer wheat (Triticum dicoccoides). Genet Resour Crop Evol 2010; 58: 11-53.
[http://dx.doi.org/10.1007/s10722-010-9581-5]

[115] Mujeeb-Kazi A, Rosas V, Roldan S. Conservation of the genetic variation of *Triticum tauschii* (Coss.) Schmalh. (*Aegilops squarrosa* auct. non L.) in synthetic hexaploid wheats (*T. turgidum* L. s.lat. x *T. tauschii*; 2n=6x=42, AABBDD) and its potential utilization for wheat improvement. Genet Resour Crop Evol 1996; 43(2): 129-34.

[http://dx.doi.org/10.1007/BF00126756]

[116] Villareal R, Sayre K, Banuelos O, *et al.* Registration of four synthetic hexaploid wheat (*Triticum turgidum*/*Aegilops tauschii*) germplasm lines tolerant to waterlogging. Crop Sci 2001; 41: 274. [http://dx.doi.org/10.2135/cropsci2001.411274x]

[117] Nakamura SI. The origin of rice cultivation in the Lower Yangtze Region, China. Archaeol Anthropol Sci 2010; 2: 107-13. [http://dx.doi.org/10.1007/s12520-010-0033-0]

[118] Bellwood P. First Farmers: The Origins of Agricultural Societies. Wiley 2004.

[119] Cowan CW, Watson PJ, Benco NL. The Origins of Agriculture: An International Perspective. University of Alabama Press 2006.

[120] Cohen DJ. The Beginnings of Agriculture in China A Multiregional View. Curr Anthropol 2011; 52: S273-93. [http://dx.doi.org/10.1086/659965]

[121] Gross BL, Zhao Z. Archaeological and genetic insights into the origins of domesticated rice. Proc Natl Acad Sci USA 2014; 111(17): 6190-7. [http://dx.doi.org/10.1073/pnas.1308942110] [PMID: 24753573]

[122] Fuller DQ. Finding Plant Domestication in the Indian Subcontinent. Curr Anthropol 2011; 52: S347-62. [http://dx.doi.org/10.1086/658900]

[123] Huang P, Molina J, Flowers JM, *et al.* Phylogeography of Asian wild rice, *Oryza rufipogon*: a genome-wide view. Mol Ecol 2012; 21(18): 4593-604. [http://dx.doi.org/10.1111/j.1365-294X.2012.05625.x] [PMID: 22646149]

[124] Rakshit S, Rakshit A, Matsumura H, *et al.* Large-scale DNA polymorphism study of *Oryza sativa* and *O. rufipogon* reveals the origin and divergence of Asian rice. Theor Appl Genet 2007; 114(4): 731-43. [http://dx.doi.org/10.1007/s00122-006-0473-1] [PMID: 17219210]

[125] Caicedo AL, Williamson SH, Hernandez RD, *et al.* Genome-wide patterns of nucleotide polymorphism in domesticated rice. PLoS Genet 2007; 3(9): 1745-56. [http://dx.doi.org/10.1371/journal.pgen.0030163] [PMID: 17907810]

[126] Londo JP, Chiang YC, Hung KH, Chiang TY, Schaal BA. Phylogeography of Asian wild rice, *Oryza rufipogon*, reveals multiple independent domestications of cultivated rice, *Oryza sativa*. Proc Natl Acad Sci USA 2006; 103(25): 9578-83. [http://dx.doi.org/10.1073/pnas.0603152103] [PMID: 16766658]

[127] Gao LZ, Innan H. Nonindependent domestication of the two rice subspecies, *Oryza sativa ssp. indica* and *ssp. japonica*, demonstrated by multilocus microsatellites. Genetics 2008; 179(2): 965-76. [http://dx.doi.org/10.1534/genetics.106.068072] [PMID: 18505887]

[128] Kovach MJ, Sweeney MT, McCouch SR. New insights into the history of rice domestication. Trends Genet 2007; 23(11): 578-87. [http://dx.doi.org/10.1016/j.tig.2007.08.012] [PMID: 17963977]

[129] Sweeney MT, Thomson MJ, Cho YG, *et al.* Global dissemination of a single mutation conferring white pericarp in rice. PLoS Genet 2007; 3(8): e133. [http://dx.doi.org/10.1371/journal.pgen.0030133] [PMID: 17696613]

[130] Li C, Zhou A, Sang T. Rice domestication by reducing shattering. Science 2006; 311(5769): 1936-9. [http://dx.doi.org/10.1126/science.1123604] [PMID: 16527928]

[131] Zhang LB, Zhu Q, Wu ZQ, *et al.* Selection on grain shattering genes and rates of rice domestication. New Phytol 2009; 184(3): 708-20. [http://dx.doi.org/10.1111/j.1469-8137.2009.02984.x] [PMID: 19674325]

[132] Asano K, Yamasaki M, Takuno S, *et al.* Artificial selection for a green revolution gene during

japonica rice domestication. Proc Natl Acad Sci USA 2011; 108(27): 11034-9. [http://dx.doi.org/10.1073/pnas.1019490108] [PMID: 21646530]

[133] Kovach MJ, Calingacion MN, Fitzgerald MA, McCouch SR. The origin and evolution of fragrance in rice (*Oryza sativa L.*). Proc Natl Acad Sci USA 2009; 106(34): 14444-9. [http://dx.doi.org/10.1073/pnas.0904077106] [PMID: 19706531]

[134] Tan L, Li X, Liu F, *et al.* Control of a key transition from prostrate to erect growth in rice domestication. Nat Genet 2008; 40(11): 1360-4. [http://dx.doi.org/10.1038/ng.197] [PMID: 18820699]

[135] Molina J, Sikora M, Garud N, *et al.* Molecular evidence for a single evolutionary origin of domesticated rice. Proc Natl Acad Sci USA 2011; 108(20): 8351-6. [http://dx.doi.org/10.1073/pnas.1104686108] [PMID: 21536870]

[136] Drummond AJ, Suchard MA, Xie D, Rambaut A. Bayesian phylogenetics with BEAUti and the BEAST 1.7. Mol Biol Evol 2012; 29(8): 1969-73. [http://dx.doi.org/10.1093/molbev/mss075] [PMID: 22367748]

[137] Drummond AJ, Rambaut A. BEAST: Bayesian evolutionary analysis by sampling trees. BMC Evol Biol 2007; 7: 214. [http://dx.doi.org/10.1186/1471-2148-7-214] [PMID: 17996036]

[138] Gross BL. Rice domestication: histories and mysteries. Mol Ecol 2012; 21(18): 4412-3. [http://dx.doi.org/10.1111/j.1365-294X.2012.05626.x] [PMID: 22967059]

[139] Jinhua X, Grandillo S, Ahn SN, *et al.* Genes from wild rice improve yield. Nature 1996; 384: 223-4. [http://dx.doi.org/10.1038/384223a0]

[140] Imai I, Kimball JA, Conway B, *et al.* Validation of yield-enhancing quantitative trait loci from a low-yielding wild ancestor of rice. Mol Breed 2013; 32: 101-20. [http://dx.doi.org/10.1007/s11032-013-9855-7]

[141] Rice L. Rice development in sub-Saharan Africa. J Sci Food Agric 2006; 86: 675-7. [http://dx.doi.org/10.1002/jsfa.2415]

[142] Ishimaru T, Hirabayashi H, Ida M, *et al.* A genetic resource for early-morning flowering trait of wild rice *Oryza* officinalis to mitigate high temperature-induced spikelet sterility at anthesis. Ann Bot (Lond) 2010; 106(3): 515-20. [http://dx.doi.org/10.1093/aob/mcq124] [PMID: 20566680]

[143] Dempewolf H, Eastwood RJ, Guarino L, *et al.* Adapting Agriculture to Climate Change: A Global Initiative to Collect, Conserve, and Use Crop Wild Relatives. Agroecol Sust Food Syst 2014; 38: 369-77. [http://dx.doi.org/10.1080/21683565.2013.870629]

[144] Badr A, Müller K, Schäfer-Pregl R, *et al.* On the origin and domestication history of Barley (Hordeum vulgare). Mol Biol Evol 2000; 17(4): 499-510. [http://dx.doi.org/10.1093/oxfordjournals.molbev.a026330] [PMID: 10742042]

[145] Komatsuda T, Maxim P, Senthil N, Mano Y. High-density AFLP map of nonbrittle rachis 1 (btr1) and 2 (btr2) genes in barley (*Hordeum vulgare* L.). Theor Appl Genet 2004; 109(5): 986-95. [http://dx.doi.org/10.1007/s00122-004-1710-0] [PMID: 15490100]

[146] Molina-Cano JL, Russell JR, Moralejo MA, Escacena JL, Arias G, Powell W. Chloroplast DNA microsatellite analysis supports a polyphyletic origin for barley. Theor Appl Genet 2005; 110(4): 613-9. [http://dx.doi.org/10.1007/s00122-004-1878-3] [PMID: 15723272]

[147] Azhaguvel P, Komatsuda T. A phylogenetic analysis based on nucleotide sequence of a marker linked to the brittle rachis locus indicates a diphyletic origin of barley. Ann Bot (Lond) 2007; 100(5): 1009-15. [http://dx.doi.org/10.1093/aob/mcm129] [PMID: 17638711]

[148] Morrell PL, Clegg MT. Genetic evidence for a second domestication of barley (*Hordeum vulgare*) east of the Fertile Crescent. Proc Natl Acad Sci USA 2007; 104(9): 3289-94. [http://dx.doi.org/10.1073/pnas.0611377104] [PMID: 17360640]

[149] Saisho D, Purugganan MD. Molecular phylogeography of domesticated barley traces expansion of agriculture in the Old World. Genetics 2007; 177(3): 1765-76. [http://dx.doi.org/10.1534/genetics.107.079491] [PMID: 17947416]

[150] Poets AM, Fang Z, Clegg MT, Morrell PL. Barley landraces are characterized by geographically heterogeneous genomic origins. Genome Biol 2015; 16: 173. [http://dx.doi.org/10.1186/s13059-015-0712-3] [PMID: 26293830]

[151] Allaby RG. Barley domestication: the end of a central dogma? Genome Biol 2015; 16: 176. [http://dx.doi.org/10.1186/s13059-015-0743-9] [PMID: 26306561]

[152] Jørgensen IH. Discovery, characterization and exploitation of Mlo powdery mildew resistance in barley. Euphytica 1992; 63: 141-52. [http://dx.doi.org/10.1007/BF00023919]

[153] Brown AH, Munday J, Oram RN. Use of Isozyme-Marked Segments from Wild Barley (*Hordeum spontaneum*) in Barley Breeding. Plant Breed 1988; 100: 280-8. [http://dx.doi.org/10.1111/j.1439-0523.1988.tb00254.x]

[154] Matus I, Corey A, Filichkin T, *et al.* Development and characterization of recombinant chromosome substitution lines (RCSLs) using *Hordeum vulgare subsp.* spontaneum as a source of donor alleles in a *Hordeum vulgare subsp.* vulgare background. Genome 2003; 46(6): 1010-23. [http://dx.doi.org/10.1139/g03-080] [PMID: 14663520]

[155] Honsdorf N, March TJ, Berger B, Tester M, Pillen K. High-throughput phenotyping to detect drought tolerance QTL in wild barley introgression lines. PLoS One 2014; 9(5): e97047. [http://dx.doi.org/10.1371/journal.pone.0097047] [PMID: 24823485]

[156] Schmalenbach I, Körber N, Pillen K. Selecting a set of wild barley introgression lines and verification of QTL effects for resistance to powdery mildew and leaf rust. Theor Appl Genet 2008; 117(7): 1093-106. [http://dx.doi.org/10.1007/s00122-008-0847-7] [PMID: 18663425]

[157] Kalladan R, Worch S, Rolletschek H, *et al.* Identification of quantitative trait loci contributing to yield and seed quality parameters under terminal drought in barley advanced backcross lines. Mol Breed 2013; 32: 71-90. [http://dx.doi.org/10.1007/s11032-013-9853-9]

[158] Friedt W, Horsley RD, Harvey BL, *et al.* Barley Breeding History, Progress, Objectives, and Technology, in Barley. Wiley-Blackwell 2010; pp. 160-220.

[159] Lakew B, Eglinton J, Henry RJ, *et al.* The potential contribution of wild barley (*Hordeum vulgare* ssp. *spontaneum*) germplasm to drought tolerance of cultivated barley (*H. vulgare* ssp. *vulgare*). Field Crops Res 2011; 120: 161-8. [http://dx.doi.org/10.1016/j.fcr.2010.09.011]

[160] Ceccarelli S, Jackson M, Ford-Lloyds B, *et al.* Drought. Plant Genetic Resources And Climate Change 2014; pp. 221-35.

[161] Savolainen O, Lascoux M, Merilä J. Ecological genomics of local adaptation. Nat Rev Genet 2013; 14(11): 807-20. [http://dx.doi.org/10.1038/nrg3522] [PMID: 24136507]

[162] Hashmi U, Shafqat S, Khan F, *et al.* Plant exomics: concepts, applications and methodologies in crop improvement. Plant Signal Behav 2015; 10(1): e976152. [http://dx.doi.org/10.4161/15592324.2014.976152] [PMID: 25482786]

[163] Pérez-de-Castro AM, Vilanova S, Cañizares J, *et al.* Application of genomic tools in plant breeding.

Curr Genomics 2012; 13(3): 179-95.
[http://dx.doi.org/10.2174/138920212800543084] [PMID: 23115520]

[164] Li JY, Wang J, Zeigler RS. The 3000 Rice Genome Project: opportunities and challenges for future rice research. Gigascience 2014; 3

[165] Teer JK, Mullikin JC. Exome sequencing: the sweet spot before whole genomes. Hum Mol Genet 2010; 19(R2): R145-51.
[http://dx.doi.org/10.1093/hmg/ddq333] [PMID: 20705737]

[166] Saintenac C, Jiang D, Akhunov ED. Targeted analysis of nucleotide and copy number variation by exon capture in allotetraploid wheat genome. Genome Biol 2011; 12(9): R88.
[http://dx.doi.org/10.1186/gb-2011-12-9-r88] [PMID: 21917144]

[167] Winfield MO, Wilkinson PA, Allen AM, *et al.* Targeted re-sequencing of the allohexaploid wheat exome. Plant Biotechnol J 2012; 10(6): 733-42.
[http://dx.doi.org/10.1111/j.1467-7652.2012.00713.x] [PMID: 22703335]

[168] Allen AM, Barker GL, Wilkinson P, *et al.* Discovery and development of exome-based, co-dominant single nucleotide polymorphism markers in hexaploid wheat (*Triticum aestivum* L.). Plant Biotechnol J 2013; 11(3): 279-95.
[http://dx.doi.org/10.1111/pbi.12009] [PMID: 23279710]

[169] Mascher M, Richmond TA, Gerhardt DJ, *et al.* Barley whole exome capture: a tool for genomic research in the genus Hordeum and beyond. Plant J 2013; 76(3): 494-505.
[http://dx.doi.org/10.1111/tpj.12294] [PMID: 23889683]

[170] Dawson IK, Russell J, Powell W, Steffenson B, Thomas WT, Waugh R. Barley: a translational model for adaptation to climate change. New Phytol 2015; 206(3): 913-31.
[http://dx.doi.org/10.1111/nph.13266] [PMID: 25605349]

[171] Esquinas-Alcázar J. Science and society: protecting crop genetic diversity for food security: political, ethical and technical challenges. Nat Rev Genet 2005; 6(12): 946-53.
[http://dx.doi.org/10.1038/nrg1729] [PMID: 16341075]

[172] Hummer KE. In the Footsteps of Vavilov Plant Diversity Then and Now. HortScience 2015; 50

[173] Sachs MM. Cereal germplasm resources. Plant Physiol 2009; 149(1): 148-51.
[http://dx.doi.org/10.1104/pp.108.129205] [PMID: 19126707]

[174] The second report on the state of the world's plant genetic resources for food and agricolture. Rome: FAO 2010.

[175] Westengen OT, Jeppson S, Guarino L. Global ex-situ crop diversity conservation and the Svalbard Global Seed Vault: assessing the current status. PLoS One 2013; 8(5): e64146.
[http://dx.doi.org/10.1371/journal.pone.0064146] [PMID: 23671707]

[176] Ramírez-Villegas J, Khoury C, Jarvis A, Debouck DG, Guarino L. A gap analysis methodology for collecting crop genepools: a case study with phaseolus beans. PLoS One 2010; 5(10): e13497.
[http://dx.doi.org/10.1371/journal.pone.0013497] [PMID: 20976009]

[177] Vincent H, von Bothmer R, Knüpffer H, *et al.* Genetic gap analysis of wild Hordeum taxa. Plant Genetic Resources 2012; 10: 242-53.
[http://dx.doi.org/10.1017/S1479262112000317]

CHAPTER 9

Domestication of New Species

Gaia Pigna[1,2] and **Piero Morandini**[1,*]

[1] *Department of Biosciences, Università degli Studi di Milano, Milano, Italy*

[2] *Department of Plant Sciences, The University of Tennessee, Knoxville, TN, USA*

Abstract: Domestication is the process through which a wild plant becomes a crop. The process is the result of the selection, either deliberate or as a byproduct of agricultural practices, of characteristics favorable to human beings. The sum of such characteristics is usually described as the 'domestication syndrome' because the types of traits selected are often shared among many different species. The most commonly selected traits are loss of seed dispersal, reduced seed dormancy, changes in growth habit, flowering time, and gigantism, all of which have an impact on morphology, reproductive strategies and, most importantly, production (yield and harvest index). Depending on the plant and its use, other traits could be selected, such as reduction or loss of toxic compounds, winter hardiness, nutritional quality, *etc*. Most domestication took place in ancient times, but there are a few examples of recent and accelerated domestication, for instance sugar beet. It is now possible to achieve the domestication of new species, based on the deliberate induction and combination of traits, using a set of approaches: classical plant breeding *via* hybridization and selection (including wide area crosses, hybrid seeds and plant cell culture), coupled with molecular tools such as Marker Assisted Selection, transgenesis, and site directed mutagenesis. Examples of interesting traits as well as candidate crops are discussed. Thus, we have the means to repeat the achievements of the early domestication wave and do even better, but this requires drastic changes in international and national regulations impacting on plant biotechnology and novel breeding techniques.

Keywords: Agriculture, Biofuel, Biomass, Biotechnology, Crop, Genetic engineering, Genetic resources, Genome, Harvest index, Marker assisted selection, Natural pesticides, Secondary metabolism, Seed dispersal, Seed dormancy.

INTRODUCTION

Agriculture has been an incredibly important invention in human history. The increased food production and reliability allowed the creation of stable settlements and denser societies. Agriculture does not rely on wild plants, but on

* **Corresponding author Piero Morandini**: Department of Biosciences, Università degli Studi di Milano, Milano, Italy; Tel: +39-02-50314816; E-mail: piero.morandini@unimi.it

plants which have been adapted to specific environments created by humans. Such 'domesticated' plants, usually called crops, present a set of specific characteristics which differentiate them from the wild counterparts. In order to replicate on new crops the achievements connected to the invention of agriculture, or to further improve existing crops, it is helpful to reflect on the nature of the domestication process and of its consequences on the organisms. Part of this chapter is devoted to these reflections, while the rest describes desirable traits, the methods to obtain them and suggestions of potentially interesting and promising crops.

1. WHAT IS DOMESTICATION? TRAITS OF THE DOMESTICATION SYNDROME

Domestication is the process driven by human selection by virtue of which a wild, spontaneous plant is turned into a crop. The concept applies also to animals, but these will not be dealt with in this chapter and we refer the reader to specific reviews [1 - 3]. The process of domestication implies the stable acquisition (and therefore the inheritance by the progeny) of a suite of traits, which are collectively defined as the 'domestication syndrome', which mark the difference between the crop and its wild ancestor(s) [4 - 7]. It is important to stress that the word syndrome (literally: running together) refers to "a group of symptoms that together are characteristic of a specific disorder, disease, or the like" (*Random House Kernerman Webster's College Dictionary)* and therefore domestication is seen as a complex illness from the point of view of the wild plant. When looked upon from the side of human beings, crops are marvelous organism on which we ultimately depend, or have depended, for most of our history, for food, feed, fiber, flower, fuel and fun (consider beverages such as wine, beer, tea or coffee, for instance).

Two Crucial Traits

These morphological and physiological differences are known since a long time, but the molecular details started to become available in the past two decades and several reviews appeared covering this subject [8 - 14]. For cereals, which represent the major and most studied group of crop plants because of their importance as a food/feed source, several genes have been identified and cloned [14]. The first crucial trait in the domestication of cereals and other grain crops is the loss of seed shattering (Fig. **1**), whose benefit is easily understood even by pre-school children, while the second most important trait is the reduction in seed dormancy.

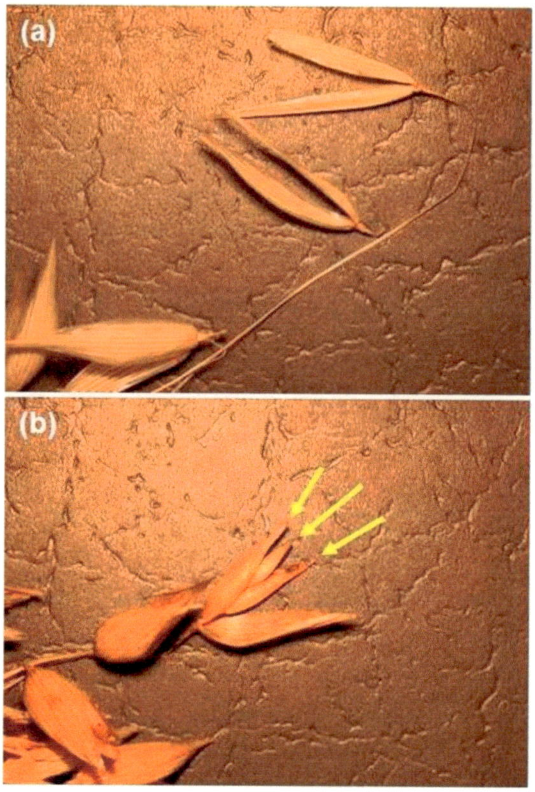

Fig. (1). Shattering behavior in wild and cultivated oat. (**a**) Wild oat spikelets at maturity are empty. (**b**) Cultivated oat spikelets retain seeds (highlighted by arrows) and only a vigorous treatment (threshing) is capable of releasing them.

The first mutations for shattering identified in molecular terms are qSH1 [15] and SH4 [16] in rice. It is remarkable that in several cases a single nucleotide change is responsible for the non-shattering trait. In the case of qSH1, the change localizes almost 12,000 bp upstream of the coding region and it is likely to affect the binding of transcription factors responsible for the activation of the gene in the pedicels, which is necessary to create the abscission layer [15]. In the case of SH4, the mutation hits a conserved Lysine residue in the DNA binding domain [16]. Interestingly, identical regulatory point mutations occurred in quite different plants [17]. Another gene (SHAT1 [18]), was identified as directly regulated by SH4 in rice and these two together influence the expression of qSH1, which, on turn, maintains SHAT1 and SH4 expression in the abscission zone. In sorghum, three independent mutations at the main shattering locus Sh1 [19] prevent seed dispersal. This gene is homologous to SH1, a minor rice QTL for shattering. In barley, the deletion of 1 or 11 bp in either of two tightly linked genes (Non-brittle rachis 1 and 2, btr1 and btr2), converted the wild type shattering behavior into a non-shattering one [20]. In wheat, the main gene conferring the non-shattering trait is called Brittle rachis (Br), but there are other important genes affecting threshability (soft glume (Sog) [21], and Tenacious glume 1 (Tg1) [22, 23]) as well as others impacting on shattering, threshability and spike compactness (*e.g.*, the Q locus [24]). Loss of seed dispersal is not restricted to domesticated cereals: it has been selected also in other crops such as in legumes, groundnut and Chile pepper, even though the genes or the tissues affected may be quite different (for soybean see [25, 26]).

The second most important trait of domestication is reduced seed dormancy. Seed dormancy is important for wild plants (Fig. **2a**) because it prevents germination soon after seed dispersal, that is, right at the end of the 'good' season in temperate climates, and spreads it over time – often years – ensuring long term survival. Any seed germinating at this stage has to face hot and dry weather in the following months, and therefore it is crucial for survival of wild plants that the seed remains dormant rather than starting its growth phase. Seed dormancy is a complex trait, with a maternal component (seed coat) and an embryo component which, of course, interact with the environment. Some wild plants require harsh treatments to rupture the seed coat to allow imbibition and germination (Fig. **2b**).

Fig. (2). Wild *Lathyrus sylvestris* seeds germinate only when scarified: (**a**) wild type seeds do not germinate, even in favorable conditions, because water uptake is prevented when the seed coat is intact. (**b**) After scarification, wild type seeds take up water, swell and germinate in a few days. Scars were created by rubbing individual seeds on glass paper. Some scars are highlighted with arrows. All seeds were transferred to moist paper at the same time.

On the contrary, crop seeds require little dormancy for a prompt germination right after sowing (Fig. **3a, b**). However, a very weak seed dormancy may translate into pre-harvest sprouting, which is a real problem for some crops, decreasing yields and affecting processing. This is most clear in viviparous mutants where seeds sprout when still on the mother plant (*e.g* [27]). Such mutations are usually lethal in the wild at the homozygous state and they can only be maintained as heterozygous. Yet, deep dormancy in a crop is not desirable because it implies erratic germination, gaps in crop rows and a delay in crop appearance that favors weed establishment and crop suppression. Therefore, crops need an optimal seed dormancy as a compromise between opposing extremes. The mutations identified so far imply a reduced or null gene function [28, 29] impacting on the various mechanisms contributing to dormancy. As before, reduced dormancy is not restricted to cereals, but it is a trait shared, for instance, by quinoa [30] and many grain legumes [31], which have thinner seed coat than their wild counterparts.

Fig. (3). Combined effect of the main domestication traits. (**a**) Kernels on a maize cob, which escaped collection by the combine harvester, germinate shortly after harvest while still on the cob, as a result of reduced seed dormancy and non-shattering; (**b**) after a few days of growth the plantlets experience strong competition for light and nutrients. (**c**) Ungerminated kernels left in the field during winter are either eaten by animals or rot because of microorganism. Image (a-b) were taken in mid-October near Chiaravalle abbey (45°25'02.8"N 9°13'59.3"E). Image (c) is from Sergnano (45°26'24.2"N 9°41'50.4"E), Crema, Italy.

We can draw a few interesting conclusions from this brief presentation of the two major domestication traits in grains: (i). the precise defect/mechanism preventing seed dispersal is of little relevance, as the important point is the result, *i.e.* seed retention. (ii). Most of the known mutations are small changes (SNP or small deletions/alterations) and usually lead to loss of function or at least reduced protein activity; it is conceivable that the frequency of such mutations was high enough in wild population for early farmers to pick them up and propagate. As selection was most certainly not at the level of individual plants, but it was done at harvest at the population level (mass selection), it probably took many generations for the selected alleles to become 'fixed' and to achieve a significant frequency in the population. (iii). The knowledge of which genes are important for domestication and the sort of mutations involved should speed up the domestication of new species (see section 4 and 5 for an in depth discussion on this issue).

Other Traits

Another interesting, yet often overlooked trait of the domestication syndrome, is a change in nutritional quality [261] and especially a reduction in natural pesticides content. For instance, bitterness represents the normal wild trait for several seeds or tissues in the Rosaceae family and it is due, for this family, to cyanogenic glucosides [32]. These compounds release hydrogen cyanide upon meeting the endogenous hydrolytic enzymes (glucosidases) after crushing of the tissue. This was known in ancient times, as Pliny the Elder mentions in *Naturalis Historia* how to prepare oil from almonds. When prepared according to the recipe, the seed cake would be free of cyanide, but not of the bitter taste [33]. This knowledge is however unknown to most urban people nowadays, as modern commercial varieties have a low content of cyanogenic glucosides and are therefore non-bitter. On the contrary, seeds of apricot, plum and peach are often quite bitter, because such seeds are not usually consumed as food as it is the case for almond. Acute poisonings and intoxications due to accidental ingestion of plants containing cyanogenic glucosides are relatively uncommon (and fatal accidents very rare), but several cases are reported in the biomedical literature. For instance, the consumption of realistic doses of bitter almonds or apricot seeds has been reported to be life-threatening [34] (see also references 2 - 6 therein) and [35, 36]. A particularly instructive case [34], allows one to estimate that five apricot seeds may be a fatal dose, if untreated, for a 30-month old female child. Similar cases are reported about consumption of cassava leaves or roots [37, 38]. There is no surprise that sweetness was an obvious target for selection, both for safety and taste issues. In general, edible crop seeds or fruits are often not only nutrient-rich, but they also have little natural pesticides and are therefore more attractive and palatable to animals and pathogens (Fig. **3c**).

There are other cases reported in the literature of wild relatives showing a higher content of natural pesticides compared to the domesticated forms. One is in cucurbits, another concerns lupin. Wild species of Cucurbita contain cucurbitacins in the fruits (as well as in leaves), while the domesticated species used as food do not [7, 39]. One must bear in mind that the cucurbitacin biosynthetic pathway is still present in domesticates, as demonstrated by sporadic cases of sudden increase in bitterness, presumably as a result of spontaneous mutation or environmental conditions [40, 41]. The lack of bitterness typical of the domesticates is due to a mutant allele of the Bitter fruit (bt) gene selected during domestication [39]. Similarly, wild lupin (*Lupinus albus*) contains more toxic alkaloids than the domesticated form (sweet lupin, still *L. albus*), which is much more susceptible to a wide range of herbivores [42]. Further example of wild-domesticates differences in natural pesticides and resistance to herbivores are mentioned in [43], who however argues that the phenomenon may not be general. See also [44] for Lima bean [45], for potato and [46] for Vigna as other examples.

Another argument for a protective function of many plant secondary compounds comes from *ad hoc* mutants with increased or decreased level of specific compounds. For instance [47], Tattersall *et al.* expressed the biosynthetic pathway of dhurrin, a cyanogenic glucoside from Sorghum, in Arabidopsis, leading to dhurrin accumulation at high levels. Not surprisingly, when this pesticide (convertible into a general metabolic poison) accumulates into Arabidopsis, resistance to herbivores increased. Conversely, when the content of an abundant natural pesticide class is specifically reduced, *e.g.* aliphatic glucosinolates in Arabidopsis, larvae of herbivores such as *Mamestra brassicae* thrive better and gain much more weight [48] when compared to wild type. The same is true for nicotine, which "functions as an efficient defense in nature" [49]. A further example is a glandless mutant cotton variety which lacked glands with their associated toxic terpenoids. This demonstrated the protective function of gossypol, but the trait turned out to be agronomically useless, as "glandless plants were extraordinarily susceptible to the attack by a host of insect pests, because they constitutively lacked protective terpenoids" ([50] and references therein).

Accumulation of secondary, toxic compounds, as well as other defenses, are often inducible traits, because there is a trade-off between defense and growth [51]. Mutants altered in defense strategies may be more palatable or less toxic (or both, as is the case for cyanogenic glucosides or cucurbitacins) to human beings, but the trait also implies that the plant is more susceptible to pests and more palatable also to animals. Therefore, other pesticides must be supplied in order to protect the plant from biotic stresses. The use of specific promoters and RNAi, to reduce the accumulation of natural pesticides only in edible parts, is a promising strategy [50, 52]. Also the engineering of more specific pesticides such as Bt cry proteins is far more sensible than that of general pesticides such as cyanogenic glucosides [47], especially if the crop is to be used for food or feed.

Other notable domestication traits present in several crops are a change in shape (growth habit) and size of the harvested part(s) of the plant, very often seed, but also, for example, inflorescence, tuber, petiole and rhizome (see [10] for an extensive list of crops and traits). Clear examples are many cereals, for instance maize and barley [13], sunflower, soybean [53], many brassicas, tomato [54, 55], pepper, eggplant, cassava, potato, pumpkin, fennel, artichoke, cardoon, sugar beet, chard, just to name a few, as well as many fruit trees and ornamental plants. An increase in the harvest index in crops (the ratio between the harvested part and the total aboveground biomass) may be achieved by different strategies: an increase in seed size, or in inflorescence size or fertility or a specific decrease in plant height. The introgression of dwarfing alleles in wheat [56] and rice [57] was one of the key components of the green revolution which increased the harvest index but also had a further positive effect by reducing lodging. The harvest index of domesticated crops is in the 0.5-0.6 range, while wild plants have values around 0.2-0.3 [58]. Sometimes the changes involve the entire plant, with underlying changes in ploidy, a trait relevant to many crops [59].

One trait connected to plant height/habit is flowering time, as it depends on the switch from vegetative to reproductive growth. Another trait is the transition from indeterminate to determinate growth, also dependent on the behavior of the apical meristem. Therefore, there is no surprise that the same genes are often involved in determining plant height, flowering time, inflorescence size and determinacy (for a review, see [12] and references therein) and that they were related to domestication. For instance, cultivated soybean flowers earlier and has larger seeds than the wild counterpart [60, 61] as additional domestication traits, beside loss of pod shattering and reduced dormancy. Also sunflower domestication required the selection of early flowering variants [62]. Changes in flowering sensitivity to photoperiod and/or to cold-induced vernalization have been very important in the adaptation of many crops to a wider range of latitudes and/or climatic conditions ([12] and references therein [63, 64]); This was true for many of the crops first domesticated in the near East when they moved from the mild and wet winters typical of the Mediterranean area to the very cold (freezing) and dry winters of northern latitudes. Also cultivated beet has a reduced photoperiod sensitivity and requires vernalization for flowering, compared to wild beet [65]. The domestication thus required the selection of a rare, partial loss-of-function allele in the BvBTC1 gene conferring reduced sensitivity to photoperiod. Vernalization overcomes the block, thus conferring bienniality to flowering [65]. This allows accumulation of large sugar quantities in the first growth year, after which the root is harvested for sugar extraction.

In some cases, the crop does not flower or the seeds do not mature in the new condition (due to changes in the length of the growing season or of the photoperiod) and the plant can complete its life cycle only when a mutation strikes some of the regulatory circuits controlling flowering. For instance, tropical maize varieties grow very tall but flower late in the season and produce little or no grain when grown in temperate environments [66]. The same must have happened to tropical rice varieties when they reached the Mediterranean region [67, 68].

Another key trait exhibited by several crops is the replacement of sexual reproduction with vegetative propagation. Several crops are propagated vegetatively, even if they have a functional sexual reproduction (seed formation). Common examples are garlic (propagated through cloves), potato, cassava, Jerusalem artichoke, which are propagated by stem (or tuber) cuttings or by rhizomes, and many fruit trees, which are propagated by grafting or through *in vitro* shoot multiplication. Some crops/cultivars have lost the ability of setting seed (*e.g.* banana/plantain because of triploidy) and are propagated by stem cuttings. Other ones have more than one strategy of vegetative propagation such as giant reed, which is sterile but can be multiplied using rhizomes or stem cuttings. All crops propagated vegetatively have fewer features of the domestication syndrome than annual seed crops because the selection is much less efficient and there is no recombination of traits.

One important key message from the analysis of the literature is that many of the changes associated with domestication are controlled by one or just a few major genes. This has interesting implications that will be discussed in section 3.

2. CONSEQUENCES OF THE SYNDROME: WHAT COMES FROM THE WILD CAN GO BACK INTO THE WILD?

Conscious or Unconscious? in any Case, A Genuine Scientific Attitude

The main traits of the domestication syndrome discussed so far, are either the 'natural' outcome of the selection due to cultivation practices or obvious deliberate (conscious) targets of selection [69]. It is often difficult to distinguish between the two, as it is obvious that the human factor and its actions are central to the development of crops. Purugganan and Fuller [70] argued that early farmers were unable to distinguish between important phenotypes, for instance, brittle and non-brittle spikes of cereals. However, it is much more reasonable to assume that late hunter-gatherers who turned into early farmers had a deep knowledge of their environment, its plants and animals. Such knowledge is an essential component for survival to contemporary hunter-gatherers [71] and it dwarfs the minimal to non-existent botanical knowledge of most western urban dwellers. For instance, the Tewa Indians of New Mexico have fifteen distinct terms for different parts of a maize plant. Many similar examples are listed by Levi-Strauss [71]. We wonder how many biology or biotechnology students could list 15 terms for different botanical parts of a maize plant. There is no reason to assume that the situation was different in the past: people ought to be able to distinguish the organisms they depended on, knowing their life cycles and the useful parts, as well as the practices and the tools to process them in order to make them edible. Levi-Strauss [71] argues "*to transform a weed into a cultivated plant, [...] to change toxic roots or seeds into foodstuffs or again to use their poison for hunting, war or ritual – there is no doubt that all these achievements required a genuinely scientific attitude, sustained and watchful interest and a desire for knowledge for its own sake. For only a small proportion of observations and experiments (which must be assumed to have been primarily inspired by a desire for knowledge) could have yielded practical and immediately useful results.*"

Turning a raw plant material, such as harvested wild wheat ears, into flour and eventually a soft bread requires a lot of trial by someone who exactly knows what he/she wants to get, but has never done it. Even more so if you are living in the Neolithic, have no idea of what you want to achieve, and you are just trying to get a better food from a bunch of hard, dry grains hidden inside their hulls. These achievements imply "*centuries of active and methodical observation, of bold hypotheses tested by means of endlessly repeated experiments*" [71], that is consciousness and deliberate action. The development of agriculture and of crops fits well into this description and early farmers must have started to deliberately select the best individuals for reproduction on the basis of vigor, color, taste, flowering time, inflorescence size, grain size, nutrient content, ease of processing or several other characteristics, for instance beauty. Of course, some traits would be the automatic result of cultivation practices, for instance reduced seed dormancy, which would be favored by the annual cycles of reaping and sowing. Quickly germinating seeds would contribute to the next generation at high frequency, while those sprouting late would not grow to maturity or be harvested. The latter may be deemed unconscious selection, but the very act of farming requires a central role for human decision-making and preferences [71, 72].

Unable to Survive

Why so few species were domesticated and why did it take such a long time? As argued by Gepts [6], some plants may be less amenable to domestication, as is the case for some animals. Sometimes potentially useful and domesticable plants simply did not live close enough to human beings to undergo the transition. In other places, they did live close enough, but there was not enough time and will to tame them, as other ones were already available. All these considerations bring us to the conclusion that many additional species could be domesticated. As we set us to discuss and describe this new endeavor, there is one important thing we need not to forget. If one reflects on the comparison between domesticated plants and wild counterparts, one clear, mostly unacknowledged, evidence comes to the mind: crops are not encountered outside cultivated fields, gardens and orchards, while wild counterparts often abound in or even suffocate agricultural landscapes. Think about all crops: cereals and cereal-like (*e.g.* quinoa, amaranth), pulses, oilseeds, tubers, vegetables, fruit trees, flowers, and fiber crops (cotton, flax, hemp...) and ask yourself if you have ever found them growing in the wild. It happens so rarely, at least to the authors, that it becomes a remarkable event. As Darwin observed in 1859, fully domesticated plants are unable to survive on their own in the wild (in his very own words: "*Very many of the most strongly-marked domestic varieties could not possibly live in a wild state*", The Origin of species, chapter I). The inability to survive without human care is the direct consequence of the domestication syndrome, most notably of the combination between the two main traits (Fig. **3a, b**). The reduction in natural pesticides also contributes because it makes seeds and plants more attractive to animals and more prone to microorganisms (Fig. **3c**). Once that a seed of an annual plant germinates at the wrong time and with strong competition among siblings, there is no way for it to conclude the life cycle (Fig. **4**). Indeed, the survival rate for maize is very low when left on its own (Fig. **5**).

Fig. (4). Voluntary maize plants in dense stands in fields near Sergnano (Crema, Italy). (**a**) seedlings in mid-October or (**b**) plants in early November. The latter have already some leaves partially burned by low temperatures and will die in a few days, well before flowering.

This plain truth is clear to farmers and gardeners, who have to fight daily against pests (weeds, pathogens, rodents, arthropods…) and abiotic stresses to make crops thrive, but hard to acknowledge by urban dwellers. Had generations of human beings not continuously done the hard work of selecting and propagating their plants, the incredible variety and richness (in terms of forms, colors, taste, scent and nutrient content) of crops and varieties would simply be unavailable today. In other words, crops are the product of a constant human activity. As a corollary, most crops/varieties would readily go extinct in one or a few seasons if humans would not keep caring for them. The firsts to go would be annual crops, then biennial and then, later on, trees and perennials. Some may eventually survive, but would become feral [262] and loose part of the traits typical of the domestication syndrome by back mutations (*e.g.* wild hexaploid wheat in Tibet, feral sunflower and feral beets in Europe). The truth is most evident with sterile crops such as triploid bananas, which have no seeds and are propagated using cuttings. If humans would not take care of them, most of these plants would die out in a few years in most locations.

Fig. (5). The survival rate of maize is very low A voluntary maize plant growing in a soybean field near Ivrea, Italy (45°28'46.4"N 7°51'54.7"E). The field was around 1 ha in size and hosted three voluntary plants in total.

The reason for this behavior is that human selection and natural selection often have opposite goals. Natural selection crucially depends on seed dispersal and seed dormancy. It also takes place in dense stands, which favors slender plants with efficient light capture, fast growth and good resistance to biotic and abiotic stresses. When grown in strong competition, resources are scarce (light, water, carbon dioxide, nutrients such as phosphate, nitrogen, sulfur, iron…) and limit growth and seed production. On the contrary, human selection favors good seed retention and prompt seed germination. Weed removal eliminates competitors for resources and favors the crop. A high harvest index means that large amounts of nutrients are removed from the soil at each harvest and soil fertility must be restored by feeding those nutrients back to the soil using various sources (manure, ash, crop waste and synthetic fertilizers). Crops must often be protected from pests and irrigated in order to give a harvest of good quantity and quality. As experienced by the real character of the book and the movie 'Into the wild' (Christopher McCandless), going back to nature is not an easy option, especially for someone raised in an urban setting. In his case, the lack of food and the perils of pristine nature (the abundant toxic substances present in a wild species mistaken as a food substitute) were fatal. For crop plants the competition with wild species is enough to bring crops to extinction.

Are the two selections always conflicting, so that a gain on one side means a loss on the other side? The question is a tough one and there is probably no simple answer. However, we have today many sophisticated tools to accelerate the domestication process while avoiding, whenever possible, or at least mitigating the conflict. Such methods are discussed in section 3 and 4.

3. LESSONS FROM OLD AND RECENT DOMESTICATION

Learning from the Past

Can we gather some lessons from the past and improve our ability to domesticate new crops in a shorter time, with better results or less efforts? To answer this question is important to remind first that domestication requires similar traits (*e.g.* loss of seed shattering) in plants of different origin or structure. Similar traits often depend on mutations at orthologous loci [73]. This means that accumulating knowledge about domestication related traits, basic plant physiology and genetics, can help a lot in the endeavour.

We also know that the amount of time required to domesticate a plant was large at the beginning (hundreds to thousands of years) but it substantially dropped in recent times. For instance, sugar beet has required less than two hundred years to achieve a remarkable improvement in sugar yield per hectare. When the crop was first cultivated for this purpose, at the beginning of the 19th century, sugar yield was around 0.7 t/ha, while nowadays average yields are in the range of 11-18 t/ha in France and Germany (see production data in the FAO-Stat database), which means yield improved around 20 times in 200 years. The components of the yield increase are diverse, but they have a strong genetic basis: improved sugar content, increased root weight of hybrid varieties and monogerm seeds. Oilseed rape is an another excellent recent example – a hybrid between two Brassica species.

In addition, one needs to acknowledge that sometimes different crops were derived from a single species or from the cross of similar species. An example of this phenomenon is again beet: sugar beets, table beets and chards are capable of cross pollination among themselves and with wild beet, from which they derive. All four plants are therefore considered members of the same species [74]. Another example is the pair globe artichoke/cultivated cardoon, both of which derive from wild cardoon, with which they are still inter-fertile [75].

It is also important to recall a third fact: the phenotypic diversity of colours, shapes and tastes within a crop is often large, despite the bottleneck of domestication has significantly reduced genetic diversity compared to wild progenitors. All these facts suggest that a strong selective pressure can create a large diversity of phenotypes/genotypes in a short time. For instance, the creation of many rice varieties adapted to higher latitudes (*e.g.* to the Mediterranean climates) has been accomplished rather quickly [67, 68]. A similar situation is true for potato in Europe: the first crop that reached the old world initially produced tubers only in short days, but clones adapted to long days appeared within 200 years [76].

A further phenomenon to take into account is that, during early domestication, only individual plants living close enough and that were compatible in some way could cross and mix their genes. Nowadays there are large collections of ecotypes and varieties for many species conserved in germplasm banks. Individual genotypes can be ordered by researchers and breeders across the world and used to create new varieties and new species. The cultivated strawberry is a perfect example of this, being the result of an accidental cross between the white-fruited Chilean strawberry [*F. chiloensis* (L.) Mill. subsp. *chiloensis f. chiloensis*] and a North American strawberry (*F. virginiana* Mill. subsp. *virginiana*) that took place in a Royal Botanical Garden in France [77]. Breeders are increasingly using specific genomic regions introgressed from wild relatives into crops to improve productivity and adaptation (*e.g.* for tomato [78, 79]).

One last interesting issue to recall here is that the selection of favourable traits has been done, for most of the time, at the population level and not of individual plants. The genetic identity of each progeny was not preserved as it is done today in modern breeding programs. Therefore it was difficult to identify favourable alleles in bulked seeds. Because of this way, it took a long time for a favourable allele to become established within a population in the early days of domestication. The ability to preserve the genetic identity allowed the selection of pure lines and their controlled mating. Nowadays, the testing of different genotypes in different environments is the key to identify favourable alleles (and combinations thereof). On the other hand, the selection of individual plants as 'founders' of new, superior cultivars, brought up the issue of seed multiplication, but allowed specific traits to become quickly established. This is most evident with transgenic traits (*e.g.* herbicide tolerance or insect resistance), for which a specific insertion event is selected and approved. Subsequently, the trait is transferred to different cultivars using classical breeding. In this way some transgenic events have been bred into hundreds or thousands of varieties (*e.g.* RR in soybean).

Modern Biotechnology As A Form Of Domestication

It is thus difficult to argue that we could not repeat once again the achievements of domestication in present times. Thanks to the powerful means and knowledge accumulated in recent decades, it is easy to predict that achievements will be even greater. What else is modern biotechnology if not a technology that allowed a wave of improved crop (and animal) domestications in a short time? Biotechnology, and especially its application of genetic engineering to crops, has created simple, highly 'rewarding' traits (HT, Bt, virus resistance....), with ample environmental and economic benefits [80 - 82]. Health benefits have been much less than anticipated because the first wave of commercial products focused much more on agronomic traits beneficial to farmers, one way or the other. It would be highly desirable that companies could deliver consumer-friendly traits, especially those related to nutrition and health, the most interesting of which are biofortified crops [83, 84]. Among these, Golden crops (biofortified with pro-vitamin A) are those with the highest potential impact in the developing world [85, 86].

The potential for further benefits is enormous, as the genes which can be incorporated into plants are not limited to those present in compatible species, but any gene conferring a useful trait can be transferred and its tissue specificity and expression level can be tuned as desired by choosing an appropriate promoter. A much more authoritative book argues that "genetic engineering can enhance crop biodiversity as a necessary adjunct to breeding" [263]. There is a valid moral justification for this approach: early farmers waited for the appropriate mutation to strike, but waiting is not imperative as we have the means and the knowledge to recreate similar changes. For instance, unlike maize, millet and wheat, there are no rice varieties able to accumulate carotenoids in the seed endosperm. We know that the carotenoid pathway is present and active in all green plants, because some of the products of this pathway, namely violaxanthin, antheraxanthin, and zeaxanthin, must be present in green tissues [87]. Question is: should one simply wait (years? Centuries? Eons?) until the desired mutation strikes and activates the pathway in rice seeds, while hundreds of thousands of children become blind or die every year? Is it not a lot more sensible to introduce the pathway into the grain by transgenesis and alleviate the sufferings now?

The potential of transgenesis has been amply demonstrated [88] by the long list of traits pursued and obtained: resistance to biotic (insects, herbicides and hence weeds, parasitic weeds, nematodes, fungi, bacteria) and abiotic stresses (viruses, ions, antibiotic or other compounds), toxin and allergen removal, improved composition, nutritional content, yield, plant shape, seedlessness, male sterility (and therefore hybrid seed production), extended shelf life, improved digestibility, processing, and many others [88 - 100]. Several of these traits were not accessible to classical breeding, nor will they be in the close future unless through trangenesis [263]. A further argument in favor of transgenesis as a form of smart breeding is that recombination is not possible over the entire genome, therefore site directed mutagenesis (SDM) and transgenesis may allow the combination of alleles that have been impossible to achieve so far. Moreover, both techniques allow the retention of varietal identity. Hence, it is especially important for vegetatively-propagated crops, but also to avoid lengthy recurrent backcrossing after mating with wild/distant relatives.

The pursuit of all these traits represents conscious form of domestication, some of which tend to counter the weakening effect of other domestication traits, which were alluded to in section 2. Replacing natural pesticides such as cyanogenic glucoside with Bt or dsRNA [92, 101] are successful attempts to make crops more resilient to pests without reverting to natural pesticides toxic to humans or domestic animals. Selectively confounding pests with homeotic mutations is a similar, much safer and/or more environmentally sustainable strategy [102].

The very same transgenic approach can be used for *de novo* domestication of new species, especially when coupled to new breeding technologies and exploiting information from high throughput techniques (*e.g.* sequencing and genotyping). These techniques and some of the most promising plants are discussed in the next sections.

Table 1. Traits of interest for domesticating new crops and the method(s) to obtain them.

Trait	Suitable crop type	Method of choice	Alternative methods
Seed retention	Grain	SDM	Transgenesis, random mutagenesis
Reduced seed dormancy	Grain	SDM	Transgenesis, random mutagenesis
Threshability	Cereals	SDM, transgenesis	Random mutagenesis

(Table 1) contd.....

Trait	Suitable crop type	Method of choice	Alternative methods
Semi-dwarfness	Cereals	SDM, transgenesis	Random mutagenesis
Pest resistance (insect)	Wide range	Transgenesis	SDM
Virus resistance	Wide range	Transgenesis	-
Fungal/bacterial resistance	Wide range	Transgenesis, wide crosses	SDM
Herbicide tolerance	Wide range	Transgenesis, SDM	Random mutagenesis
Apomixis	Grain	Transgenesis	SDM
Parthenocarpy	Fruit	Transgenesis, SDM	Random mutagenesis
Male sterility / hybrid seed	Many	Transgenesis, SDM	Random mutagenesis
Somatic embryogenesis	Many	Transgenesis	Cell/tissue culture
Oil composition	Oil	Transgenesis, SDM	Random mutagenesis
Sugar content	Sugar	Transgenesis	Wide crosses
Yield	All	Wide crosses, transgenesis	SDM
Polyploidy/wide hybrids	Many	Wide crosses, cell culture	Transgenesis, mutagenesis
Shelf life	Many	Transgenesis	SDM

4. OLD AND NEW TRAITS & APPROACHES FOR NEW CROPS

As was discussed in section 1, the domestication of a plant usually requires few, crucial mutations, many of which are recessive. This means that new domestication attempts can exploit once again spontaneous mutation and subsequent selection of useful genotypes. However, we have nowadays the possibility to increase greatly the frequency of mutations by several means and to examine large number of plants with automatic, high-throughput and sophisticated analyses, thus shortening the time required to identify useful traits. More recently, several techniques for site-directed mutagenesis, gene deletion/ inactivation and gene replacement in plant genomes became available (see below). These are potent tools for crop improvement and accelerated novel crop domestication.

Thus, domesticating novel crops should be much easier today, given the deep knowledge of physiological and genetic processes involved in plant growth and development, as well as the differences between crops and their wild relatives. It should even be possible, at least in a number of cases, to predict which genes have to be mutated and the sort of mutation necessary to achieve the desired traits. As an exercise, let us try to discuss which traits would be most desirable (Table **1**) and which plants would be good candidates to domesticate first (Table **2**). Let us begin with the traits.

Table 2. – Examples of plants potentially interesting or already under domestication. [a] "-" indicates that the species is wild and no significant domestication was achieved; "-/+" means is being attempted and "+"achieved at least for some trait; "++" means an advanced stage of domestication.

Plant name	Full Latin name	Domestication stage[a]	Crop type	Products	Targets	Ref
Physic nut	*Jatropha curcas*	-/+	Oilseed	Oil, protein	Reduce/ remove toxins	[209, 210]
Field pennycress	*Thlaspi arvense*	-/+	Oilseed	Oil, protein	Oil yield	[212]
Giant miscanthus	*Miscanthus × giganteus*	-/+	Biomass	Lignocellulose	Biomass yield, digestibility	[232]
Giant reed	*Arundo donax*	-	Biomass	Lignocellulose	Biomass yield, digestibility	[238], this chapter
Blueberry	*Vaccinium corymbosum*	++	Fruit	Berries	Fruit size	[194]
Macadamia	*Macadamia integrifolia*	+	Oilseed	Nuts	Yield,	[195]
Potato bean	*Apios americana*	+	Starchy tuber	Starch, protein	Yield, more compact	[190]
Weeping grass	*Microlaena stipoides*	-/+	Cereal grain	Starchy seeds	Yield, grain size	[186]
Switchgrass	*Panicum virgatum*	-/+	Biomass	Lignocellulose	Biomass yield, digestibility	[223]

Two Plus Two Fundamental Traits

Seed Dispersal And Seed Dormancy

Turning an undomesticated plant into a grain or an oil crop, requires the two fundamental traits mentioned in section 1: seed retention and reduced seed dormancy; for a cereal crop, threshability [21] and reduced height [56] would be desirable additions. Target genes for loss of ear/pod shattering have been identified in several species and sometimes the mutations strike the same genes or even the same nucleotide in different species [12, 14, 17, 19, 25, 73]. Such genes and positions would then be good candidates for accelerated domestication. The mean of choice would be site directed mutagenesis using CRISPR/Cas9, because many mutations of domestication are single/few nucleotide changes. Changes in flowering time or inflorescence size would also be easily achievable [12] and transgenesis would be the method of choice for this trait. If transgenesis keeps being rejected or opposed in a few countries, then SDM could be used, at least in some cases. This is a pity, as for some traits there are already proof of principles (*e.g.*, loss of shattering in Brassica [103 - 105]): so there is no scientific justification to reject a proven approach in favor of unproved ones, especially if the trait achieved is the same and it confers a disadvantage or no advantage compared to wild relatives (see next paragraph).

Herbicide Tolerance And Mitigation

Another desirable trait, especially for annual crops, is herbicide tolerance. It simplifies weed management and, in certain cases, it allows fighting parasitic weeds [106] which are serious pest in several places. This successful trait was achieved not only trough transgenic technology [107], but also through spontaneous mutation as well as random and site directed mutagenesis. Indeed, non-transgenic HT crops [108] have been and are still grown on millions of hectares: rice, maize, soybean, sunflower, lentil, just to name the most important ones.

This does not mean that the products of genetic engineering are without problems, but these are not peculiar to the technology. For instance, gene flow from HT crops takes place also with non-transgenic varieties obtained through mutagenesis [108]. Indeed, imidazolinone-resistance has already escaped into con-specific weedy rice in several countries (Costa Rica, Brasil, US) [109] and recently also in northern Italy [110]. Similarly, any pesticide (herbicide, insecticide, fungicide, *etc.*) or approach will select resistant individuals, if given enough time or a population large enough. Thus, the appearance of resistant weeds is not a surprise, but rather a predictable outcome, especially when good agricultural practices such as crop rotation or change of active principles are not followed. However, it must be pointed out that gene flow, even though it increases the competitiveness of weeds, it does so in cultivated fields. It is therefore not an ecological but rather an agronomic problem. In fact, extensive gene flow to weedy relatives makes the specific herbicide useless, but it has no effect on wild environments, as herbicides are not sprayed outside cultivated fields and anthropic areas (*e.g.* roadsides). Therefore, gene flow from crops to wild or weedy relatives does not confer any advantage to a plant in the wild and, since it comes from a crop, it is much more likely to confer a disadvantage because of the genetic burden represented by the domestication alleles. Another example of superior versatility and predictability of phenotypes from transgenic approaches is the possibility of mitigating gene flow from HT crops by coupling the resistance with mitigator genes [111 - 113]. In short, a mitigator gene is an allele typical of domestication (*e.g.*, semi-dwarfness, non-shattering or similar) which is normally present in a crop but absent in a wild/weedy relative. By transforming the construct conferring HT with a tightly linked mitigator gene, it is possible to confer a reduced fitness to individuals where gene flow has occurred. Every time the HT character is passed onto the progeny, also the mitigator, penalty-conferring gene is passed onto the same individual, preventing thus a weedy, invasive behaviour to the hybrid. Only when a crossing-over event separates the tolerance gene from the mitigator gene –an extremely rare event as one is next to the other – does gene flow create a weed more difficult to eliminate from fields. Mitigation has proven to be an effective strategy to avoid technology loss and a better alternative to the various containment strategies. However mitigation is impossible with non-transgenic crops.

Pest Resistance: Not Only Insects

The success of Bt technology as an effective and quite durable insect control strategy is beyond doubt [101], but we must get ready with new solutions when pest will evolve resistance to Bt toxins. One option is to look for new biological pesticides whose synthesis can be engineered directly into plants. With such strategy, the delivery of insecticidal compounds (peptides or else) is much more efficient than with sprays and hits only those insects that not only feed on the crop, but which are also sensitive to the compound. One possibility is the accumulation of toxic metabolites [114], that is, small molecules synthesized through specific biochemical pathways, the other one is toxic peptides. Most of these metabolites/peptides are of course natural compounds (*e.g.* spinosad, benzoxazinoids, Cry proteins) whose metabolic pathways or coding regions are known, so the method of choice for transferring them to other species would be transgenesis or by mean of wide crosses. It would be however nonsensical to engineer the accumulation of compounds such as cyanogenic glucosides which upon herbivory produce hydrogen cyanide, a compound of very general toxicity. On the contrary, the engineering of compounds with selective toxicity, such as Bt Cry proteins, is wiser, even though the source of the gene may be a very distant organism and transgenesis the only method available for the transfer. Another strategy is to recruit natural enemies of pests (parasitoids) by attracting them with volatile signal molecules, as already demonstrated for many plants [114 - 116]. It must be noted however that some signals may attract also other pests, so a compromise may be needed [117]. A third possibility is to induce endogenous defences using signal molecules [118]. Yet another strategy has recently become available: the use of dsRNA targeting essential genes of the pest [91, 93]. A technical advance, namely the accumulation of the dsRNA in the plastid rather than in the cytosol, has recently improved the efficacy of the approach [92]. Similar strategies can be discussed for pests such as nematodes, fungi and bacteria [89, 93, 119 - 125]. For several of these approaches, but in particular for virus resistance, the transgenic approach is easy to implement and effective, often it is the only viable strategy against viruses. The knowledge of pest life cycles may allow intelligent solutions and the avoidance of pesticides altogether [102], be they natural or novel. Whatever the strategy, there is always a dilemma between fitness and yield, as there is always a trade-off between resource allocation for growth or defence.

Perenniality

Perenniality is a trait of interest from several perspectives [126 - 128]. Perennial crops prevent soil erosion, improve soil structure, are more efficient in water and nutrient use and allow carbon sequestration. Perennial crops are often more rustic and show early vigor in the season; this translates into an efficient crop establishment and weed growth suppression. They improve sustainability by reducing tillage, expenditure for seed, fertilizer, and labour. Perenniality seems to involve, at least for the limited experience acquired so far, a yield penalty, but it is unclear whether this is a result of a competition in resource allocation (more carbon allocated to the root) or simply the lack of prolonged selection [129]. As stated by DeHaan and van Tassel [130] "*rapid domestication will likely require genes with large effect that are expected to exhibit strong pleiotropy and epistasis*". Thus they are most likely due to rare, dominant, gain-of-function mutations, which are more easily achieved by transgenesis, *e.g.* by overexpressing regulatory genes in new tissues or at a different time. The capacity to shift allocation from vegetative to reproductive structures may be a possible breeding target, as two perennial species were found with a harvest index greater than 0.5 [131] that is typical of highly domesticated crops. Since several crops have perennial relatives, both among the cereals and the legumes [52, 126, 128], there are several groups investigating the possibility of incorporating the trait in crops. The transfer of perenniality and other complex traits such as C4 photosynthesis or symbiosis with N2-fixing bacteria would be remarkable achievements, but these will require a better understanding of the phenomena and the coordinate expression of many genes. As such, they are not short terms goals, even though the knowledge has tremendously improved in past years. Time will tell whether transgenesis is the only way forward.

Composition: Oil, Sugar, Protein And More

If an oil crop were the target for accelerated domestication, then changes in oil composition for the different applications (food, feed, chemical precursor or fuel) and/or protein quality would be highly desirable, besides the obvious fundamental traits discussed above. Successful approaches to achieve a substantial change in amino acid or fatty acid composition were discussed at length in recent reviews [96, 132 - 135, 264], but most of these strategies are dependent on transgenesis and require manipulation of the supply of fatty acids as well as on the stimulation of their incorporation into triacylglycerols [96,136]. Many wild plants produce and accumulate unusual fatty acids, with double or triple bonds, epoxides or rings at specific positions [137]. Several of these unusual fatty acids are very interesting as chemicals feedstocks for the production of precursors for industrial applications, from nylon synthesis to lubricants and detergents [138 - 141] or for nutraceutical products, for instance by increasing the content of healthy fatty acids and reducing those less healthy [134, 141]. Random mutagenesis and RNAi approaches have also been used to achieve some of these traits. The production of many other compounds in crops was achieved or enhanced especially by means of transgenesis [96]. The use of a single gene, for instance, in sugarcane allowed the doubling of sugar content [142].

Seeds Without Fertilization, Sterile Plants And Other Forms Of Propagation

Apomixis (that is reproduction without meiosis and fertilisation) would be another remarkable achievement for crop plants, as it would mean fixation of superior genotypes in outcrossing species and possibly also the maintenance of hybrid vigour through generations, thus avoiding the need for producing F1 seeds every year. There are several examples of plants showing this behaviour and they are heavily investigated (for recent reviews, see [143 - 145]. The combination of several mutants allowed the creation of apomictic seeds (*e.g.* [146]) in model plants. Three genetically independent genes seem to be required for apomixis, so it is likely that a combination of site directed mutations and the use of tissue specific expression of RNAi could achieve the result. However, the transformation of a single gene from *Pennisetum squamulatum* was recently reported to induce apomixis in pearl millet [147]. Another recent paper [148] reports that ectopic overexpression of *BELL1* (an homeobox gene) triggers embryo formation in *Physcomitrella patens* leading to fully functional moss sporophytes.

Hybrid seeds breeding for crop production is very relevant for some plants. They can be easily produced when a plant is an outcrosser, when a plant has separate male and female flowers (on the same or different plants) or when the mother plant is male sterile and must therefore be pollinated by another individual to bear seeds. Seeds of crops such as maize, sugar beet, sunflower, onion and many vegetable crops are often or mainly available as F1 hybrid seeds which are produced by exploiting male sterile mutants (*e.g.* sugar beet) or mechanical removal of the male flowers (maize). Other crops (*e.g.* rice or wheat) have little share of hybrid seed because they normally self-pollinate and sterility systems to block self-pollination are inefficient. Many have created transgenic plants with inducible male sterility which would facilitate hybrid seed production [149 - 152]. In other cases, single mutation achieved with SDM may allow the creation of efficient male-sterile/restorer lines, but this will probably be far more complicate to achieve compared to transgenesis.

Fig. (6). Vegetative propagation of *Kalanchoe delagoense* (Mother of millions). The tip of each leaf bears several plantlets which are the result of a developmental program resembling zygotic embryogenesis.

Another interesting characteristic shared by many plants is the ability to produce new plantlets directly on the leaves (for an example, see Fig. **6**).

The elucidation of the molecular mechanisms in some species [153] suggests that it may be possible to reproduce the phenomenon in crops. If the approaches are replicable and applicable to a wide range of crops, then a new industrial field is going to thrive.

Fruits Without Seeds

Fruit formation can be uncoupled from fertilization and seed development, as demonstrated by seedless fruits (parthenocarpy), which is another interesting phenomenon occurring in more than 400 species. Breeders are trying to obtain more seedless fruits [144, 154, 155] mainly because it is a desirable trait for consumers, but also because partenocarpic fruits have a longer shelf-life. Fruit development is usually dependent on seed formation as seeds are a major source of the phytohormone auxin that promotes fruit growth. The trait is therefore obtained by expressing the auxin synthesising gene iaaM under the control of the DefH9 promoter, which is specifically active in the ovule and the placenta and in the tissues derived from them. This approach has worked in tomato, eggplant, cucumber and raspberry [156] while at least two other transgenic approaches did work for tomato [157, 158] and there is good reason to believe that they should work in other species as well.

Polyploids And Wide Range Hybrids

Polyploidization – whole genome multiplication by combining the same or different genomes – is another interesting feature shared by some important crops: wheat, potato, sugarcane, banana, oilseed rape, triticale and cotton. Often both autopolyploid and allopolyploid are accompanied by wide adaptability, heterosis, robustness and higher yields when compared to the parents (for a review see [58, 159]). Some crops were created from scratch by researchers, such as triticale, whose history begins in 1875 with a cross between wheat and rye performed by a Scottish botanist. Today it grows on 4 M ha, mainly in Europe [160]. Polyploidy is being recognized as an important mechanism for formation of new species and its pathways are better understood and reproduced in the lab [161]. It is now proposed that asexual rather than sexual pathways have been a major source of new polyploid species. The use of grafting in conjunction with cell culture techniques should facilitate the creation of new species by asexual means, however the strategy requires a way to select for the rare events [161].

Wide crosses involve the formation of inter-specific or intergeneric hybrids, that is from different but similar species (within the same genus) or from species of different genera that do not normally mate [159, 162 - 166]. The molecular identification of the Ph1 (Pairing homoeologous 1) gene of *Aegilops speltoides* [167, 168] and the availability of Ph1 mutants, should allow the introgression of many more genes from wild and cultivated relatives, because it is now possible to overcome the block on recombination imposed by the Ph1 gene [169]. The broadening of the gene pool available for breeding is therefore an imprecise, but welcome, substitute of transgenesis. It requires low cost cell culture methods (embryo rescue, somatic cell hybridization, organ regeneration, somatic embryogenesis...) which are affordable to many laboratories and present no opposition from environmental groups and the society at large, at least for now. Indeed, the experience of the past decades with classical methods supports the notion that there have been few problems in manipulating plant genes [170 - 174], but the more so should be expected for more precise and predictable methods.

Exploiting Genome Sequences

Genome sequences are being revealed at an increased rate. It took more than 10 years to get the Arabidopsis genome, but nowadays new sequencing technologies allow the sequencing of much more complex genomes in a fraction of the time and at a fraction of the cost. In the Phytozome database [175] there were almost 25 complete plant genomes available in 2011, while the number of total plant genome sequences is now (2016) around 80, mostly referring to crop species for obvious reasons. The availability of a genome sequence facilitates a number of different experimental approaches, such as mapping insertional mutants, candidate gene identification, cloning and primer design. The use of genome sequence information however finds many applications in crop improvement and breeding [176]. For instance, once that a crop reference genome is made available, the resequencing of other varieties of the same crop or of wild relatives allows the identification of millions of SNPs or thousands of microsatellites that can be used as genetic markers in breeding programs. Once a desirable trait is mapped onto the genome, then more specific genetic markers can be designed to identify individuals bearing favorable alleles long before the favorable trait is expressed in a segregating population. With the availability of genetic markers tightly linked to the gene(s) of interest, it is possible to accelerate breeding programs (the approach is usually called marker assisted selection) and reduce the number of plants to be examined at the phenotypic level. If the resequencing of many varieties is coupled to a phenotypic analysis, then it is possible to do a whole genome association study in order to identify the genetic basis of the phenotypic variations. The identification of millions of SNPs for a species, allowed the development of chips to detect thousands or hundreds of thousands of SNPs with a single hybridization experiment. Thus any variety can be genotyped at all these loci for a small cost. The drop in sequencing costs has made the SNP genotyping by microarray almost obsolete: several approaches are now possible thanks to NGS technologies [177]. NGS allows sequencing different varieties or individuals at a low coverage to identify the genotype of each chromosomal segment. This approach allows SNP discovery and genotyping to happen at the same time; it is expected that mapping by sequencing will become the standard in the future.

The application of NGS is useful for sequence-based association studies [178 - 180], to identify regions of low diversity due to domestication for unraveling the history of a crop and the genomic regions involved [2, 181 - 183], to identify specific genes in a wild relative [184, 185] or for accelerating the domestication of new species [186, 187].

5. NOVEL CROPS: PARADIGMATIC CANDIDATES AND RELEVANT TRAITS

Some examples of plants that could undergo domestication in the near future, or that are already undergoing it, are presented below. The list does not pretend to be exhaustive, rather the contrary, as one can imagine. Many other plants may be suitable for domestication, but no one has ever tried! There are centers dedicated to candidate and emerging crops (*e.g.* Center for New Crops & Plant Products, https://hort.purdue.edu/newcrop/default.html, see also Table S1 from ref [10].), and a book series as well [188], which list hundreds of species, with ample bibliography and other information; a further authoritative book provides several compelling examples of crops and traits [263] and we refer readers to these sources. The crops proposed here are summarized in Table **2** and discussed below. The examples provided want only to be a paradigm for some crop types (cereal, starchy tuber, edible tree/shrub, oil seed, biomass) with the indication of the sort of traits needed and the combination of technologies best suited to achieve them. Some emphasis is placed on biofuel crops for which an accelerated domestication is very much desirable. We shall be happy to receive suggestions for inclusion of other crops in our list, as well as corrections for mistakes.

Paradigmatic Crop Types: Food Crops

Weeping grass (Microlaena stipoides)

This plant is a tetraploid relative of cultivated rice native to Australia. It is a hardy perennial evergreen plant with seeds similar, albeit smaller, to rice. It is being targeted for domestication by random mutagenesis [186] and genome sequencing both of wild type and mutant lines. Several interesting phenotypes related to domestication have been identified [185, 186], most notably several non-shattering or reduced shattering mutants (Fig. **7**).

Fig. (7). A non-shattering mutant of *Microlaena stipoides*. (Image courtesy of Dr. Chivers) (www.nativeseeds.com.au).

While the random mutagenesis seems to work well even in a tetraploid plant, one wonders if using site directed mutagenesis would be more efficient and less laborious for the creation of specific phenotypes, particularly for a close relative of rice where many domestication-related genes have been identified and the relative mutations characterized in detail.

Potato bean (Apios Americana)

It is a nitrogen-fixing legume vine (Fig. **8a**) native to eastern North America which was a significant food source for native North Americans. The starchy and protein-rich tubers found along the rhizome are the main target for harvest (Fig. **8b**). They have been mostly collected from the wild and therefore the plant underwent little or no domestication. However also the seeds, which have a better protein content, can be used as food, but pods tend to break open at maturity. A breeding program for the deliberate domestication of this plant started in the '80s at Louisiana State University (Baton Rouge, LA) to create plants with a significant tuber yield in a single season [189]. A collection of Apios is still maintained and analyzed at Iowa State University [190]. *De novo* sequencing of transcripts allowed the identification of polymorphic markers for building a high-density linkage map [191]. Given these premises, it would be sensible to target a few traits, such as indehiscent pods (shattering resistance), longer pods, larger seeds and a more compact growth habit. Mutations of this kind are already characterized or at least identified in other species ([25, 26, 61, 180, 181, 183, 192, 193]). Making the potato bean a compact plant means that it could grow untrellised and that the tubers would be more closely-spaced, reducing the work at harvest. For some of the traits, random mutagenesis appears to be a sensible approach, while for those mutations already characterized in similar species, SDM or transgenesis would be quicker.

Fig. (8). *Apios americana.* (**a**) Starchy tubers, image courtesy of A. Haines, New England, Wild Flower Society. (**b**) Inflorescence (image courtesy of Daniela Longo) (www.actaplantarum.org).

Blueberry (Vaccinium corymbosum)

Vaccinium corymbosum is a shrub native to North America. Berries were picked since ancient times from the wild, but its domestication begins just in 1911, when the botanist Frederick Coville from the U.S. Dept. of Agriculture demonstrated that blueberries require moist, very acidic soils for germination and growth. Next, he tackled breeding, demonstrating that blueberry is not self-fertile and thus it requires cross-pollination. He crossed different wild plants and created the first hybrids. He also found a way to propagate them vegetatively, allowing the production of thousands of identical plants from a single individual. At this point fruit production could start on a large-scale [194] and become a viable business. Since blueberry is already at an advanced stage of domestication, we propose to follow a similar approach with other berries that could thus be brought into cultivation. The main steps are essentially the same followed by Coville plus a few additions: isolation of ecotypes with good characteristics, mutagenesis and selection for a few key traits (reduced fruit shattering, compact growth habit, yield, fruit quality and resistance to pathogen), hybrid production and vegetative multiplication.

Macadamia (Macadamia tetraphylla)

Macadamia are nut bearing trees of the Australian subtropical rainforest belonging to the Proteaceae family. Two *Macadamia* species underwent domestication recently: *Macadamia integrifolia* and *M. tetraphylla* and their hybrids [195]. The development of methods for vegetative propagation was a major event in the domestication process. The development of an efficient and reproducible regeneration system for macadamia allows the multiplication on a large scale of elite cultivars [196]. The identification, using next-generation sequencing, of 12 polymorphic microsatellite markers [197] will provide the tools for discrimination among macadamia cultivars and the selection of genotypes in breeding programs. Further domestication could target growth habit, yield (which depend on several variables, among which the age of first crop, inflorescence size and number) and nut size and shape [195]. There is little information from other species about relevant domestication genes, but a few candidate genes could be tested by transgenesis, such as general regulators of flowering time and plant architecture (*e.g.* FT, TFL) or resistance to pathogens and pests.

Recent advances in biotechnology made it possible to reduce the number and the duration of breeding cycles required for improving tree crops. On one side, rapid-cycle breeding reduces the time required by different means. Once a cross has been performed, it is possible to induce a rapid germination of seeds, which, once mature, are usually dormant for many months. One possibility is the so called embryo-rescue technique, which envisages removing immature embryos from seeds before dormancy is imposed and culturing them *in vitro* under conditions that promote seedling development. The other possibility is to promote germination of seeds by treatment with phytohormones or nitric oxide (NO). Using a rapid-cycle approach, the length of the breeding cycle was reduced from the usual 8 years to just 3 in a breeding program for the improvement of cassava carotenoid content [198].

Another approach is to reduce the juvenile phase by ectopic expression of flowering inducing activities (*e.g.* FT, LEAFY, APETALA or a MADS box protein [199 - 201]; see van Nocker *et al.* [202] for a review), which also bypasses the chilling requirement for flowering. For instance, the overexpression of the BpMADS4 gene from the European silver birch shortens the breeding cycle to one year [201]. The transgene can be subsequently segregated out of the plant, while maintaining the gene of interest (*e.g.*, pathogen resistance) introgressed through the cross, in order to avoid the excessive regulatory burden imposed on transgenic events. The FT protein is the flowering signal traveling from leaves to shoot apex; this simple biological fact suggests the use of stable transgenic lines expressing FT as a rootstock to drive flowering in a grafted scion. In this way flowering is induced but there is no need for transgene segregation at the end of the process.

The other important biotechnological revolution accelerating tree breeding is MAS, which reduces the number of cycles required and the number of plants to be examined by identifying the plants with the most favorable genotypes at an early stage and discarding the other ones. The screening can be done at the seedling stage and therefore long before the trait of interest is phenotypically revealed. The availability of thousands, if not millions, of genetic markers makes it easy to find always one or a few tightly linked to the genes/alleles of interest. By tracing the presence of the markers, each favorable gene/allele can be followed in breeding programs. A MAS pre-screen employing a few markers can be used to eliminate the undesired genotypes lacking critical traits (for instance pest and disease resistance). Then a denser and costly analysis follows to identify the superior genotypes (*e.g* [203]; see [202] for other references). The two approaches (rapid cycling and MAS) synergize and they significantly shorten the breeding of better tree crops.

Paradigmatic Crop Types: Biofuel Crops

The last decade has seen a rush to obtain energy and fuels from renewable sources. This has been translated into changes in policies (particularly subsidies), research funding, technology development and land use to accelerate the transition. In particular, biofuel crops have gained much hectarage at the expense of food/feed crops. Often the surface was employed for the same crops (mainly maize and, on a lower scale, soybean, rapeseed and sugarcane), what has changed was the use of the primary product (*e.g.* corn). If edible crop products such as corn starch or rapeseed oil would be used for biofuel production, one source calculated [204] that a substantial part of US agricultural land would be needed just to meet the 15% target of blending with biofuel, let alone if the target is complete replacement. At present, more than 30% of the corn grown in the US is converted into ethanol and this is regarded as one major factor for recent food price spikes [205]. One big problem with this approach is that most biofuel production is not (highly) efficient: estimates of the Energy Return On Energy Investment (EROEI) index can vary from 5-8 for ethanol from sugarcane to 3 for biodiesel from soybean and they dangerously approach 1 or even drop below 1 for ethanol from corn or sugar beet. There are significant variations in the reported values, depending on how the calculations are made, but the take-home-message remains valid: the majority of first generation biofuels are not efficient.

The conflict between food and fuel is not an all-or-none dilemma, because byproducts of fuel production are still going to become feed (*e.g.* dry distillers grain or soybean meal) and they have a high value, giving a substantial contribution to profitability of the whole process. However, there is definitely a conflict and the tendency is to try reducing it. Therefore, biofuels need to be obtained from non-food crops (possibly using land unsuitable for food crops) or from inedible part of food crops (*e.g.* cereal straw) that have a negative value. Lignocellulosic biomasses are already produced as agricultural byproducts at a world annual rate of 2 billion tons, but their conversion into fuel is inefficient– around 20-25% at best with present technologies– and expensive, and therefore it would meet only a small part of the world demand; as an example consider that Italy alone requires 100 Mt of mineral oil each year. Thus plant biologists, geneticists and agronomists have to develop better, more productive second generation energy crops able to grow on soil unsuitable for food production. We discuss below some candidate 'second generation' biofuel crops: two candidates for oil production and three for biomass. In all cases we reiterate the plea by Gressel [206] that all useful technologies are needed, if we are going to improve sustainability of renewable fuel production and at the same time achieve the ambitious targets set by international agreements.

According to a source [207], a good second generation biomass crop candidate should be "*high yielding, fast growing,* [...] *and requiring relatively small energy inputs for its growth and harvest*". How good are the biofuel crops in development according to such criteria?

Physic Nut (Jatropha curcas)

One crop repeatedly mentioned as promising, but in dire need of domestication, is *Jatropha curcas*, a stem-succulent deciduous shrub/tree of the Euforbiaceae family. Several sites claim great virtues of the plant, according to which *Jatropha* is a pest resistant plant, able to grow on land unsuitable for food production, with little water and fertilizer, but with high yields. The seeds are indeed rich in oil (35%), which can be converted into biodiesel, and protein. However, this crop has several problematic sides, the most important being the presence of toxic proteins and metabolites: curcin, which is a toxalbumin similar to the highly poisonous ricin (but 'only' half as toxic), as well as phorbol esters (diterpenoid alkaloids) which are known skin tumor promoters, for which there is no known antidote (for references see those cited in [206]). The cultivation and processing of such a toxic material is likely to create a series of problems and risks. The best approach would be to eliminate both toxic proteins and metabolites, which are quite easy goals to achieve when using transgenic approaches or alternatively SDM. In this way the resulting protein meal obtained after oil extraction would be a high value byproduct that could be used as feed. Thus an accelerated domestication of *Jatropha* would require several traits: non-shattering, dwarfing and suppressed branching to improve the harvest index and facilitate fruit harvesting, respectively, and finally yet very important, suppression of curcin and phorbol esters production. However, it is easy to predict that the plant would then become much more prone to insect pests and would therefore require the introduction of other form of resistance (*e.g.* Bt). Maghuly and Laimer [208] also suggest the following additional traits: a reduction in polyunsaturated fatty acids, to improve biofuel quality, and the modification from continuous to more seasonal fruiting, to reduce harvest costs. Many groups have published sequencing programs to elucidate genome structure and its expressed fraction from various tissues (see [208] for references). These data facilitate approaches such as tilling mutagenesis for a quick mutant identification, SNP discovery and candidate gene identification [209]. The accelerated domestication will require also a set of molecular tools to accelerate breeding, such as genetic markers, and explore the genetic diversity and the extent of outcrossing [209, 210]. Several other shrubs share similar characteristics with *Jatropha* and represent likewise attractive target for domestication [206, 211].

Field Pennycress (Thlaspi arvense)

A potential oil-rich crop is field pennycress, a member of the Brassicaceae family and a relative of rapeseed and Arabidopsis. The domestication of field pennycress was discussed in detail by Sedbrook *et al.* [212], with ample details on the accessions, genome and transcriptome sequencing projects [187], molecular tools (transformation, markers) and attractive gene targets. This plant has many attractive traits typical of wild species (high seed oil content, fallow land-tolerant, extreme cold-tolerant, requires little agricultural inputs...), but obviously lacks all the traits typical of domesticated Brassicaceae and namely: reduced fruit shattering and seed dormancy, large seed size, low erucic acid content, more compact growth, synchronized flowering. Given the closeness to Arabidopsis it is easy to identify gene candidates for many traits, for instance SHATTERPROOF1 and 2 for reduced shattering. Readers are referred to Sedbrook *et al.* [212] for a list of traits and potential candidates. Mutants can be created by random mutagenesis and then identified using TILLING, or by transgenesis.

Paradigmatic Crop Types: Biomass Crops

According to Lewandowski *et al.* [213], the best candidates for biomass production in temperate climates are the following rhizomatous, perennial grasses: miscanthus, reed canarygrass, giant reed and switchgrass. We discuss three of these candidates below, with a particular emphasis on giant reed. We refer the reader to recent reviews on the parallel between domestication of food and energy crops [214, 215].

Panicum Virgatum (switchgrass)

Switchgrass (*Panicum virgatum* L.) is a perennial, warm season C4 grass native to North America [216] (Fig. **9**) which has been used as a hay and forage crop for 70 years. Switchgrass taxonomy recognizes two distinct ecotype categories: lowland and uplands that are, respectively, tetra and (mainly) octoploids, and are regarded to be reproductively isolated [217]. The plant begun to be considered as a promising energy crop only recently. At the end of the '80s, the US department of energy (DOE) recognized the potential of switchgrass for biofuel production and developed a project for the intensive study of the crop [218, 219], but the domestication level of switchgrass is still unsatisfactory [220].

The use of EST-SSR molecular marker analysis suggests a great genetic variability among switchgrass germplasm especially within population that represents a potential resource for domestication [221]. The efforts for the development of an improved crop have mostly focused on obtaining higher biomass yield and lower cell wall recalcitrance [222]. Besides traditional breeding, biotechnological approaches have been recently used to accelerate the domestication process. These involve the introduction of foreign genes by transgenesis (with *Agrobacterium tumefaciens* as a vector) to increase the release of fermentable sugars. For example, the overexpression of gibberellin 2-oxidase (GA2ox) results in a plant with modified plant architecture and reduced lignin content [223, 224]. Another involved the silencing of one of the key genes implicates in the monolignol biosynthetic pathway, 4-coumarate: coenzyme A ligase [225], leading to a reduction in lignin content, increased saccharification efficiency, but the same biomass yields.

Fig. (9). Switchgrass (*Panicum virgatum*) grown in greenhouse.

Miscanthus

Miscanthus x giganteus is a perennial C4 grass regarded as promising candidate because of a high productivity and low nutritional requirements [226, 227]. It is a naturally occurring sterile hybrid between the diploid *Miscanthus sinensis* and the tetraploid *Miscanthus sacchariflorus* and it is therefore a triploid. Due to its sterility, it is propagated through rhizomes or *in vitro*. Although the first test of Miscanthus as an energy crop took place more than 30 years ago in Europe [226], the domestication process has just begun [228]. The high natural variation and adaptive plasticity found in the *Miscanthus* genus suggest that there is large room for improvement in the development of hybrid, more productive genotypes [229]. Therefore, Robson *et al.* [228] suggest a two-step strategy as it is done for maize or switchgrass: first the best genotypes are identified within each group and then crosses are performed between individuals belonging to different groups to exploit heterosis. There were also differences between plants at different ploidy levels, with triploid and tetraploid hybrids having the highest yield. To accelerate the process of domestication, the introduction of morphometric selection may be needed in order to rapidly identify high yielding genotypes [228, 230] and the underlying genes/QTLs. Loci previously mapped in sorghum and other cereals can be quickly tested to see if they affect also the same traits in Miscanthus. Once candidate genes are identified, selection could be done using MAS. Transgenic approaches would be useful to add traits such as pest resistance and herbicide tolerance, or to test genes that proved to augment biomass in other species. A particularly important trait to improve is cold tolerance, if the crop is going to be used in northern latitudes [231]. Engler and Jakob [232] discuss at length all the traits desirable for improved Miscanthus genotypes. Transgenesis will be one of the pillar to obtain them, together with SDM and molecular breeding. The genome sequence has recently become available [233 - 234] as well as a genetic map [235]. The combination of many biotechnology tools with plant breeding could make cultivation of Miscanthus profitable in a few years. Other relatives among rhizomatous grasses could be explored (*e.g.* Pampas grass, see Fig. **10**).

Fig. (10). Pampas grass (*Cortaderia selloana*) growing in the Aosta valley (45°35'50.8"N 7°48'11.2"E). It is a perennial rhizomatous grass similar to Miscanthus and switchgrass.

Giant Reed (Arundo donax)

Arundo donax is also a perennial rhizomatous grass (Fig. **11a - d**). However, it is a C3 plant, well adapted to temperate and subtropical climates, and with an excellent productivity (30-40 t dry matter/ha/yr) [236 - 239]. Its culms can grow several meters (Fig. **11a**). It is a very hardy plant, as it requires little fertilizer, it is quite competitive against weeds because it sprouts early in the season from rhizomes (Fig. **11d**) and resists well to pests.

Being perennial, the production can continue for several years, even without addition of fertilizers, and it is able to grow even in semi-dry conditions and saline soils, albeit with lower yields. For an extensive report on the biology of *Arundo*, see [237].

A. donax is a probable allotetraploid or a triploid derived from two similar but different species [240 - 242]. It is a vigorous plant with a large (ca. 0.5 m) inflorescence, which however does not produce seeds, apparently because of defects in ovule and pollen development after meiosis [240]. It is propagated only vegetatively (*via* rhizome or culm fragments), often through human intervention.

Fig. (11). *Arundo donax*. (**a**) A single *Arundo* clump near Chiaravalle abbey, Milan, Italy. (**b**) Inflorescences at various stages of development. (**c**) A young plant. (d) A sprouting rhizome.

Once established it tends to grow as monocultural stands (Fig. **12**) and may become invasive, primarily in riparian habitats where dispersal is mediated by transport of plant parts due to floods [237]. The reproductive biology and the genetics of *Arundo* have profound implications for breeding strategies. The lack of sexual reproduction means that one cannot rely on crosses and segregation to identify new genetic combinations more suitable for human purposes. Any mutation, be it spontaneous or induced, would not be able to achieve homozygosity and therefore only dominant or semi-dominant mutations, which are rare, would be expressed at the phenotypic level. Moreover, its more-than-diploid nature means that it would be even more difficult to knock out genes by classical means than in a sterile diploid plant. On the contrary, new traits will be easy to obtain through transgenesis, by either expressing new proteins or silencing endogenous ones, since both approaches are likely to be dominant or semi-dominant. The same is possibly true for SDM approaches, as the CRISPR/Cas9 approach is reported to target more than one allele.

Fig. (12). A large mature stand of *Arundo donax* near Oristano, Sardinia (39°52'16.8"N 8°37'18.5"E).

Improving the Plant

The production of ethanol from *Arundo* becomes competitive, at present biomass yields and technologies, only if the oil price is above 70$/barrel. This is therefore a system worth of further development, which could be done on several fronts (*e.g.* agronomic, process engineering, *etc.*). The most obvious one to pursue is genetics: giant reed is an undomesticated crop in terms of biomass production. Therefore there should be room for improvements, similarly to what has happened for many of the crops domesticated during the course of history, both in terms of production (yield) and suitability for conversion into fuel [214 - 215, 232].

Factors Limiting Productivity

Other non-perennial crops such as Sorghum (a C4 plant) are being used for biomass production and have comparable growth rates, but the actual final biomass quantity is substantially lower than that of *Arundo* (Fig. **13**) [243]. This is due to the fact that sorghum emerges later in the season and it has a smaller LAI, especially in the early phases. The overall energy balance is therefore much more favorable to *Arundo*, because sorghum needs to be sown every year and has higher input requirements (tilling, water, fertilizer…).

It is known that aboveground productivity of *Arundo* is maximal early in the season, but decreases towards the end of the summer [243]. What is the cause for this 'summer slump' (Fig. **13**) and the inefficient use of intercepted radiation? It is unlikely to be due to abiotic factors (N2, water, light…), as it is observed in different years, despite the application of fertilizers (nitrogen and phosphorous).

Several interpretations are possible, but if one excludes unlikely abiotic factors (heat and/or water stress), then the most likely factor is the reduction in endogenous demand of photosynthate [96, 244]. There is strong evidence that sink strength is a major determinant of plant productivity. As stated by McCormick [244], a "*reduced sink demand will lead to the accumulation of sucrose in the phloem and, consequently, regulate* [inhibit] *sucrose production at the source...*". Once that the growth program is brought to completion with inflorescence development, grain filling and maturation, then the additional sunlight cannot be transformed into biomass. Evidence for this is, for instance, a strong increase in harvested biomass for late flowering or non-flowering genotypes in sorghum, maize, miscanthus and rice [245 - 248] as well as early experiments with sink removal in maize [249], just to cite a few examples. In addition, grain filling seems to be a major drive for sink strength [250]. Transformation of sugarcane with a single bacterial gene led to almost doubling the sugar content [142, 251]; the easiest interpretation is that an increase in demand (sink strength) drives an increase in assimilation.

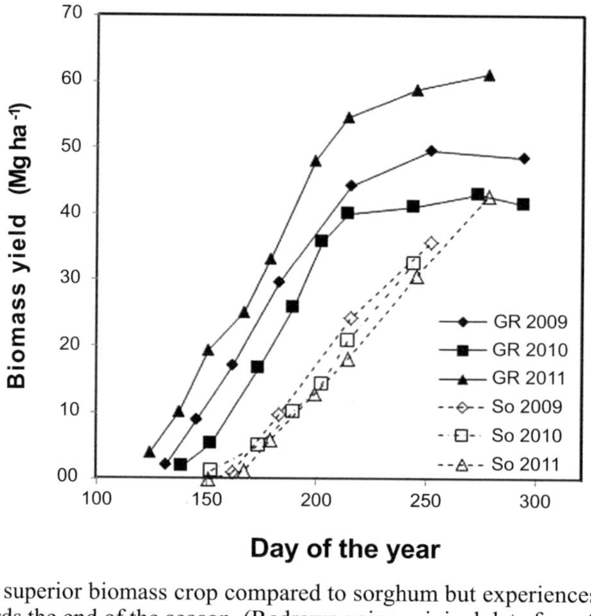

Fig. (13). *Arundo donax* is a superior biomass crop compared to sorghum but experiences a summer slump, a reduced rate of growth towards the end of the season. (Redrawn using original data from Ceotto *et al.*, 2013).

In *Arundo*, sink strength decreases towards the end of the summer because the main aboveground sink, that is the developing vegetative tip, is becoming thinner and thinner and it is eventually replaced by the inflorescence. The small diameter of the tip late in the summer and the lack of seed filling in the inflorescence translate into a weak sink strength compared to the developing vegetative shoot early in the season. An alternative explanation is that assimilation is not reduced, but that translocation to belowground organs (rhizomes or roots) increases, but this is less likely, as new rhizomes are continuously produced and they do not seem to enlarge just at the end of the season. If the summer slump is caused by a reduced demand, then the inhibition of flowering should bring an extension of the vegetative growth phase and an increase biomass.

Thus an increase in biomass seems a feasible target and this should be achieved by delaying flowering as well as, possibly, by accumulation of soluble sugars (similarly to sugarcane) and by triggering seed development even in the absence of meiosis (apomixis, see above).

Factors Limiting Conversion

Other possible targets for an accelerated domestication of *Arundo* are an improvement in the conversion into biofuel or a reduction in inputs (*e.g.* enzymes). It is known that the conversion into ethanol by saccharification and fermentation is still inefficient: the conversion rate is now around 25% of the input dry biomass, while the theoretical maximum is around 40%. This is due to the natural recalcitrance of the plant cell walls to enzymatic hydrolysis [252]. At the same time, the process requires substantial inputs: the energy for pretreatments and other chemical processes (inhibitor removal and end product purification), and the enzymes for the hydrolysis of sugar polymers. The accessibility of cellulose and emicellulose to the hydrolytic enzymes could be improved by expressing proteins which modify pectins [253], while the expression of thermotolerant enzymes, inactive during growth but active during saccharification, would translate into a higher conversion rate or reduced conversion costs. Enzyme costs still represent a significant share of the total costs [254], even though their price dropped considerably in recent years and their actual weight depends on the feedstock as well as the pretreatment.

Improving Transformation

One major technical gap for *Arundo* is the inability to obtain stable transformants. No such protocol has been published so far, although there are single publications describing, respectively, the transient transformation of calli [255], the regeneration of plants from embryogenic calli derived from immature inflorescences [256], as well as the *in vitro* propagation and regeneration from axillary meristems [257]. The last group achieved also the selection of transformed calli with hygromycin, but they were unable to regenerate whole plants [258]. It is thus crucial to set up a reliable protocol for the stable transformation of *Arundo*. Two papers appeared on the sequencing of RNAs from different stages and organs [259, 260] and these data will supply endogenous sequences for candidate genes.

CONCLUDING REMARKS

The domestication syndrome of many crops did evolve over thousands of generations as desirable traits were selected for in the agricultural environment and became fixed within crop genomes. The overall time required to domesticate a species has decreased since the earliest domestication events [10], but it is a never-ending process and one can recognize several phases. The first one entails a fairly rapid fixation through selection of key traits, then a spread of the crop with a long period of minor modifications, variety 'development' and a yield stasis due to inefficient farmer selection. The last, ongoing phase began with the application of modern science (genetics, breeding and more recently cell and molecular biology, among others) which brought spectacular improvements in many traits (yield, resistance, composition…).

Today we have a far better understanding on how plants grow and develop as well as the means to tame more precisely growth and development or to transfer traits across organisms, in order to get more out of our crops. Thanks to the improved understanding, this last phase of domestication promises to accelerate in the near future, bringing new crops into cultivation and further improving existing ones. This large potential however will not be realized without drastic changes in the international and national regulations impacting on transgenesis and novel breeding techniques [86, 261]. The decision to regulate the products of novel breeding techniques, and in particular of a powerful technique such as CRISPR/Cas9, in the same way as it is done for the product of transgenesis, would weaken our ability to (further) domesticate crops in order to feed humans and animals more sustainably.

LIST OF ABBREVIATIONS

Bt	Bacillus thuringensis
Cas9	CRISPR Associated protein 9
CRISPR	Clustered, Regularly Interspaced, Short Palindromic Repeats
HT	Herbicide Tolerant
MAS	Marker Assisted Selection
NGS	Next Generation Sequencing
QTL	Quantitative Trait Loci
RR	Roundup Ready
SDM	Site Directed Mutagenesis
SNP	Single Nucleotide Polymorphism
TILLING	Targeting Induced Local Lesions IN Genomes
wt	wild type

CONFLICT OF INTEREST

The authors confirm that they have no conflict of interest to declare for this publication.

ACKNOWLEDGEMENTS

The picture of the *Apios americana* tubers was kindly provided by Lee Reich, the inflorescence is from Daniela Longo (through the actaplantarum website). The picture of *Microlaena stipoides* was kindly provided by Ian Chivers. Thanks also to Enrico Ceotto for providing the data for Figure 13, to Simona Masiero for help on selected topics, to Jonathan Gressel for suggestions and corrections, to David Horner for polishing the language, to Luca Mizzi for improving picture quality and to Frances Shapter for help on getting the *Microlaena* picture.

This paper is dedicated to Jonathan Gressel who has been a constant source of inspiration with his writing and lectures.

REFERENCES

[1] Larson G, Fuller DQ. The evolution of animal domestication. Annu Rev Ecol Evol Syst 2014; 45: 115-36.
[http://dx.doi.org/10.1146/annurev-ecolsys-110512-135813]

[2] Wang GD, Xie HB, Peng MS, Irwin D, Zhang YP. Domestication genomics: evidence from animals. Annu Rev Anim Biosci 2014; 2: 65-84.
[http://dx.doi.org/10.1146/annurev-animal-022513-114129] [PMID: 25384135]

[3] Larson G, Burger J. A population genetics view of animal domestication. Trends Genet 2013; 29(4): 197-205.
[http://dx.doi.org/10.1016/j.tig.2013.01.003] [PMID: 23415592]

[4] Harlan JR. Agricultural origins: centers and noncenters. Science 1971; 174(4008): 468-74.
[http://dx.doi.org/10.1126/science.174.4008.468] [PMID: 17745730]

[5] Harlan JR. Crops and man Madison: American Soc. Madison, WI: Agronomy 1992.

[6] Gepts P. Crop domestication as a long-term selection experiment. Plant Breed Rev 2004; 24: 1-44.

[7] Pickersgill B. Domestication of plants in the Americas: insights from Mendelian and molecular genetics. Ann Bot (Lond) 2007; 100(5): 925-40.
[http://dx.doi.org/10.1093/aob/mcm193] [PMID: 17766847]

[8] Doebley JF, Gaut BS, Smith BD. The molecular genetics of crop domestication. Cell 2006; 127(7): 1309-21.
[http://dx.doi.org/10.1016/j.cell.2006.12.006] [PMID: 17190597]

[9] Gross BL, Olsen KM. Genetic perspectives on crop domestication. Trends Plant Sci 2010; 15(9): 529-37.
[http://dx.doi.org/10.1016/j.tplants.2010.05.008] [PMID: 20541451]

[10] Meyer RS, DuVal AE, Jensen HR. Patterns and processes in crop domestication: an historical review and quantitative analysis of 203 global food crops. New Phytol 2012; 196(1): 29-48.
[http://dx.doi.org/10.1111/j.1469-8137.2012.04253.x] [PMID: 22889076]

[11] Meyer RS, Purugganan MD. Evolution of crop species: genetics of domestication and diversification.

Nat Rev Genet 2013; 14(12): 840-52.
[http://dx.doi.org/10.1038/nrg3605] [PMID: 24240513]

[12] Lenser T, Theißen G. Molecular mechanisms involved in convergent crop domestication. Trends Plant Sci 2013; 18(12): 704-14.
[http://dx.doi.org/10.1016/j.tplants.2013.08.007] [PMID: 24035234]

[13] Sakuma S, Salomon B, Komatsuda T. The domestication syndrome genes responsible for the major changes in plant form in the Triticeae crops. Plant Cell Physiol 2011; 52(5): 738-49.
[http://dx.doi.org/10.1093/pcp/pcr025] [PMID: 21389058]

[14] Sang T. Genes and mutations underlying domestication transitions in grasses. Plant Physiol 2009; 149(1): 63-70.
[http://dx.doi.org/10.1104/pp.108.128827] [PMID: 19126696]

[15] Konishi S, Izawa T, Lin SY, *et al.* An SNP caused loss of seed shattering during rice domestication. Science 2006; 312(5778): 1392-6.
[http://dx.doi.org/10.1126/science.1126410] [PMID: 16614172]

[16] Li C, Zhou A, Sang T. Rice domestication by reducing shattering. Science 2006; 311(5769): 1936-9.
[http://dx.doi.org/10.1126/science.1123604] [PMID: 16527928]

[17] Arnaud N, Lawrenson T, Østergaard L, Sablowski R. The same regulatory point mutation changed seed-dispersal structures in evolution and domestication. Curr Biol 2011; 21(14): 1215-9.
[http://dx.doi.org/10.1016/j.cub.2011.06.008] [PMID: 21737279]

[18] Zhou Y, Lu D, Li C, *et al.* Genetic control of seed shattering in rice by the APETALA2 transcription factor shattering abortion1. Plant Cell 2012; 24(3): 1034-48.
[http://dx.doi.org/10.1105/tpc.111.094383] [PMID: 22408071]

[19] Lin Z, Li X, Shannon LM, *et al.* Parallel domestication of the Shattering1 genes in cereals. Nat Genet 2012; 44(6): 720-4.
[http://dx.doi.org/10.1038/ng.2281] [PMID: 22581231]

[20] Pourkheirandish M, Hensel G, Kilian B, *et al.* Evolution of the grain dispersal system in barley. Cell 2015; 162(3): 527-39.
[http://dx.doi.org/10.1016/j.cell.2015.07.002] [PMID: 26232223]

[21] Sood S, Kuraparthy V, Bai G, Gill BS. The major threshability genes soft glume (sog) and tenacious glume (Tg), of diploid and polyploid wheat, trace their origin to independent mutations at non-orthologous loci. Theor Appl Genet 2009; 119(2): 341-51.
[http://dx.doi.org/10.1007/s00122-009-1043-0] [PMID: 19421730]

[22] Dvorak J, Deal KR, Luo MC, You FM, von Borstel K, Dehghani H. The origin of spelt and free-threshing hexaploid wheat. J Hered 2012; 103(3): 426-41.
[http://dx.doi.org/10.1093/jhered/esr152] [PMID: 22378960]

[23] Faris JD, Zhang Z, Chao S. Map-based analysis of the tenacious glume gene Tg-B1 of wild emmer and its role in wheat domestication. Gene 2014; 542(2): 198-208.
[http://dx.doi.org/10.1016/j.gene.2014.03.034] [PMID: 24657062]

[24] Simons KJ, Fellers JP, Trick HN, *et al.* Molecular characterization of the major wheat domestication gene Q. Genetics 2006; 172(1): 547-55.
[http://dx.doi.org/10.1534/genetics.105.044727] [PMID: 16172507]

[25] Dong Y, Yang X, Liu J, Wang BH, Liu BL, Wang YZ. Pod shattering resistance associated with domestication is mediated by a NAC gene in soybean. Nat Commun 2014; 5: 3352.
[http://dx.doi.org/10.1038/ncomms4352] [PMID: 24549030]

[26] Funatsuki H, Suzuki M, Hirose A, *et al.* Molecular basis of a shattering resistance boosting global dissemination of soybean. Proc Natl Acad Sci USA 2014; 111(50): 17797-802.
[http://dx.doi.org/10.1073/pnas.1417282111] [PMID: 25468966]

[27] Porch TG, Tseung CW, Schmelz EA, Settles AM. The maize Viviparous10/Viviparous13 locus encodes the Cnx1 gene required for molybdenum cofactor biosynthesis. Plant J 2006; 45(2): 250-63. [http://dx.doi.org/10.1111/j.1365-313X.2005.02621.x] [PMID: 16367968]

[28] Sugimoto K, Takeuchi Y, Ebana K, *et al.* Molecular cloning of Sdr4, a regulator involved in seed dormancy and domestication of rice. Proc Natl Acad Sci USA 2010; 107(13): 5792-7. [http://dx.doi.org/10.1073/pnas.0911965107] [PMID: 20220098]

[29] Liu S, Sehgal SK, Lin M, *et al.* Independent mis-splicing mutations in TaPHS1 causing loss of preharvest sprouting (PHS) resistance during wheat domestication. New Phytol 2015; 208(3): 928-35. [http://dx.doi.org/10.1111/nph.13489] [PMID: 26255630]

[30] Bruno MC. A morphological approach to documenting the domestication of *Chenopodium* in the Andes. In: Zeder MA, Bradley DG, Emshwiller E, Smith BD, Eds. Documenting domestication: new genetic and archaeological paradigms. Berkeley: University of California Press 2006; pp. 32-45.

[31] Lush WM, Evans LT. The seed coats of cowpeas and other grain legumes: structure in relation to function. Field Crops Res 1980; 3: 267-86. [http://dx.doi.org/10.1016/0378-4290(80)90034-9]

[32] Sánchez-Pérez R, Jørgensen K, Olsen CE, Dicenta F, Møller BL. Bitterness in almonds. Plant Physiol 2008; 146(3): 1040-52. [http://dx.doi.org/10.1104/pp.107.112979] [PMID: 18192442]

[33] Pliny the Elder (77) Chapter VII. In Naturalis Historia, Liber XV 2016; 26. http://penelope.uchicago.edu/Thayer/L/Roman/Texts/Pliny_the_Elder/15*.html (Latin version) or http://www.masseiana.org/pliny.htm#BOOK%20XV (English) both accessed on Jan 2016

[34] Nader R, Mathieu-Daudé JC, Deveaux M, Faure K, Hayek-Lanthois M, de Haro L. Child cyanide poisoning after ingestion of bitter almonds. Clin Toxicol (Phila) 2010; 48(6): 574-5. [http://dx.doi.org/10.3109/15563650.2010.492351] [PMID: 20560788]

[35] Sanchez-Verlaan P, Geeraerts T, Buys S, *et al.* An unusual cause of severe lactic acidosis: cyanide poisoning after bitter almond ingestion. Intensive Care Med 2011; 37(1): 168-9. [http://dx.doi.org/10.1007/s00134-010-2029-8] [PMID: 20845029]

[36] Lasch EE, El Shawa R. Multiple cases of cyanide poisoning by apricot kernels in children from Gaza. Pediatrics 1981; 68: 5-7.

[37] Ruangkanchanasetr S, Wananukul V, Suwanjutha S. Cyanide poisoning, 2 cases report and treatment review. J Med Assoc Thai 1999; 82 (Suppl. 1): S162-7. [Case Reports]. [PMID: 10730537]

[38] Ozun J, Glaizal M, Flechelles O, *et al.* Un cas d'intoxication pédiatrique en Martinique par ingestion de feuilles crues de manioc. Ann Toxicol Anal Clin 2014; 3: 169-71.

[39] Shang Y, Ma Y, Zhou Y, *et al.* Plant science. Biosynthesis, regulation, and domestication of bitterness in cucumber. Science 2014; 346(6213): 1084-8. [http://dx.doi.org/10.1126/science.1259215] [PMID: 25430763]

[40] Herrington ME. Intense Bitterness in Commercial Zucchini. Cucurbit Genetics Cooperative Report 1983; 6: 75-6. (http://www.cuke.hort.ncsu.edu/cgc/cgc06/cgc6-38.html accessed on June 2016)

[41] Browning S, Hodges L. Bitterness in zucchini, squash and cucumber. Cucumber Crop Information 2005. (http://www.cuke.hort.ncsu.edu/cucurbit/cuke/cukehndbk/cukebitterness.html acc. Aug 2016)

[42] Wink M. Plant breeding: importance of plant secondary metabolites for protection against pathogens and herbivores. Theor Appl Genet 1988; 75: 225-33. [http://dx.doi.org/10.1007/BF00303957]

[43] Turcotte MM, Turley NE, Johnson MT. The impact of domestication on resistance to two generalist herbivores across 29 independent domestication events. New Phytol 2014; 204(3): 671-81.

[http://dx.doi.org/10.1111/nph.12935] [PMID: 25039644]

[44] Vanderborght T. Evaluation of Phaseolus vulgaris wild types and weedy forms. Plant Genet Resour Newsl 1983; 54: 18-25.

[45] Sotelo A, Contreras E, Sousa H, Hernandez V. Nutrient composition and toxic factor content of four wild species of Mexican potato. J Agric Food Chem 1998; 46: 1355-8. [http://dx.doi.org/10.1021/jf970773f]

[46] Koona P, Osisanya EO, Lajide L, *et al.* Assessment of chemical resistance of wild and cultivated Vigna species to the brown pod-bug *Clavigralla tomentosicollis* Stal (Hem., Coreidae). J Appl Entomol 2003; 127: 293-8. [http://dx.doi.org/10.1046/j.1439-0418.2003.00755.x]

[47] Tattersall DB, Bak S, Jones PR, *et al.* Resistance to an herbivore through engineered cyanogenic glucoside synthesis. Science 2001; 293(5536): 1826-8. [http://dx.doi.org/10.1126/science.1062249] [PMID: 11474068]

[48] Beekwilder J, van Leeuwen W, van Dam NM, *et al.* The impact of the absence of aliphatic glucosinolates on insect herbivory in Arabidopsis. PLoS One 2008; 3(4): e2068. [http://dx.doi.org/10.1371/journal.pone.0002068] [PMID: 18446225]

[49] Steppuhn A, Gase K, Krock B, Halitschke R, Baldwin IT. Nicotines defensive function in nature. PLoS Biol 2004; 2(8): E217. [http://dx.doi.org/10.1371/journal.pbio.0020217] [PMID: 15314646]

[50] Sunilkumar G, Campbell LM, Puckhaber L, Stipanovic RD, Rathore KS. Engineering cottonseed for use in human nutrition by tissue-specific reduction of toxic gossypol. Proc Natl Acad Sci USA 2006; 103(48): 18054-9. [http://dx.doi.org/10.1073/pnas.0605389103] [PMID: 17110445]

[51] Cipollini D, Walters D, Voelckel C. Costs of resistance in plants: From theory to evidence. In: Voelckel C, Jander G, Eds. Annual Plant Reviews 47: Insect-Plant Interactions. Chichester, UK: John Wiley & Sons Ltd 2014; pp. 263-307. [http://dx.doi.org/10.1002/9781118829783.ch8]

[52] Palle SR, Campbell LM, Pandeya D, *et al.* RNAi-mediated Ultra-low gossypol cottonseed trait: performance of transgenic lines under field conditions. Plant Biotechnol J 2013; 11(3): 296-304. [http://dx.doi.org/10.1111/pbi.12013] [PMID: 23078138]

[53] Sherman-Broyles S, Bombarely A, Powell AF, *et al.* The wild side of a major crop: soybeans perennial cousins from Down Under. Am J Bot 2014; 101(10): 1651-65. [http://dx.doi.org/10.3732/ajb.1400121] [PMID: 25326613]

[54] Frary A, Nesbitt TC, Grandillo S, *et al.* fw2.2: a quantitative trait locus key to the evolution of tomato fruit size. Science 2000; 289(5476): 85-8. [http://dx.doi.org/10.1126/science.289.5476.85] [PMID: 10884229]

[55] Bai Y, Lindhout P. Domestication and breeding of tomatoes: what have we gained and what can we gain in the future? Ann Bot (Lond) 2007; 100(5): 1085-94. [http://dx.doi.org/10.1093/aob/mcm150] [PMID: 17717024]

[56] Peng J, Richards DE, Hartley NM, *et al.* Green revolution genes encode mutant gibberellin response modulators. Nature 1999; 400(6741): 256-61. [http://dx.doi.org/10.1038/22307] [PMID: 10421366]

[57] Spielmeyer W, Ellis MH, Chandler PM. Semidwarf (sd-1), green revolution rice, contains a defective gibberellin 20-oxidase gene. Proc Natl Acad Sci USA 2002; 99(13): 9043-8. [PMID: 12077303]

[58] Evans LT. Crop evolution, adaptation, and yield. Cambridge, UK: Cambridge Univ. Press 1993.

[59] Renny-Byfield S, Wendel JF. Doubling down on genomes: polyploidy and crop plants. Am J Bot

2014; 101(10): 1711-25.
[PMID: 25090999]

[60] Liu B, Fujita T, Yan ZH, Sakamoto S, Xu D, Abe J. QTL mapping of domestication-related traits in soybean (Glycine max). Ann Bot (Lond) 2007; 100(5): 1027-38.
[http://dx.doi.org/10.1093/aob/mcm149] [PMID: 17684023]

[61] Zhou Z, Jiang Y, Wang Z, *et al.* Resequencing 302 wild and cultivated accessions identifies genes related to domestication and improvement in soybean. Nat Biotechnol 2015; 33(4): 408-14.
[http://dx.doi.org/10.1038/nbt.3096] [PMID: 25643055]

[62] Blackman BK, Strasburg JL, Raduski AR, Michaels SD, Rieseberg LH. The role of recently derived FT paralogs in sunflower domestication. Curr Biol 2010; 20(7): 629-35.
[http://dx.doi.org/10.1016/j.cub.2010.01.059] [PMID: 20303265]

[63] Liu H, Liu H, Zhou L, *et al.* Parallel domestication of the Heading Date 1 gene in cereals. Mol Biol Evol 2015; 32(10): 2726-37.
[http://dx.doi.org/10.1093/molbev/msv148] [PMID: 26116860]

[64] Yang S, Weers BD, Morishige DT, Mullet JE. CONSTANS is a photoperiod regulated activator of flowering in sorghum. BMC Plant Biol 2014; 14: 148.
[http://dx.doi.org/10.1186/1471-2229-14-148] [PMID: 24884377]

[65] Pin PA, Zhang W, Vogt SH, *et al.* The role of a pseudo-response regulator gene in life cycle adaptation and domestication of beet. Curr Biol 2012; 22(12): 1095-101.
[PMID: 22608508]

[66] White WG, Moose SP, Weil CF, *et al.* Tropical maize: exploiting maize genetic diversity to develop a novel annual crop for lignocellulosic biomass and sugar production. In: Buckeridge MS, Goldman GH, Eds. Routes to cellulosic ethanol, Part 3. New York: Springer 2011; pp. 167-79.

[67] Gómez-Ariza J, Galbiati F, Goretti D, *et al.* Loss of floral repressor function adapts rice to higher latitudes in Europe. J Exp Bot 2015; 66(7): 2027-39.
[PMID: 25732533]

[68] Shrestha R, Gómez-Ariza J, Brambilla V, Fornara F. Molecular control of seasonal flowering in rice, arabidopsis and temperate cereals. Ann Bot (Lond) 2014; 114(7): 1445-58.
[PMID: 24651369]

[69] Zohary D. Unconscious selection and the evolution of domesticated Plants. Econ Bot 2004; 58: 5-10.

[70] Purugganan MD, Fuller DQ. Archaeological data reveal slow rates of evolution during plant domestication. Evolution 2011; 65(1): 171-83.
[PMID: 20666839]

[71] Levi-Strauss C. The savage mind. Chicago: The University of Chicago Press 1966; p. 9.

[72] Abbo S, Lev-Yadun S, Gopher A. The Human Mind as a common denominator in plant domestication. J Exp Bot 2014; 65(8): 1917-20.
[http://dx.doi.org/10.1093/jxb/eru068] [PMID: 24638899]

[73] Paterson AH, Lin YR, Li Z, *et al.* Convergent domestication of cereal crops by independent mutations at corresponding genetic Loci. Science 1995; 269(5231): 1714-8.
[http://dx.doi.org/10.1126/science.269.5231.1714] [PMID: 17821643]

[74] Mangin B, Sandron F, Henry K, *et al.* Breeding patterns and cultivated beets origins by genetic diversity and linkage disequilibrium analyses. Theor Appl Genet 2015; 128(11): 2255-71.
[http://dx.doi.org/10.1007/s00122-015-2582-1] [PMID: 26239407]

[75] Gatto A, De Paola D, Bagnoli F, Vendramin GG, Sonnante G. Population structure of *Cynara cardunculus* complex and the origin of the conspecific crops artichoke and cardoon. Ann Bot (Lond) 2013; 112(5): 855-65.
[http://dx.doi.org/10.1093/aob/mct150] [PMID: 23877076]

[76] Simmonds NW, Ed. Evolution of crop plants. London, UK: Longman 1976; pp. 1-339.

[77] Liston A, Cronn R, Ashman TL. Fragaria: a genus with deep historical roots and ripe for evolutionary and ecological insights. Am J Bot 2014; 101(10): 1686-99. [http://dx.doi.org/10.3732/ajb.1400140] [PMID: 25326614]

[78] Gur A, Zamir D. Unused natural variation can lift yield barriers in plant breeding. PLoS Biol 2004; 2(10): e245. [http://dx.doi.org/10.1371/journal.pbio.0020245] [PMID: 15328532]

[79] Zamir D. Improving plant breeding with exotic genetic libraries. Nat Rev Genet 2001; 2(12): 983-9. [http://dx.doi.org/10.1038/35103590] [PMID: 11733751]

[80] Barfoot P, Brookes G. Key global environmental impacts of genetically modified (GM) crop use 19962012. GM Crops Food 2014; 5(2): 149-60. [http://dx.doi.org/10.4161/gmcr.28449] [PMID: 24637726]

[81] Raymond Park J, McFarlane I, Hartley Phipps R, Ceddia G. The role of transgenic crops in sustainable development. Plant Biotechnol J 2011; 9(1): 2-21. [http://dx.doi.org/10.1111/j.1467-7652.2010.00565.x] [PMID: 21040386]

[82] Brookes G, Barfoot P. Global income and production impacts of using GM crop technology 19962013. GM Crops Food 2015; 6(1): 13-46. [http://dx.doi.org/10.1080/21645698.2015.1022310] [PMID: 25738324]

[83] Pérez-Massot E, Banakar R, Gómez-Galera S, *et al.* The contribution of transgenic plants to better health through improved nutrition: opportunities and constraints. Genes Nutr 2013; 8(1): 29-41. [http://dx.doi.org/10.1007/s12263-012-0315-5] [PMID: 22926437]

[84] Sanahuja G, Farré G, Berman J, *et al.* A question of balance: achieving appropriate nutrient levels in biofortified staple crops. Nutr Res Rev 2013; 26(2): 235-45. [http://dx.doi.org/10.1017/S0954422413000176] [PMID: 24134863]

[85] Beyer P. Golden Rice and Golden crops for human nutrition. N Biotechnol 2010; 27(5): 478-81. [http://dx.doi.org/10.1016/j.nbt.2010.05.010] [PMID: 20478420]

[86] Potrykus I. Unjustified regulation prevents use of GMO technology for public good. Trends Biotechnol 2013; 31(3): 131-3. [http://dx.doi.org/10.1016/j.tibtech.2012.11.008] [PMID: 23273413]

[87] Niyogi KK. Photoprotection revisited: genetic and molecular approaches. Annu Rev Plant Physiol Plant Mol Biol 1999; 50: 333-59. [http://dx.doi.org/10.1146/annurev.arplant.50.1.333] [PMID: 15012213]

[88] Murphy DJ. Overview of Applications of Plant Biotechnology. Handbook of Plant Biotechnology. In: Christou P, Klee H, Eds. Handbook of Plant Biotechnology. New York: Wiley & Sons 2004; pp. 35-48. [http://dx.doi.org/10.1002/0470869143.kc002]

[89] Jo KR, Kim CJ, Kim SJ, *et al.* Development of late blight resistant potatoes by cisgene stacking. BMC Biotechnol 2014; 14: 50. [http://dx.doi.org/10.1186/1472-6750-14-50] [PMID: 24885731]

[90] Jones JD, Witek K, Verweij W, *et al.* Elevating crop disease resistance with cloned genes. Philos Trans R Soc Lond B Biol Sci 2014; 369(1639): 20130087. [http://dx.doi.org/10.1098/rstb.2013.0087] [PMID: 24535396]

[91] Kola VS, Renuka P, Madhav MS, Mangrauthia SK. Key enzymes and proteins of crop insects as candidate for RNAi based gene silencing. Front Physiol 2015; 6: 119. [http://dx.doi.org/10.3389/fphys.2015.00119] [PMID: 25954206]

[92] Zhang J, Khan SA, Hasse C, Ruf S, Heckel DG, Bock R. Pest control. Full crop protection from an

insect pest by expression of long double-stranded RNAs in plastids. Science 2015; 347(6225): 991-4. [http://dx.doi.org/10.1126/science.1261680] [PMID: 25722411]

[93] Saurabh S, Vidyarthi AS, Prasad D. RNA interference: concept to reality in crop improvement. Planta 2014; 239(3): 543-64. [http://dx.doi.org/10.1007/s00425-013-2019-5] [PMID: 24402564]

[94] Koch A, Kogel KH. New wind in the sails: improving the agronomic value of crop plants through RNAi-mediated gene silencing. Plant Biotechnol J 2014; 12(7): 821-31. [http://dx.doi.org/10.1111/pbi.12226] [PMID: 25040343]

[95] Morandini P. Rethinking metabolic control. Plant Sci 2009; 176(4): 441-51. [PMID: 26493133]

[96] Morandini P. Control limits for accumulation of plant metabolites: brute force is no substitute for understanding. Plant Biotechnol J 2013; 11(2): 253-67. [PMID: 23301840]

[97] Wang P, Dudareva N, Morgan JA, Chapple C. Genetic manipulation of lignocellulosic biomass for bioenergy. Curr Opin Chem Biol 2015; 29: 32-9. [PMID: 26342806]

[98] Poovaiah CR, Nageswara-Rao M, Soneji JR, Baxter HL, Stewart CN Jr. Altered lignin biosynthesis using biotechnology to improve lignocellulosic biofuel feedstocks. Plant Biotechnol J 2014; 12(9): 1163-73. [PMID: 25051990]

[99] Nicolia A, Manzo A, Veronesi F, Rosellini D. An overview of the last 10 years of genetically engineered crop safety research. Crit Rev Biotechnol 2014; 34(1): 77-88. [PMID: 24041244]

[100] Newell-McGloughlin M. Nutritionally improved agricultural crops. Plant Physiol 2008; 147(3): 939-53. [PMID: 18612071]

[101] Koch MS, Ward JM, Levine SL, Baum JA, Vicini JL, Hammond BG. The food and environmental safety of Bt crops. Front Plant Sci 2015; 6: 283. [PMID: 25972882]

[102] Kater MM, Franken J, Inggamer H, *et al.* The use of floral homeotic mutants as a novel way to obtain durable resistance to insect pests. Plant Biotechnol J 2003; 1(2): 123-7. [http://dx.doi.org/10.1046/j.1467-7652.2003.00013.x] [PMID: 17147749]

[103] Hossain S, Kadkol GP, Raman R, *et al.* Breeding *Brassica napus* for shatter resistance. In: Abdurakhmonov IY, Ed. Plant Breeding. Rijeka, Croatia: Intech Open Access Publisher 2012; pp. 313-32. [http://dx.doi.org/10.5772/29051]

[104] Kord H, Shakib AM, Daneshvar MH, *et al.* RNAi-mediated down-regulation of SHATTERPROOF gene in transgenic oilseed rape. 3 Biotech 2015; 5: 271-7.

[105] Østergaard L, Kempin SA, Bies D, Klee HJ, Yanofsky MF. Pod shatter-resistant Brassica fruit produced by ectopic expression of the FRUITFULL gene. Plant Biotechnol J 2006; 4(1): 45-51. [http://dx.doi.org/10.1111/j.1467-7652.2005.00156.x] [PMID: 17177784]

[106] Gressel J, Valverde BE. The other, ignored HIV — highly invasive vegetation. Food Security 2009; 1: 463-78.

[107] Rao VS. Transgenic Herbicide Resistance in Plants. Boca Raton, FL: CRC Press 2014.

[108] Tan S, Evans RR, Dahmer ML, Singh BK, Shaner DL. Imidazolinone-tolerant crops: history, current status and future. Pest Manag Sci 2005; 61(3): 246-57. [PMID: 15627242]

[109] Gressel J, Valverde BE. A strategy to provide long-term control of weedy rice while mitigating herbicide resistance transgene flow, and its potential use for other crops with related weeds. Pest Manag Sci 2009; 65(7): 723-31.
[PMID: 19367567]

[110] Busconi M, Rossi D, Lorenzoni C, Baldi G, Fogher C. Spread of herbicide-resistant weedy rice (red rice, *Oryza sativa* L.) after 5 years of Clearfield rice cultivation in Italy. Plant Biol (Stuttg) 2012; 14(5): 751-9.
[PMID: 22443148]

[111] Al-Ahmad H, Gressel J. Mitigation using a tandem construct containing a selectively unfit gene precludes establishment of *Brassica napus* transgenes in hybrids and backcrosses with weedy *Brassica rapa*. Plant Biotechnol J 2006; 4(1): 23-33.
[PMID: 17177782]

[112] Al-Ahmad H, Galili S, Gressel J. Poor competitive fitness of transgenically mitigated tobacco in competition with the wild type in a replacement series. Planta 2005; 222(2): 372-85.
[PMID: 15931502]

[113] Rose CW, Millwood RJ, Moon HS, *et al.* Genetic load and transgenic mitigating genes in transgenic *Brassica rapa* (field mustard) x *Brassica napus* (oilseed rape) hybrid populations. BMC Biotechnol 2009; 9: 93.
[PMID: 19878583]

[114] Birkett MA, Pickett JA. Prospects of genetic engineering for robust insect resistance. Curr Opin Plant Biol 2014; 19: 59-67.
[PMID: 24747775]

[115] Tamiru A, Bruce TJ, Woodcock CM, *et al.* Maize landraces recruit egg and larval parasitoids in response to egg deposition by a herbivore. Ecol Lett 2011; 14(11): 1075-83.
[PMID: 21831133]

[116] Köllner TG, Held M, Lenk C, *et al.* A maize (E)-beta-caryophyllene synthase implicated in indirect defense responses against herbivores is not expressed in most American maize varieties. Plant Cell 2008; 20(2): 482-94.
[PMID: 18296628]

[117] Fantaye CA, Köpke D, Gershenzon J, Degenhardt J. Restoring (E)-β-Caryophyllene production in a non-producing maize line compromises its resistance against the fungus *Colletotrichum graminicola.* J Chem Ecol 2015; 41(3): 213-23.
[PMID: 25893788]

[118] Erb M, Veyrat N, Robert CA, *et al.* Indole is an essential herbivore-induced volatile priming signal in maize. Nat Commun 2015; 6: 6273.
[PMID: 25683900]

[119] Eschen-Lippold L, Landgraf R, Smolka U, *et al.* Activation of defense against Phytophthora infestans in potato by down-regulation of syntaxin gene expression. New Phytol 2012; 193(4): 985-96.
[PMID: 22243492]

[120] Grover A, Gowthaman R. Strategies for development of fungus-resistant transgenic plants. Curr Sci 2008; 84: 330-40.

[121] Collinge DB, Lund OS, Thordal-Christensen H. What are the prospects for genetically engineered, disease resistant plants? Eur J Plant Pathol 2007; 121: 217-31.

[122] Collinge DB, Jørgensen HJ, Lund OS, Lyngkjaer MF. Engineering pathogen resistance in crop plants: current trends and future prospects. Annu Rev Phytopathol 2010; 48: 269-91.
[PMID: 20687833]

[123] Wulff BB, Horvath DM, Ward ER. Improving immunity in crops: new tactics in an old game. Curr Opin Plant Biol 2011; 14(4): 468-76.

[PMID: 21531167]

[124] Song J, Bradeen JM, Naess SK, *et al.* Gene RB cloned from *Solanum bulbocastanum* confers broad spectrum resistance to potato late blight. Proc Natl Acad Sci USA 2003; 100(16): 9128-33. [PMID: 12872003]

[125] Newhouse AE, Polin-McGuigan LD, Baier KA, *et al.* Transgenic American chestnuts show enhanced blight resistance and transmit the trait to T1 progeny. Plant Sci 2014; 228: 88-97. [PMID: 25438789]

[126] Glover JD, Reganold JP, Bell LW, *et al.* Agriculture. Increased food and ecosystem security *via* perennial grains. Science 2010; 328(5986): 1638-9. [PMID: 20576874]

[127] Kantar MB, Tyl CE, Dorn KM, *et al.* Perennial grain and oilseed crops. Annu Rev Plant Biol 2016; 67: 703-29. [http://dx.doi.org/10.1146/annurev-arplant-043015-112311] [PMID: 26789233]

[128] Van Tassel DL, DeHaan LR, Cox TS. Missing domesticated plant forms: can artificial selection fill the gap? Evol Appl 2010; 3(5-6): 434-52. [PMID: 25567937]

[129] Vico G, Manzoni S, Nkurunziza L, Murphy K, Weih M. Trade-offs between seed output and life span - a quantitative comparison of traits between annual and perennial congeneric species. New Phytol 2016; 209(1): 104-14. [PMID: 26214792]

[130] DeHaan LR, Van Tassel DL. Useful insights from evolutionary biology for developing perennial grain crops. Am J Bot 2014; 101(10): 1801-19. [PMID: 25326622]

[131] Bell LW, Ryan MH, Bennett RG, Collins MT, Clarke HJ. Growth, yield and seed composition of native Australian legumes with potential as grain crops. J Sci Food Agric 2012; 92(7): 1354-61. [PMID: 22083564]

[132] Vanhercke T, Wood CC, Stymne S, Singh SP, Green AG. Metabolic engineering of plant oils and waxes for use as industrial feedstocks. Plant Biotechnol J 2013; 11(2): 197-210. [PMID: 23190163]

[133] Wenefrida I, Utomo HS, Linscombe SD. Mutational breeding and genetic engineering in the development of high grain protein content. J Agric Food Chem 2013; 61(48): 11702-10. [PMID: 23869957]

[134] Haslam RP, Ruiz-Lopez N, Eastmond P, Moloney M, Sayanova O, Napier JA. The modification of plant oil composition via metabolic engineeringbetter nutrition by design. Plant Biotechnol J 2013; 11(2): 157-68. [PMID: 23066823]

[135] Bates PD, Stymne S, Ohlrogge J. Biochemical pathways in seed oil synthesis. Curr Opin Plant Biol 2013; 16(3): 358-64. [PMID: 23529069]

[136] Bates PD, Johnson SR, Cao X, *et al.* Fatty acid synthesis is inhibited by inefficient utilization of unusual fatty acids for glycerolipid assembly. Proc Natl Acad Sci USA 2014; 111(3): 1204-9. [PMID: 24398521]

[137] Voelker T, Kinney AJ. Variations in the biosynthesis of seed storage lipids. Annu Rev Plant Physiol Plant Mol Biol 2001; 52: 335-61. [PMID: 11337402]

[138] Thelen JJ, Ohlrogge JB. Metabolic engineering of fatty acid biosynthesis in plants. Metab Eng 2002; 4(1): 12-21. [PMID: 11800570]

[139] Bansal S, Durrett TP. *Camelina sativa*: An ideal platform for the metabolic engineering and field production of industrial lipids. Biochimie 2016; 120: 9-16.
[PMID: 26107412]

[140] Liu J, Rice A, McGlew K, *et al.* Metabolic engineering of oilseed crops to produce high levels of novel acetyl glyceride oils with reduced viscosity, freezing point and calorific value. Plant Biotechnol J 2015; 13(6): 858-65.
[http://dx.doi.org/10.1111/pbi.12325] [PMID: 25756355]

[141] Napier JA, Haslam RP, Beaudoin F, Cahoon EB. Understanding and manipulating plant lipid composition: Metabolic engineering leads the way. Curr Opin Plant Biol 2014; 19: 68-75.
[http://dx.doi.org/10.1016/j.pbi.2014.04.001] [PMID: 24809765]

[142] Mudge SR, Basnayake SW, Moyle RL, *et al.* Mature-stem expression of a silencing-resistant sucrose isomerase gene drives isomaltulose accumulation to high levels in sugarcane. Plant Biotechnol J 2013; 11(4): 502-9.
[http://dx.doi.org/10.1111/pbi.12038] [PMID: 23297683]

[143] Hand ML, Koltunow AM. The genetic control of apomixis: asexual seed formation. Genetics 2014; 197(2): 441-50.
[PMID: 24939990]

[144] Ronceret A, Vielle-Calzada JP. Meiosis, unreduced gametes, and parthenogenesis: implications for engineering clonal seed formation in crops. Plant Reprod 2015; 28(2): 91-102.
[PMID: 25796397]

[145] Barcaccia G, Albertini E. Apomixis in plant reproduction: a novel perspective on an old dilemma. Plant Reprod 2013; 26(3): 159-79.
[PMID: 23852378]

[146] Marimuthu MP, Jolivet S, Ravi M, *et al.* Synthetic clonal reproduction through seeds. Science 2011; 331(6019): 876.
[PMID: 21330535]

[147] Conner JA, Mookkan M, Huo H, Chae K, Ozias-Akins P. A parthenogenesis gene of apomict origin elicits embryo formation from unfertilized eggs in a sexual plant. Proc Natl Acad Sci USA 2015; 112(36): 11205-10.
[PMID: 26305939]

[148] Horst NA, Katz A, Pereman I, Decker EL, Ohad N, Reski R. A single homeobox gene triggers phase transition, embryogenesis and asexual reproduction. Nat Plants 2016; 2: 15209.
[PMID: 27250874]

[149] Chakraborty A, Mitra J, Bhattacharyya J, *et al.* Transgenic expression of an unedited mitochondrial orfB gene product from wild abortive (WA) cytoplasm of rice (*Oryza sativa* L.) generates male sterility in fertile rice lines. Planta 2015; 241(6): 1463-79.
[PMID: 25754232]

[150] Kempe K, Rubtsova M, Gils M. Split-gene system for hybrid wheat seed production. Proc Natl Acad Sci USA 2014; 111(25): 9097-102.
[PMID: 24821800]

[151] Jiang J, Mou T, Yu H, Zhou F. Molecular breeding of thermo-sensitive genic male sterile (TGMS) lines of rice for blast resistance using Pi2 gene. Rice (N Y) 2015; 8: 11.
[PMID: 25844116]

[152] Hawkes T, Pline-Srnic W, Dale R, *et al.* D-glufosinate as a male sterility agent for hybrid seed production. Plant Biotechnol J 2011; 9(3): 301-14.
[PMID: 20678098]

[153] Garcês HM, Koenig D, Townsley BT, Kim M, Sinha NR. Truncation of LEAFY COTYLEDON1 protein is required for asexual reproduction in Kalanchoë daigremontiana. Plant Physiol 2014; 165(1):

196-206.
[PMID: 24664206]

[154] Molesini B, Pandolfini T, Rotino GL, Dani V, Spena A. Aucsia gene silencing causes parthenocarpic fruit development in tomato. Plant Physiol 2009; 149(1): 534-48.
[PMID: 18987210]

[155] Østergaard L, Pandolfini T, Molesini B, Spena A. Parthenocarpy in Crop Plants. In: Østergaard L, Ed. Ann Plant Rev Vol 38: Fruit development and seed dispersal. Oxford, UK: Wiley-Blackwell 2010; pp. 326-45.

[156] Pandolfini T. Seedless fruit production by hormonal regulation of fruit set. Nutrients 2009; 1(2): 168-77.
[PMID: 22253976]

[157] Martínez-Bello L, Moritz T, López-Díaz I. Silencing C19-GA 2-oxidases induces parthenocarpic development and inhibits lateral branching in tomato plants. J Exp Bot 2015; 66(19): 5897-910.
[PMID: 26093022]

[158] Mounet F, Moing A, Kowalczyk M, *et al.* Down-regulation of a single auxin efflux transport protein in tomato induces precocious fruit development. J Exp Bot 2012; 63(13): 4901-17.
[http://dx.doi.org/10.1093/jxb/ers167] [PMID: 22844095]

[159] Mason AS, Batley J. Creating new interspecific hybrid and polyploid crops. Trends Biotechnol 2015; 33(8): 436-41.
[http://dx.doi.org/10.1016/j.tibtech.2015.06.004] [PMID: 26164645]

[160] Randhawa HS, Bona L, Graf RJ. Triticale breeding- progress and prospect. In: Eudes F, Ed. Triticale. New York: Springer International Publishing 2015; pp. 15-32.

[161] Fuentes I, Stegemann S, Golczyk H, Karcher D, Bock R. Horizontal genome transfer as an asexual path to the formation of new species. Nature 2014; 511(7508): 232-5.
[PMID: 24909992]

[162] Goodman RM, Hauptli H, Crossway A, Knauf VC. Gene transfer in crop improvement. Science 1987; 236(4797): 48-54.
[PMID: 17759205]

[163] Friebe B, Jiang J, Raupp WJ, *et al.* Characterization of wheat-alien translocations conferring resistance to disease and pests: current status. Euphytica 1996; 91: 59-87.

[164] Sharma HC. How wide can a wide cross be? Euphytica 1995; 82: 43-64.

[165] Kuzmanović L, Gennaro A, Benedettelli S, Dodd IC, Quarrie SA, Ceoloni C. Structural-functional dissection and characterization of yield-contributing traits originating from a group 7 chromosome of the wheatgrass species *Thinopyrum ponticum* after transfer into durum wheat. J Exp Bot 2014; 65(2): 509-25.
[PMID: 24319256]

[166] Faure N, Serieys H, Cazaux E, Kaan F, Bervillé A. Partial hybridization in wide crosses between cultivated sunflower and the perennial Helianthus species *H. mollis* and *H. orgyalis.* Ann Bot (Lond) 2002; 89(1): 31-9.
[PMID: 12096817]

[167] Bhullar R, Nagarajan R, Bennypaul H, *et al.* Silencing of a metaphase I-specific gene results in a phenotype similar to that of the Pairing homeologous 1 (Ph1) gene mutations. Proc Natl Acad Sci USA 2014; 111(39): 14187-92.
[PMID: 25232038]

[168] Moore G. The control of recombination in wheat by Ph1 and its use in breeding. Methods Mol Biol 2014; 1145: 143-53.
[PMID: 24816666]

[169] Rey MD, Calderón MC, Prieto P. The use of the ph1b mutant to induce recombination between the chromosomes of wheat and barley. Front Plant Sci 2015; 6: 160.
[PMID: 25852713]

[170] Bradford KJ, Van Deynze A, Gutterson N, Parrott W, Strauss SH. Regulating transgenic crops sensibly: lessons from plant breeding, biotechnology and genomics. Nat Biotechnol 2005; 23(4): 439-44.
[PMID: 15815671]

[171] Parrott W. Genetically modified myths and realities. N Biotechnol 2010; 27(5): 545-51.
[PMID: 20609417]

[172] Morandini P. Inactivation of allergens and toxins. N Biotechnol 2010; 27(5): 482-93.
[PMID: 20601271]

[173] Parrott WA, Jez JM, Hannah LC. To be or not to be transgenic. Nat Biotechnol 2012; 30(9): 825-6.
[PMID: 22965047]

[174] Ladics GS, Bartholomaeus A, Bregitzer P, *et al.* Genetic basis and detection of unintended effects in genetically modified crop plants. Transgenic Res 2015; 24(4): 587-603.
[PMID: 25716164]

[175] Goodstein DM, Shu S, Howson R, *et al.* Phytozome: a comparative platform for green plant genomics. Nucleic Acids Res 2012; 40(Database issue): D1178-86.
[http://dx.doi.org/10.1093/nar/gkr944] [PMID: 22110026]

[176] Bolger ME, Weisshaar B, Scholz U, Stein N, Usadel B, Mayer KF. Plant genome sequencing - applications for crop improvement. Curr Opin Biotechnol 2014; 26: 31-7.
[http://dx.doi.org/10.1016/j.copbio.2013.08.019] [PMID: 24679255]

[177] Huang X, Han B. Natural variations and genome-wide association studies in crop plants. Annu Rev Plant Biol 2014; 65: 531-51.
[http://dx.doi.org/10.1146/annurev-arplant-050213-035715] [PMID: 24274033]

[178] Huang X, Wei X, Sang T, *et al.* Genome-wide association studies of 14 agronomic traits in rice landraces. Nat Genet 2010; 42(11): 961-7.
[PMID: 20972439]

[179] Desta ZA, Ortiz R. Genomic selection: genome-wide prediction in plant improvement. Trends Plant Sci 2014; 19(9): 592-601.
[PMID: 24970707]

[180] Zhang J, Song Q, Cregan PB, Jiang GL. Genome-wide association study, genomic prediction and marker-assisted selection for seed weight in soybean (*Glycine max*). Theor Appl Genet 2016; 129(1): 117-30.
[PMID: 26518570]

[181] Han Y, Zhao X, Liu D, *et al.* Domestication footprints anchor genomic regions of agronomic importance in soybeans. New Phytol 2016; 209(2): 871-84.
[PMID: 26479264]

[182] Qin C, Yu C, Shen Y, *et al.* Whole-genome sequencing of cultivated and wild peppers provides insights into Capsicum domestication and specialization. Proc Natl Acad Sci USA 2014; 111(14): 5135-40.
[PMID: 24591624]

[183] Li YH, Zhao SC, Ma JX, *et al.* Molecular footprints of domestication and improvement in soybean revealed by whole genome re-sequencing. BMC Genomics 2013; 14: 579.
[PMID: 23984715]

[184] Henry RJ. Next-generation sequencing for understanding and accelerating crop domestication. Brief

Funct Genomics 2012; 11(1): 51-6.
[PMID: 22025450]

[185] Malory S, Shapter FM, Elphinstone MS, Chivers IH, Henry RJ. Characterizing homologues of crop domestication genes in poorly described wild relatives by high-throughput sequencing of whole genomes. Plant Biotechnol J 2011; 9(9): 1131-40.
[PMID: 21762354]

[186] Shapter FM, Cross M, Ablett G, *et al.* High-throughput sequencing and mutagenesis to accelerate the domestication of *Microlaena stipoides* as a new food crop. PLoS One 2013; 8(12): e82641.
[PMID: 24367532]

[187] Dorn KM, Fankhauser JD, Wyse DL, Marks MD. A draft genome of field pennycress (*Thlaspi arvense*) provides tools for the domestication of a new winter biofuel crop. DNA Res 2015; 22(2): 121-31.
[PMID: 25632110]

[188] Peter KV. Underutilized and underexploited horticultural crops. New Indian Publishing Agency 2007; 1-5.Pitam Pura, New Delhi, India

[189] Reynolds BD, Blackmon WJ, Wickremesinhe E, *et al.* Domestication of *Apios americana.* In: Janick J, Simon JE, Eds. Advances in new crops. Portland, OR: Timber Press 1990; pp. 436-42.

[190] Belamkar V, Wenger A, Kalberer SR, *et al.* Evaluation of phenotypic variation in a collection of *Apios americana*: an edible tuberous legume. Crop Sci 2015; 55: 712-26.

[191] Singh J, Kalberer SR, Belamkar V, *et al.* A transcript-based linkage map in *Apios americana* ("Potato Bean") and synteny analysis with other warm season legumes. Plant and Animal Genome XXIV Conference. January 08-13, 2016; San Diego, CA, USA. 1020.

[192] Kongjaimun A, Kaga A, Tomooka N, Somta P, Vaughan DA, Srinives P. The genetics of domestication of yardlong bean, *Vigna unguiculata* (L.) Walp. ssp. unguiculata cv.-gr. sesquipedalis. Ann Bot (Lond) 2012; 109(6): 1185-200.
[PMID: 22419763]

[193] Isemura T, Kaga A, Tomooka N, Shimizu T, Vaughan DA. The genetics of domestication of rice bean, *Vigna umbellata.* Ann Bot (Lond) 2010; 106(6): 927-44.
[PMID: 20880934]

[194] Schneider SM, Finley JW. Blueberry growing comes to the National Agricultural Library. Agricultural Research Magazine 2011; 59: 14-6.

[195] Hardner CM, Peace C, Lowe AJ. Genetic resources and domestication of *Macadamia.* In: Janick J, Ed. Horticultural Reviews. United States: Wiley-Blackwell 2009; pp. 1-126.

[196] Mulwa RM, Bhalla PL. *In vitro* plant regeneration from immature cotyledon explants of *macadamia* (*Macadamia tetraphylla* L. Johnson). Plant Cell Rep 2006; 25(12): 1281-6.
[PMID: 16858554]

[197] Nock CJ, Elphinstone MS, Ablett G, *et al.* Whole genome shotgun sequences for microsatellite discovery and application in cultivated and wild Macadamia (Proteaceae). Appl Plant Sci 2014; 2(4): 2.
[PMID: 25202615]

[198] Ceballos H, Morante N, Sánchez T, *et al.* Rapid Cycling Recurrent Selection for Increased Carotenoids Content in Cassava Roots. Crop Sci 2013; 53: 1-10.
[http://dx.doi.org/10.2135/cropsci2013.02.0123]

[199] Peña L, Martín-Trillo M, Juárez J, Pina JA, Navarro L, Martínez-Zapater JM. Constitutive expression of Arabidopsis LEAFY or APETALA1 genes in citrus reduces their generation time. Nat Biotechnol 2001; 19(3): 263-7.
[http://dx.doi.org/10.1038/85719] [PMID: 11231561]

[200] Flachowsky H, Le Roux PM, Peil A, Patocchi A, Richter K, Hanke MV. Application of a high-speed breeding technology to apple (*Malus × domestica*) based on transgenic early flowering plants and marker-assisted selection. New Phytol 2011; 192(2): 364-77.
[PMID: 21736565]

[201] Weigl K, Wenzel S, Flachowsky H, Peil A, Hanke MV. Integration of BpMADS4 on various linkage groups improves the utilization of the rapid cycle breeding system in apple. Plant Biotechnol J 2015; 13(2): 246-58.
[PMID: 25370729]

[202] van Nocker S, Gardiner SE. Breeding better cultivars, faster: applications of new technologies for the rapid deployment of superior horticultural tree crops. Hortic Res 2014; 1: 14022.
[PMID: 26504538]

[203] Kumar S, Garrick DJ, Bink MC, Whitworth C, Chagné D, Volz RK. Novel genomic approaches unravel genetic architecture of complex traits in apple. BMC Genomics 2013; 14: 393.
[http://dx.doi.org/10.1186/1471-2164-14-393] [PMID: 23758946]

[204] Chisti Y. Biodiesel from microalgae. Biotechnol Adv 2007; 25(3): 294-306.
[PMID: 17350212]

[205] Martin MA. First generation biofuels compete. N Biotechnol 2010; 27(5): 596-608.
[PMID: 20601265]

[206] Gressel J. Transgenics are imperative for biofuel crops. Plant Sci 2008; 174: 246-63.

[207] Waclawovsky AJ, Sato PM, Lembke CG, Moore PH, Souza GM. Sugarcane for bioenergy production: an assessment of yield and regulation of sucrose content. Plant Biotechnol J 2010; 8(3): 263-76.
[PMID: 20388126]

[208] Maghuly F, Laimer M. *Jatropha curcas*, a biofuel crop: functional genomics for understanding metabolic pathways and genetic improvement. Biotechnol J 2013; 8(10): 1172-82.
[PMID: 24092674]

[209] Carels N. Towards domestication of *Jatropha curcas*. The Integration of Sciences. In: Bahadur B, Sujatha M, Carels N, Eds. Jatropha, Challenges for a New Energy Crop. New York: Springer 2013; pp. 263-99.

[210] Achten WM, Nielsen LR, Aerts R, *et al.* Towards domestication of *Jatropha curcas*. Biofuels 2010; 1: 91-107.

[211] Azam MM, Waris A, Nahar NM. Prospects and potential of fatty acid methyl esters of some non-traditional seed oils for use as biodiesel in India. Biomass Bioenergy 2005; 29: 293-302.

[212] Sedbrook JC, Phippen WB, Marks MD. New approaches to facilitate rapid domestication of a wild plant to an oilseed crop: example pennycress (*Thlaspi arvense* L.). Plant Sci 2014; 227: 122-32.
[PMID: 25219314]

[213] Lewandowski I, Scurlock JM, Lindvall E, Christou M. The development and current status of perennial rhizomatous grasses as energy crops in the US and Europe. Biomass Bioener 2003; 25: 335-61.

[214] Allwright MR, Taylor G. Molecular Breeding for Improved Second Generation Bioenergy Crops. Trends Plant Sci 2016; 21(1): 43-54.
[PMID: 26541073]

[215] Sang T. Toward the domestication of lignocellulosic energy crops: learning from food crop domestication. J Integr Plant Biol 2011; 53(2): 96-104.
[PMID: 21261812]

[216] Parrish DJ, Fike JH. The biology and agronomy of switchgrass for biofuels. Crit Rev Plant Sci 2005; 24: 423-59.

[217] Vogel KP. Switchgrass. In: Moser LE, Sollenberger L, Burson B, Eds. Warm-season (C4) grasses. Madison, WI, USA, ASA Monogr 45 ASA, CSSA and SSSA: Madis American Society of Agronomy 2004; pp. 561-88.

[218] Lowenberg-DeBoer J, Cherney JH. Biophysical simulation for evaluating new crops: the case for switchgrass for biomass energy feedstock. Agric Syst 1989; 29: 233-46.

[219] McLaughlin SB, Kszos LA. Development of switchgrass (*Panicum virgatum*) as a bioenergy feedstock in the United States. Biomass Bioenergy 2005; 28: 515-35.

[220] Antizar-Ladislao B, Turrion-Gomez JL. Second-generation biofuels and local bioenergy systems. Biofuels. Bioprod Bioref 2008; 2: 455-69.

[221] Narasimhamoorthy B, Saha MC, Swaller T, Bouton JH. Genetic diversity in switchgrass collections assessed by EST-SSR markers. BioEnergy Res 2008; 1: 136-4.

[222] Lemus R, Brummer EC, Moore KJ, *et al.* Biomass yield and quality of 20 switchgrass populations in southern Iowa, USA. Biomass Bioenergy 2002; 23: 433-42.

[223] Wuddineh WA, Mazarei M, Zhang J, *et al.* Identification and overexpression of gibberellin 2-oxidase (GA2ox) in switchgrass (*Panicum virgatum* L.) for improved plant architecture and reduced biomass recalcitrance. Plant Biotechnol J 2015; 13(5): 636-47.
[PMID: 25400275]

[224] Do PT, De Tar JR, Lee H, Folta MK, Zhang ZJ. Expression of ZmGA20ox cDNA alters plant morphology and increases biomass production of switchgrass (*Panicum virgatum* L.). Plant Biotechnol J 2016; 14(7): 1532-40.
[PMID: 26801525]

[225] Xu B, Escamilla-Treviño LL, Sathitsuksanoh N, *et al.* Silencing of 4-coumarate:coenzyme A ligase in switchgrass leads to reduced lignin content and improved fermentable sugar yields for biofuel production. New Phytol 2011; 192(3): 611-25.
[PMID: 21790609]

[226] Lewandowski I, Clifton-Brown JC, Scurlock JM, Huisman W. *Miscanthus*: European experience with a novel energy crop. Biomass Bioenergy 2000; 19: 209-27.

[227] Liu W, Yan J, Li JQ, Sang T. Yield potential of *Miscanthus* energy crops in the Loess Plateau of China. GCB Bioenergy 2012; 4: 545-54.

[228] Robson P, Jensen E, Hawkins S, *et al.* Accelerating the domestication of a bioenergy crop: identifying and modelling morphological targets for sustainable yield increase in *Miscanthus.* J Exp Bot 2013; 64(14): 4143-55.
[PMID: 24064927]

[229] Yan J, Chen W, Luo F, *et al.* Variability and adaptability of *Miscanthus* species evaluated for energy crop domestication. Glob Change Biol Bioenergy 2012; 4: 49-60.
[http://dx.doi.org/10.1111/j.1757-1707.2011.01108.x]

[230] Slavov GT, Nipper R, Robson P, *et al.* Genome-wide association studies and prediction of 17 traits related to phenology, biomass and cell wall composition in the energy grass *Miscanthus sinensis.* New Phytol 2014; 201(4): 1227-39.
[PMID: 24308815]

[231] Sage RF, de Melo Peixoto M, Friesen P, Deen B. C4 bioenergy crops for cool climates, with special emphasis on perennial C4 grasses. J Exp Bot 2015; 66(14): 4195-212.
[PMID: 25873658]

[232] Engler D, Jakob K. Genetic Engineering of Miscanthus. In: Paterson AH, Ed. Genomics of the Saccharinae Plant Genetics and Genomics: Crops and Models. New York: Springer 2012; Vol. 11: pp. 255-301.

[233] Ma XF, Jensen E, Alexandrov N, *et al.* High resolution genetic mapping by genome sequencing

reveals genome duplication and tetraploid genetic structure of the diploid *Miscanthus sinensis*. PLoS One 2012; 7(3): e33821.
[PMID: 22439001]

[234] Swaminathan K, Alabady MS, Varala K, *et al.* Genomic and small RNA sequencing *of Miscanthus x giganteus* shows the utility of sorghum as a reference genome sequence for Andropogoneae grasses. Genome Biol 2010; 11(2): R12.
[PMID: 20128909]

[235] Swaminathan K, Chae WB, Mitros T, *et al.* A framework genetic map for *Miscanthus sinensis* from RNAseq-based markers shows recent tetraploidy. BMC Genomics 2012; 13: 142.
[PMID: 22524439]

[236] Angelini L, Ceccarini L, Bonari E. Biomass yield and energy balance of giant reed (*A. donax*) cropped in central Italy as related to different management practices. Eur J Agron 2005; 22: 375-89.

[237] Giessow JH, Casanova J, Leclerc R, *et al.* Arundo donax distribution and impact report. California Invasive Plant Council 2011 March; At http://www.cal-ipc.org/ip/research/arundo/ (accessed June 2016)

[238] Pilu R, Manca A, Landoni M. *Arundo donax* as an energy crop: pros and cons of the utilization of this perennial plant. Maydica 2013; 58: 54-9.

[239] Corno L, Pilu R, Adani F. *Arundo donax* L.: a non-food crop for bioenergy and bio-compound production. Biotechnol Adv 2014; 32(8): 1535-49.
[PMID: 25457226]

[240] Mariani C, Cabrini R, Danin A, *et al.* Origin, diffusion and reproduction of the giant reed (*Arundo donax* L.): A promising weedy energy crop. Ann Appl Biol 2010; 157: 191-202.

[241] Hardion L, Verlaque R, Rosato M, Rosselló JA, Vila B. Impact of polyploidy on fertility variation of Mediterranean *Arundo* L. (Poaceae). C R Biol 2015; 338(5): 298-306.
[PMID: 25908118]

[242] Hardion L, Verlaque R, Baumel A, *et al.* Revised systematics of Mediterranean *Arundo* (Poaceae) based on AFLP fingerprints and morphology. Taxon 2012; 61: 1217-26.

[243] Ceotto E, Di Candilo M, Castelli F, *et al.* Comparing solar radiation interception and use efficiency for the energy crops giant reed (*Arundo donax* L.) and sweet sorghum (*Sorghum bicolor* L. Moench). Field Crops Res 2013; 149: 159-66.

[244] McCormick AJ, Watt DA, Cramer MD. Supply and demand: sink regulation of sugar accumulation in sugarcane. J Exp Bot 2009; 60(2): 357-64.
[PMID: 19050062]

[245] Murphy RL, Klein RR, Morishige DT, *et al.* Coincident light and clock regulation of pseudoresponse regulator protein 37 (PRR37) controls photoperiodic flowering in *sorghum.* Proc Natl Acad Sci USA 2011; 108(39): 16469-74.
[PMID: 21930910]

[246] Colasanti J, Yuan Z, Sundaresan V. The indeterminate gene encodes a zinc finger protein and regulates a leaf-generated signal required for the transition to flowering in maize. Cell 1998; 93(4): 593-603.
[PMID: 9604934]

[247] Jensen E, Robson P, Norris J, *et al.* Flowering induction in the bioenergy grass *Miscanthus sacchariflorus* is a quantitative short-day response, whilst delayed flowering under long days increases biomass accumulation. J Exp Bot 2013; 64(2): 541-52.
[PMID: 23183254]

[248] Wu C, You C, Li C, *et al.* RID1, encoding a Cys2/His2-type zinc finger transcription factor, acts as a master switch from vegetative to floral development in rice. Proc Natl Acad Sci USA 2008; 105(35): 12915-20.

[PMID: 18725639]

[249] Ceppi D, Sala M, Gentinetta E, Verderio A, Motto M. Genotype-dependent leaf senescence in maize : inheritance and effects of pollination-prevention. Plant Physiol 1987; 85(3): 720-5. [PMID: 16665767]

[250] Preuss SB, Meister R, Xu Q, *et al.* Expression of the *Arabidopsis thaliana* BBX32 gene in soybean increases grain yield. PLoS One 2012; 7(2): e30717. [PMID: 22363475]

[251] Wu L, Birch RG. Doubled sugar content in sugarcane plants modified to produce a sucrose isomer. Plant Biotechnol J 2007; 5(1): 109-17. [PMID: 17207261]

[252] Himmel ME, Ding SY, Johnson DK, *et al.* Biomass recalcitrance: engineering plants and enzymes for biofuels production. Science 2007; 315(5813): 804-7. [PMID: 17289988]

[253] Lionetti V, Francocci F, Ferrari S, *et al.* Engineering the cell wall by reducing de-methyl-esterified homogalacturonan improves saccharification of plant tissues for bioconversion. Proc Natl Acad Sci USA 2010; 107(2): 616-21. [PMID: 20080727]

[254] Klein-Marcuschamer D, Oleskowicz-Popiel P, Simmons BA, Blanch HW. The challenge of enzyme cost in the production of lignocellulosic biofuels. Biotechnol Bioeng 2012; 109(4): 1083-7. [PMID: 22095526]

[255] Dhir S, Knowles K, Pagan CL, Mann J. Optimization and transformation of *A. donax* using particle bombardment. Afr J Biotechnol 2010; 9: 6460-9.

[256] Marton L, Czako M. Method for micropropagation of monocots based on sustained totipotent cell cultures. US 2011; p. 7863046.

[257] Takahashi W, Takamizo T, Kobayashi M, Ebina M. Plant regeneration from calli in giant reed (*A. donax*). Grassland Sci 2010; 56: 224-9.

[258] Takahashi W, Takamizo T. Molecular breeding of grasses by transgenic approaches for biofuel production. In: Ozden Çiftçi Y, Ed. Transgenic plants - Advances and limitations. Rijeka: InTech 2012; pp. 91-116.

[259] Sablok G, Fu Y, Bobbio V, *et al.* Fuelling genetic and metabolic exploration of C3 bioenergy crops through the first reference transcriptome of *Arundo donax* L. Plant Biotechnol J 2014; 12: 554-67.

[260] Barrero RA, Guerrero FD, Moolhuijzen P, *et al.* Shoot transcriptome of the giant reed, *Arundo donax.* Data Brief 2015; 3: 1-6. [PMID: 26217707]

[261] McHughen A. Introduction to the GM crops special issue on biosafety, food and GM regulation. GM Crops Food 2012; 3(1): 6-8. [PMID: 22614635]

[262] Gressel JB, Ed. Crop Ferality and Volunteerism. Boca Raton, Florida: CRC Press 2005.

[263] Gressel JB. Genetic glass ceilings - transgenics for crop biodiversity. Baltimore, MD: Johns Hopkins University Press 2008.

[264] Sands DC, Morris CE, Dratz EA, Pilgeram A. Elevating optimal human nutrition to a central goal of plant breeding and production of plant-based foods. Plant Sci 2009; 177(5): 377-89. [PMID: 20467463]

CHAPTER 10

Breeding for Drought Stress Resistance in Plants

Massimo Galbiati*

Department of Biosciences, Università degli Studi di Milano, Milano, Italy

Abstract: Most cropping environments worldwide are suboptimal for plant growth and reproduction. Unfavorable environmental conditions prevent crops from attaining their full yield potential. Abiotic stresses including heat, cold, drought, salinity, and flooding have a major impact on world agriculture, reducing by 50% the average yield for most crop plants. Among these, water scarcity is the major factor limiting the expansion of agriculture and the single leading cause of crop losses worldwide. With the impact of climate change and demographic growth looming, stress-tolerant varieties and climate-resilient crops emerge as relevant and necessary targets to ensure global food security and to improve sustainability in agriculture. Breeding of crops with enhanced stresses tolerance has been particularly compelling, as the related agronomic traits are complex and quantitative in nature, often associated to several loci exhibiting additive effects. Advances in plant genomics have greatly contributed to dissect such complex traits, unraveling the mechanisms underlying the plant response to stress, and opening unprecedented avenues in breeding improved varieties against unfavorable environmental conditions. This chapter will focus on recent successful stories in molecular breeding and biotechnological strategies for crop improvement against abiotic stress, with particular emphasis on drought tolerance.

Keywords: Climate change, Crop biotechnology, Crop breeding, Crop yield, Drought, Drought-resistance, Food security.

INTRODUCTION

Plant growth and productivity are constantly challenged by stressful conditions originating from the interaction with pathogenic insects and microorganisms (biotic stress) and from the physical environment (abiotic stress). Altogether abiotic factors, including drought, salinity, heat, cold, flood and heavy metals, account for up to 80% of annual yield losses in agriculture [1]. Among them, drought is thus far the single leading cause of crop losses worldwide [2]. Nearly 40% of the world's land surface experience precipitation that is largely insufficient for agricultural practices. Areas subjected to predictable 'dry seasons',

* **Corresponding author Massimo Galbiati**: Department of Biosciences, Università degli Studi di Milano, Milano, Italy; Tel: +39.02.50315012; Fax: +39.02.50315044; E-mail: massimo.galbiati@unimi.it

Roberto Pilu and Giuseppe Gavazzi (Eds.)

as many developing regions in Africa and Asia, can suffer annual yield loss up to 50% or more [3]. Importantly, high yielding cropping environments can also endure periods of water scarcity of variable duration. According to the "US Department of Agriculture" (USDA), the drought encountered in the Corn Belt region in the summer of 2012, reduced maize yields of 21% compared with the 2009-2011 mean levels (http://quickstats.nass.usda.gov/). Ongoing climatic changes are expected to further limit the availability of water for agricultural needs. Global warming will enhance evapotranspiration losses and decrease rainfall in many agricultural contexts, severely affecting staple crops that are essential to food-security [4, 5]. The increased frequency of concurrent dry and warm conditions will also threat food production in the most developed regions of the planet, as demonstrated by the acute environmental, agricultural and economic impacts associated with the 2012–2014 drought in California [6].

In this scenario, access to irrigation is becoming increasingly important to sustain crop productivity and food production. In the period 1998-2002, irrigated cereals yielded 442 Mg km^{-2} on average, whereas rainfed cereals only attained 266 Mg km^{-2} [7]. Data indicate that "without irrigation, global cereal production on irrigated land is anticipated to decrease by 47%, corresponding to a 20% loss of total cereal production" [7]. Currently, agriculture uses over 70% of the total human consumption of fresh water, with an increase of over 6-fold in the past 100 years [8]. Following this pattern, agricultural demands for fresh water are predicted to double by 2030 [9]. Yet, sustainability issues question any additional expansion of crop irrigation. Globally, we experience severe shortages of freshwater, lowering of underground water reservoirs and competition for dwindling resources [10]. Further, with increasing irrigation comes the threat of salinization of field soil, a problem that is already affecting crop productivity in many regions of the world [11].

Under rainfed conditions, the adoption of suitable water management strategies can significantly improve agricultural yields, especially where water limitations seem to be strongest. For instance, the introduction of relatively inexpensive small-scale water harvesting methods (*i.e.* terracing, dams and ditches), together with conservation tillage in several parts of Africa, has increased yield stability and raised plant productivity from 1 ton per hectare to 3–4 tons [12]. Improved management coupled with a better deployment of existing crop varieties is expected to help in closing yield gaps across many developing regions of the planet. Concurrently, the continued development of high yielding/stress-tolerant varieties and of climate resilient crops is an urgent and essential target to reduce crop losses and increase potential yields into the future. Indeed, breeding approaches to select crop plants with enhanced yield under stress have thus far met with limited success. This is in part due to the quantitative nature of tolerant traits, with low heritability and significant genotype-by-environment (G×E) interaction [13]. Additionally, breeding for drought resistance is hindered by the plant phenology (*e.g.* reproductive phases are generally more sensitive to drought compared with vegetative development) and by the simultaneous exposure of the plant to concurrent stress conditions (*e.g.* drought often occurs in conjunction with high temperatures and high irradiance) [14].

This chapter will critically address the main questions that challenge the breeding of stress-tolerant traits in crops, while providing examples of success stories in delivering drought-resistant varieties to farmers worldwide.

WHAT IS "DROUGHT"? WHAT IS "DROUGHT RESISTANCE"?

The definitions of "drought" and "drought resistance" are not necessarily univocal. The perception of drought varies substantially among agronomists, meteorologists, hydrologists and economists (see Table **1** for definition of drought and drought-related traits). What is relevant in a cropping environment is the amount of moisture present in the soil, readily accessible to plants at a particular developmental stage. Drought develops when soil moisture is insufficient to cope with the requirement of the crop. In natural environments, drought resistance is essentially associated with the ability of plants to survive dry conditions, regardless of their capability to bear fruits and seeds. Clearly, plant survival *per se* is a trait of little value in an agricultural context. Rather, drought resistance in crops must be considered in terms of **yield** in relation to a limiting water supply.

Even brief periods of moderate drought can significantly affect crop yields, as most land plants, crops persistently lose water from their surfaces due to disequilibrium with the atmosphere. Under fully hydrated conditions, plant water potential (Ψw) for mesophyte vegetation, ranges from -0:3 to -0:5 MPa [15]. Only at very high relative humidity (RH=99.6%) the air water potential (Ψwv=-0.54 MPa) is in equilibrium with plant tissues. Minor reductions in RH result in significant decreases in Ψwv. Even at 99% RH, Ψwv is far below equilibrium, reaching -1:36 MPa. At 90% RH, Ψwv matches-14:2 MPa, whereas at RH below 50% it drops to less than -93:6 MPa [16]. As evident, even under the most favorable cropping conditions plants are constantly exposed to an exceptionally water-demanding environment. When this demand is not met by adequate supply from the soil, plants undergo the net loss of water to the atmosphere with the consequent decline of Ψw and relative water contents (RWC). Under these conditions, the closure of the stomatal pores drastically reduces transpiration and CO_2 uptake and growth is rapidly inhibited. If the water deficit persists, plants experience premature leaf senescence, wilting, desiccation and ultimately death [17].

The extent, both in terms of magnitude and timing, to which plants can sustain water loss, defines the level of their **drought resistance**. Plants can resist drought by: (i) **dehydration avoidance**, (ii) **dehydration tolerance**, or (iii) **drought escape** [22] (Table **1**). **Dehydration avoidance** refers to the ability of plants to maintain high water status and cellular hydration under drought conditions. The vast majority of land plants, including crops, can be considered drought avoiders. By this mechanism cellular functions are marginally exposed to tissue dehydration. Crops avoid dehydration by enhancing water uptake from the soil, by reducing water loss by transpiration, and by preserving a relatively high level of cellular hydration even in the presence of low Ψw [19].

Dehydration tolerance refers to the capacity of maintaining plant functions in a dehydrated state and it is sometimes considered as a second line of defense behind dehydration avoidance [19]. While specialized plant organs, including pollen and seeds, can endure severe stress, tolerance mechanisms are rare in crop plants. Extreme dehydration tolerance is essentially limited to the so-called "resurrection plants", which are capable of outliving considerable water loss from vegetative tissues [20]. Finally, **drought escape** refers to the attitude of developing rapidly and setting seeds before water shortage becomes too severe [21]. In nature, this response in mainly associate to acceleration of flowering and represents a key adaptive trait, which can guarantees reproductive success under unfavorable conditions [22].

Table 1. Drought-related terms definition

Term	Definition
Drought	"Extended period of deficient precipitation compared to the statistical multi-year average for a region that results in water shortage" [18].
Meteorological Drought	"When precipitation departs from the long-term normal" [18].
Agricultural drought	"When there is insufficient soil moisture to meet the needs of a particular crop at a particular time" [18].
Hydrological drought	"When deficiencies occur in surface and subsurface water supplies" [18].
Dehydration	"The loss of water from a cell occurring during drought or water deficit" [19].
Desiccation	"The process whereby all free water is lost from the protoplast" [19].
Drought avoidance	"Avoiding the impact of drought by utilizing adaptations that limit the perception of water deficit by the protoplasm" [19].
Drought tolerance	"The ability to withstand suboptimal water availability by utilizing adaptations that permit metabolism to occur at low water potential" [19].
Drought escape	"The ability of plants to complete their life cycle before stress conditions lead to lethality" [20].

(Table 1) contd.....

Term	Definition
Desiccation tolerance	The ability "to lose all free water from the protoplasm and recover normal metabolism upon rewatering" [19].
Water use efficiency	"The ratio of CO_2 assimilation into the photosynthetic biochemistry to water lost, *via* transpiration, through the stomata" [21].

WHAT TRAIT BREEDING FOR?

Recent advances in plant genomics and genetics, coupled with the development of high-throughput phenotyping platforms and modeling methods, greatly advanced our understanding of the molecular basis of plant responses and plant resistance to drought stress. Large-scale screens identified several drought-responsive genes whose role in stress adaptation has been determined at the molecular and cellular level [23]. Comprehensive analyses of large datasets revealed the involvement of common mechanisms of drought resistance in both dicotyledonous and mono-cotyledonous plants [24]. These include activation of signal transduction pathways, regulation of gene expression, functional protection of proteins, accumulation of osmolytes, synthesis of protective antioxidants and production of scavanger molecules [14]. Yet, there is a striking contrast between the enormous amount of data generated in the laboratory and the limited number of field-testing of improved crops for better performance under stress [25].

Several approaches have been employed to improve drought resistance in crops, ranging from classical breeding, to marker assisted selection (MAS) and genetic engineering (GM). Independently of the breeding strategy adopted, the clear identification of target drought-resistance traits still remains the primary and, perhaps most difficult, tusk for breeders. Adaptive traits and related QTLs or genes, which influence yield under stress, can be essentially classified as **constitutive** or **drought-responsive** [26]. The former are expressed under both stressed and well-watered conditions, whereas the latter are only expressed in response to water shortage.

Yield-Related Constitutive Traits

Paradoxically, the most efficient breeding strategy to increase yield under stress has been the selection for genotypes, which guarantee high yield under optimal growing conditions. Indeed, most of the progress achieved by breeders during the last century derives from the selection of constitutive traits affecting dehydration avoidance [27]. Among yield-related constitutive traits, **plant height**, **grain number**, **grain filling** and **flowering time** have played a key role in sustaining plant productivity in unfavorable cropping environments. Paradigmatic is the example of the semidwarf wheat varieties empirically selected for their superior yield potential under favorable conditions. In comparison with taller varieties specifically developed for dryland conditions, the semidwarf genotypes demonstrate an indisputable yield superiority over a wide range of water-limited environments [28]. Interestingly, mutations in the gibberellic acid-insensitive dwarfing genes *Reduced Hight B1b* and *D1b* (*Rht-B1b* and *–D1b*), also known as the "Green Revolution genes", not only decrease **plant height** but also improve **spike fertility**, thus augmenting **grain number** [29]. Enhanced spike fertility, mainly associated with higher survival of floret primordia is one of most important constitutive trait introgressed in modern wheat varieties leading to higher yield under both optimal cropping conditions and moderate water shortage [30].

Corn hybrids, developed for high yielding environments have also significantly contributed to enhance stress tolerance in this relevant crop. Two main constitutive traits are associated with increased productivity and drought resistance in modern hybrids: (i) the enhanced leaf photosynthesis during **grain-filling**, which results in increased dry matter accumulation, and (ii) the larger **number of kernels** derived from improved partitioning of nutrients to developing ears [31]. Interestingly, a recent and extensive analysis of rice pyramiding lines (PLs) identified efficient partitioning as a key mechanism contributing to both improved yield potential and drought resistance. As a result, the best performing PLs showed heavier grain weight and higher spike fertility under control and stress conditions [32].

Flowering time represents a major constitutive trait targeted by breeders for crop adaptation to water-limited conditions [33]. Constitutive "early flowering" varieties with a shorter vegetative development have been successfully introduced in regions, which experience recurrent terminal droughts at the end of the cropping season. In these environments, water is generally available during vegetative development and early reproduction. Nevertheless, as the cropping season proceeds, precipitations become very scarce and the reproductive cycle is often completed under severe water shortage. Under these circumstances, early flowering allows plants to anticipate the onset of drought, thus ensuring yield. Earliness has been extensively exploited in major cereals cultivated in the Mediterranean regions and Australia, and is becoming increasingly important for several minor crops (*e.g.*, sorgum, pearl millet, chickpea, groundnut), broadly cultivated under rainfed conditions [34]. Ongoing research at the "International Crops Research Institute for the Semi-Arid Tropics" (ICRISAT) indicates that short-cycle chickpea varieties (70–90 days to maturity) have significantly contributed to secure yields in the mild-winter regions of peninsular India [35]. Super-early cultivars have also been developed, allowing for a further extension of chickpea cultivation in the region [36]. Similarly, breeders at the "International Rice Research Centre" (IRRI) have developed an early flowering rice variety targeted for the rainfed upland areas of eastern India, severely affected by terminal droughts [37]. The drought-tolerant breeding line, IR 74371-70-1-1, commonly known as *Sahbhagi Dhan* ("rice developed through collaboration"), has been released in 2009 and rapidly adopted by farmers in South Asia, where it can deliver a yield advantage of over one ton per hectare [38]. Yet, under optimal growing conditions constitutive early flowering can result in yield penalty, as compared to late flowering. For instance, in many Mediterranean regions, years with adequate precipitations throughout the main cereal-growing season alternate with years characterized by rather severe terminal droughts, as well as with years with early drought occurring during vegetative growth or flowering [17]. The fine-tuning of flowering time is thus essential to synchronizing reproductive development with the most favorable conditions occurring in a specific environment. In cereals, breeding for flowering time traits mainly exploited the broad natural allelic variation occurring at the major flowering loci, *VERNALIZATION 1-3* (*VRN1–3*) and *PHOTOPERIOD-1* (*Ppd1*) [39]. Several additional loci and QTLs involved in the control of flowering time have been reported for barley and wheat, which represent an, as yet, unexploited source of breeding targets [40].

Drought-Responsive Traits

From an historical perspective, breeding strategies based on the identification of relevant drought-responsive traits and their introgression in high yielding elite germoplasm have only marginally contributed to the development of tolerant crops. These simplistic strategies did not take into proper account the trade off between yield and resistance observed in most crops. Indeed, selection for sustained yield under water-limited conditions mostly resulted in dehydration-tolerant phenotypes (*e.g.* reduced growth, reduced leaf area, enhanced stomatal closure) that are hardly reconcilable with high yield potential. Nevertheless, evidence indicates that high yield potential and stress resistance are not mutually exclusive when specific tolerant-traits are selectively introduced into elite germoplasm [27]. The genetic improvement of **water use efficiency** (WUE) in wheat represents a successful example for breeding of adaptive physiological traits into high yield genotypes. WUE refers to the amount of water exchanged for unit of production. As breeding target WUE is expressed in terms of harvested product attained from the total amount of water supplied to plants throughout the cropping season (**agronomic WUE**). Selection for higher WUE under limited water conditions has frequently led to a genetic shift towards small plants with decreased transpiration and reduced biomass, which result in severe yield penalty under less severe stress conditions [41]. At the leaf level, WUE reflects the moles of carbon gained in photosynthesis (A) per unit of water lost by transpiration (T) (**instantaneous WUE** or **A/T**). Different technologies, ranging from gas exchange to carbon isotope discrimination ($\Delta^{13}C$), have been developed to estimate A/T under field condition, allowing for large-scale screens of breeding lines. The $\Delta^{13}C$ method has been particularly effective in assisting the selection of wheat genotypes characterized by high A/T values [42]. The subsequent identification of the QTLs associated with this physiological trait and their introgression into elite material finally resulted in the release of "Drysdale" and "Rees", two high-yielding and **drought-tolerant wheat** cultivars [43]. As expected both cultivars showed a significant yield advantage over other wheat varieties in the drought-prone and low-yielding environments (c. 2–3 t ha^{-1}) of Southern Australia. Most importantly, "Drysdale" and "Rees" did not carry yield penalty in the more favorable cropping regions of Western Australia nor in the high-yield environments of Eastern Australia (c. 6 t ha^{-1}) [43]. From the pioneering work of Farquhar and colleagues, advances in high-throughput phenotyping (phenomics), Marker-Assisted Selection (MAS) and genomic analyses, greatly facilitated the selection of physiological traits for sustained yield under water shortage, the identification of sources of genetic variation for these traits and the transfer of desirable alleles into high yielding genotypes. Managed-environment (ME) technologies have been established, allowing for reliable measurement of drought-related physiological traits in hundreds of genotypes under appropriate field conditions [44]. Precision phenotyping of several individuals and meta-analysis of large datasets allow dissecting the overall plant tolerance into single causative traits that can be independently evaluated. Mathematical methods have been developed for modeling the comparative advantage of a given trait or allele on crop yield in a multiplicity of environmental scenarios that are difficult to handle experimentally [45]. Importantly, model-assisted phenotyping greatly assists the statistical analysis of the complex Genotype x Environment x Management (GxExM) interactions, thus enabling the assessment of a large network of field conditions [46]. Finally, key genomic technologies including next-generation sequencing (NGS) methods, genotyping-by-sequencing (GBS) approaches and genome-wide association studies (GWAS), significantly advanced the definition of the genetic architecture of physiological traits, providing a vast array of molecular markers, which facilitate mapping of selected traits and assist their introgression and pyramiding into elite germoplasm.

The recently released AQUAmax®**drought-tolerant maize** hybrids (Pioneer), represents a successful case study of advanced breeding for stress tolerance in elite germoplasm [47]. AQUAmax® hybrids are the result of a decadal targeted drought breeding effort which integrated state-of-the-art genomic, phenotyping and modeling technologies with germoplasm selection and the clear definition of the product targets. Multiple testing under field conditions, demonstrated a significant yield improvement over standard drought-tolerant corn hybrids, in a wide range of target environments with different management practices [47]. In

2014 AQUAmax® hybrids were planted on about 10 million acres in the U.S., up from about 7 million in 2013. Remarkably, during the extreme drought in 2012 which affected many corn growing regions, when most hybrids yielded 40 bushel per acre on average, AQUAmax® hybrids yielded up to 70 bushels per acre [48]. Clearly, the long-term genetic gain offered by these hybrids indicates the feasibility of breeding of high yielding genotypes with superior performance under both optimal and unfavorable cropping conditions.

WHAT CONTRIBUTION FROM PLANT BIOTECHNOLOGY?

Biotechnology traits were first introduced in agriculture nearly 20 years ago to enhance tolerance to herbicide and insects. The first generation of biotech crops greatly contributed to improve yield protection while reducing the amount and the cost of chemical inputs [49]. Since the release of the first herbicide- and insect-resistant varieties, both industry and academia demonstrated a rising interest in drought tolerance as a biotechnological target. A total of 117 field trials for GM-drought resistant crops were approved in 2013 by U.S. regulatory authorities alone, up from just 29 in 2004 [50].

Biotechnological strategies for plant stress tolerance mainly rely on the ectopic expression of genes involved in different aspects of cellular homeostasis. These include genes coding for: (i) osmoprotectants, molecular chaperones or antioxidants, (ii) signal transduction components, such as kinases, phosphatases and protein modification enzymes, in addition to (iii) stress-responsive regulatory genes (*e.g.* transcription factors), which simultaneously coordinate multiple stress responses [51]. Recent evidence also indicates a role for non-coding RNAs in the plant adaptation to drought, thus providing exciting novel targets for engineering stress responses in crops [52]. Along the path to discovery of new target genes, significant progress has been achieved in the development of useful *cis*-acting regulatory regions, which can precisely modulate the temporal and spatial expression of the transgenes. As opposite to constitutively active promoters (*e.g.* *35SCaMV*), tissue-specific or stress-regulated promoters allow for proper expression of stress-related genes, avoiding unwanted side effects on plant growth and productivity. In particular, regulatory modules, which integrate tissue /cellular specificity with activation by environmental stress (*e.g.* drought) and/or internal signals (*e.g.* ABA) can be particularly effective in tailoring the plant response to the prevailing climatic conditions [53].

Despite advances in the identification of lead genes and suitable promoters, progress towards commercialization of biotech tolerant traits has been exceedingly slow. Several factors hampered the translation of results obtained from model species in the laboratory to field crops. Most stress-related candidate genes reported in the literature, have only been evaluated in transgenic model plants with poorly designed experiments, often lacking detailed characterization of yield parameters [54]. Even when a single transgene can confer tolerance in greenhouse experiments, multiple adaptive traits and coordinated mechanisms of drought resistance are often necessary to provide economic yield benefit in cropping environments.

As for conventional breeding, the development of drought-resistant biotech varieties requires extensive field experiments, performed at a large number of sites in target environments over a period of years, and involves the coordination of large multidisciplinary teams of scientists. Clearly, financial and technical resources for testing drought efficacy in the field are limited to most researchers, whereas the actual strict governmental regulations and the very high cost for the environmental release of GM crops often preclude field-testing for most public research institutes [55]. Not surprisingly, the breeding of stress-resistant biotech crops rapidly moved out form universities and public programs into the private sector. It is interesting to note that, in the past 15 years, over 60% of U.S. field trial permits or notifications for stress tolerance were granted to a single multinational company. Two-thirds of such studies involved corn as a target crop, whereas a smaller percentage involved cotton or soybean [50].

Several innovative approaches have been employed to engineering stress tolerance in crops and many GM products are in the research and development pipeline of agbiotech companies [24, 49, 51, 56, 57]. Here we focus on the limited number of stress-resistant GM crop currently released on the market or nearing the final commercial development (Table **2**).

The First Drought-Resistant Biotech Crop

The first biotech trait for drought resistance approved for cultivation is represented by the series of Genuinity® DroughtGard® maize products, commercially released on the U.S.A. market in 2012. The DroughtGard® technology entails the expression of the bacterial **COLD SHOCK PROTEIN B (CspB)** from *Bacillus subtilis* in maize [58]. CSPs proteins are ubiquitously found in all living organisms and are well known to act as RNA chaperones [74]. In bacteria they play an important role in sustaining active growth after stress acclimation by favoring gene transcription and mRNA translation [75]. In a preliminary experiment, Castiglioni and colleagues, demonstrated that constitutive over-expression of *CspB* in Arabidopsis and rice could enhance tolerance to cold, heat and drought stress in greenhouse experiments [58]. Transgenic expression of *CspB* in corn was subsequently demonstrated to improve both vegetative growth and kernel yield under controlled water deficit in field tests. Grain harvest data collected from four field sites revealed that, under stress, yield of the transgenic plants were significantly greater than control lines (7.5% increase on average), with the best performing line showing improvements up to 20% [58]. The *CspB* transgene was successively deployed in three different commercial hybrid backgrounds to investigate the stability of the yield protection trait. Remarkably, all the *CspB* hybrids consistently out-yielded the non-transgenic controls in both vegetative and reproductive stress treatments across four years of testing, at five replicated locations [58]. Comparisons of yield performance in *CspB* and control hybrids cultivated at 12 low yielding dryland-cropping environments without supplemental water, further confirmed the yield benefits provided by the *CspB* trait under prevailing agronomic practices. Importantly, improvements in stress tolerance associated with *CspB* expression did not results in yield penalty in high-yielding environments [58]. Similarly to the previously described drought-tolerant conventional corn hybrids [31], the increased **kernel number** per ear is the primary contributor to yield protection under stress in the lead *CspB* line (MON 87460) [59]. Following drought treatment at the silking stage, MON87460 plants show a concurrent reduction of leaf growth and increase of ear size, compared with an isogenic control hybrid. Analysis of CspB expression along the longitudinal axis of the maize leaf revealed that the recombinant protein preferentially accumulates in the active cell division zone located at the leaf base [59]. A model has been proposed in which the CspB protein functions by transiently down-regulating global gene transcription and translation in actively dividing cells, thus contributing to reduce leaf growth under stress [59]. The transitory reduction in leaf growth observed in MON 87460 is a typical avoidance response, which helps to limit water use and to reduce the impact of stress at silking. This in turn leads to more efficient ear partitioning and ear growth, resulting in augmented ear size. The enhanced sink activity exerted by larger ears can provide an adequate supply of nutrients to prevent the stress-induced abortion of a large number of kernels [59]. The Genuinity® DroughtGard® technology was deregulated by the "U.S. Department of Agriculture" (USDA) in December 2011. In the two succeeding years, MON87460 hybrids were only made available to selected farmers located in the drought-prone region of the western Great Plains. During this period, over 2,500 farmers participating in the trial reported reliable yield increase during drought over competitive drought-tolerant conventional hybrids (on average 5 bushel per acre) [50]. In 2014, Genuinity® DroughtGard® hybrids were planted by U.S. corn growers on 11.5 million acres, out-yielding conventional hybrids in a broader range of dry regions across the country [48]. Five consecutive years of commercial cultivation of the first biotech drought resistant crop clearly demonstrated that water use and yield-related tolerant traits could be efficiently engineered in high-yielding germoplasm without affecting yield potential under favorable environmental conditions.

Table 2. Transgenic drought tolerant crops on the market or under commercial evaluation

Developer	Crop	Transgene	Mechanism	Tolerance	Status	Field trial results	Reference
Monsanto	Corn	Constitutive expression of *COLD SHOCK PROTEIN B* (*CspB*) from *B. subtilis*	Stabilization of RNA	Drought	Approved for cultivation in USA (2014)	Yield advantage under drought (+5 bushel/acre)	[58, 59]
PT Perkenbunan Nusatara XI; Ajinomoto	Sugarcane	Constitutive expression of *betA* from *R. meliloti*	Synthesis of the osmoprotectant glycine betaine	Drought	Approved for cultivation in Indonesia (2013)	20-30% higher sugar yield during drought	[50]
Verdeca	Soybean	Constitutive expression of *Hahb-4* from sunflower	Downregulation of ethylene synthesis and siganling	Drought Multiple stresses	Field trials (Argentina, USA)	Yield advantage under drought (+15-27%)	[60]
Performance Plants	Canola, Corn, Petunia, Rice	Stress-inducible RNAi-mediated suppression of plant farnsyltransferase *era1*	Enhanced stomatal closure	Drought	Licenced to different biotech complanies	Yield advantage under drought in canola (+26%)	[61, 62]
DuPont Pioneer	Corn	Constitutive expression of an *ACC SYNTHSASE* (*ACS*) RNA interference construct	Downregulation of ethylene synthesis	Drought N-deficiency	Field trials (USA, Chile)	Yield advantage under drought (+9.3 bushel/acre) and N-deficiency (+7.1 bushel/acre)	[63]
Syngenta	Corn	Floem-specific overexpression of rice *TREHALOSE-6-PHOSPHATE PHOSPHATASE* (*TPP*)	Treahalose pathway	Drought	Field trials (USA)	Yield advantage under mild drought (+9-49%) and severe drought (31-123%)	[64]

(Table 2) contd.....

Developer	Crop	Transgene	Mechanism	Tolerance	Status	Field trial results	Reference
Arcadia Biosciences	Rice, Canola, Cotton, Peanut	Stress- and senescence- induced expression of *isopentenyltransferase* (*ipt*) from Agrobacterium,	Upregulation of cytokinin synthesis	N-deficiency Drought Multiple stresses	Field trials (USA)	Yield advantage under N-deficiency, drought stress, multiple stress (+12-18%)	[65 - 68]
Japan International Research Centre for Agricultural Sciences	Wheat, Soybean,	Stress-inducible expression of *DREB1A* transcription factor	Regulation of transcription		Field trials (CYMMIT, IRRI, ICRISAT)	Variable	[69, 70]
Japan International Research Centre for Agricultural Sciences	Rice, Peanuts	Stress-inducible expression of *DREB1A* transcription factor	Regulation of transcription	Drought	Field trials (India)	Variable	[71, 72]
Agricultural Genetic Engineering Research Institute (Egypt)	Wheat	Constitutive expression of the LATE EMBRIOGENESIS ABUNDANT PROTEIN gene *HVA1* from barley	LEA protein accumulation	Drought	Field trials (Egypt)	Not released	[73]

Drought-Resistant Sugarcane

A drought-tolerant sugarcane (*Saccharum officinarum* L.) variety has been recently released for cultivation in Indonesia. In May 2013, the "National Genetically Modified Product Biosafety Commission", approved a commercial variety expressing a bacterial enzyme involved in the synthesis of **glycine betaine (GB)** [76]. GB is a compatible solute, occurring in some but not all higher plants where it acts as an osmoprotectant. GB accumulates in response to different environmental cues and helps cells to acclimate to stress conditions [77]. Even if the details of the protective role of GB have not been completely elucidates, evidence indicates a function in the stabilization of enzymes and proteins and in the preservation of membrane integrity [78].

In bacteria GB is synthesized from the precursor choline by a two-step process, involving the enzymes, CHOLINE DEHYDROGENASE (CDH) and BETAINE ALDEHYDE DEHYDROGENASE (BADH). In plants, the synthesis of GB occurs in the chloroplast. The first oxygenation step is catalyzed by CHOLINE MONOOXYGENASE (CMO), followed by BADH-mediated dehydrogenation [77]. First attempts to engineering GB metabolism in plants, by expressing either the bacterial (*E. coli*) or plant biosynthetic genes produced very modest results. The amount of GB observed in early transgenics was significantly lower (< 5 µ mol/g fresh weight) than the concentrations detected in species, which naturally produce GB (4–40 µ mol/g fresh weight) [77]. The poor synthesis of GB suggests that other factors, other than CDH/CMO and BDH activity, might represent the limiting steps in the pathway. The lack of adequate concentration of endogenous choline [79] and the rate of import in the chloroplast have been proposed as major limitations to GB accumulation in transgenic plants [80]. Despite initial difficulties, scientists at the Indonesian research institute "PT Perkebunan Nusantara XI" succeeded in producing a GB-accumulating biotech sugarcane variety by introducing the CDH-coding gene *betA* from the symbiotic nitrogen-fixing bacterium *Rhizobium meliloti*. In Indonesia, nearly 40% of the cultivated sugarcane is rainfed and suffers from water deficit due to limited precipitations during the four-month wet season. Under these conditions, the *betA* transgenic variety produces 20–30% more sugar compared with the conventional parental lines [50]. Commercial planting of the sugarcane GM-variety, the second drought-resistant crop ever to be released on the market, started in 2014.

Drought-Resistant Biotech Soybean

In April of 2015, the first biotech stress tolerance trait in soybean was released for cultivation in Argentina [50]. The trait, developed by Verdeca LLC a joint venture between Bioceres S.A. and Arcadia Biosciences, Inc., is based on the constitutive overexpression of the **homeobox-leucine zipper transcription factor HAHB-4** from sunflower. The *HAHB-4* gene was first identified based on its enhanced expression following ABA treatment or exposure to water deficit [81]. Ectopic expression of *HAHB-4* in Arabidopsis resulted in enhanced tolerance to water stress, associated with improved development and higher survival rates compare with untransformed plants. Importantly, under either normal or drought conditions, the *HAHB-4* lines produced the same seed weight as wild-type plants under normal growth conditions, suggesting a possible contribution of the transgene to yield protection under stress. Developmental analysis of *HAHB-4* plants revealed a significant delay in the onset of the **senescence** program, a developmental defect often associated to alterations in **ethylene metabolism** [81]. Transcriptomic analyses confirmed that genes related to ethylene synthesis and signal transduction were significantly downregulated in plant expressing *HAHB-4*, as compared with the wild-type [82]. Consistently, transgenic plants showed reduced sensitivity to external ethylene as demonstrated by the complete suppression of the triple response of dark-grown seedlings, in addition to delayed senescence [82]. These findings highlight a novel cross-talk mechanism between drought responses and ethylene signaling, and establish a role for *HAHB-4* as a key negative regulator of such a cross-talk. In this perspective, the ectopic expression of *HAHB-4* causes a reduction in ethylene accumulation and the inhibition of ethylene signaling in plants exposed to drought stress. This determines a significant delay of ethylene-associated senescence processes allowing leaves to maintain active photosynthesis for longer periods to produce osmoprotectants and other metabolites that improve stress tolerance [82].

Additional evidence indicates a possible role for *HAHB-4* in mediating the plant response to herbivorous insects [83]. Transcript levels of the gene were strongly enhanced when sunflower leaves were defoliated by herbivores or mechanically wounded. Expression of *HAHB-4* in stable transgenic Arabidopsis or maize plants resulted in in upregulation of defense-related genes and increased synthesis of protective compounds, including jasmonic acid (JA) and JA-isoleucine [83]. Noteworthy, growth of larvae from *Spodoptera littoralis* or *Spodoptera frugiperda* fed on leaves from *HAHB-4* transgenic corn was inhibited compared with that of larvae fed on untransformed control lines [83]. Even if the protective role of *HAHB-4* against plant feeding pathogens has not yet been assessed under field conditions, these results indicate *HAHB-4* as a valuable biotech target to enhance both biotic- and abiotic-stress responses in crops.

Six consecutive years of field trials at multiple sites in Argentina and the United States indicated that the *HAHB-4* trait ensures up to 15% yield increase over comparable conventional varieties when exposed to multiple stress conditions. *HAHB-4* soybean varieties are being developed by "Verdeca LLC" in collaboration with "Tropical Melhoramento e Genética Ltda". Primary target for these new varieties are growers in Argentina and Brazil, two of the largest soybean producers worldwide. Commercial cultivation of the first drought tolerant *HAHB-4* varieties is expected to begin in 2017 [50].

WHO BREEDING FOR?

Advancements in breeding technologies have delivered promising solutions to counteract the detrimental effects of the ongoing climatic changes on global crop yield and food security. Until now, biotech and conventional stress-tolerant crop varieties have been made available to farmers in the most developed and high-yielding areas of the planet. However, the need and the urgency of extending the available technologies to smallholders farmers in drought-prone regions in developing countries are becoming increasingly evident. Specific selection strategies targeted to low-yielding cropping environments worldwide are also highly desirable to enhance yield in food insecure countries [84]. Importantly, climate models indicate that the poorer region of the world will be mostly exposed to the detrimental effects of global warming and changes in rainfall. By 2080, the average yield declines could be as severe as 23% for South Asia, 17% percent for East Asia and the Pacific and 15% for Sub-Saharan Africa [85].

International public-private partnerships as the **"Water-Efficient Maize for Africa" (WEMA)** project (http://wema.aatf-africa.org/) can play a pivotal role in transferring novel crop technologies to developing countries, contributing to build local capacities to foster further development and innovation. WEMA is coordinated by the "African Agriculture Technology Foundation" (AATF) and involves agricultural researcher centers in Uganda, Kenya, Mozambique, Tanzania and South Africa in addition to international academic partners and multinational companies. Aim of the project is the selection of drought-tolerant and insect-protected maize optimized for the Sub-Saharan Africa, where over 300 million people rely on corn as their principal source of calories [86]. In this area, the vast majority of corn is grown under rainfed conditions and its productivity is severely affected by frequent droughts. In recent years, prolonged periods without rain prevented maize from growing in Kenya, causing severe food shortages for over 4 million people [87].

The WEMA project entails the use of conventional breeding, marker-assisted breeding, and biotechnology to transfer drought-resistant traits into local corn genotypes. Breeding programs within WEMA build upon germoplasm and technologies freely contributed by the participating private companies. In 2014, AATF announced results from the first harvest of WE1101 DroughtTEGO™, one of WEMA's drought-tolerant corns developed through conventional breeding.

Initial plantings of DroughtTEGO™ in Kenya resulted in significantly higher seeds production compared with other local or commercial corn hybrids [86]. Among the 39 experimental sites surveyed, DroughtTEGO™ yielded on average 4.5 tons per hectare, a remarkable result when compared to the national average of 1.8 tons per hectare. Up to date, 25 DroughtTEGO™ hybrids have received approval for commercialization in Kenya, Tanzania and Uganda by the local authorities [86].

The biotech strategy exploited by the project is centered on the *CspB* drought-tolerant trait (MON87460), developed by Monsanto and freely licensed to WEMA. *CspB* hybrids have been tested for six consecutive years in five experimental sites across Kenya, Uganda and South Africa, providing an average yield increase of 8–14% for *CspB* hybrids compared with non-transgenic control hybrids [86]. The commercial development of the biotech trait still awaits years of field trials to introgress the event into genetic backgrounds suitable for cultivations in the region and to comply with the regulatory requirements of the partner countries. The first release of WEMA biotech drought-tolerant corn is expected in 2016-2017 in South Africa where the biosafety review process is nearing completion [50].

Finally, it is important to emphasize that, even with the involvement of large multinational companies, all the conventional and biotech corn-tolerant hybrids developed within WEMA, will be freely released to smallholder farmers directly by local seed companies.

CONCLUDING REMARKS

Demographic expansion and environmental changes raise great concerns about global food security and prompt new urgency into the development of durable high-yielding/stress-tolerant crop varieties. Further developments to accelerate the rate of genetic gain in crop breeding for stress tolerance will require:

i. the implementation of large-scale phenotyping and genotyping methods, to fully exploit the available genetic diversity in breeding germoplasm;
ii. the deployment of multidisciplinary approaches integrating conventional breeding with molecular breeding and genetic engineering;
iii. the improved understanding of the cross-talk between different stress-response pathways, to identify molecular targets and traits conferring tolerance to multiple stress conditions which simultaneously occur in the field (*e.g.* drought and high temperature);
iv. the effective and timely delivering of breeding technologies in developing and food insecure countries, where they are most needed and where the most yield gain is expected to occur;

v. the combination of efforts from academia and industry to improve translation of fundamental research into concrete solutions in agriculture;
vi. the integration of improved stress-tolerant genotypes with the most appropriate management practices for water-limited cropping environments.

CONFLICT OF INTEREST

The author confirms that author has no conflict of interest to declare for this publication.

ACKNOWLEDGEMENTS

This work was in part supported by "Fondazione Umberto Veronesi per il Progresso delle Scienze", Milano, Italy (project "Sustainability in Agriculture" - AGRISOST).

REFERENCES

[1] Khan NA, Singh S. Abiotic Stress and Plant Responses. New Delhi: I. K. International Publishing House 2008.

[2] Araus JL, Slafer GA, Royo C, Serret MD. Breeding for Yield Potential and Stress Adaptation in Cereals. Crit Rev Plant Sci 2008; 27: 377-412. [http://dx.doi.org/10.1080/07352680802467736]

[3] Bray E, Bailey-Serres J, Weretilnyk E. Responses to abiotic stresses. In: Buchanan B, Gruissem W, Jones R, Eds. Biochemistry and Molecular Biology of Plants. American Society of Plant Physiologists 2000; p. 1160.

[4] Fu R. Global warming-accelerated drying in the tropics. Proc Natl Acad Sci USA 2015; 112(12): 3593-4. [PMID: 25770219]

[5] Lobell DB, Burke MB, Tebaldi C, Mastrandrea MD, Falcon WP, Naylor RL. Prioritizing climate change adaptation needs for food security in 2030. Science 2008; 319(5863): 607-10. [http://dx.doi.org/10.1126/science.1152339] [PMID: 18239122]

[6] Diffenbaugh NS, Swain DL, Touma D. Anthropogenic warming has increased drought risk in California. Proc Natl Acad Sci USA 2015; 112(13): 3931-6. [http://dx.doi.org/10.1073/pnas.1422385112] [PMID: 25733875]

[7] Siebert S, Döll P. Quantifying blue and green virtual water contents in global crop production as well as potential production losses without irrigation. J Hydrol (Amst) 2010; 384: 198-217. [http://dx.doi.org/10.1016/j.jhydrol.2009.07.031]

[8] FAO. The State of the World's Land and Water Resources for Food and Agriculture. Rome: FAO 2011.

[9] Blatt M, Chaumont F, Farquhar G. Focus on water. Plant Physiol 2014; 164: 1553-5. [http://dx.doi.org/10.1104/pp.114.900484]

[10] Cominelli E, Galbiati M, Tonelli C, Bowler C. Water: the invisible problem. Access to fresh water is considered to be a universal and free human right, but dwindling resources and a burgeoning population are increasing its economic value EMBO Rep 2009; 10(7): 671-6. [http://dx.doi.org/10.1038/embor.2009.148] [PMID: 19543229]

[11] Botella MA, Rosado A, Bressan RA, Hasegawa PM. Plant adaptive responses to salinity stress. In:

Jenks MA, Hasegawa PM, Eds. Plant Abiotic Stress Biological Sciences Series. Oxford, UK: Blackwell Publishing Ltd 2005.
[http://dx.doi.org/10.1002/9780470988503.ch3]

[12] Rockström J, Falkenmark M. Semiarid crop production a hydrogeological perspective: gap between potential and actual yields. Crit Rev Plant Sci 2000; 19: 319-46.
[http://dx.doi.org/10.1080/07352680091139259]

[13] Cattivelli L, Rizza F, Badek F-W, Mazzucotelli E, Mastrangelo A, Francia E, *et al.* Drought tolerance improvement in crop plants: An integrated view from breeding to genomics. Field Crops Res 2008; 105: 1-14.
[http://dx.doi.org/10.1016/j.fcr.2007.07.004]

[14] Fleury D, Jefferies S, Kuchel H, Langridge P. Genetic and genomic tools to improve drought tolerance in wheat. J Exp Bot 2010; 61(12): 3211-22.
[http://dx.doi.org/10.1093/jxb/erq152] [PMID: 20525798]

[15] Boyer JS. Leaf enlargement and metabolic rates in corn, soybean, and sunflower at various leaf water potentials. Plant Physiol 1970; 46(2): 233-5.
[http://dx.doi.org/10.1104/pp.46.2.233] [PMID: 16657441]

[16] Nobel P. Biophysical Plant Physiology and Ecology. New York: W.H. Freeman 1983.

[17] Cattivelli L, Rizza F, Badeck F-W, Mazzucotelli E, Mastrangelo AM, Francia E, *et al.* Drought tolerance improvement in crop plants: an integrated view from breeding to genomics. Field Crops Res 2008; 105: 1-14.
[http://dx.doi.org/10.1016/j.fcr.2007.07.004]

[18] FAO. Drought Facts - Food and Agriculture Organization of the United Nations. 2013.

[19] Wood AJ. Eco-physiological adaptations to limited water environments. In: Jenks MA, Hasegawa PM, Eds. Plant Abiotic Stress Biological Sciences Series. Oxford, UK: Blackwell Publishing Ltd 2005.
[http://dx.doi.org/10.1002/9780470988503.ch1]

[20] Cockram J, Jones H, Leigh FJ, *et al.* Control of flowering time in temperate cereals: genes, domestication, and sustainable productivity. J Exp Bot 2007; 58(6): 1231-44.
[http://dx.doi.org/10.1093/jxb/erm042] [PMID: 17420173]

[21] Bacon M. Water use efficiency in plant biology. In: Ma B, Ed. Water use efficiency in plant biology Biological Sciences Series 12. Oxford, UK: Blackwell Publishing Ltd 2004; pp. 1-22.

[22] Levitt J. Responses of plants to environmental stresses. New York: Academic Press 1972.

[23] Cominelli E, Conti L, Tonelli C, Galbiati M. Challenges and perspectives to improve crop drought and salinity tolerance. N Biotechnol 2013; 30(4): 355-61.
[http://dx.doi.org/10.1016/j.nbt.2012.11.001] [PMID: 23165101]

[24] Todaka D, Shinozaki K, Yamaguchi-Shinozaki K. Recent advances in the dissection of drought-stress regulatory networks and strategies for development of drought-tolerant transgenic rice plants. Front Plant Sci 2015; 6: 84.
[http://dx.doi.org/10.3389/fpls.2015.00084] [PMID: 25741357]

[25] Reynolds M, Tuberosa R. Translational research impacting on crop productivity in drought-prone environments. Curr Opin Plant Biol 2008; 11(2): 171-9.
[http://dx.doi.org/10.1016/j.pbi.2008.02.005] [PMID: 18329330]

[26] Tuberosa R. Phenotyping for drought tolerance of crops in the genomics era. Front Physiol 2012; 3: 347.
[http://dx.doi.org/10.3389/fphys.2012.00347] [PMID: 23049510]

[27] Blum A. Drought resistance, water-use efficiency, and yield potential—are they compatible, dissonant, or mutually exclusive? Aust J Agric Res 2005; 56: 1159-68.
[http://dx.doi.org/10.1071/AR05069]

[28] Richards R. Breeding and selecting for drought resistance in wheat Drought resistance in crops with emphasis on rice. Los Banos, Laguna, Philippines: INTERNATIONAL RICE RESEARCH INSTITUTE 1982.

[29] Alvaro F, Isidro J, Villegas D. Moral LGd, C Royo C. Old and modern durum wheat varieties from Italy and Spain differ in spike components. Field Crops Res 2008; 106: 86-93. [http://dx.doi.org/10.1016/j.fcr.2007.11.003]

[30] Slafer G, Calderini D, Miralles D, Dreccer M. Preanthesis shading effects on the number of grains of three bread wheat cultivars of different potential number of grains. Field Crops Res 1994; 36: 31-9. [http://dx.doi.org/10.1016/0378-4290(94)90050-7]

[31] Tollenaar M, Lee E. Dissection of physiological processes underlying grain yield in maize by examining genetic improvement and heterosis. Maydica 2006; 51: 399-408.

[32] Guan YS, Serraj R, Liu SH, *et al.* Simultaneously improving yield under drought stress and non-stress conditions: a case study of rice (Oryza sativa L.). J Exp Bot 2010; 61(15): 4145-56. [http://dx.doi.org/10.1093/jxb/erq212] [PMID: 20660496]

[33] Richards R. Physiological traits used in the breeding of new cultivars for water-scarce environments. Agric Water Manage 2006; 80: 197-211. [http://dx.doi.org/10.1016/j.agwat.2005.07.013]

[34] Serra R, Bidinger FR, Chauhan YS, Seetharama N, Nigam SN, Saxena NP. Management of Drought in ICRISAT Cereal and Legume Mandate Crops. 2003.

[35] Kumar J, Sethi S, Johansen C, Kelley T, Rahman M, Rheenen Hv. Potential of short-duration varieties in chickpea. Indian J Dryland Agr Res 1996; 11: 26-9.

[36] Kumar J, van Rheenen HA. A major gene for time of flowering in chickpea. J Hered 2000; 91(1): 67-8. [http://dx.doi.org/10.1093/jhered/91.1.67] [PMID: 10739130]

[37] Eisenstein M. Plant breeding: Discovery in a dry spell. Nature 2013; 501(7468): S7-9. [http://dx.doi.org/10.1038/501S7a] [PMID: 24067764]

[38] Reyes L. Making rice less thirsty. Rice Today 2009; 8(3): 12-5.

[39] Jung C, Müller AE. Flowering time control and applications in plant breeding. Trends Plant Sci 2009; 14(10): 563-73. [http://dx.doi.org/10.1016/j.tplants.2009.07.005] [PMID: 19716745]

[40] Cockram J, Jones H, Leigh FJ, *et al.* Control of flowering time in temperate cereals: genes, domestication, and sustainable productivity. J Exp Bot 2007; 58(6): 1231-44. [http://dx.doi.org/10.1093/jxb/erm042] [PMID: 17420173]

[41] Blum A. Effective use of water (EUW) and not water-use efficiency (WUE) is the target of crop yield improvement under drought stress. Field Crops Res 2009; 112: 119-23. [http://dx.doi.org/10.1016/j.fcr.2009.03.009]

[42] Farquhar G, Ehleringer J. KT KH. Carbon isotope discrimination and photosynthesis. Annu Rev Plant Physiol Plant Mol Biol 1989; 40: 503-37. [http://dx.doi.org/10.1146/annurev.pp.40.060189.002443]

[43] Condon AG, Richards RA, Rebetzke GJ, Farquhar GD. Breeding for high water-use efficiency. J Exp Bot 2004; 55(407): 2447-60. [http://dx.doi.org/10.1093/jxb/erh277] [PMID: 15475373]

[44] Turner NC, Blum A, Cakir M, Steduto P, Tuberosa R, Young N. Strategies to increase the yield and yield stability of crops under drought – are we making progress? Funct Plant Biol 2014; 41: 1199-206. [http://dx.doi.org/10.1071/FP14057]

[45] Tardieu F, Tuberosa R. Dissection and modelling of abiotic stress tolerance in plants. Curr Opin Plant

Biol 2010; 13(2): 206-12.
[http://dx.doi.org/10.1016/j.pbi.2009.12.012] [PMID: 20097596]

[46] Cooper M, van Eeuwijk FA, Hammer GL, Podlich DW, Messina C. Modeling QTL for complex traits: detection and context for plant breeding. Curr Opin Plant Biol 2009; 12(2): 231-40.
[http://dx.doi.org/10.1016/j.pbi.2009.01.006] [PMID: 19282235]

[47] Cooper M, Gho C, Leafgren R, Tang T, Messina C. Breeding drought-tolerant maize hybrids for the US corn-belt: discovery to product. J Exp Bot 2014; 65(21): 6191-204.
[http://dx.doi.org/10.1093/jxb/eru064] [PMID: 24596174]

[48] Minford M. New hybrids cut yield losses during drought and optimize available water for higher yields. Corn+Soybean Digest [Internet] 2015.

[49] Deikman J, Petracek M, Heard JE. Drought tolerance through biotechnology: improving translation from the laboratory to farmers fields. Curr Opin Biotechnol 2012; 23(2): 243-50.
[http://dx.doi.org/10.1016/j.copbio.2011.11.003] [PMID: 22154468]

[50] Waltz E. Beating the heat. Nat Biotechnol 2014; 32(7): 610-3.
[http://dx.doi.org/10.1038/nbt.2948] [PMID: 25004222]

[51] Yang S, Vanderbeld B, Wan J, Huang Y. Narrowing down the targets: towards successful genetic engineering of drought-tolerant crops. Mol Plant 2010; 3(3): 469-90.
[http://dx.doi.org/10.1093/mp/ssq016] [PMID: 20507936]

[52] Zhang B. MicroRNA: a new target for improving plant tolerance to abiotic stress. J Exp Bot 2015; 66(7): 1749-61.
[http://dx.doi.org/10.1093/jxb/erv013] [PMID: 25697792]

[53] Rusconi F, Simeoni F, Francia P, *et al.* The Arabidopsis thaliana MYB60 promoter provides a tool for the spatio-temporal control of gene expression in stomatal guard cells. J Exp Bot 2013; 64(11): 3361-71.
[http://dx.doi.org/10.1093/jxb/ert180] [PMID: 23828545]

[54] Blum A. Genomics for drought resistance – getting down to earth. Funct Plant Biol 2014; 41(11): 1191-8.
[http://dx.doi.org/10.1071/FP14018]

[55] Fedoroff NV, Battisti DS, Beachy RN, *et al.* Radically rethinking agriculture for the 21st century. Science 2010; 327(5967): 833-4.
[http://dx.doi.org/10.1126/science.1186834] [PMID: 20150494]

[56] Rothstein SJ, Bi YM, Coneva V, Han M, Good A. The challenges of commercializing second-generation transgenic crop traits necessitate the development of international public sector research infrastructure. J Exp Bot 2014; 65(19): 5673-82.
[http://dx.doi.org/10.1093/jxb/eru236] [PMID: 24948680]

[57] Langridge P, Reynolds MP. Genomic tools to assist breeding for drought tolerance. Curr Opin Biotechnol 2015; 32: 130-5.
[http://dx.doi.org/10.1016/j.copbio.2014.11.027] [PMID: 25531270]

[58] Castiglioni P, Warner D, Bensen RJ, *et al.* Bacterial RNA chaperones confer abiotic stress tolerance in plants and improved grain yield in maize under water-limited conditions. Plant Physiol 2008; 147(2): 446-55.
[http://dx.doi.org/10.1104/pp.108.118828] [PMID: 18524876]

[59] Nemali KS, Bonin C, Dohleman FG, *et al.* Physiological responses related to increased grain yield under drought in the first biotechnology-derived drought-tolerant maize. Plant Cell Environ 2015; 38(9): 1866-80.
[http://dx.doi.org/10.1111/pce.12446] [PMID: 25210866]

[60] Manavella PA, Dezar CA, Ariel FD, Drincovich MF, Chan RL. The sunflower HD-Zip transcription factor HAHB4 is up-regulated in darkness, reducing the transcription of photosynthesis-related genes.

J Exp Bot 2008; 59(11): 3143-55.
[http://dx.doi.org/10.1093/jxb/ern170] [PMID: 18603614]

[61] Wang Y, Beaith M, Chalifoux M, *et al.* Shoot-specific down-regulation of protein farnesyltransferase (alpha-subunit) for yield protection against drought in canola. Mol Plant 2009; 2(1): 191-200. [http://dx.doi.org/10.1093/mp/ssn088] [PMID: 19529821]

[62] Wang Y, Ying J, Kuzma M, Chalifoux M, Sample A, McArthur C, *et al.* Molecular tailoring of farnesylation for plant drought tolerance and yield protection. The Plant J : For Cell Mol Biol 2005; 43(3): 413-24.
[http://dx.doi.org/10.1111/j.1365-313X.2005.02463.x]

[63] Habben JE, Bao X, Bate NJ, *et al.* Transgenic alteration of ethylene biosynthesis increases grain yield in maize under field drought-stress conditions. Plant Biotechnol J 2014; 12(6): 685-93. [http://dx.doi.org/10.1111/pbi.12172] [PMID: 24618117]

[64] Nuccio ML, Wu J, Mowers R, *et al.* Expression of trehalose-6-phosphate phosphatase in maize ears improves yield in well-watered and drought conditions. Nat Biotechnol 2015; 33(8): 862-9. [http://dx.doi.org/10.1038/nbt.3277] [PMID: 26473199]

[65] Kuppu S, Mishra N, Hu R, *et al.* Water-deficit inducible expression of a cytokinin biosynthetic gene IPT improves drought tolerance in cotton. PLoS One 2013; 8(5): e64190. [http://dx.doi.org/10.1371/journal.pone.0064190] [PMID: 23675526]

[66] Qin H, Gu Q, Zhang J, *et al.* Regulated expression of an isopentenyltransferase gene (IPT) in peanut significantly improves drought tolerance and increases yield under field conditions. Plant Cell Physiol 2011; 52(11): 1904-14.
[http://dx.doi.org/10.1093/pcp/pcr125] [PMID: 21920877]

[67] Reguera M, Peleg Z, Abdel-Tawab YM, Tumimbang EB, Delatorre CA, Blumwald E. Stress-induced cytokinin synthesis increases drought tolerance through the coordinated regulation of carbon and nitrogen assimilation in rice. Plant Physiol 2013; 163(4): 1609-22.
[http://dx.doi.org/10.1104/pp.113.227702] [PMID: 24101772]

[68] Kant S, Burch D, Badenhorst P, Palanisamy R, Mason J, Spangenberg G. Regulated expression of a cytokinin biosynthesis gene IPT delays leaf senescence and improves yield under rainfed and irrigated conditions in canola (Brassica napus L.). PLoS One 2015; 10(1): e0116349.
[http://dx.doi.org/10.1371/journal.pone.0116349] [PMID: 25602960]

[69] Kidokoro S, Watanabe K, Ohori T, Moriwaki T, Maruyama K, Mizoi J, *et al.* Soybean DREB1/CBF-type transcription factors function in heat and drought as well as cold stress-responsive gene expression. The Plant journal : for cell and molecular biology 2015; 81(3): 505-18.

[70] Pellegrineschi A, Reynolds M, Pacheco M, Brito RM, Almeraya R, Yamaguchi-Shinozaki K, *et al.* Stress-induced expression in wheat of the Arabidopsis thaliana DREB1A gene delays water stress symptoms under greenhouse conditions. Genome / National Research Council Canada = Genome / Conseil national de recherches Canada 2004; 47(3): 493-500.

[71] Bhatnagar-Mathur P, Devi MJ, Reddy DS, *et al.* Stress-inducible expression of At DREB1A in transgenic peanut (Arachis hypogaea L.) increases transpiration efficiency under water-limiting conditions. Plant Cell Rep 2007; 26(12): 2071-82.
[http://dx.doi.org/10.1007/s00299-007-0406-8] [PMID: 17653723]

[72] Ravikumar G, Manimaran P, Voleti SR, *et al.* Stress-inducible expression of AtDREB1A transcription factor greatly improves drought stress tolerance in transgenic indica rice. Transgenic Res 2014; 23(3): 421-39.
[http://dx.doi.org/10.1007/s11248-013-9776-6] [PMID: 24398893]

[73] Sivamani E, Bahieldin A I, Wraith JM, *et al.* Improved biomass productivity and water use efficiency under water deficit conditions in transgenic wheat constitutively expressing the barley HVA1 gene. Plant Sci 2000; 155(1): 1-9.

[http://dx.doi.org/10.1016/S0168-9452(99)00247-2] [PMID: 10773334]

[74] Horn G, Hofweber R, Kremer W, Kalbitzer HR. Structure and function of bacterial cold shock proteins. Cell Mol Life Sci 2007; 64(12): 1457-70. [http://dx.doi.org/10.1007/s00018-007-6388-4] [PMID: 17437059]

[75] Barria C, Malecki M, Arraiano CM. Bacterial adaptation to cold. Microbiology 2013; 159(Pt 12): 2437-43. [http://dx.doi.org/10.1099/mic.0.052209-0] [PMID: 24068238]

[76] Pocard JA, Vincent N, Boncompagni E, Smith LT, Poggi MC, Le Rudulier D. Molecular characterization of the bet genes encoding glycine betaine synthesis in Sinorhizobium meliloti 102F34. Microbiology 1997; 143(Pt 4): 1369-79. [http://dx.doi.org/10.1099/00221287-143-4-1369] [PMID: 9141699]

[77] Chen TH, Murata N. Enhancement of tolerance of abiotic stress by metabolic engineering of betaines and other compatible solutes. Curr Opin Plant Biol 2002; 5(3): 250-7. [http://dx.doi.org/10.1016/S1369-5266(02)00255-8] [PMID: 11960744]

[78] Papageorgiou GC, Murata N. The unusually strong stabilizing effects of glycine betaine on the structure and function of the oxygen-evolving Photosystem II complex. Photosynth Res 1995; 44(3): 243-52. [http://dx.doi.org/10.1007/BF00048597] [PMID: 24307094]

[79] Huang J, Hirji R, Adam L, *et al.* Genetic engineering of glycinebetaine production toward enhancing stress tolerance in plants: metabolic limitations. Plant Physiol 2000; 122(3): 747-56. [http://dx.doi.org/10.1104/pp.122.3.747] [PMID: 10712538]

[80] Nuccio ML, McNeil SD, Ziemak MJ, Hanson AD, Jain RK, Selvaraj G. Choline import into chloroplasts limits glycine betaine synthesis in tobacco: analysis of plants engineered with a chloroplastic or a cytosolic pathway. Metab Eng 2000; 2(4): 300-11. [http://dx.doi.org/10.1006/mben.2000.0158] [PMID: 11120642]

[81] Dezar C, Gago GM, González DH, Chan RL. Hahb-4, a sunflower homeobox-leucine zipper gene, confers drought tolerance to Arabidopsis thaliana plants. Transgenic Res 2005; 14: 429-40. [http://dx.doi.org/10.1007/s11248-005-5076-0] [PMID: 16201409]

[82] Manavella PA, Arce AL, Dezar CA, Bitton F, Renou JP, Crespi M, *et al.* Cross-talk between ethylene and drought signalling pathways is mediated by the sunflower Hahb-4 transcription factor. Plant J: for cell and molecular biology 2006; 48(1): 125-37. [http://dx.doi.org/10.1111/j.1365-313X.2006.02865.x]

[83] Manavella PA, Dezar CA, Bonaventure G, Baldwin IT, Chan RL. a sunflower HD-Zip protein, integrates signals from the jasmonic acid and ethylene pathways during wounding and biotic stress responses. The Plant journal : for cell and molecular biology 2008; 56(3): 376-88.

[84] Tester M, Langridge P. Breeding technologies to increase crop production in a changing world. Science 2010; 327(5967): 818-22. [http://dx.doi.org/10.1126/science.1183700] [PMID: 20150489]

[85] Hallegatte S, Bangalore M, Bonzanigo L, Fay M, Kane T, Narloch U, *et al.* Shock Waves: Managing the Impacts of Climate Change on Poverty. Washington: World Bank Group 2016.

[86] AATF. Adding Value: Making a Difference Annual Report 2014. Nairobi, Kenya: African Agricultural Technology Foundation 2015.

[87] Rockström J, Falkenmark M. Agriculture: increase water harvesting in Africa. Nature 2015; 519(7543): 283-5. [http://dx.doi.org/10.1038/519283a] [PMID: 25788079]

CHAPTER 11

Genetic Strategies to Improve Resistance to Biotic Stresses in Plants

Stefano Sangiorgio[1] and **Mario Motto**[2,*]

[1] *Department of Agricultural and Environmental Sciences – Production, Landscapes, Agroenergy, Università degli Studi di Milano, Milano, Italy*

[2] *Fondazione Istituto Tecnico Superiore per le nuove Tecnologie della Vita, Bergamo, Italy*

Abstract: The long-term target of improving crops resistance to biotic stresses is a familiar goal for breeders. Plants ought to constantly protect themselves *versus* aggressions from a wide spectrum of organisms that include viruses, bacteria, oomycetes, fungi, insects and other herbivores, and weeds. In this chapter attention will be given to depict a picture on the genetic and molecular mechanisms that plants have promoted to recognize and react to invasion by numerous parasites (pathogens and pests). These topics include non-host resistance, constitutive barriers, and race-specific resistance. The chapter also examines current progresses in clarifying the structure and molecular devices developed by plants to neutralize pathogen and pest aggressions. Moreover, it takes a look with aspects experienced in breeding for resistance to relevant biotic stress factors. Major considerations in breeding for resistance to pathogens, insect pests, and weeds, traditional sources of resistance or other possible strategies, such as mutation breeding, genetic manipulations, and molecular strategies to develop crops more resistant to parasites are also explored.

Keywords: Defense mechanisms, Genetic basis of resistance, Pathogenesis related proteins, Signal transduction network, Transgenic plants.

INTRODUCTION

Biotic stresses, the damage caused by plant pathogen, insect, and weed pests, have a negative impact on productivity of our crops by reducing yield and quality [1]. It is estimated that 35% of crop production, on a global scale, is annually lost to pre-harvest biotic stresses, with an additional 6 to 20% of losses due to post-harvest events [2]. A survey of the potential and actual yield injuries attributable to biotic constraints in important crop plants is shown in Table **1**. This information indicate that there is remarkable deficit between potential and realized crop productions.

* **Corresponding author Mario Motto**: Fondazione Istituto Tecnico Superiore per le nuove Tecnologie della Vita, Bergamo, Italy; Tel: +39 0350789106; Fax: +39 0350789107; E-mail: mariomotto@libero.it

Roberto Pilu and Giuseppe Gavazzi (Eds.)

Table 1. Survey of deficit between potential and realized crop yields due to fungal and bacterial pathogens, viruses, animal pests, and weeds including the efficacy of the used pest control activities in various crops (*e.g.* maize, wheat, rice, barley, potatoes, soybean, sugar beet, and cotton). Modified from Oerke and Dehne [3].

	Pests and pathogens				
	Fungi and bacteria	**Viruses**	**Animal pests**	**Weeds**	**Total**
Loss potential(%)[a]	14.9	3.1	17.6	31.8	67.4
Actual losses (%)[a]	9.9	2.7	10.1	9.4	32.0
Efficacy(%)[b]	33.8	12.9	42.4	70.6	52.5

[a] As percentage of attainable yields; [b]As a percentage of potential prevented yields loss

Plant protection has a cardinal function in assuring crop yield performances against plant pathogens, animal pests, and weeds. This relationship is illustrated by a 15–20-fold annual increase in the volume of chemicals (pesticides) employed on a global scale [3]. However, plants are able to counteract parasite damages by several genetically inherited mechanisms, acting at the morphological, physiological, biochemical, cellular, and molecular levels. Therefore, the introduction of genetic resistance or tolerance into plants to the plethora of biotic stresses that severely damage our crops, is an important goal for scientists. Besides, this strategy has showed up relevant consequences for both growers and the seed and agrochemical industrial sectors [4]. Notably, genetically resistant or tolerant crops able to neutralize pest attacks have various benefits over the employment of pesticides or additional procedures to manage biotic stresses. These advantages are reflected by the following arguments: i) economic savings of the costs of pesticide treatments, ii) seed of resistant varieties generally costs to growers no more that the susceptible varieties, iii) although the resistance does not fully protect the crop, partial resistance may conduct to a sizeable decrease in the amount of pesticides needed to provide a tolerable control. Evidence suggests that these benefits depend on simple genetic stability, insignificant expenses after varieties are produced, and a remarkable effectiveness. The principal drawback of genetic resistance in plants to neutralize biotic stress factors is due to the issue that selection pressure is focused on parasite populations: the development of individuals with inherited mechanisms to breakdown plant resistance are favored within these populations. Thereby, it is obvious that this occurrence is restricting the temporal length of resistance performance in crops.

Typically, microbial organisms causing diseases are referred as pathogens, and herbivorous insects, mammals, and birds are termed pests. In this chapter we take a close look at the importance of genetic, biochemical, and molecular processes by which plants protect themselves from diseases and damages caused by plant pathogens, insects, and weed pests. Additionally, breeding strategies devoted to the development of tolerant or resistant plants are also highlighted.

PLANT PATHOGENS

The many organisms that cause infectious diseases and damages in plants include fungi, oomycetes, bacteria, viruses, and nematodes. A comprehensive description of individual diseases and the methods used in their control is outside the objectives of this chapter. A number of books and reviews have been written in this field to which the reader is addressed for a more detailed illustration [5]. For the sake of brevity, specific pathogens or the diseases and damages they cause are herein mentioned without further explanation.

The main findings emerging from those publications that may worth noting indicate that:

i. Several pathogens are specialized to growth on a specific plant species and cannot strike and produce disease in other plants. Others can devastate numerous, frequently unrelated, plant species. To exploit a distinct species as nutriment, a pathogen must be competent to defeat the species defense systems. Nearly all plant species are resistant to the majority of pathogens.
ii. Pathogens can enter into plants through several routes, such as direct penetration *via* intact surfaces, entry *via* natural opening (*e.g.* stomata), or *via* opportunist entry represented by existing wounds or cracks on the plant surface.
iii. After the pathogens have entered into the plant, three major colonization tactics are used by these organisms to take advantage of the host plant as a nutritional substrate for their growth and development. Essentially, either these organisms parasitize the vital plant to pick up nourishments (biotrophic lifestyle) or they destroy the plant tissues that are infected and use up nutrients from the non-longer alive tissues (necrotrophic lifestyle). Hemibiotrophs embrace both lifestyles, shifting from a biotrophic stage at the starting of the infection to a necrotrophic lifestyle as pathogenesis advances.
iv. Pathogenesis describes the series of phases concerning host and pathogen interaction (*e.g.* infection, colonization and plant pathogen reproduction) to the progress of the whole syndrome.
v. Recent evidence indicates that fungal pathogen employed sex pheromone receptors for perceiving chemotropically host plant signals in an intricate environment medium like the soil [6].
vi. A pathogen race that induces disease is named virulent. Its favorable outcome may depend from different elements that include: i) very quick and elevated rate of reproduction throughout the central growing season for plants; ii) high performance dispersal system and long-standing survival ability; (iii) great

effectiveness to induce genetic variability during haploid phase and successive sexual reproduction.

vii. Pathogens give raise to a variety of symptoms on infected plants. These symptoms are, from time to time, peculiar for a specific pathogen, permitting it to be detected by visible symptoms, although distinct pathogens can generate analogous effects. Symptoms of the diseases caused by the pathogens on plants result from various factors such as: disruption of respiration, photosynthesis, translocation of nutrients, transpiration, and other aspects affecting plant growth and development.

Plant Defense Mechanisms

Plants are sedentary entities that are permanently subjected to invasion by several pathogens. However, plants are armed with various intricate, highly effective defense mechanisms in response to challenge by microbial pathogens that are able in preventing, delaying or mitiging damages induced by pathogens. A given plant species is normally resistant to most pathogens and susceptible to only a few pathogens. Disease resistance is determined by an elaborate, multilayered system of defense mainly due to both preformed defense mechanisms and their innate immune system that are able to block potential pathogen invasion [7].

Preformed Defense

Preformed defenses are among the most relevant components of non host resistance – *i.e.* an ample array of plant defense that confers immunity to all individuals of a plant species - against all races of a microorganism that is pathogenic to different plant species. These defenses include both physical restriction and chemical inhibition of pathogens. For example, the cuticle layer, the epidermis, and the cell wall are natural obstacles present in all plant species. Additionally, the low frequency of natural openings and the high cuticular wax on leaf surfaces act as physical barriers for the entry of bacterial pathogens and their access to cellular nutrients [8]. However, most fungal pathogens can penetrate the cuticle by mechanical rupture and secretion of cutinases that hydrolyse cutin polyesters [9]. Notably these last authors have indicated that i) cuticle breakdown products are recognized by invading fungi and promote developmental processes throughout pathogenesis, ii) cutin monomers (*e.g.* n,16-hydroxypalmitic acid) form potential cues distinguished by plants; they are among the primary elicitors developed concurrently with infection and promote a complex of reactions that frequently culminates in resistance *versus* necrotrophs.

An additional physical barrier to counteract pathogen attacks is given by the plant cell wall. This wall provides both a structural and a protection element against abiotic and biotic stress agents and is controlled by a specific cell wall integrity maintenance mechanism [10]. The cell wall is also implicated in perception of different stimuli, such as pathogen presence. Survival of the complete plant might rely on a prompt downstream signal transduction mechanism, acting from the external to the inside of the cell surface, to promote antimicrobial responses [11]. In this context it was found, for example, that Arabidopsis plants are able to challenge pathogen penetration into cells by several non host fungal pathogens (predominantly powdery mildews) after the expression of penetration (*PEN*) genes, which provide a rapid and efficient protection system at the level of cell wall component [12].

Besides physical barriers, plants constantly synthesize a collection of secondary metabolites (*e.g.* phenolics, terpenoids saponins, steroids, and glucosinalates), and antifungal proteins [13]. Several of them function as antimicrobial molecules, generically called phytoanticipins. These chemical agents, able to inhibit pathogen growth, may exist in plant cells either in biologically active shapes or stored as inert precursor compounds. These last precursors are subsequently modified to their active molecular shapes by plant host enzymes as a reaction to pathogen aggression or tissue injuries. For example, in *Brassicaceae,* specifically in Arabidopsis (*Arabidopsis thaliana*), sulfur-and nitrogen-containing secondary metabolite compounds (*e.g.* glucosinolates and their derivatives) are distantly localized *in planta* from the enzymes that convert them to yield toxic cyanide or isocyanates, thiocyanates, and nitriles, after wounding [14, 15]. Collectively, it was found that these chemicals have a function in the plant defense arsenal towards an ample variety of pathogens [16, 17].

Inducible Plant Defence

In addition to preformed defenses, plants have elaborated a battery of sophisticated vigilance mechanisms to block potential pathogen invasion and rapidly react before pathogens have a chance to produce severe injuries, reflected by losses in plant performances [1].

From genetic and molecular point of views, the role of inducible post-infection plant defenses to pathogen attack has been well described. There are a number of recent reviews to which the reader is referred for detailed information [1, 18 - 21]. Briefly: i) interactions between pathogens and their host plants can be described as composed of four stages, reflecting a series of events in the evolution of host-pathogen relationships. These are depicted by the so called 'zigzag model' of events in coevolution of host plants and their pathogens: whereas a pathogen arrives in contact with a potential host plant, surface molecules on the pathogen are recognized by receptor in the plant cell activating defense responses; ii) plants do not have a specialized immune cell system or organs; therefore every individual cell exhibits the potential ability to elicited immune reactions autonomously; iii) the outcome of a resistance response depends on its rate of reaction, which is guaranteed by an effective perception and signal transduction systems; iv) plant innate immune recognition system produces both local and systemic responses; this permits a plant to disable invasion by pathogens, both in a very quick and restricted modality or on an broad time scale by temporal and spatial processes; v) the presence of plant basal defense mechanisms provides selective pressure on potential pathogen genetic variants that produce proteins and other molecules, defined as effector molecules. These enable pathogens to interact with plants and inhibit those mechanisms, allowing the pathogen to assault plants, which in the literature is popular as effector-triggered susceptibility mechanism.

The innate immune system can occur *via* various mechanisms that include: i) pathogen/microbial-associated molecular pattern (P/MAMP hereafter referred as PAMP)-triggered immunity (PTI), ii) effector-triggered immunity (ETI), iii) systemic acquired resistance (SAR) also called induce systemic resistance (ISR), and iv) noncoding small RNAs (sRNAs or RNAi) [22]. A general diagrammatic picture of these plant defensive mechanisms is given in Fig. (**1**).

Notably all of these defensive mechanisms are more efficient than preformed defence. Therefore, understanding their mechanism of action is important to engineer long lasting disease-resistant crops in response to challenge by potential microbial pathogens [1, 7]. PTI and ETI can be indicated as branches (sub-networks) of the resistance signalling pathway. Downstream components are shared by PTI and ETI, and several of their outputs differ in speed and amplitude rather than qualitatively. In fact, PTI cannot entirely prevent pathogen colonization, but can postpone pathogen invasion [18]. On the other hand, ETI can be considered as a strengthen and a long-lasting PTI mechanism that involves the evolution of SAR and a fast, local programmed cell death (PCD), usually specify as hypersensitive response (HR) Therefore, PTI and ETI are considered as primary and secondary innate immunity system, respectively [23].

PAMP -Triggered Immunity (PTI)

The majority of microbial agents include surface molecules (frequently cell wall items) that are detected or recognized by receptors on plant cells, eliciting the activation of defences in a complex referred to as basal defence mechanisms or PTI [21]. This is a primordial type of plant immunity that grants enough protection to host non-adapted pathogens, providing basal resistance. It is regulated by cell surface local pattern recognition receptors (PRRs)-like kinases that recognizes various danger signals during pathogen infection represented by PAMPs and elicit PTI.

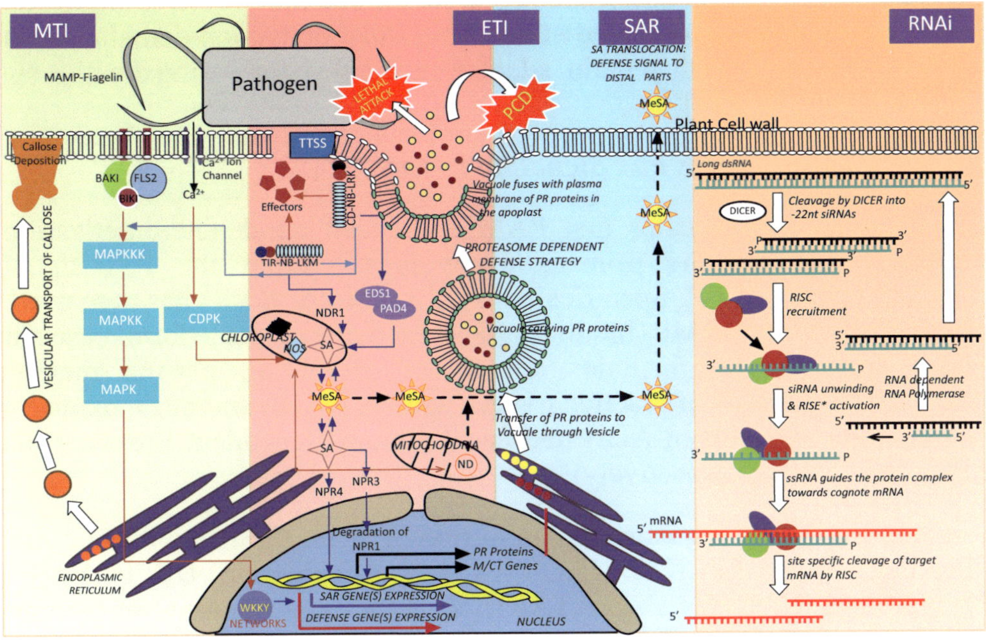

Fig. (1). Schematic representation of plant immunity mechanisms: (a) PAMP-triggered immunity (PTI); (b) effector-triggered immunity (ETI); (c) systemic acquired resistance (SAR); (d) gene silencing. ETI representation contains the features operating in cell autonomous immune system organized on membrane fusion; this is able to neutralize intercellular bacteria and stimulating local necrosis (*i.e.* hypersensitive response-programmed cell death) (Figure adapted from Muthamilarasan and Prasad, 2013, [22]).

PAMPs are small, widely different molecules (*e.g.* elongation factors, flagellins quorum-sensing proteins, peptidoglycans from bacteria or chitin from fungal cell wall) [24]. They are mostly conserved microbial substances and invariant among diverse groups of microorganisms and excreted on the surface by a category of pathogens functioning as elicitors.

Typically PRRs (*e.g.* FLS2, EFR, Xa21), or surface receptors, directly sense PAMPs, promoting downstream signalling events, including defence gene induction [25]. PRRs occur in multigene families such as those for the *Crinkler (CRN)* members of the *Phytophtora genus* [18]. Notably from a structural point of view PRRs comprises receptor-like kinases (RLK) and receptor-like proteins (RLP) components, respectively. RLKs are located in plasma membrane and are formed by a presumed extracellular ligand-binding domain associated to an individual transmembrane domain, and to an intracellular serine/threonine kinase domain [21, 26]. Evidence shows that plant RLKs are structurally correlated to animal receptor-tyrosine kinases (RTKs) and that the biggest and most variable (L-type LecRK) class is lineage-specific [27]. Similarly, RLPs are formed by an extracellular and a membrane-spanning domain, but they are missing a kinase domain. Because they are deprived of an intracellular activation domain, therefore they need to interact with certain adapter molecules for acting on the signal transduction pathway [28].

Functional Characteristics and Signaling Events

Functional evidence indicates that PAMPs induced rapid heteromerization of PRRs that represents the primary event in triggering the PTI signaling downstream response [29]. Specifically, once a receptor (PPR) has perceived the pathogen, following PAMP binding, PRRs activate a multiform cascade of signaling episodes that include: ion fluxes promoting plasma membrane depolarization, synthesis of reactive oxygen species (ROS) and nitric oxide (NO), and stimulation of Mitogen-Activated and Calcium-Dependent Protein Kinases (MAPKs and CDPKs, respectively).

The signaling episodes regulate also transcription factor (TF) activities of distinct families that include: WRKY, MADS (MCM1, AGAMOUS, DEFICIENS, and SRF) box and NAC (NAM, ATAF1, 2, CUC2) promoting an extensive transcriptional reprogramming connected to plants defense responses [31]. Interestingly the WRKY TFs were reported to perform contrasting (positive and negative) functions in regulating plant defence responses [30]. It is predicted that peculiar members of the WRKY family may be potentially utilized to develop plants equipped with durable transgenic disease resistance [31, 32].

Mechanisms controlling the expression of WRKY activity by MAPK proteins are especially well substantiated. It was found that MAPKs play a function not exclusively in the phosphorylation of substrates, but additionally in the sequestration and release of TFs, which permits passage to target promoters [33]. In other circumstances, triggering of WRKY TFs *via* MAPKs can be generated by phosphorylation-induced structural modifications. Additional research in this field have also indicated that other functions, such as epigenetic alterations, embracing DNA methylation and histone changes, play a significant role in PTI signaling events [34]. For example, two Arabidopsis *histone deacetylase (HDAC)* genes, namely *HDA19* and *SRT2*, are involved in regulating disease defense networks. In this context it was also found that cytosine DNA methylation, especially in the promoter region, frequently shows a repressive function in varying gene expression as a reaction to stresses. Furthermore, ubiquitination has a central function in cell signaling that controls distinct processes [35]. These involve protein degradation and immunological response exploiting operational proteasomes and protein targeting systems [35].

The activity of the TFs is also coordinated by a blend of signaling phytohormones, including salicylic acid (SA), ethylene (ET), jasmonic acid (JA), and abscisic acid (ABA [36]). The interaction and the fine tuning among these hormones and other plant hormones, like auxins (AXs), gibberellins or brassinosteroids (BRs), orchestrate the induction of previously cited defense mechanisms and permit to address immune responses *versus* the peculiar invader [37]. Several of the defense responses precisely rely on the activity of these plant hormones [36, 37]. More recently, research has highlighted that on the expression of plant defense mechanisms, a role is played by light cues recognized by photoreceptors [38]. This result provides new information into the mechanisms that plants employ to drive adaptive solutions in cases they are attacked by diversified biotic stress organisms. Collectively, plants possesses an intricate arsenal of defense reaction pathway to contrast pathogen invasions or able to confine, at least partially, the growth and diffusion of pathogens in plant tissues.

Defence Response

Gene activation, involved in plant defense mechanisms, culminates with the deposition of various enzymes and stress-specific metabolites [23, 39]. Notably this response is mediated by i) pathogenesis-related (PR) proteins such as hydrolytic enzymes (β- 1,3-glucanases and chitinases), deteriorating microbial cell walls, cationic defensins degrading pathogen membranes, peroxidases, inhibitors of proteinase or lipid-transfer proteins; ii) molecules possessing antimicrobial activity like phytoalexins; iii) lignin and callose accumulated in the cell wall active in reinforcing its protection iv) production of ROS having a signaling function and driving antimicrobial response, and v) closure of stomatal elements.

Typically the final responses of PTI lead to the successive layer of defense responses *via* activation of HR, a confined type of PCD limiting pathogen diffusion. It was also reported an additional strengthening of the cell walls, and synthesis of antimicrobial chemical compounds (*e.g.* defense proteins and phytoalexins) [40].These events proceed in both susceptible and resistant plant-pathogen interactions, probably with different velocity and intensity rates.

Effector-Triggered Immunity (ETI)

Although the PTI defence reactions may be strong enough to block infection before the invader microbe becomes established, proliferant pathogens have evolved a repository of toxins and effector proteins (*i.e.* protein transferred inside plant cells by a pathogen) that play a pivotal function in the modification and suppression of plant defence processes [7]. Specifically, presence of effectors compounds that enable propagating colonization by the pathogen determinates a selective pressure on the plant populations. This event favour in plant the selection of genetic variants that can detect pathogen effectors and respond to their presence by activating additional defence mechanisms [41]. In this scenario, it was found that those variant plants possess receptors for the effector molecules, encoded by genes, which are familiar as resistance genes *(R)*. This modality of resistance is referred as *R*-gene mediated defence or effector-triggered immunity (*ETI*). This system of resistance is able to elicite PCD and qualitative resistance mechanisms [41] that are stronger than does PTI [23].

Research in this field of disease resistance has been largely reviewed [*e.g.* 23, 25, and 42]. The main findings emerging from these studies generally indicate that two central genes are mandatory for expression of resistance: the *R*-gene in the host plant cell and the related avirulence (*AVR*) effector gene in the pathogen organism. Individual *R*-gene provides resistance to pathogen strains possessing the corresponding *AVR* effector gene: briefly, the efficiency of *R*-genes is pathogen strain-dependent.

Usually AVR effectors are produce and release inside the host to favor the pro infection advancement and therefore are central players in developing pathogenicity. The capacity of the pathogen to neutralize resistance evolves from mutations concerning the *AV*R gene. This event is dependent in the absence of identification of AVR effector by the corresponding *R*-gene product [43]. Several *AVR* genes have been analyzed from distinct plant pathogens. They typically encode small proteins, usually cysteine-rich, and generally show lack of homology to familiar proteins in database collections [44 and reference therein]. Effector genes usually manifest a coordinated expression during plant infection that is modulated by epigenetic mechanisms [44]. Elusion of ETI following mutation is a favorable mechanism: cytoplasmic effector arsenals are largely redundant and any single effector is usually needles [43].

Similarly *R*-genes have been identified and molecularly characterized from several plant species [13, 23]. They encode receptor proteins, generally nucleotide-binding leucine-rich repeat (NLRs) that are able to distinguish, either directly or indirectly, AVR proteins synthetized by pathogens [45, 46]. In addition to R- NLRs, another class of plant disease resistance proteins, *i.e.* transcription activator-like effector (TALE), connected with executor type *R*-genes, has been shown to contrast pathogen virulence functions and open an alternative way to counteracts pathogen proliferation in plants (see [47] for a recent review).

R- NLRs Genes: Structural Characteristics, Diversity, and Evolution

Plants are equipped with a wide collection of *R*-genes that *via* distinct recognition peculiarities are switched on in response to a plethora of microbial pathogens. The robust phenotypes and tangible variation occurring at *R*-loci have been useful to molecular biologists for cloning the *R*-genes and for studying their fine mechanisms of action [25, 42].

Studies have indicated that *R*-genes in plants form highly amplified gene families, ranging from 50 in *Cucumis sativus* [48] to 653 in *Oryza sativa* [49]. Moreover, these investigations have highlighted that among the eight families of *R*-genes, currently identified, the most recurrent one is typified by a distinct NLR amino acid sequence motif, implicated in the recognition of a microbial pathogen and its associated roles [42]. Further observations on the structural motif indicate that R-proteins are formed by a multi-domain architectural element with a C-terminal LRR and a central nucleotide-binding domain (NB-ARC). In the majority of examples, R-proteins also accommodate at their N-terminus a coiled-coil (CC) domain or a domain with homologies to the TOLL/interleukin-1 receptor (TIR) element [50]. In Fig. (**2**) is given a graphical representation of various domains of plant disease R-proteins.

Further research in this field has highlighted that *R-NLR*s genes are located either as single loci or clustered in the genome as a result of tandem and segmental duplications [25]). These clusters are either homogeneous, comprising closely related sequences, or heterogeneous, including distantly related sequences and sometimes both TNL- and CNL-encoding genes. In this respect, it was suggested that *R* clusters favor a fast evolution of these genes facilitated by recombination mispairings, thus providing new *R*-gene sequence variants that may encode R-protein with modified specificities or have an unexpected expression profile.

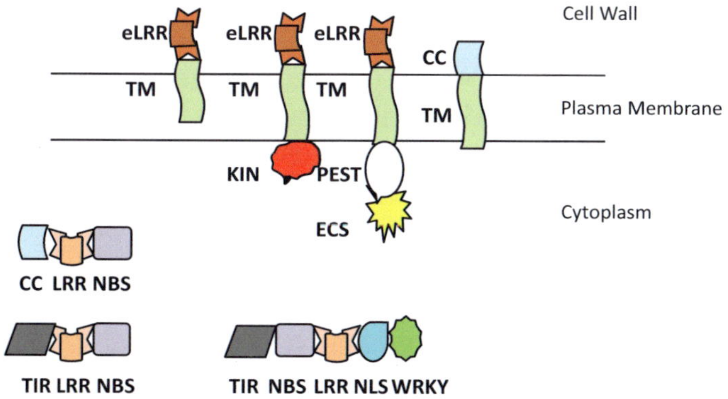

Fig. (2). Various domain regions of plant disease R-proteins. LRR: Leucine Rich Repeat, eLRR: Extracellular LRR, NBS: Nucleotide Binding Site, CC: Coiled Coil, TIR: Toll-Interleukin Receptor, NLS: Nuclear Localization Signal, WRKY: DNA binding Domain (60 amino acids), TM:Trans-membrane Domain, KIN: Kinase Domain, PEST: Degradation Domain (proline, glycine, serine, threonine), ECS: Endocytosis Cell Signaling Domain (Redraw from Buscaill and Rivas, 2014 [31]).

The rate of evolution of *R*-genes differs widely in agreement with subfamilies [7]: some evolve comparatively slowly, while others display classical characteristics of very quick evolution. Accordingly, Yang and coworkers have reported that a different collection of plant *R*-genes, for very quick evolving pathogens, takes place not exclusively within plant species, but also among species [51]. Moreover, the previous workers have indicated the existence of a functional homology among NLR proteins, experiencing a very quick evolution, in various plant species. In conclusion, it is suggested that the utilization of this variability is a valuable source of *R*-genes for contrasting rice blast diseases and potentially additional plant pathogens, and eventually may represent a choice to be exploited in map-based cloning strategies.

Another mechanism that increases diversity in R-protein is due to the fact that numerous *R*-genes undergo alternative splicing (AS) [52]. In this respect, it was suggested that in the finely-tuned mechanism of R-protein activation, the truncated isoforms produced by AS can cooperate in stimulating plant disease resistance. This activation function either by abolishing the unfavorable regulation of immunity initiation, or by straight enrollment into the effector-triggered signaling network. Although, other information has indicated the functional implication of AS in sustaining plant biotic stress responses, this phenomenon remains to a great extent an unexplored research topic. A number of attractive issues concerning AS mechanism active on *R*-genes, particularly influencing its functional activities and regulation, will demand innovative tools and further research activity to untangle them.

R-NLRs Genes: Recognition of Effectors and Downstream Signaling Mechanism

As above indicated plant disease *R*-gene products belonging to the NLR family may directly or indirectly perceived and identity pathogen effectors [13, 23, 45, 46]. Direct interaction proposes that the pathogen AVR effectors directly interact with plant R-proteins to make them ineffective (gene-for-gene hypothesis: *i.e.* for individual host plant *R*-gene there is frequently a matching pathogen *AVR* gene) [53]. Actually a number of examples of matching *R*- and *AVR*-genes have been documented and, in most instances, single NLR proteins mediate AVR effector identification, and thereby conferring resistance to microbial pathogens [7, 54]. Classical model of this type of interactions have been reported in tomato, wherever Pto interacts with AvrPto gene product of *Pseudomonas syringae* [55] and in the rice blast pathosystem (*Magnaporthe* pathosystem), where Pita interacts with Avr-Pita [56].

Relatively to the indirect recognition, it is postulated that the NLR reacts to changes promoted by a plant protein which is activated by the AVR-protein. This plant protein may be represented by either i) an operational target of the effector, in this circumstance it is termed a "guardee": it is guarded by the NLR (Guard Model [13]), ii) a mimic of the operational target and in this case is termed a "decoy" (or co-effector) specialized in attracting effector (Decoy Model [45]). Validations supporting the Guard Model have increased during the past decade with information regarding guarded effector targets, like the claimed guardees RCR3 of tomato. This is an excreted papain-like cysteine protease that in tomato is believed to be guarded by the R-protein Cf-2 [7]. Notably Cf-2 imparts resistance to the leaf mold pathogen, *Cladosporium fulvum,* acting *via* detection of the fungal protease inhibitor AVR2 [57]. This physically interacts with RCR3 and ultimately suppresses it [57]. The Decoy Model is compatible with more recent results about effector activities and processes underlying pathogen recognition by plants. The model is corroborated by various examples about effector recognition mechanisms observed on effector perception studies concerning Pto, Bs3, RCR3, and RIN4 elements [45]. More recently Cesari *et al.* [46] have described an "Integrated Decoy" Model addressed to the functional role of receptor aggregates. In this model, a plant protein targeted by an effector has been replicated and joined to a representative of the NLR pair, whereas it acts as a bait to activate defense signaling mechanisms by the second NLR at the time of effector binding. In Fig. (**3**) is given a schematic representation of the various interaction models mentioned above between R- and effector-proteins.

Downstream Signaling Mechanisms

Identification of the *AVR* gene product through R-protein brings about an intricate cascade of signal transduction mechanisms similar to the above-mentioned processes involved in PTI. In fact, like PTI, ETI is generally correlated with the accumulation of ROS, modifications in ion channels with an influx of Ca^{2+} and H^+ and outpouring of K^+ ions. These events are triggering distinct families of defense-related genes, such as those encoding various classes of pathogenesis-related (PR) proteins [23]. This network of signaling events in ETI is connected with HR, a dramatic local defence process, promoting PCD at the site of infection as above indicated for PTI [23].

Transcription Activator-Like Effector (TALE) Family

TALEs are a subset of type III effector proteins that are translocated from pathogenic bacteria (*e.g.* genus *Xanthomonas* and *Ralstonia*) within plant host cells (see [47], for a recent review). Once inside the cells TALEs penetrate into the nucleus. In the nuclei they function as TFs to modulate host gene expression.

The modulation is achieved by binding to specific promoter site elements, designated as effector binding elements (EBEs) of host resistance (*R*)-genes, accompanied with a reprogramming of the plant transcriptome. These genes are known as executor resistance (*E*) genes, or susceptibility *(S)* genes, because they are able of activating resistance reactions or promoting disease susceptibility, respectively [58 - 61].

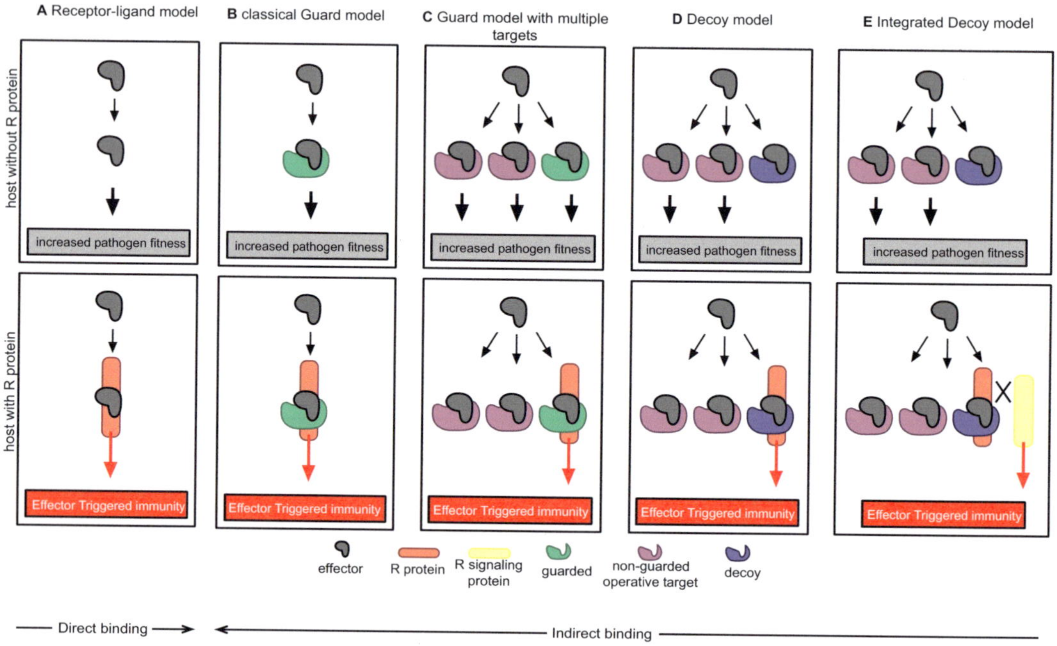

Fig. (3). Schematic comparisons of the Receptor- ligand Model or direct recognition (A) Guard, Decoy, and Integrated Decoy models. The standard Guard Model (B) is confronted with a modified Guard Model in which the effector binds diversified plant proteins (C), the Decoy Model (D), and the Integrated Decoy model (E). Effectors are represented in gray, operative effector targets in purple, guardee in green, decoy in blue, and the R- protein in orange and for signaling in yellow (Modified from Cesari *et al.*, 2014).

Typically, TALEs retain a multiform conserved architecture that is shaped by: i) an N- terminus domain that is indispensable for the type III secretion system (TTSS)-dependent translocation of effectors inside cells of the host plant, ii) a central repeat domain that is formed by a variable score of amino acid repeats (34-35 repeats), able to confer a high binding specificity, and iii) a C-terminus domain that include the nuclear localization signals and an acidic activation domain promoting transcription [62]. Notably, as above mentioned the binding specificity is driven by the variable score of repeats of the central amino acids domain. In a single repeat, a pair of variable residues, located at sites 12 and 13 of this domain (jointly named the repeat-variable di-residue (RVD), are usually linked with a diverse nucleotide to delineate the length and sequence of the EBE

[47, 62]. A schematically representation of TAL effector is depicted in Fig. (**4**). Notably, this modularity in structure makes TAL effectors to be remodeled and thus this feature is an important tool for genome editing, systems biology, and other DNA targeting applications [63]. The expression of the executor *R*-genes (*E*S), similarly to *AVR/R* gene interactions, causes PCD or a HR in plants, therefore limiting pathogen diffusion and favoring disease resistance phenotypes [64, 65].

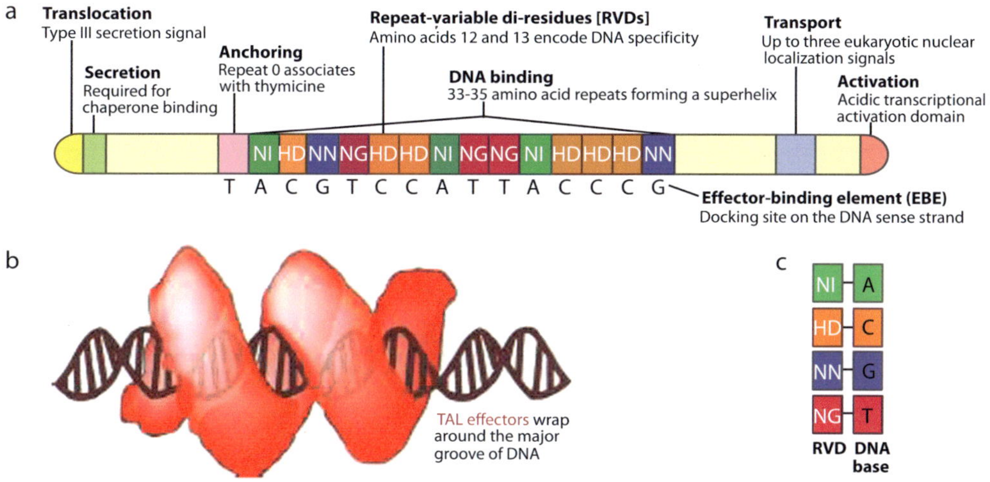

Fig. (4). Structure of transcription activator-like (TAL) effectors. (*a*) TAL effectors consist of three functional domains. The N terminus is required for secretion and translocation into the host. The central domain consists of repeats formed by amino acid repeat-variable di-residues (RVDs) that associate with a specific DNA sequence. The C terminus confers transcriptional activation of open reading frames. (*b*) The tertiary structure of TAL effector proteins consists of a right-handed super helix those winds around the major groove of double-stranded DNA, interacting with the sense strand bases. (*c*) A minimal set of RVDs and their DNA base preference are shown. This set can be used to generate a custom TALE that is able to bind to any sequence of choice (From Schornack *et al.*, 2013 [62]).

Systemic Acquired Resistance (SAR) and Induced Systemic Resistance (ISR) Processes

As previously herein reported the systems of signaling PTI and ETI complexes are connected with the HR, a strong local defence process, leading to PCD at the of infection site: HR is an extremely orchestrated (both temporally and spatially) defence response system; this system engages local cell death (*e.g.* HR, deeply confined deposition of phenolic aggregates and cell-wall expansions in cells circumventing the area of cell death) and a more distal activation of general induced defence mechanisms (*e.g.* systemic acquired resistance or SAR), which exclude a progression of infections on distal sectors of the plant) for a reviewed see [66]. Moreover there is information indicating that a local HR can also 'immunize' plants against future infection. In fact, localized exposure of plants to certain pathogen may induce subsequent resistance to a large array of virulent pathogen agents in distant, non-infected plant tissues [67]. Research has revealed that induced resistance in plants can be spliced into two classes: systemic acquired resistance (SAR) and induced systemic resistance (ISR) (Fig. **5**). SAR is a reaction to pathogens, whilst ISR is activated by beneficial microorganisms and triggers a smaller detrimental defence reaction to provide the plant to contrast potential pathogen invasion [66]. Notably SAR implies the production of a mobile signal(s) in the initial infected tissues (within 4–6 hours), its rapid translocation to systemic uninfected tissues, and preparation of the systemic tissue for subsequent infections from related/unrelated pathogens [68]. On plant tissues, it is commonly visible by a local necrosis, expression of genes encoding PR protein, and demands for the activity of the SA pathway. On the other hand, ISR is repeatedly promoted by plant growth-promoting rhizobacteria [69]. However, it is not connected with necrosis and embraces the JA and ET pathways [70]. Besides to the individual plant, SAR also appears trans/generationally beneficial, whereas an immune 'memory' is transmitted to the next generation ([71, 72]. For the onset of SAR and ISR networks, the cardinal signalling compounds are represented by SA and ET associated with JA, respectively [73, 74]. SAR gives rise to the activation of a 'defence-ready' status [68].

The signalling pathways of this phenomenon constitute an intricate regulatory network which is finely tuned by hormones [37, 75]. Abscisic acid, notably, has been reported to be a central regulator of SA signalling, but additionally pressures from ET/JA signalling are important in the pathogens responses as above mentioned [76 - 78]. Moreover, it was found that SAR is largely interdependent on the activity of specific genes, such as the *Nonexpressor of PR genes1 (NPR1)* identified as a central regulator of SA-induced responses [79, 80]. Specifically, once inside the nucleus, NPR1 provides immunity *via* a transcriptional network, involving transcriptional activators (*e.g.,* TGA3) and repressors (*e.g.,* WRKY70), resulting in the extensive trigger of antimicrobial *PR* genes [80, 81]. Although, the process how NPR1 regulates transcription is incompletely unclear, it was recently reported that sumoylation of NPR1 changes its association from transcriptional repressors to transcriptional activators [82]. Moreover, E3 ubiquitin ligases promote, transiently, defence gene expression.

Priming

The priming phenomenon (*i.e.* physiological event that allows cells to react to very small amounts of a stimulus in a quicker and strong way in comparison to non-primed cells) was proven to be a crucial defensive system in different kinds of systemic plant immunity mechanisms [83]. Notably priming agents include natural (*e.g.* vitamins, chitosans, oligogalacturonides, volatile organic compounds) or synthetic (*e.g.* β-amino-butyric acid, para-aminobenzoic acid, azelaic acid, AZA) compounds and the general activation of wounding processes.

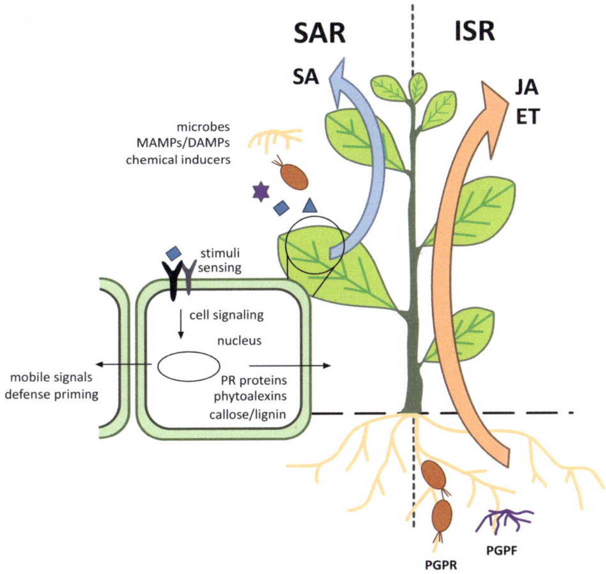

Fig. (5). Draft of distinct classes of systemic resistance complexes... The systemic acquired resistance (SAR, the left draft region is activated following pathogen invasion, foliar applications of plants with MAMPs, DAMPs, plant phyto-hormones or chemical activators. Following the recognition by a sensor molecule, the plant cell activates multiple signalling and defence responses, such as synthesis of pathogenesis-related (PR) proteins, antimicrobial phytoalexins or cell wall reinforcement with callose or lignin. Local defence reactions are accompanied by the synthesis of mobile cues that are moved through xylem and prime distal plant regions for defence compounds aggregation. Induced systemic resistance (ISR, the right draft region) can be activated by root colonization *via* plant-growth-promoting rhizobacteria (PGPR) or fungi (PFPF). ISR is SA-independent network, controlled primarily *via* jasmonic acid (JA) and ethylene (ET). Alike SAR and ISR stimulate a reinforcement of cell walls and a rapid plant defence response, following biotic and abiotic stress agents (From Burketova *et al.*, 2015 66).

In spite of the fact that this event has been familiar since decades, the molecular background of priming is largely unclear. However, current evidence has shown that in primed plants, crucial steps are potentially correlated with chromatin modifications [83]. These variations in chromatine are expressed as methylation and acetylation of histones, occurring on defence gene promoters, and are accompanied by an extensive accumulation of inactive MPK3 and MPK6 signalling proteins, besides to an increased quantity of SA and AZA compounds [83 - 85]. Furthermore, the last cited authors proposed that priming agents, in addition to provide new disease control approaches, fitting environmental regulations, are helpful devices to gain insights into the multiform mechanisms controlling the inducible plant response.

RNAi-Mediated Defence Against Pathogens

In addition to the HR response, plants can defend themselves against pathogens by other mechanisms including a sophisticated genetic defensive system termed RNA silencing or RNA interference (RNAi). The benefit of RNAi as a new gene device to contract fungal, viral, and bacterial infections in plants is due to the fact that it controls gene expression, including mRNA degradation, repression of translation, and chromatin remodeling, *via* the action of small non-coding RNAs (sRNAs). Obviously the flexibility of this strategy provides a powerful tool to analyze gene function and to bioengineering resistance against damages induced by plant pathogens.

Notably, silencing is activated following the formation of the double stranded (ds) sRNAs influencing RNA degradation of the transgene and of the correspondent endogenous genes [86, 87]. Typically the silencing action usually starts with the cleavage of a dsRNA precursor into shorter sequences, 21–26 nucleotides in length, by an enzyme named Dicer [88, 89]. These fragmented RNAs, familiar as small interfering RNA (siRNA) and microRNA (miRNA), drive the RNA-induced silencing complex (RISC) to break down the complementary target mRNA sequences [90]. The finding of dsRNA as the promoter of RNA silencing in plants has stimulated the growth of a more powerful strategy for engineering virus resistance in crops [22]. Furthermore, it is worth noting that the RNA silencing signals diffuse from cell to cell *via* plasmodesmata connections and across extensive intervals through the phloem mimicking the SAR phenomenon previously described [91].

A number of researches regarding sRNAs from plants and pathogens have recently disclosed the function of noncoding sRNAs in plant immunity process [89]. For example, in Arabidopsis, accumulation of the 222 miRNA, miR393, after perception of flg22, drives an unfavourable regulation of transcripts for F-box auxin receptors and a suppression of the auxin signalling mechanism [92]. This event enhances resistance to bacterial pathogens and, therefore, may contribute to reinforce the defence pathogen strategy provided by the PTI system. Similarly, from rice studies it was reported, following infection with stripe virus, five over-expressed and two down-expressed miRNAs [93]. Moreover, further research has also documented that, in agreement to protein effector genes, several of these sRNAs are generated from TE regions located in the plant genome [89]. These molecules are probably cooperating for a speedily evolving virulence and a host adaptation processes.

Collectively, these studies indicated that in higher plants RNA silencing operates as an innate antiviral defensive mechanism. The development of effective RNAi are powerful to i) exploit *in vitro* assays, ii) detect and study RNA silencing mutants, and iii) analyse viral-encoded silencing suppressors. En mass they provide an attractive and exciting investigation field to develop strategies to contrast infections caused by pathogens.

Genetics and Biotechnology Strategies for Improving Disease Resistance

Typically in plants two principal classes sustaining disease resistance have been described: (i) complete resistance controlled by an individual gene (mono-genically) and (ii) incomplete resistance controlled by a complex of genes (polygenically), each of these one is contributing a partial effect. In their deep forms, these kinds of resistance are evident and quickly detected [94, 95]. A plethora of terms have been employed to indicate this division, such as horizontal against vertical, complete against incomplete, major-gene against minor-gene and narrow-spectrum against broad spectrum. or gene-for-gene resistance' or 'race specific resistance'. This variety of terms indicates the spectrum of interests and assumptions elaborated by the various workers. However, it has also increased a factor of complication in the literature: some terms are adopted in distinct manners by various scientists. Therefore, according to St Clair [95] herein the terms 'qualitative disease resistance' (QDR) and 'quantitative disease resistance' (QRL) will be utilized to indicate the corresponding disease resistance event, namely QDR or QRL.

Qualitative Disease Resistance (QDR)

Qualitative resistance generally donates a great degree of resistance. It is typically race specific, sustained by single, frequently dominant, acting genes (*R*-genes) of major effect [96, 97]. Moreover these genes possess high penetrance and tend to be stable across diverse environmental conditions. This class of resistance genes operates at the recognition phase of plant-pathogen interaction, usually *versus* biotrophic pathogens, where the expression of resistance is frequently connected with a type of PCD. A common example is that reported in Brassica, concerning the blackleg resistance gene (*Rlm1),* a race-specific *R*-genes, which confers resistance to a fungal pathogen (*Leptosphaeria maculans)* [98]. Furthermore, research has identified at least fifteen genes supporting resistance to nematodes in different crops [99]. For example, two genes equipped with nematodes resistance have been isolated and cloned, according to their chromosomal position, and identified through genetic complementation. The first gene, termed $Hs1^{pro\text{-}1}$, from sugar beet provides resistance to the beet cyst nematode (*Heterodera schachtii*). The second gene, named *Mi-1,* induces a hypersensitive reaction to tomato roots, following infection with *Meloidogyne* spp. Similarly, resistance to soybean mosaic virus (SMV) is attributable to various genes (*i.e. Rsv1, Rsv*3 and *Rsv4*) located at three different loci in soybean germplasm [97]. Therefore, pyramiding of *Rsv* resistance genes from distinct loci in soybean genome has been advanced as a promising strategy for developing a long life or durable and large interval of resistance to all races of SMV.

Disease resistance of genetically homogeneous lines, holding an individual source of resistance, is often broken down by novel pathogenic strains when planted on large-scale and for a prolong time-period. For instance in Brassica, *Rlm1*-mediated resistance gene evolved into feebleness in protecting plants [98]. This event was seen in as short period of time: within 3 years of commercial release of the cultivar that incorporated *Rlm1*. To bypass breakdown of resistance induced by individual *R*-gene, different approaches, such as composite crosses, synthetics, and multiline breeding, have been exploited [100]. These interventions were made with the purpose of stacking or pyramiding multiple *R* genes *via* traditional breeding or marker-assisted selection (MAS) schemes into a valuable genetic background. Such strategies permit to distinct *R*-gene alleles to be concentrated as *R*-gene stacks, producing plant varieties with an ample range of *R*-genes for avoiding pathogen evolution. Notably this approach also up hold the ability of providing a longer lasting or a more ample range of resistance *versus* pathogens. The probability of pathogen to erase resistance at two or more additional genes is definitively inferior in comparison to individual gene. However, pyramiding of the *R*-genes, using conventional breeding approaches, may not always be possible, or it may be laborious and time-consuming. This is due to the transfer of un-desirable genes (linkage drag) along with *R*-genes that need to be removed by backcrossing and to the presence of dominance and epistatic effects [97]. In this context, it was suggested that the cooperation of genomics and molecular-based breeding tools for assembling disease resistance plant varieties, with gene-based MAS, is a powerful procedures for an efficient selection program. Improvement of disease resistance has been experimentally proved in various crops, including blast disease resistance in rice [101]. It should also worth noting that not all monogenic *R*-genes confer complete immunity: single *R*-genes may as well as determine intermediate or partial levels of resistance. For example, in wheat monogenic rust resistance genes showed a large spectrum of interaction phenotypes [102]. In addition, various genes conferring resistance to pathogens are recessive, especially those contrasting virus infections [103]. Interestingly, recessive resistance genes are disease-susceptible and resistance may reflect a lack of an interaction required for susceptibility. This is in agreement with the event of allelic variants at the *eIF4E* locus that promotes resistance in a number of plant species to potyviruses [104].

Lack of a required molecular activity may also lead to resistance. For instance *mlo* is a durable, recessive gene that supports a large spectrum resistance *versus* all races of powdery mildew in barley and other plant species [105]. It was shown that wild-type MLO is a major unfavorable regulator of plant defenses [106]. Molecularly is formed by a plasma-membrane protein involved in calmodulin binding that accumulates at sites of fungal penetration [107]. Loss of MLO activity, in both Arabidopsis and barley, promotes cell wall depositions, increases H_2O_2 accumulation, and arrests cell entry for all pathogen genotypes [108].

In addition to *R*-genes, further mechanisms have been connected with qualitative resistance, such as detoxification processes concerning fungal toxins such as the case of *Hm1* and *Hm2* loci in maize [1]). In this case *Hm1* gene, protects plants against a lethal leaf blight and ear rot damages racked up by *Cochliobolus carbonum* race 1 (CCR1). Specifically *Hm1 locus* induces resistance against CCR1 pathosystem *via* inactivating HC-toxin, a host-specific toxin definitively requested by the pathogen to colonize maize tissues. This inactivation is moderated by an NADPH-dependent reductase, called HC-toxin reductase (HCTR), which is encoded by the *Hm1* gene. Similarly to *Hm1*, *Hm2*, activates protection against CCR1. However, differently to *Hm*1, which contributes completely dominant resistance during all stages of plant development, *Hm2*-conferred resistance is partially dominant. It turns completely effective exclusively at plant maturity. As anticipated, *Hm2* is a duplicate of *Hm1*, although the protein it encodes is heavily truncated in comparison with HM1. The ability of *Hm2* in protecting plants against CCR1 rises gradually over time, varying from having small or no influence in seedling tissues up to conferring full immunity at anthesis. The developmentally specified phenotype of *Hm2* is not determined at the transcriptional level: the degree of expression of this gene, either taking place constitutively or going through ample and transient induction in response to infection, does not vary with the age of plants. Nevertheless, the *Hm2* transcript is much more represented in homozygous *Hm2* plants for this gene in comparison with plants that bears a single copy of the gene. This finding indicate a transcriptional support for the dosage-dependent nature of *Hm2*. Altogether, this information advocates that distinct mechanisms appear to influence the developmentally program of the *Hm1* phenotype against the partially dominant resistance phenotype of *Hm2*.

Another example is represented by the rice recessive resistance gene *xa13*. This is an allele of *Os-8N3*, which confers resistance to bacterial blight [1]. This last gene is activated transcriptionally by *Xanthomonas oryzaen pv. oryzae (Xoo*) races that is able to express the TALE PthXo1 compounds. The *xa*13 gene possesses a mutated effector-binding sequence in its promoter region that abolishes PthXo1 binding and generates plants resistant to PthXo1 races of the bacterium. This result is also corroborated by the evidence that the *Os-8N3* is necessary to reveal the susceptibility pattern [1].

Quantitative Disease Resistance (QRL)

This kind of disease protection is determined by numerous loci, typically indicated as polygenes or quantitative resistant loci (QRLs). In fact, they do not fit with an elementary Mendelian mechanism of inheritance. QRLs tend to include phenotypes that vary from partial to full resistance. Furthermore, they are efficient against multiple strains of a pathogen, are sensitive to environmental conditions, and are typically determined by several to many genes with small additive effects [95, 109, 110].

A growing amount of research was carried out on the detection and analysis of QRLs for partial and frequently long-lasting or durable resistance [95, 96]. All together these findings have provided details that several of these QRLs are encoding genes precisely engaged in resistance. Examples include fusarium head blight in wheat, late blight in tomato, leaf rust in barley, white mold in common bean, southern and northern leaf blight in maize, as well as resistance to tanspot and rust in wheat [95, 105]. For instance, resistance to white mold (causal agent of *Sclerotinia sclerotiorium*) in common bean is marked off by more than 10 independent QRLs with major-effects, located on various bean linkage groups [111, 112]. Moreover, the last cited authors have employed MAS backcrossing procedures to transfer a QRL located on linkage group B7 into bean cultivars. The inbred lines derived from these backcrosses, tested in replicated field trials and greenhouse disease evaluations, decreased disease damage severity (-15%), advising that transfer of individual QRL significantly decrease disease injures in distinct genetic backgrounds. Moreover Poland [109], have listed, from research reports available in literature, twenty-five QRL-mapping studies carried out for diseases in rice, forty-three for maize, and thirteen for Arabidopsis. It is predictable that in the near future, meta-QTL analyses -*i.e.* tools for integrating information from multiple QTLs studies [113]- and genome-wide association studies -GWAS to identified marker-trait associations [114]-, combined with the increasing number of genomic sequences and RNA-seq studies, will provide more detailed insights regarding polygenic resistance as well as candidate genes and markers useful in MAS [96] For instance, GWAS procedures have already permitted to identify marker associations with disease resistance for southern and northern leaf blight in maize [115, 116], as well as for rust and tanspot in wheat [117, 118]. Collectively, this evidence should cut down the drawbacks of integrating polygenic resistance into breeding schemes and might help in assembling major genes for resistance with other agronomical relevant genes [95]. Efficient genotyping on a global scale may also permit genomic selection of complex polygenic traits, such as adult plant resistance to wheat rust, without requiring genetic dissection of the phenotype [119]. This should facilitate the use of tractable population sizes. Notably, *via* the cloning of QRLs it should also be feasible to clarify the mechanisms supporting resistance and eventually concentrate these genes with divergent and complementing mechanisms of action. This approach appears particularly useful to develop efficiently long-term pathogen resistance crops. In this respect, it is interesting to note that, although durability (*i.e.* resistance that persists efficient when employed in an extensive cultivated area over a long time period in environments supportive to disease diffusion) of QDR has been often claimed, the experimental evidence in favor of this assumption remains poor: findings are largely variable among plant-pathosystems leaving the question yet unsolved [110]. Nevertheless, as more durable resistance genes will be cloned and their mechanisms is understood, it would be feasible to address selection pressure on those genes that recognize and react to vital components of pathogenicity and thereby arresting pathogen infection.

QRL-mapping investigations for disease resistance genes have also provided evidence in support of the occurrence of several disease resistance (MDR) genes [109]. Findings, suggestive of the occurrence of functionally variable MDR genes, have been reported by Wisser and coworkers using a set of different maize inbred lines, characterized by a high gene level of linkage disequilibrium (a necessity for mapping resolution) [110a]. They have found that a *glutathione S-transferase (GST)* gene manifested sizeable favorable genetic associations for resistance against three distinct foliar fungal pathogens. These fungi, belonging to paramount damaging diseases in global maize cultivation, are represented by: (i) southern leaf blight (SLB), produced by *Cochliobolus heterostrophus*; (ii) gray leaf spot (GLS), generated by *Cercospora zeae-maydis*, and *Cercospora zeina*; and (iii) northern leaf blight (NLB)), promoted by *Setosphaeria turcic*a.

Mutagenesis

Induced mutations, including that underlying disease resistance, have successfully assisted the developing of improved and new cultivars [120]. A notable example is given by the barley *CNGC4* gene. This gene is encoding a disease resistance-related cyclic nucleotide-gated ion channel component [121]. In particular the barley mutant allele *nec1,* a variant due to a frame-shift mutation of CNGC4, manifests a clear necrotic phenotype and constitutively over-expresses a pathogen related protein (*i.e.* PR-1). Furthermore, *nec1* mutant plants produce in their tissues a large level of SA and H_2O_2 when confronted to its parental variety named Parkland. Further investigations, concerning *nec1* disease resistance plants, showed, when assay at high inoculum spore densities, a favorable influence of *nec1* mutation on non-host resistance *versus* the bacterium *Pseudomonas syringae pv. tomato* (*Pst*) [122]. By contrast when the plants were assayed with a normal *Ps*t inoculum concentration, *nec1* resistance plants did not differ from the wild type ones. The same authors provided evidence that the *nec1* mutant is also able to overexpress the *BI-1* and *MLO* genes supporting other processes promoting resistance of the race non-specific fungus *Blumeria graminis f. sp. hordei* (*Bgh*). This is the causal agent causing powdery mildew in barley. It was speculated that the *nec1* mutant alter induction of SAR production, probably *via* a SA signaling activation process. This can implement resistance to contrast invasion of several additional types of pathogens.

Novel *R*-gene alleles, able to promote new recognition specificities, may be developed *via* directed mutagenesis [123] or *via* domain swapping interventions among *R*-genes [124]. Mutations created within the potato *R*-gene, *RX*, which is able to promote resistance against the potato virus X (PVX), gave rise to new alleles. These were shown to be able to combat positively the infection caused by additional strains of the PVX virus and by a virus of poplar trees (*Populus* spp.), *i.e.,* poplar mosaic virus [123].

Additionally mutagenesis, including TILLING (Targeting Induced Local Lesions In Genome), has become a powerful device in gene discovery and functional analyses in reverse genetics. In particular, TILLING enables screening of mutations at the molecular level [125]. In this respect, a well-documented example for TILLING procedures was the identification of mutants conferring resistance to *Fusarium sp.* These fungi are of concern not exclusively for relevant losses in the production of major crops (*i.e.* maize, wheat, barley, oats) but, notably, because any species of fungi, engaged in infection development, may synthesize a plethora of mycotoxins, poising serious threats to human and animal health [126]. Although there is genetic variability in host susceptibility *versus Fusarium* species, no main disease resistance gene has been presently detected [127]. For example, in oats variation in resistance to fusarium head blight disease in breeding populations was not large enough to detect and thus select lines possessing a clear resistant to this specific disease [128]. In this respect, Sikora *et al.* [125], assaying an oat TILLING population, have detected three lines that were less infected, at high inoculum spores density, in comparison with the principal resistant profitable cultivar and showed a better resistance when compared to the original Belinda variety. Collectively, these results reveal that phenotypic screening of mutagenized populations is useful to detect multiple features, such as pathogen resistance, whenever the screening protocols are thoroughly designed. Finally for the enhancement of disease resistance, the development of spontaneous mutations may also by exploited by other distinct mutagenesis tools such as: virus promoting gene silencing, RNA-mediated interference, and insertional mutagenesis using transposon tagging [125].

Alternative Strategies to Improve Resistance

Transgenic Prospects

The transgenic technology is an appropriate solution in circumstances where sufficient levels of resistance to pathogens are not available in a specific plant species or alternatively in its wild relatives and germplasm resources. It is also useful in the case that resistance is laborious to transfer by traditional breeding procedures, on account of either reproductive isolation of the species or linkage with additional unwanted characteristics.

Several genes were used for generating transgenic crop varieties (or genetically modified, GM varieties), provided with resistance features for contrasting damages induced by the invasion of fungal, viral, and bacterial pathogens [1, 129].

It was evident from these results that attempts employed for developing the first generation of transgenic disease resistant crops were essentially based upon the use of single genes, including *R*-genes or genes that encode antimicrobial proteins (*e.g.* chitinases) and additional antimicrobial proteins (PR proteins) that are directly involved in conferring resistance [129]. Nevertheless among the genes conferring disease resistance to transgenic plants, none of them turn to practical account from a commercial point of view. Similar strategies have been exploited to produce GM modified nematode resistance plants. These strategies, based on the overexpression of cysteine proteinase inhibitors (cystatins), compounds able to inhibit intestinal digestion of nematodes by impairing their dietary protein intake from the plant tissues, were effective against a range of these pests with different tactics of parasitism [130].

One of the more serious practical problems found with these transgenes is that the activity of the antifungal or antibacterial proteins is altered by the endogenous defense process innate in wild-type plants: the newly synthesized proteins have to match themselves with the native or endogenous defense substances produced by plants. According to Chen and coworkers [129] to overcome these difficulties, the next-generation of transgenic plants should be directed in manipulating signal-transduction networks. In agreement with the previous cited authors, this type of intervention devoted to control the induced defense mechanisms should result more valuable for contrasting pathogen invasions. However, further research is needed to clarify in details the complexity of the signaling pathways. These includes a large number of complex factors such as hormones, defense modulators, and TFs, as well as their interactions with abiotic stress agents *(i.e.* drought, heat, cold, salt) and the biology of the organisms infecting plants. Progresses in sequencing crop genomes, combined with extensive gene expression and functional gene investigations, would certainly advance the production of transgenic disease-resistant crops. Notably, approaches exploiting the general defense pathways (PCDs) and the antifungal protein overexpression should improve our understanding of plant defensive repertoire. This will supply information on the way how to complement defensive arsenals or calibrate the expression of genes supporting them. Additionally, investigations on pathogen-responsive promoters, will also give insights to develop a non-pathogen-specific hypersensitive reaction in plants following pathogen invasions. Moreover, it is also projected that some non-host resistance genes might be discovered. These can be subsequently evaluated for their long lasting resistance or durability.

From the most outstanding examples of transgenic disease resistance plants, commercially exploited, is given by virus resistant papaya and summer squash. An attempt applied to develop virus-resistant plants was to transfer genes, from the pathogen itself inside the plant genome, producing the so called pathogen-derived resistance (PDR) [97]. The principal widely used intervention was to express the virus coat protein (CP), which is implicated several processes concerning viral biology - *e.g.* encapsidation, virus replication, cell to cell or systemic movement, vector transmission, symptom development- in transgenic plants. Practically, the expression of viral genes breaks up infection or symptom evolution induced by viruses. One of the most successful implemented GM project in this direction is the Rainbow papaya resistant plants to Papaya Ringspot Virus *(*PRSV) [131]. The planting of these transgenic trees essentially recovered the papaya industry in Hawaii from the danger caused by dreaded ring spot disease. Interestingly, these GM varieties possesses a gene that encodes a PRSV *(papaya ringspot virus*) coat protein, an approach mimicking the development of cross protection.. In the normal phenomenon of cross protection, infection by a moderate strain of a virus, promotes resistance to successive infection caused by a more virulent strain [97]. This approach has been expanded to other plants, including wheat, soybean, rice, tomato, peanut, and plum tree with favorable results, without reaching, however, a commercial breakthrough.

Beside CP mRNA, other strategies, based on satellite (sat) RNA, deficient interfering RNA, non-coding regions of viral genome, viral sequences in sense or antisense orientation or in double stranded forms, were also found to implement RNA mediated resistance. Interestingly, these RNAs are acting by degradation of viral RNA through the siRNA pathway [97]. RNAi-based transgenic defenses were also proven effective to control infection by nematodes [130]. Another form of PDR was the introduction of viral replicase genes (or RNA-dependent RNA polymerase genes), which likely function *via* post-transcriptional gene silencing. This intervention has been adopted to bring about resistance to different diseases in crops, such as potato leafroll virus in potato, barley yellow dwarf virus in oats, spherical virus in rice, wheat streak mosaic virus in wheat, and cucumber mosaic virus in tomato [97]. RNAi-based transgenic defenses strategies were also found effective to control infection by nematodes [130]. Similarly, short peptides mediated resistance were credited to bypass synthesis of an elevated amount of transgenic mRNA that might activate post transcriptional gene silencing and suppress protein expression [97]. However this approach provides a poor level of broad spectrum resistance [130a]. Furthermore, to induce protection of plants *versus* more than one virus, ribosomal inactivating proteins (RIPs), have been synthesized in transgenic plants. Notably, RIPs are powerful inhibitors of protein synthesis. Moreover, depending on the plant species, from which they derive, RIPs exhibits several degrees of toxicity against distinct viral hosts. For example, poke weed antiviral protein (PAP) promotes resistance to PVX and PVY in transgenic potatoes, while PAPII induces resistance to TMV and PVX. Another RIP, termed b-32, identified in maize endosperm was found to decrease fungal infections in tobacco, wheat, and maize plants [132]. Recent evidence has highlighted that in tomato enrichment with anthocyanin compounds can reduce the susceptibility of fruit to the postharvest, necrotrophic pathogen, *Botrytis cinerea,* associated with an extension of the shelf life of fruits of tomato [133]. It was proposed that anthocyanins enhance the anti-oxidant ability of plants, and therefore decrease the oxidative injuries that take place during over-ripening and pathogen infection stages.

Even if, the above cited strategies gives a certain degree of resistance to one or more families of viruses, none of them is able to induce immunity. More recently, approaches to develop transgenic plants resistant to viruses *via* the plant innate immune system have been appointed. These include the 2,5A oligoadenylate (OAS) antiviral pathway activated *via* interferons in mammalian cells [134]. The OAS pathway represents an innate immunity complex that reacts to the PAMPs, similarly as virus double stranded RNA (dsRNA) molecules. In this context, research has pointed out evidence indicating that the introduction of the OAS complex in tobacco developed a broad spectrum of virus resistance in transgenic plants when compared to controls [135, 136]. Visibly the resistance reaction of transgenic tobacco plants is analogous to the HR response, where the virus infection is confined or restricted into very narrow necrotic lesion zones. Finally, the 2-5A pathway is also able of triggering defense-related genes and SAR in transgenic tobacco plants that were CMV-infected [137].

Omics Strategies

Next-generation sequencing and genome engineering, or as a matter of choice genome editing, tools are among the most fasting growing sectors of biotechnology in recent years. These technologies can increase our knowledge regarding crucial biological mechanisms and will permit to design new alternatives for crop improvement. The very quick development of these tools has also propelled a new generation of genomic technologies. These integrate diversified strategies to functionally typifying and managing aspects related to the manipulation of genomic information.

In the current resistance selection programs, effectors are coming out as devices to increase the speed and advance detection, functional analysis, and selection and breeding of genes underlying resistance to pathogens. In this context genome-wide database concerning effectors have recently become usable for a number of pathogens [138]. This information has admitted to establish a high-throughput functional genomics approach, or effector-based breeding, termed "effectromics". This technology may provide favorable information extremely useful to enhance disease resistance in crops. In particular, the last cited authors have suggested that effectromics might contribute several advancements in this area of plant breeding activity such as: i) rapid identification of novel *R*-genes belonging to the NLR family ii) prevention of redundant breeding or cloning efforts; iii) support in "Mendelization" of field resistance; and iv) expansion in the specificity of effector recognition of a distinct *R*-gene. Interestingly inspection of effector allelic variability into pathogen populations may help in *R*-gene classification available in agriculture and to engineering new *R*-genes with a broaden interval in pathogen recognition. For example, Hummels *et al.* [139] have added new TALE binding sites, located in the promoter region of a rice gene that confers resistance to a strain-specific pathogen, promoting bacterial blight in this crop. The resulting rice transgenic plants equipped with this modified promoter were made effective against additional strains and against bacterial leaf streak. Notably, this is another rice pathogen for which no major gene resistance has been yet detected and consequently this strategy is highly suggestive. Similarly the recent evolution in developing pathogen resistance is represented by targeted genome-editing tools [140]. They include zinc-finger nucleases (ZFNs), transcription activator-like effector nucleases (TALENs), and the clustered, regularly interspaced, short palindromic repeats (CRISPR)/Cas techniques. Interestingly, these technologies supply stimulating potentiality to manage the problems of remarkably specific genome changes with extreme effectiveness and precision [141]. Notably these techniques take advantage of sequence-specific designer nucleases that are able to alter targeted loci. This circumstance permit to generate small insertions and deletions, insertion of new DNA, as well as the substitution of single alleles. For example, the TALEN technology was successfully used to edit the promoter of the rice S-gene *Os11N3*, which contains an effector-binding element for AvrXa7 of *X. oryzae pv. oryza*e [142]. Thereby, gene editing makes feasible to modify exclusively the *S-gene* function without disturbing other biological roles. Similarly Wang *et al.* [65], using TALEN and CRISPR/Cas9 technologies, successfully introduced targeted mutations in the three homeoalleles (*i.e. TaMLO-A1, TaMLO-B1 and TaMLO-D1*) encoding the MLO protein in hexaploid bread wheat. The resulting induced mutations of the *TaMLO* homeologues, evaluated in transgenic wheat plants, promote a heritable broad-spectrum resistance to powdery mildew generated by *Blumeria graminis f. sp. Tritici*. This fungus is one of the principal devastating plant pathogens globally diffuse in cultivated fields.

PLANT INSECT PESTS

Insects produce large losses in crops, both directly and by facilitating infection of pathogens. In particular, they are responsible for two major kinds of damages to growing crops: direct feeding on above-ground and below-ground plant components. The amount and quality of damages switches dramatically based on the feeding tactic used by insects [143]. Additionally, it was indicated that insects may also create damages to plants following egg depositions (oviposit) inside plant regions. Massive oviposition into stalks may generate mortality or devastation of stalks or branches, whilst oviposition on fruits can develop in deformed or aborted fruits. Insects able to form galls provoke their host plants to develop abnormally. Based on the insect species, the gall formation may be accelerated by feeding or by oviposition (egg-deposition) within plant tissues or organs. Furthermore, various insects, particularly sap-suckers, are indirectly responsible to transmit plant diseases from one host to another [144]. Insects that deliver plant diseases are named vectors. The majority of the plant diseases delivered by insects sum up plant viruses, although there are models from additional plant pathogen categories (*e.g.* bacteria, fungi, or nematodes). Viruses transmitted by insect are either stylet-borne (adhering in a specific way to insect mouthparts) or contained in the saliva of insect (circulative viruses). This virus-insect interaction is largely specific, although viruses can infect a large array of plants. Notably, insects employ different feeding tactics to pick up nutrient substances from all above (vegetative) and below-ground (root) plant components and bring about mechanical injuries on plants zones.

Plant Defense Mechanisms

Maxwell and Jennings [145] have defined insect resistance as "*those heritable characteristics possessed by the plant which influence the ultimate degree of damage done by insects*". Resistance is a comparative concept: spatial and temporal variations in the environment influence its expression and/or effectiveness. Furthermore, host-plant resistance can be the consequence of a series of interactions between plants and insects. This event may bias the choice of plants as hosts and the impacts of plants regarding survival and population increase of insects.

Generally, the development of plants resistant to particular insect pests is obtained by selection procedures to breed resistance traits. These naturally plant's defensive features are related with various mechanisms that include: *antibiosis* (*i.e.* morphological characteristics produce unfavorable effect on the insect behavior, including growth rate, development and reproductive progress), *antixenosis* (*i.e.* morphological traits manifest unfavorable effects on the insect feeding preference, including selection of host plant, oviposition and feeding tactics), and *tolerance (i.e.* plants are able to withstand or recover from insect damage) [146]. In spite of these inherited traits, tolerance is more deeply affected by variation due to environmental conditions rather than non-preference and antibiosis. Moreover, the age, dimension, and common vitality of the plant and dimension of the insect-resistant population can heavily modify the level of tolerance of plants to damages brought about by insects.

Host Plant Defenses against Insects

To manage with the devastating damages caused by attacking insects, plants have elaborated a sophisticated repertory of defense mechanisms. Typically, plant resistance to insects is dependent on two types of response that involve direct and/or indirect defense strategies [147]. These types of response can be, constitutively orchestrated or induced following herbivore aggression, respectively.

Direct Defense Mechanisms

Direct defense strategies include structural barriers, morphological traits for physical defense (*e.g.* spines, spikes, and trichomes, epicuticular wax coatings, wax crystals, tissue persistence, in addition to secretory complexes and pipes for the transit of lattices or resins (reviewed by [148, 149]). Furthermore, they also comprise various aggregates operating in chemical defense, such as secondary metabolites. These chemical are characterized by low molecular weight molecules, and defensive peptides and proteins, generally with a high molecular weight, that are very effective in contrasting insect aggressions. Although, several defensive chemical compounds are likely toxic to the plants that synthesize them, the accumulation of relatively beneficial precursors that are switched on by herbivory agents is a reiterated topic of research in plant biology.

Interestingly, secondary metabolites consist of a wide spectrum of compounds varying from terpenoids, alkaloids, tannins, furanocoumarins, phenolics, glucosinolates, cyanogenic glycosides, saponins, and benzoaxinoids. This ample variability of molecules derives, to some degree, from an evolutionary mechanism promoted by selection processes designed to upgrade chemical defenses for contrasting microbial and herbivorous assaults [148, 150]. For example, many terpenoids and alkaloids play a significant role in plant defense [151]. For instance, in cotton derivatives of sesquiterpene promote resistance to several insects by deterring insect feeding on plants. Azadirachtin, a derivatives of squalene from *Azadirach indica,* is one of the most effective recognized deterrents to insect feeding. Specific plants reshape squalene to form molecules that are able to mimic the insect molting hormone ectysone. These plant ectysones disorganize the larval molting phases of insects nourishing on plants.

Similarly benzoxazinoids (BXs), another class of secondary metabolites, are powerful to promote plant defenses and allelopathy [152]. These chemicals are produce in an ample supply in grasses, including the principal crops (*i.e.* wheat and maize). They are produced during the seedling stage and deposited as glucosides inside the vacuoles [153]. A specific glycosidase, situated in the chloroplast, catalyzes the production of the toxic aglucon after a cell is damaged and decomposed. Specifically, glicosides such as DIBOA [2,4-dihydroxy- 2H-1,4-benzoxazin-3-one and its C-7-methoxy] are deposited as inactive DIBOA-glucoside, and enzymatically switch on to generate toxic breakdown compounds. Their derivatives, termed DIMBOAs, are the principal examples of BXs in plants. DIMBOAs were found to act as an inhibitor of various important compounds involved in insect metabolism, such as α-chymotrypsin, aphid cholinesterase, and plasma membrane H+-ATPase [153]. The relationships between DIMBOA concentration and degree of protection given by the insect feeding injury traits were particularly evaluated [153]. The findings of these studies pointed out that DIMBOA may function as a feeding obstacle and decrease insect larvae viability [153]. Therefore this evidence has suggested a feasible utilization of this relationship exploitable in breeding programs devoted to develop maize plants with enhanced insect resistance [153].

Plants also accumulate large amounts of distinct polyphenolic acids, besides their glycosides and esters. These chemical molecules are involved in two defense mechanisms, namely the phenolic reinforcement of cell walls and the detrimental result of fiber concentration on insect growth [154]. Free phenols, prevalently 4-coumaric and ferulic acids, were suggested as favorable agents in advancing resistance in maize to contrast ECB and weevil (*Sitophilus zeamais*), and pink stalk borer (*Sesamia nonagrioides*) infestations. Moreover, some interest has been recently focused on plant resistance characteristics that are dependent on the biosynthesis of small lipophilic molecules (SLMs) [155]). Interestingly, SLMs are opening new types of strategies in this modality of specific breeding programs. However, these approaches are more challenging than the genetic engineering strategies currently adopted.

Besides potential synergistic response, metabolic variability in toxin synthesis by single plants may contribute defense mechanisms against a number of herbivorous insects with diverse feeding habits or resistance networks. Current research on glucosinolates shows how natural selection for a diverse pattern of secondary metabolites has supplied a range of defensive specificity.

Glucosinolates, mainly found in the Brassicacae, are stored in the plants separately for enzymes that break down them [156]. Hydrolyses of these compounds, providing molecules that are disgusting or toxic to herbivores, take place only in plants that are mechanically injured. Herbivores that may tolerate, breaking down products of glucosinolates, frequently avoid plant that synthesizes them. For instance, white butterfly of cabbage is attracted to deposit eggs on leaves of *Brassica* spp. by glucosinolates presence.

Approximately forty different glucosinolates have been detected in *A. thaliana*, and more than one hundred breakdown compounds are probably generated, following activation due to myrosinase enzymes. Experimental trials, using four insect herbivores, demonstrated that tryptophan-derived indole and methionine-derived aliphatic glucosinolates exhibit distinct effects on *Hemiptera* and *Lepidoptera* larvae [157]. Indole glucosinolates, able to break down in the deficiency of the activating enzyme myrosinase [158], promote a more excellent defense for contrasting *Myzus persicae* damages in comparison with the more stable aliphatic glucosinolate compounds [159]. The majority of genes, implicated for the synthesis of glucosinolates, have been identified in the model plant *A. thaliana* and other *Cruciferae* [160]. As an example of how this type of information, concerning biochemical pathways that may be exploited to modify positively plant immunity to herbivory, has been reported in A. *thaliana.* Plants of this species were engineered with genes, encoding three different enzymes found in grain sorghum that is able to synthesize the cyanogenic glycoside dhurrin. The use of these transgenes significantly improves resistance to damages induced by the yellow-striped flea beetle (*Phyllotreta nemorum* [161]).

A number of plants deposited in their own tissues high amounts of certain non-protein amino acids [150]. Some of these molecules sharply resemble protein amino acids and thus cannot by identified form them by animal tRNA synthases with the correct amino acids during translation processes involved in protein synthesis. The tRNAs of herbivores that eat plants including non-protein amino acids thereby turn into pressed with these molecules instead that with the right amino acids, driving to distortion of protein synthesis or unsettled proteins. For example, canavanine, which resembles arginine, is deposited largely in jackbean (*Canavalia)* seeds and is erroneously taken up as arginine by the arginil tRNA synthase of animal organisms. In spite of this arginil tRNA synthase of jackbean plants can single out arginine from canavanine and thus the plant is undisturbed by its own toxin.

Finally there is evidence that, in response to injuries caused by herbivorous insects, plants may also release a specific mixture of volatile organic compounds (VOCs) [162]. These compounds are represented by an ample set of low molecular weight terpenoids and green-leaf volatiles [162]. It was suggested that VOCs can protect plants by repelling, deterring, and exhibiting toxicity to herbivorous attackers.

Defensive Proteins

Studies have shown that a large array of plant defensive proteins is implied in plant defense mechanisms to counteract insect herbivores development [163]. The best known example of plant proteins clearly engaged in defense networks is given by lectins, inhibitors of proteolytic enzymes, RIPs, chitinases, and glycohydrolases [148, 164]).

Protease inhibitors (PIs) are produced and stored in plants seeds and tubers [165, 166]. The expression of several *PI* genes is promoted in response to mechanical wounding or insect injuries [165]. PIs, which damage different families of digestive proteases in the insect midgut, have been analyzed in detail for their function in the active defense reaction [167, 168]. Inhibition of gut proteases by PIs generates amino acid shortcomings that negatively influence insect growth and development [169, 170].

The defensive role of several PIs to contrast insect pests, directly or by expression in transgenic plants, has been investigated in trials with various lepidopteran [166] and hemipteran insects [167]. The performance of PIs as a defense is frequently beaten by adaptive capacity of insects to synthesized digestive proteases that are either unresponsive to the host plant counterpart of PIs or that are able to suppress PIs activities [171, 172]). However, advances in genome sequencing have allowed identification of a large number of PIs and other defense components that are produced in plants following herbivore damages [148].

Plant lectins represent carbohydrate-binding (glyco) proteins, ever present in nature, and exhibit a protective role against a plethora of insect pests [173, 174]. One of the most remarkable features of lectins is their ability to survival in the digestive apparatus of herbivores [174]. This event confers to lectines an outstanding insecticidal capacity. It was also found that lectins operate as antinutritional and/or toxic compounds by binding to membrane glycosyl groups. These mechanisms protect the digestive tract by promoting a range of damaging systemic responses.

Additionally the plant's defensive protein repository involves enzymes that disorganize the digestive physiology of insects and other properties of food assimilation of these organisms. Representatives of the cysteine protease family of enzymes, for instance, includes the chitin-rich peritrophic membrane that have a role in protecting the gut epithelium [175, 176]. Furthermore, it was found that oxidative enzymes, like polyphenol oxidase and lipoxygenase, covalently alter dietary protein by the formation, respectively, of reactive o-quinones and lipid peroxides [148, 177]. This impairs the nutrient uptake by insects and, in turn, their survival. Accordingly, the discovery of novel defensive proteins may be accelerated by proteomic studies carried out on gut content and feces of insect herbivores. The application of this intervention to contrast larvae of tomato-reared *Manduca sexta* has allowed detecting isoforms of arginase and threonine deaminase [178]. These enzymes, able to downgrade the essential amino acids arginine and threonine, respectively, in the midgut of lepidopterans, turned out to be another favorable strategy usable against these insects.

Indirect Defenses

Herbivore-induced plant volatiles are among the principal substances operating as signals in indirect defense mechanisms. Besides gaseous phyto-hormones (*e.g.* ET) triggered by herbivorous insect species, plants synthesize and emit a specific blend of VOSs in response to damages from insects through the expression of specific sets of genes [179]. VOCs are produced specifically in response to herbivore attacks and are popular as herbivore-induced plant volatiles (HIPVs). These compounds come forth from the exposed herbaceous regions (most likely leaves and flowers) and supply air-born cues by attracting carnivorous arthropod species (predators and parasitoids) or deterring ovipositing female individuals from their host plants [180].

The induced volatiles are prospected to accomplish a range of physiological and ecological roles, including attraction of natural enemies of herbivores, and/or engagement of predators and parasitoids, besides to act as airborne signals and reallocation of resources upon attack [180]. The specificity of this interaction has been elegantly exemplified by Degenhardt *et al.* [181]. They showed that herbivore-induced expression of a *TPS23* (*terpene synthase 23*) gene in a maize strain promoted the synthesis of the volatile substance, namely (E)-β-caryophyllene, that induced a substantial attraction of the natural competitors of larvae of western corn rootworm (*Diabrotica virgifera virgifera*), a severe root pest in maize cultivation, in comparison with a maize strain that did not trigger *TPS23* expression. The results of this research suggested that plant VOCs can be manipulated to improve the performance of biological control elements opening the door for a novel ecologically sound pest management approach.

Inducible Defenses and Signal Transduction Pathway

Orchestration of defensive frameworks to contrast attacking insects is achieved from interactions between the plant and herbivore-derived elicitors and effectors. These events proceed with a very quick activation of complex plant signaling networks. A schematic representation of these events elaborate by Arimura *et al.* [147a] is shown in Fig. (**6**).

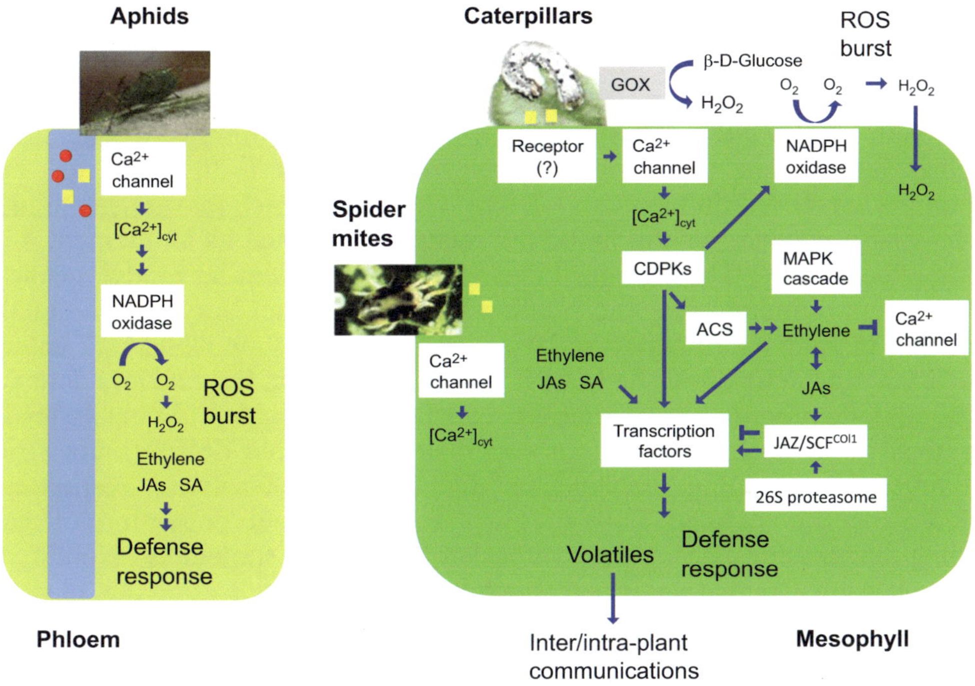

Fig. (6). Cartoon of the signaling cascades for plant defense reactions to chewing (caterpillars) and sap-sucking organisms (aphids and spider mites). Arrows and bars represent positive and negative interactions, respectively. The general picture can vary in specific plant taxa. However, generally, masticating organisms trigger JA-dependent defense reactions, while sap-sucking organisms often trigger SA-dependent defense reactions. Circles in red and squares in yellow are molecules representing oral elements of organisms (effectors and elicitors, respectively). Outlines: ACS, 1-aminocyclopropane-1-carboxylate (ACC) synthase; CDPKs, Ca^{2+}-dependent protein kinases; GOX, glucose oxidase; JAs, jasmonates; MAPK, mitogen-activated protein kinase; ROS, reactive oxygen species; SA, salicylic acid. (From Arimura *et al.*, 2011 [147a]).

Below are briefly given some key features regarding the activation of inducible defenses. Recent reviews have described these features in considerably more detail [*e.g.,* 148-149, 164]. In analogy with PAMPs, herbivore-associated molecular patterns (HAMPs) were suggested as a term to include all herbivore-derived signaling substances that can be recognized by host plants and thus elicit defense responses. Evidence indicates that HAMPs comprise multiple molecules, such as fatty acid–amino acid conjugates (FACs), inceptins, caeliferins, oligouronides, β-glucosidase, glucose oxidase, and lipases. Signal transduction networks perceive the signaling molecules, like those described above, and transmit the signals to the nuclear genomic equipment. This is realized *via* an exhaustive pathway of interacting networks downstream of the sensor/receptor mechanisms. In this respect, there are several components that are implicated in the systemic induction of defense reactions. They comprise system in peptides, oligogalacturonides, formed by polygalacturonase systemically activated following wounding, and jasmonates [149]. It was documented that systemins – a small protein of eigthteen amino acid residues- can promote responses at extremely low concentration. It is produced by cleavage for the C-terminus of prosystemin at the site of wounding. Moreover systemins, move across the wounded leaf within (~30 min a short time period (~30 min) and other unwounded leaves within hours (~ 1-2 hrs). Once they reach leaves systemins are able to bind a plasma membrane-bound receptor, detected as an element of the LRR leucine-rich Ser/Thr receptor kinase family. With similar model, occurring at a specific wounding site, the binding triggers distinct very quick signaling events that include membrane depolarization, a raised of cytosolic Ca^{2+} amounts and of ROS, and activation of a MAPK signaling cascade. Beyond any doubt, the MAPK cascade promotes the synthesis of JA hormones, which ultimately leads to the activation of defense genes. In conclusion, inducible defenses are strictly coordinated by signaling mechanisms that control downstream responses to herbivore-specific signals. This involves transcriptional activation of genes encoding enzymes sustaining the synthesis of defensive molecules. Notably this response is similar to that described above for pathogens indicating that plants use the same pathway to contrast pest attacks.

Genetic Bases for Insect Resistance

Studies indicate that there are a relative small number of single dominant genes in various plant species conferring resistance to several insect orders [182]. Individual genes are known to provide resistance to a single insect species or to strictly allied insect species occurring within the same genus.

Unlike plant–pathogen mechanism for gene-for-gene interactions, there is a poor knowledge about *R*-genes engaged in plant–insect interactions. These are related to the patterns of how plants perceive the products released by invading insects and what are the downstream reactions that are turned on. The Mi gene from tomato is one of the few examples to illustrate these systems. This gene was initially discovered as a dominant gene in conferring resistance to a root-knot nematode (*Meloidigyne incognita*). Its molecular cloning indicated that is a representative of the NBS-LRR repeat family of R- genes [183]. After cloning, further investigations showed that it is arranged at the same locus as that earlier acknowledge as *Meu-1*, contributing resistance to various biotypes of the potato aphid (*Macrosiphum euphorbiae*). Another NLR protein that is encoded by the melon VAT gene provides improved resistance to both cotton aphid (*Aphis gossypii*) and prevents the diffusion of plant viruses *via* this aphid species [184]. In agreement with plant defense mechanisms to contrast pathogen attacks, these results provide evidence suggesting the existence of a gene-for-gene interaction linking plants and insects. Nevertheless, the postulated AVR proteins in the saliva of aphids have not yet been discovered. Correspondently, studies on nematodes have detected approximately fifteen genes that are able to promote resistance to nematodes in several crops [185]. For instance two genes conferring nematodes resistance were cloned, by using chromosomal position strategies, and identified through genetic complementation tools. The first gene, termed *Hs1-pro-1* from sugar beet, provides a remarkable resistance to the beet cyst nematode (*Heterodora schachtii*). The second gene, named *Mi-1*, is involved in the hypersensitive response of tomato roots following attack of Meloidogyne spp.

Future research should be devoted to i) clarify the resistance pathway operating downstream of insect *R*-genes and ii) discover the insect-derived molecules (effector proteins), believed to exist in the insect saliva, that are activating the resistance response of plants following insect infestations [186, 187].

Gene Mapping and Molecular Markers

QTLs impressing insect-resistance have been mapped in different plant species [188 - 190]. For any individual character that was evaluated, the number of QTLs discovered ranges from one to ten, and the proportion of variability accounted for each individual QTL varies from 1.3% to 58%. A common illustration is emphasized by QTLs governing the accumulation of methylketones and acylsugars in wild relatives of tomato plants [191, 192]. Furthermore, several studies were accomplished on the variability in the synthesis of glucosinolates and QTLs sustaining their production [193]. Though several QTLs determining variability in insect resistance were detected, the inherent molecular networks at the basis of these QTLs continue to be amply unexplained.

Besides to their benefit as selectable tools to accelerate breeding procedures, molecular markers may be in addition applied to upgrade our knowledge regarding the processes elaborated by plant to withstand with insect attacks. It is advanced that helpful information might be gained into the type of resistance, by i) mapping of QTLs, encoding for distinct plant physical and/ or biochemical traits correlated with insect resistance, and ii) comparing the chromosomal positions of these QTLs with those discovered for the phenotypic manifestation of resistance to a pest species. Frequently, this information may have consequences, for a fundamental as well as practical viewpoint, exploitable in developing insect-resistant crop varieties more thoroughly. The benefits of these strategies are documented by studies conducted by Byrne and collaborators [194]. Notably they have exploited molecular markers and QTL mapping tools to untangle the genetic determinants conferring resistance to maize plants against corn earworm (CEW) (*Helicoverpa zea)* larvae. These organisms determine a straightforward production loss, often accompanied by the growth of kernel-rotting fungi that further depress grain yields. Additionally, sizeable unfavorable relationships were found between maysin amounts in fresh silks and development of CEW larvae in dried-silk in biological trials [195]. Due to the fact that C-glycosyl flavones are produces through a ramification of the well-known flavonoid synthetic pathway, Byrne *et al.* [194] have speculated that several loci are thus involved in this pathway. These genetic determinants might account for a relevant percentage of the measurable variation observed in maysin amount deposited in plant tissues, and by stretching out, with resistance to CEW. These loci were proposed as "candidate genes" for QTL studies that need to be performed to pull off their architecture and function by the use of contrasting insect infestation treatments.

Breeding Approaches for Enhancing Plant Pest Resistance

Conventional Strategies

A principal target in agriculture is to produce a broad-spectrum and a durable resistance to contrast insect pests damages. Although major genes for resistance to several pests have been detected, the majority of these genes induce plant resistance individually to a distinct insect species or biotype. Furthermore, as mentioned above, this resistance is frequently not very long lasting in time, reducing the effectiveness of breeding programs in developing resistant crops.

Susceptibility plant genes, a part from *R*-gene family, can be very effective in pest manipulation. Various herbivorous insects possess the capacity to inhibit active defense reactions of plant tissues to their own benefit. It is hypotheses that these insects activate specific plant genes, the so-called susceptibility genes, to abolish or avoid the defense mechanisms that contrast insect development [196, 197]. These events result in the so called host plant susceptibility. Clearly knockdown of these susceptibility genes can be beneficial in controlling plant-manipulating insects. This type of approach has been found to increase resistance by significantly contrasting the growth of specific pathogen agents [198, 199], featuring is performance in disease management applications. Thereby, studying the promising to exploit susceptibility genes for controlling insect pest's merits further research in this field.

To achieve an extension of durability of resistance, dependent by major genes, an appealing complementary approach would be to integrate direct resistance with indirect defense attributes [200, 201]. This intervention can rule out or delay direct resistance disruption. The potential practical exploitation of indirect defense strategy in agriculture is endorsed by a recent investigation indicating a robust escalation in the rate of predation. These were shown as a consequence in herbivore-induced variation of volatile compounds secreted in nature by *Nicotiana attenuate* plants [202]. Integrating these two mechanisms of resistance in plants is apparently puzzling. However it is expected that this strategy will be beneficial in term of novel resistance repertoires and thus increasing durability phenomenon of plant resistance. Moreover there are only a poor number of investigations pointing out genetic differences for indirect resistance attributes, such as the effects of HIPVs about the attraction of natural antagonists [203 - 205]. Therefore our knowledge on the molecular basis of this resistance category is currently primitive and further research is required. Durable insect pest management is generally dependent on integrations of techniques such as the biological control and the host plant resistance. This combination of distinct tools into integrated pest management (IPM) [206] calls for a specific strategy where the components are favorably balanced. Integrating direct and indirect defense strategies in crops may be a helpful advancement to the future growth of IPM tools [207].

An alternative strategy for accomplishing a superior durable varietal resistance is the application selective breeding procedures, by exploiting QTLs sustaining a specific category of insect resistance (*e.g.* antibiosis, antixenosis), or tolerance. In this respect, Alam and Cohen [208], in a QTL study, concerning six traits related with rice resistance to the brown planthopper, identified a total of seven QTLs, out of which one was most importantly related with antixenosis and a second one with tolerance. The majority of other QTL studies related with insect resistance, carried out till now, have evaluated chemical or morphological antibiotic resistance traits, or rating of plant damages scored under free-choice environments [190]. However, these results may be biased by all modalities of resistance...

Transgenic Strategies for Insect Pest Control

Plant scientists have used genetic engineering techniques to produce insect resistant varieties in a number of crops; including maize, cotton, and soybean that can synthesize autonomously foreign antifeedants or insecticidal proteins under field conditions. Contrasting insect pest damages by introducing resistance to Bt toxins in transgenic crops, *via* "Bt technology", has been an exceptional goal. This has permitted to increase crop productivity and economic benefits to farmers. Bt crops, first commercialized in 1996, are currently cultivated on approximately 55 million hectares on a global scale [209]. Fig. (**7**) gives an example of the effectiveness of this technology.

Fig. (7). Infestation of transgenic Bt soybean plants by using velvet bean caterpillar. The non-transgenic plants (control) on the right clearly exhibit large defoliation damages.

The Bt toxins represent compounds generated by the common soil bacteria *Bacillus thuringiensis* (Bt) [210]. These bacteria are that is able to synthesize crystalline proteins throughout the sporulation stage. The Bt proteins possess an outstanding insecticidal activity, at very low concentration levels, and induce exiguous or no injuries to the majority of non-target living beings including helpful insects, wild-life, and above all humans [211]. Several strains of *B. thuringiensis* have been characterized which are toxic to distinct insect families, whereas the proteinaceous toxins is operating in the midgut of the digestive apparatus of insects [212, 213]. Specifically, the BtCRY1a toxin, affecting *Lepidoptera*, was proved a powerful agent for contrasting a number of stem-boring insect larvae, particularly in maize. It is worth noting that in this crop, the Corn Borers (*Sesamia nonagrioides* and *Ostrinia nubilalis)* are among the principal dangerous pests for this cultivation, provoking sizeable yield losses worldwide. The *CRY3a* gene, derived from a different *B. thuringiensis* strain, encodes a Bt toxin active *versus Coleoptera*, and is employed widely in cotton to contrast the growth of Cotton Bollworm (*Helicoverpa armigera*) larvae. It is also worth noting that, whilst the familiar Bt toxins utilized in plants are active *versus* chewing insects, exist an ample category of insects, such as sucking insects (*e.g.*

leaf hoppers and aphids), *versus* which no natural Bt toxins are recognized. In this respect to overcome this drawback have been generated a set of synthetic or chimeric toxins [214]. These interventions will expand the spectrum of these toxins to additional insect families.

In spite of the fact that Bt crops have been largely planted since 1996, there have been no cases of a pest evolving resistance [215]. However, as a dedicated criterion, various approaches devoted to insect resistance management have been suggested as a bundle for Bt crop cultivations. These approaches involve i) refuge strategy (*e.g.* provide pests a refuge whereas they may feed on plants devoid of toxins, thus maintaining Bt susceptible resistance alleles inside the insect population), ii) gene pyramiding, and iii) an elevated dosage of the protein in the plant tissues to avoid any insects to escape from the Bt cultivated field [215]. Furthermore, as an option to the use of *CRY* genes, there are other strategies based, for example, on the adoption of genes encoding different classes of proteins exhibiting insecticidal activities. From this point of view are already been discovered a number of proteins, like lectins, proteinase inhibitors, and a-amylase inhibitors. For example, plant lectins were employed to control sap-sucking insects included in the *Homoptera* order, which involves several the principal destructive pests globally diffuse [216, 217]. The lectin effects is due to restriction of nutrient uptake or perturbation of midgut cells by activating endocytosis process and likely distinct toxic metabolites existent in the midgut [218]. Research on rice pointed out that several of these chemical candidates show up encouraging and thereby may contribute a valid option to the Bt strategy [219]. Notably, in maize the transgenic expression of wheat oxalate oxidase greatly upgrade phenolic levels in the plant tissues, essentially ferulic acid. Field trials indicated that the transgenic maize plants displayed a superior resistance to ECB in comparison to their non-transgenic plant versions. It was indicated that transgenic oxalate oxidase elicits defense reactions by production of H_2O_2 and triggering jasmonic acid signaling pathway [220]. More recently, Schellenberger and colleagues, by analysing soil samples have identified a small protein (~ 10Kda), produced by *Pseudomonas chlororaphis*, that killed rootworm larvae of *Diabrotica virgifera* virgifera LeConte, a devastating coleoteran crop pest (corn rootworm) in North America and Europe maize cultivations [220a]. Interestingly, these workers have shown that maize plants engineered to synthesize this protein suffered meager injury from the rootworm, in spite of whether or not the pest was sensitive to Bt proteins.

An additional attractive viewpoint concerning the advancement of insect-resistant plants is given by RNAi technology [221]. This technology, initially characterized in *Caenorhabditis elegans* [222], has come out as a current strategy to fight insects, and particularly to direct resistance improvement on specific insect pests.

In this respect, gene knockdown of several insects has been worked out through supply of dsRNA, involving insects belonging to various orders like *Hymenoptera, Coleoptera, Diptera*, and *Lepidoptera* [221]. Results reported by a number of workers [223, 224] are promising. In these studies, by adopting plant-mediated RNAi tools, it was possible to knock down the *cytochrome P450* (*CYP6AE14*), *ecdysone receptor* (*EcR*), and *hunchback* (*HB*) genes, that are able to fight against various insect species (*e.g. Helicoverpa armigera, Myzus persicae*, and *Spodoptera exigua*). However, this methodology is yet in an early stage of development and being carefully studied by various research teams.

HERBICIDE-TOLERANT CROPS

Evidence indicates that weeds are a primary constraint in crop yield and consequently they need to be controlled for several reasons. As a matter of fact, they compete with crops resources, including sunlight, water, and soil nutrients, and producing sizeable crop yield losses. Additionally, they may condition the velocity and efficiency of harvesting operations and contaminate with seeds of poisonous weeds, crop seeds during harvesting.

An agronomical practice to regulate diffusion of weeds is to distribute herbicides these are chemicals that kill plants and are commercially exploited to destroy weeds. Obviously, the challenges to farmers are to suppress the weeds without killing crops The utilization of herbicides is accepted as the most economical and efficient agricultural method in weed control [225]. Incorrect herbicide application may produce crop damages, inadequate control of weed, and development of herbicide resistant weeds, environmental pollution, or health hazards.

Several of the most recent herbicides integrate extensive weed killing ability with little- or no-environmental persistence. Among the input traits provided to crop growers by breeding, herbicide resistance has been the primary feature largely exploited in developing contemporary crop varieties. Therefore, a central request in weed control concerns the discovery of selective herbicides for their application in crop managements.

Mechanisms of Action and Metabolism

There are several classes of herbicides that kill plants in distinct modes. The capacity of an herbicide to suppress specific plant species without damaging others is termed selectivity. Herbicides that suppress or kill the development of the majority of plant species are almost nonselective. Nonselective herbicide application is restricted to circumstances where control of all plant species is applicable, or the herbicide action is designed on target weeds and aside from cultivated plants. Moreover, a herbicide need to fitting a number of conditions to be useful such as i) arrive in touch with the specific weed, ii) absorption by the weed, ii) transfer to the place of action in the weed, and accumulation of appropriate amounts at the place of action to suppress or destroy the development of target plants. Weed control is inadequate except that these conditions are satisfied.

Interestingly herbicide site of action indicates the peculiar biochemical or bio-physical mechanisms acting in the plant that the herbicide breaks up to inhibit plant growth and development. This includes photosynthesis, pigment biosynthesis, mitosis, or synthesis of essential amino acids (reviewed by Délye *et al.* [225]). Based on their modality of action, herbicides have been classified in eighteen by the *Herbicide Resistance Action Committee* (HRAC; ttp://www.hrac-global.com). In Fig. (**8**) is given a graphical representation regarding cell targets of herbicide action and herbicide categorization according to HRAC.

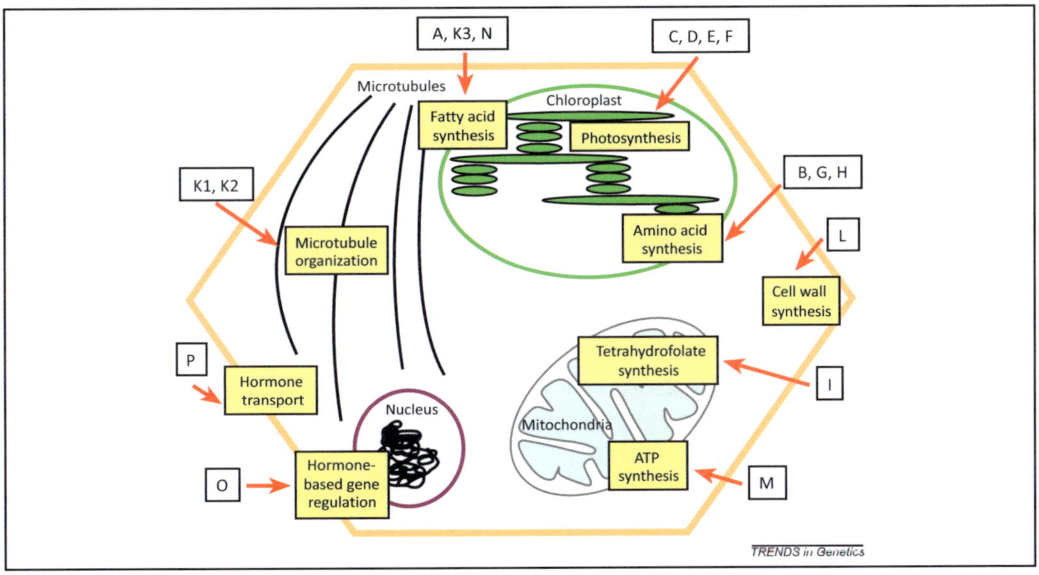

Fig. (8). Cell targets of herbicide action and herbicide categorization by their modality of action according to the HRAC (From Délye *et al.*, 2013 [225]).

Knowledge on the mechanisms or modality of action and metabolism of different herbicides has been often reviewed [226, 227]). Briefly herbicide resistance in crop plants is generally provided by biochemical processes analogous to those that produce herbicide resistance in weeds: i) deficiency in herbicide uptake and activity at its place of action ; ii) very fast metabolic detoxification of the herbicide allowing plants to survive with slight or no irreversible consequences after the herbicide use. Notably, metabolic rates may change in relation with plant growth phases and environments. This is one motive why crop safety and weed control effectiveness may occasionally differ severely in cool environments in comparison with more optimal growing locations. Collectively the previous considerations above mentioned has allowed detecting several herbicide-tolerant target enzymes, commonly occurring in distinct plant species or microorganisms, in addition to various herbicide-modifying enzymes that promoted the growth of herbicide-tolerant plants.

Genetic Bases for Herbicide Resistance

Studies carried out with a large number weed and crop species have indicated that herbicide target-site resistance (TSR) is generally strictly herbicide-specific and widely subject to monogenic inheritance [225]. For example, for the herbicides represented by sulfonylurea (SU) and imidazolinone (IMI), tolerant plants have been discovered in many weeds (> 80 species) and crop plants [228]. Notably SU and IMI herbicides are peculiar and powerful interceptors of acetolactate synthase (ALS, also commonly referred as acetohydroxyacid synthase), therefore inhibiting the synthesis of the branched-chain amino acids (*i.e.* valine, leucine and isoleucine). This inhibition drives to development of herbicide-susceptible genotypes characterized by the very fast arrest of plant cell division and consequently weeds growth. Usually, *R* genes for ALS-inhibiting herbicides manifest an inheritance model due to partial or complete dominance mechanisms. However in same examples the level of resistance was influenced, in specific genetic backgrounds, by a second genetic factor. In this respect, Tan and coworkers [228] argued that although resistance to ALS-inhibiting herbicides is largely due to heritable determinants, the phenotypic result of ALS mutations are frequently influenced by additional factors (*e.g.* herbicide uptake, its metabolism, and the level of translocation rate of the herbicide to its site of action. Moreover, molecular studies on mutant alleles of *ALS* genes have revealed, with some peculiarity, that resistances to ALS-inhibiting herbicides, in different susceptible species, are typically originated by point mutations in genes encoding ALS [228, 229]; this specificity decrease the reactivity of the enzyme to herbicide blockage.

Investigations were also executed on inheritance of resistance dependent on e non-target-site (NTRS) attributes (*i.e.* a more generic adaptive reaction to herbicides in comparison to TSR), promoting a mechanism of cross-resistance to herbicides with several types of action). Recently Duhoux and Delye [230] have employed a substantiated reference gene array to check the potential function of five cytochrome *P450* genes, in providing NTSR to herbicides able to repress ALS activity. Their results have indicated that resistance to NTSR is poly-genetically controlled and was likely evolved from pre-existing stress reaction pathways. It was also suggested that 'omics' strategies may permit to elucidate the genetic makeup of intricate processes. These aspects concerning resistances, occurrence, selection, and diffusion of herbicide resistance in weed populations may simply be completely resolved by concentrating our attention on evolutionary fluxes and implementing complementary modeling attempts.

Breeding Strategies

The breeding of crops that are resistant to herbicides is a relatively new technology designed to improve weed control in agricultural fields. Herbicide-resistant crops are made resistant by either transgene technology or by selection in cell or tissue culture for mutations that confer herbicide resistance.

Conventional Methods

Two strategies can be exploited. The first is the design of specific chemicals with broad selectivity for crops. This approach, however, is time consuming and expensive: the resulting crops may be uneconomical for grower uses or may also increase the already increasing chemical load in the environment. Moreover, it has become increasingly difficult to discover new herbicides and even harder to come up with a new one that has a novel mode of action [228]. The second and more frequent approach to crop herbicide selectivity is the development of crop cultivars with tolerance to the already existing effective broad-spectrum herbicides. This may allow us to increase the crop alternatives in which they can be exploited. For example in common sunflower populations with cross-resistance to both SU and IMI herbicides were first discovered in fields that had been repetitiously treated with herbicides for seven-eight years [230]. Genes for resistance to ALS-inhibiting herbicides in sunflower have been introgressed *via* backcrosses from resistant wild populations into elite inbred lines. This has been made with the purpose of developing and deploying herbicide resistant cultivars and hybrids. Traditionally, sunflower producers have had few herbicides for controlling broadleaf weeds. Tolerance to ALS-inhibiting herbicides has greatly increased the spectrum of herbicides for controlling broadleaf weeds in sunflower. Moreover, triazine tolerant oilseed rape varieties were developed through introgression of the triazine tolerant weed biotype cytoplasm into the triazine susceptible oilseed rape [231]. The transfer was accomplished through a combination of backcrossing and cytogenetic selection, in which the triazine tolerant weed biotype served as the donor and a susceptible canola variety served as the recurrent pollen parent.

Several variant of *ALS* genes conferring IMI tolerance were discovered in plants through mutagenesis and selection, and were used to create IMI tolerant crops such as maize, wheat, rice, oilseed rape, and sunflower [228]. These crops were bred using conventional breeding methods and commercialized as Clearfield® crops since 1992. Because all commercialized IMI-tolerant crops have been developed through the selection of spontaneous or mutagen-induced mutations in ALS, these crops are non–transgenic. To overcome these limitations of both the transgenic and mutagenesis approaches described above, it has been suggested that direct modification of an endogenous target gene is a major goal in crop molecular breeding [232]. This can be obtained *via* homologous recombination-dependent gene targeting or genome editing tools.

Using cell culture techniques, it was produced a maize hybrid that is resistant to an SU herbicide, sethoxidim [228]. In this study, a mutant cell line (S2) was identified, following continuous culture of maize embryo tissues, grown under high sethoxidim selection pressure. Plants regenerated from this somaclonal mutant line were found to contain a form of the enzyme ALS (target of SUs/IMIs), which was insensitive to the herbicide. This resistance was subsequently transferred to a commercial hybrid by backcrossing the S2 line with both of its parental lines. Further investigations showed that the sethoxidim tolerance was inherited as a single partially dominant allele. Similarly, it was selected a glyphosate-tolerant sugar cane cell line in liquid medium containing 0.8 mM glyphosate [233]. The regenerated plants were able to tolerate a concentration of glyphosate, up to five-fold, that needed that killed plants from unselected cells. Cell culture under lethal concentrations of certain herbicides also results in gene amplification in surviving cells. This is t leading to resistance through the overproduction of enzymes targeted by herbicides. For example, a petunia cell line with resistance to glyphosate was selected in this manner. Moreover, plants regenerated from them survived lethal levels of glyphosate [234]. This resistance was found to be due to an amplification of the gene encoding 5-enolpyruvyl-shikimate-3-phosphate (EPSP) synthase that caused its overproduction in the cells.

Transgenic Strategy

Herbicide tolerance is the best familiar output trait in profit-making transgenic plants [235]. This strategy includes the detection of a herbicide resistance gene from a plant or microorganism, its selection and modification for a powerful plant expression (if the gene is of microbial nature) and its successive transfer, reliable integration and expression in the cells of the target crop species. The majority of genes, encoding for commercial herbicide resistance in crops, were detected and cloned from herbicide down-grading soil microorganisms.

Tolerance to herbicide in plants, obtained by genetic transformation, is contributed by one or a combination of various mechanisms including: i) addition of a gene(s) encoding for the synthesis of a herbicide detoxifying enzyme(s); ii) addition of a of gene(s) encoding for the production of a herbicide insensitive variant of a normal functional enzyme or over expression of the genes encoding for the synthesis of a herbicide target enzyme so that the normal metabolic functionality is further maintained in the plant, although a part of the enzyme is repressed; iii) manipulation of the target enzyme of herbicide in a molecular form so that the herbicide compound does not adhere to it; and iv) genetic engineering of an efficient efflux of the herbicide from the plant cells.

Jointly eleven crop varieties, furnished with herbicide-resistance, are commercially cultivated worldwide. In this context, varieties having resistance to glyphosate (*e.g.* active ingredient in Roundup™) and glufosinate (*e.g.* active ingredient in Basta ™) are the herbicides mostly employed to control weeds. For instance, important crops such as maize, soybean, cotton, sugar beet, and canola are commercially on market as glyphosate-resistant varieties. Notably, several of them are currently extensively cultivated in numerous nations, where they absorb the majority of the 181.5 millions of hectares planted with GM-crops on a global scale [235].

Glyphosate (N-phosphonomethyl-glycine) is nonselective and the number-one selling herbicide in the world. It actively rivals in the chloroplast with the phosphoenolpyruvate (PEP) compound at the EPSPS enzyme-binding site [236]. This behavior generates a repression of the shikimate biosynthetic pathway, which involved the synthesis of essential aromatic amino acids (*e.g.* tryptophan, tyrosine, and phenylalanine), as well as other relevant plant metabolic substrates. Resistant plants (*e.g.* Roundup™ Readymaize) have a modified *EPSPS* gene sequence, derived from maize, associated to a constitutive actin promoter region derived from rice [237]. The altered EPSPS protein, encoded by this transgene, showed a very little affinity for glyphosate. Moreover, it varied from the original EPSPS peptide simply for two amino acid residues [237], involving a mutation at amino acid 106, which has been described in a number of glyphosate resistant weeds [238]. The relatively slow modality of action and physio-chemical attributes promote glyphosate translocation across plants and its concentration in the key growing points of the plant tissues, where phytotoxicity takes place. The ample range of weeds controlled by glyphosate have made its application in crop weed control an attractive solution.

Glufosinate (Bialaphos or Basta) is a natural compound isolated from two species of *Streptomyces* fungi [239]. Resistance to glufosinate is attributable to a metabolic inactivation of the source molecule by either of two homologous enzymes, namely phosphinothricin N-acetyltransferase (PAT) or bialaphos N-acetyltransferase (BAR), that are effective to catalyze the acetylation of glufosinate [239]. Both genes encoding these enzymes were detected and cloned from the soil microorganisms. Specifically *PAT* gene was isolated from *Streptomyces viridochromogenes,* while the *BAR* one was isolated from *Streptomyces hygroscopicus*. In Fig. (**9**) is given an example of the effectiveness of glufosinate.

Fig. (9). Field trial 'Basta' resistant spring oilseed rape: in foreground = wild-type cultivar 'Drakar'; in background = doubled haploids of cultivar 'Drakar' expressing the *bar* gene (courtesy of Leemans, Plant Genetic Systems N.V., Gent, Belgium).

Cotton and soybean farmers, who are concerned with problems to manage glufosinate resistant plants, can speedily use glufosinate-resistant crop cultivars and the application of glufosinate. "Dual stack"crop varieties, combining resistance to glufosinate as well as to glyphosate, are currently in the market and used in various crops (*e.g.* cotton, soybeans, and maize). This permit to farmers options in choosing between two large spectrum herbicides and additionally a range of selective herbicides for varying their strategies used in weed management.

Other applications concerning transgenic approaches were devoted to develop of auxin-herbicide-tolerant crops. An example is given by auxin herbicide-tolerant crops achieved by expressing bacterial aryloxyalkanoate dioxygenase (AAD) [240]. AADs efficiently decompose 2,4-D and other largely applied auxin herbicides. Maize plants expressing an *AAD (i.e. AAD-1)* gene were tolerant to aryloxy-phenoxypropionate herbicide applications. The enzymatic peptide (*i.e.* 4-hydroxyphenyl pyruvate dioxygenase (HPPD), encoded by this gene, modifies 4-hydroxyphenyl pyruvate to homogentisate, a crucial step in biosynthesis of plastoquinone. Three families of chemical compounds (*i.e.* isoxazoles, triketones, and callistemones) are HPPD-inhibiting herbicides. The suppression of the activity of HPPD produces a bleaching syndrome on more recent zones of plant growth. This effect is mediated *via* an indirect suppression of carotenoid biosynthesis, likely due to the need of plastoquinone as cofactor of phytoene desaturase (PDS) [240]. Although this information indicates the latest type of action detected for a herbicide, similar tests are actively performed currently to identify new herbicide molecules applicable in crop managements [241].

Concern with the impact of GM crop varieties on the environment, have recently stimulated public opinion to promote research on this topic. Among the potential impacts, the 'escape' of the transgene, either *via* spreading of the crop plants away the cultivated land or *via* hybridization process with wild relatives, might increase the so called "weediness" hazard. In several nations the extensive and concentrate application of glyphosate for weed management has turned up with the selection of herbicide-resistant weeds [242]. Presently, the occurrence of glyphosate-resistant populations has been reported in a number of weed plants. Moreover, various mechanisms conferring resistance to this herbicide have been described in weeds genetically equipped with glyphosate resistance. It is suggested that with the application of different weed management strategies, farmers will maintain the benefit of herbicide resistance characteristics and herbicide methodologies, and assist in preserving cost-effective and environmentally sustainable crop yield systems for future needs [243].

CONCLUDING REMARKS

Plants cultivated in the fields are permanently exposed to a broad spectrum of biotic stresses. One of the principal demands for plant breeding is upgrading resistance to the abundance of parasites that wrech crop productivity. Therefore, the advancement of innovative approaches to neutralize damages induced by parasites, depending on a plant's own protection systems, is a central point for sustainable and cost-effective crop production systems. Furthermore, these strategies are also respectful of environments and permit to reduce health hazards by avoiding pesticide inputs or other techniques to control biotic stresses.

Efforts to reach these objectives are based on detection in plants of genetic variability that underlay key features – *i.e.* morphological, structural, and biochemical- and to operate on intricate characteristics - related to plant growth and development or tolerances to abiotic stress agents - generally in growing environments that are greatly changeable and unreliable.

Studies on the genetic control affecting disease, insect, and weeds resistance in plants and host-pest interactions, has dramatically accelerate the performance of managing pests resistance genes in breeding projects. This has brought about the adoption of high-yielding genetically resistant crop varieties that were and are cultivated across extensive land area. However, in this scenario more aspects need to be understood, including those acting at the interface between genetic resistance and crop physiology. For example, the characterization of *R*-genes and their corresponding *AVR*-genes has disclosed an intricate puzzle of networks, affecting both the makeup and the modality in which gene products interact to promote resistance. Obviously these aspects indicate that our knowledge on the processes conferring resistance to pests is still unsatisfactory. For example, several breaks still persist, such as those concerning defense signal transduction networks. These points should be bypassed before applying cleaver approaches triggering resistance pathways. Likewise, genetic mapping of plant variants that genetically modify insect herbivore resistance, or possible reactions from purified insect elicitors, might provide the discovery of formerly unfamiliar plant defense processes.

An additional aspect worth to be mentioned is related to major genes and QTLs involved in plant resistance against several classes of pathogens and insect pests. Although most of these genetic determinants in numerous plant genomes have been located on genetic maps, the benefit of this knowledge for their applications in MAS selection projects has not still been fully exploited. Moreover, success in growing transgenic crops may profit from information about the signal transduction networks that control pathogenesis. This is particularly accurate for host spectrum of resistance and the presence of a broad array of appropriate genes that may be exploited to improve plant resistance. Additionally genetic engineering approaches need detailed knowledge on the functions and behaviors of these genes. This may enable a retailoring the utilization of genetic resources existing in plants.

It is expected that genomics might also supply productive devices able to i) accelerate the identification of key topics above indicated ii) facilitate the functional investigation of genes sustaining plant resistance and susceptibility, iii) elucidate signaling circuits and metabolic networks involved in plant resistance, and iv) identify crucial genes, acting as regulators, that may be targeted either for silencing or overexpression; this with the scope to analyze the function of these mechanisms in plant-pests interactions.

In general, gains in plant genetics and genomics are cooperating in speeding up gene discovery useful in crop improvement, including biotic stresses. In this panorama, numerous novel genomic tools - including next generation sequencing, high-throughput marker genotyping- may be applied not uniquely for generating high density genetic and physical maps, but also for producing transcriptome or sequence data sets [244]. These strategies integrated with –omics methodologies (*e g.* transcriptomics, metabolomics, and proteomics) have been turned up, as compelling devices, for explaining genome variability - at DNA, RNA, and protein levels- occurring in crop plants. They are also powerful for discovering genes or genomic regions implicated in expression of characteristic(s) that are relevant in upgrading plant resistance to biotic stresses. Furthermore, a set of innovative genetic tools, notably devices collectively termed as 'genome editing', have been recently worked out, enabling scientists to perform targeted modifications in the genomes with high efficiency [140, 245]. These changes may cover various possibilities such as i) adding or removing DNA at a distinct site of the plant genome, or ii) substituting a specified fragment of DNA, like those concerning *R*-genes with an unlike one, thereby expanding the potential genetic arsenal of weapons to contrast efficiently pest diseases and damages. In conclusion the prospects for developing resistant crop varieties to biotic stresses is encouraging as an enviromentally satisfactory ground for integrating programs of disease and pest control are available for improving current and future crop production systems.

CONFLICT OF INTEREST

The authors confirm that they have no conflict of interest to declare for this publication.

ACKNOWLEDGEMENTS

We apologize to all those whose contributions we did not include in this chapter due to space constraints, personal preferences, or bona fide oversight.

REFERENCES

[1] Dangl JL, Horvath DM, Staskawicz BJ. Pivoting the plant immune system from dissection to deployment. Science 2013; 341(6147): 746-51. [http://dx.doi.org/10.1126/science.1236011] [PMID: 23950531]

[2] Popp J, Pető K, Nagy J. Pesticide productivity and food security. A review. Agron Sustain Dev 2013; 33: 243-55. [http://dx.doi.org/10.1007/s13593-012-0105-x]

[3] Oerke EC. Crop losses to pests. J Agric Sci 2006; 144: 31-43. [http://dx.doi.org/10.1017/S0021859605005708]

[4] McGrann GR, Stavrinides A, Russell J, *et al.* A trade off between mlo resistance to powdery mildew and increased susceptibility of barley to a newly important disease, Ramularia leaf spot. J Exp Bot 2014; 65(4): 1025-37. [http://dx.doi.org/10.1093/jxb/ert452] [PMID: 24399175]

[5] Lucas J. Plant Pathology and Plant Pathogens:. 3rd ed. John Wiley & Sons 2009; p. 274.

[6] Turrà D, El Ghalid M, Rossi F, Di Pietro A. Fungal pathogen uses sex pheromone receptor for chemotropic sensing of host plant signals. Nature 2015; 527(7579): 521-4. [http://dx.doi.org/10.1038/nature15516] [PMID: 26503056]

[7] Jones JD, Dangl JL. The plant immune system. Nature 2006; 444(7117): 323-9. [http://dx.doi.org/10.1038/nature05286] [PMID: 17108957]

[8] Yeats TH, Rose JK. The formation and function of plant cuticles. Plant Physiol 2013; 163(1): 5-20. [http://dx.doi.org/10.1104/pp.113.222737] [PMID: 23893170]

[9] Serrano M, Coluccia F, Torres M, LHaridon F, Métraux JP. The cuticle and plant defense to pathogens. Front Plant Sci 2014; 5: 274. [http://dx.doi.org/10.3389/fpls.2014.00274] [PMID: 24982666]

[10] Hamann T. Plant cell wall integrity maintenance as an essential component of biotic stress response mechanisms. Front Plant Sci 2012; 3: 77. [http://dx.doi.org/10.3389/fpls.2012.00077] [PMID: 22629279]

[11] Łaźniewska J, Macioszek VK, Kononowicz AK. Plant-fungus interface: The role of surface structures in plant resistance and susceptibility to pathogenic fungi. Physiol Mol Plant Pathol 2012; 78: 24-30. [http://dx.doi.org/10.1016/j.pmpp.2012.01.004]

[12] Langenbach C, Campe R, Schaffrath U, Goellner K, Conrath U. UDP-glucosyltransferase UGT84A2/BRT1 is required for Arabidopsis nonhost resistance to the Asian soybean rust pathogen Phakopsora pachyrhizi. New Phytol 2013; 198(2): 536-45. [http://dx.doi.org/10.1111/nph.12155] [PMID: 23356583]

[13] Dangl JL, Jones JD. Plant pathogens and integrated defence responses to infection. Nature 2001; 411(6839): 826-33. [http://dx.doi.org/10.1038/35081161] [PMID: 11459065]

[14] Morrissey JP, Osbourn AE. Fungal resistance to plant antibiotics as a mechanism of pathogenesis. Microbiol Mol Biol Rev 1999; 63(3): 708-24. [PMID: 10477313]

[15] Idnurm A, Howlett BJ. Pathogenicity genes of phytopathogenic fungi. Mol Plant Pathol 2001; 2(4): 241-55. [http://dx.doi.org/10.1046/j.1464-6722.2001.00070.x] [PMID: 20573012]

[16] Bednarek P. Chemical warfare or modulators of defence responses - the function of secondary metabolites in plant immunity. Curr Opin Plant Biol 2012; 15(4): 407-14. [http://dx.doi.org/10.1016/j.pbi.2012.03.002] [PMID: 22445190]

[17] Bednarek P. Sulfur-containing secondary metabolites from *Arabidopsis thaliana* and other Brassicaceae with function in plant immunity. ChemBioChem 2012; 13(13): 1846-59. [http://dx.doi.org/10.1002/cbic.201200086] [PMID: 22807086]

[18] Stam R, Mantelin S, McLellan H, Thilliez G. The role of effectors in nonhost resistance to filamentous plant pathogens. Front Plant Sci 2014; 5: 582. [http://dx.doi.org/10.3389/fpls.2014.00582] [PMID: 25426123]

[19] Wiesel L, Newton AC, Elliott I, *et al.* Plants have evolved a battery of defense mechanisms to protect themselves against pathogen attack. Front Plant Sci 2014; 21: 655. [PMID: 25484886]

[20] Zipfel C. Plant pattern-recognition receptors. Trends Immunol 2014; 35(7): 345-51. [http://dx.doi.org/10.1016/j.it.2014.05.004] [PMID: 24946686]

[21] Bigeard J, Colcombet J, Hirt H. Signaling mechanisms in pattern-triggered immunity (PTI). Mol Plant 2015; 8(4): 521-39. [http://dx.doi.org/10.1016/j.molp.2014.12.022] [PMID: 25744358]

[22] Muthamilarasan M, Prasad M. Plant innate immunity: an updated insight into defense mechanism. J

Biosci 2013; 38(2): 433-49.
[http://dx.doi.org/10.1007/s12038-013-9302-2] [PMID: 23660678]

[23] Dodds PN, Rathjen JP. Plant immunity: towards an integrated view of plant-pathogen interactions. Nat Rev Genet 2010; 11(8): 539-48.
[http://dx.doi.org/10.1038/nrg2812] [PMID: 20585331]

[24] Yamaguchi Y, Huffaker A. Endogenous peptide elicitors in higher plants. Curr Opin Plant Biol 2011; 14(4): 351-7.
[http://dx.doi.org/10.1016/j.pbi.2011.05.001] [PMID: 21636314]

[25] Hofberger JA, Zhou B, Tang H, Jones JD, Schranz ME. A novel approach for multi-domain and multi-gene family identification provides insights into evolutionary dynamics of disease resistance genes in core eudicot plants. BMC Genomics 2014; 15: 966.
[http://dx.doi.org/10.1186/1471-2164-15-966] [PMID: 25380807]

[26] Böhm H, Albert I, Fan L, Reinhard A, Nürnberger T. Immune receptor complexes at the plant cell surface. Curr Opin Plant Biol 2014; 20: 47-54.
[http://dx.doi.org/10.1016/j.pbi.2014.04.007] [PMID: 24835204]

[27] Hofberger JA, Nsibo DL, Govers F, Bouwmeester K, Schranz ME. A complex interplay of tandem- and whole-genome duplication drives expansion of the L-type lectin receptor kinase gene family in the *brassicaceae.* Genome Biol Evol 2015; 7(3): 720-34.
[http://dx.doi.org/10.1093/gbe/evv020] [PMID: 25635042]

[28] Yang B, Sugio A, White FF. Os8N3 is a host disease-susceptibility gene for bacterial blight of rice. Proc Natl Acad Sci USA 2006; 103(27): 10503-8.
[http://dx.doi.org/10.1073/pnas.0604088103] [PMID: 16798873]

[29] Wu S, Shan L, He P. Microbial signature-triggered plant defense responses and early signaling mechanisms. Plant Sci 2014; 228: 118-26.
[http://dx.doi.org/10.1016/j.plantsci.2014.03.001] [PMID: 25438792]

[30] Pandey SP, Somssich IE. The role of WRKY transcription factors in plant immunity. Plant Physiol 2009; 150(4): 1648-55.
[http://dx.doi.org/10.1104/pp.109.138990] [PMID: 19420325]

[31] Buscaill P, Rivas S. Transcriptional control of plant defence responses. Curr Opin Plant Biol 2014; 20: 35-46.
[http://dx.doi.org/10.1016/j.pbi.2014.04.004] [PMID: 24840291]

[32] Sarris PF, Duxbury Z, Huh SU, *et al.* A plant immune receptor detects pathogen effectors that target WRKY transcription factors. Cell 2015; 161(5): 1089-100.
[http://dx.doi.org/10.1016/j.cell.2015.04.024] [PMID: 26000484]

[33] Ishihama N, Yoshioka H. Post-translational regulation of WRKY transcription factors in plant immunity. Curr Opin Plant Biol 2012; 15(4): 431-7.
[http://dx.doi.org/10.1016/j.pbi.2012.02.003] [PMID: 22425194]

[34] Ding B, Wang GL. Chromatin *versus* pathogens: the function of epigenetics in plant immunity. Front Plant Sci 2015; 6: 675.
[http://dx.doi.org/10.3389/fpls.2015.00675] [PMID: 26388882]

[35] Marino D, Peeters N, Rivas S. Ubiquitination during plant immune signaling. Plant Physiol 2012; 160(1): 15-27.
[http://dx.doi.org/10.1104/pp.112.199281] [PMID: 22689893]

[36] Glazebrook J. Contrasting mechanisms of defense against biotrophic and necrotrophic pathogens. Annu Rev Phytopathol 2005; 43: 205-27.
[http://dx.doi.org/10.1146/annurev.phyto.43.040204.135923] [PMID: 16078883]

[37] Robert-Seilaniantz A, Grant M, Jones JD. Hormone crosstalk in plant disease and defense: more than just jasmonate-salicylate antagonism. Annu Rev Phytopathol 2011; 49: 317-43.

[http://dx.doi.org/10.1146/annurev-phyto-073009-114447] [PMID: 21663438]

[38] Ballaré CL. Light regulation of plant defense. Annu Rev Plant Biol 2014; 65: 335-63. [http://dx.doi.org/10.1146/annurev-arplant-050213-040145] [PMID: 24471835]

[39] Boller T, Felix G. A renaissance of elicitors: perception of microbe-associated molecular patterns and danger signals by pattern-recognition receptors. Annu Rev Plant Biol 2009; 60: 379-406. [http://dx.doi.org/10.1146/annurev.arplant.57.032905.105346] [PMID: 19400727]

[40] Fujita M, Fujita Y, Noutoshi Y, *et al.* Crosstalk between abiotic and biotic stress responses: a current view from the points of convergence in the stress signaling networks. Curr Opin Plant Biol 2006; 9(4): 436-42. [http://dx.doi.org/10.1016/j.pbi.2006.05.014] [PMID: 16759898]

[41] Deslandes L, Rivas S. Catch me if you can: bacterial effectors and plant targets. Trends Plant Sci 2012; 17(11): 644-55. [http://dx.doi.org/10.1016/j.tplants.2012.06.011] [PMID: 22796464]

[42] Gururani MA, Venkatesh J, Upadhyaya CP, *et al.* Plant disease resistance genes: current status and future directions. Physiol Mol Plant Pathol 2012; 78: 51-65. [http://dx.doi.org/10.1016/j.pmpp.2012.01.002]

[43] Schneider DJ, Collmer A. Studying plant-pathogen interactions in the genomics era: beyond molecular Kochs postulates to systems biology. Annu Rev Phytopathol 2010; 48: 457-79. [http://dx.doi.org/10.1146/annurev-phyto-073009-114411] [PMID: 20687834]

[44] Soyer JL, El Ghalid M, Glaser N, *et al.* Epigenetic control of effector gene expression in the plant pathogenic fungus *Leptosphaeria maculans*. PLoS Genet 2014; 10(3): e1004227. [http://dx.doi.org/10.1371/journal.pgen.1004227] [PMID: 24603691]

[45] van der Hoorn RA, Kamoun S. From Guard to Decoy: a new model for perception of plant pathogen effectors. Plant Cell 2008; 20(8): 2009-17. [http://dx.doi.org/10.1105/tpc.108.060194] [PMID: 18723576]

[46] Cesari S, Bernoux M, Moncuquet P, Kroj T, Dodds PN. A novel conserved mechanism for plant NLR protein pairs: the integrated decoy hypothesis. Front Plant Sci 2014; 5: 606. [http://dx.doi.org/10.3389/fpls.2014.00606] [PMID: 25506347]

[47] Zhang J, Yin Z, White F. TAL effectors and the executor R genes. Front Plant Sci 2015; 6: 641. [http://dx.doi.org/10.3389/fpls.2015.00641] [PMID: 26347759]

[48] Wan H, Yuan W, Bo K, Shen J, Pang X, Chen J. Genome-wide analysis of NBS-encoding disease resistance genes in *Cucumis sativus* and phylogenetic study of NBS-encoding genes in Cucurbitaceae crops. BMC Genomics 2013; 14: 109. [http://dx.doi.org/10.1186/1471-2164-14-109] [PMID: 23418910]

[49] Shang J, Tao Y, Chen X, *et al.* Identification of a new rice blast resistance gene, Pid3, by genomewide comparison of paired nucleotide-binding siteleucine-rich repeat genes and their pseudogene alleles between the two sequenced rice genomes. Genetics 2009; 182(4): 1303-11. [http://dx.doi.org/10.1534/genetics.109.102871] [PMID: 19506306]

[50] Takken FL, Goverse A. How to build a pathogen detector: structural basis of NB-LRR function. Curr Opin Plant Biol 2012; 15(4): 375-84. [http://dx.doi.org/10.1016/j.pbi.2012.05.001] [PMID: 22658703]

[51] Yang S, Li J, Zhang X, *et al.* Rapidly evolving R genes in diverse grass species confer resistance to rice blast disease. Proc Natl Acad Sci USA 2013; 110(46): 18572-7. [http://dx.doi.org/10.1073/pnas.1318211110] [PMID: 24145399]

[52] Yang S, Tang F, Zhu H. Alternative splicing in plant immunity. Int J Mol Sci 2014; 15(6): 10424-45. [http://dx.doi.org/10.3390/ijms150610424] [PMID: 24918296]

[53] Flor HH. Current status of the gene-for-gene concept. Annu Rev Phytopathol 1971; 9: 275-96.

[http://dx.doi.org/10.1146/annurev.py.09.090171.001423]

[54] Bernoux M, Ellis JG, Dodds PN. New insights in plant immunity signaling activation. Curr Opin Plant Biol 2011; 14(5): 512-8.
[http://dx.doi.org/10.1016/j.pbi.2011.05.005] [PMID: 21723182]

[55] Scofield SR, Tobias CM, Rathjen JP, *et al.* Molecular basis of gene-for-gene specificity in bacterial speck disease of tomato. Science 1996; 274(5295): 2063-5.
[http://dx.doi.org/10.1126/science.274.5295.2063] [PMID: 8953034]

[56] Jia Y, McAdams SA, Bryan GT, Hershey HP, Valent B. Direct interaction of resistance gene and avirulence gene products confers rice blast resistance. EMBO J 2000; 19(15): 4004-14.
[http://dx.doi.org/10.1093/emboj/19.15.4004] [PMID: 10921881]

[57] Rooney HC, Vant Klooster JW, van der Hoorn RA, Joosten MH, Jones JD, de Wit PJ. Cladosporium Avr2 inhibits tomato Rcr3 protease required for Cf-2-dependent disease resistance. Science 2005; 308(5729): 1783-6.
[http://dx.doi.org/10.1126/science.1111404] [PMID: 15845874]

[58] Gu K, Yang B, Tian D, *et al.* R gene expression induced by a type-III effector triggers disease resistance in rice. Nature 2005; 435(7045): 1122-5.
[http://dx.doi.org/10.1038/nature03630] [PMID: 15973413]

[59] Yang S, Feng Z, Zhang X, *et al.* Genome-wide investigation on the genetic variations of rice disease resistance genes. Plant Mol Biol 2006; 62(1-2): 181-93.
[http://dx.doi.org/10.1007/s11103-006-9012-3] [PMID: 16915523]

[60] Kay S, Hahn S, Marois E, Hause G, Bonas U. A bacterial effector acts as a plant transcription factor and induces a cell size regulator. Science 2007; 318(5850): 648-51.
[http://dx.doi.org/10.1126/science.1144956] [PMID: 17962565]

[61] Antony G, Zhou J, Huang S, *et al.* Rice *xa13* recessive resistance to bacterial blight is defeated by induction of the disease susceptibility gene Os-11N3. Plant Cell 2010; 22(11): 3864-76.
[http://dx.doi.org/10.1105/tpc.110.078964] [PMID: 21098734]

[62] Schornack S, Minsavage GV, Stall RE, Jones JB, Lahaye T. Characterization of AvrHah1, a novel AvrBs3-like effector from *Xanthomonas gardneri* with virulence and avirulence activity. New Phytol 2008; 179(2): 546-56.
[http://dx.doi.org/10.1111/j.1469-8137.2008.02487.x] [PMID: 19086184]

[63] Doyle EL, Stoddard BL, Voytas DF, Bogdanove AJ. TAL effectors: highly adaptable phytobacterial virulence factors and readily engineered DNA-targeting proteins. Trends Cell Biol 2013; 23(8): 390-8.
[http://dx.doi.org/10.1016/j.tcb.2013.04.003] [PMID: 23707478]

[64] Tian W, Cao Y, Ismael A, Stone D, Liang J. Roles of regulated internalization in the polarization of cell surface receptors. Conf Proc IEEE Eng Med Biol Soc 2014; 2014: 1166-9.
[PMID: 25570171]

[65] Wang CL, Qin TF, Yu HM, *et al.* The broad bacterial blight resistance of rice line CBB23 is triggered by a novel transcription activator-like (TAL) effector of *Xanthomonas oryzae* pv. oryzae. Mol Plant Pathol 2014; 15(4): 333-41.
[http://dx.doi.org/10.1111/mpp.12092] [PMID: 24286630]

[66] Burketova L, Novakova M, Sasek V, *et al.* Leptosphaeria maculans effector AvrLm4-7 affects SA- and ET-signalling and H_2O_2 accumulation in Brassica napus.. Mol Plant Phatol 2015.
[http://dx.doi.org/DOI: 10.1111/mpp.12332]

[67] Pieterse CM, Zamioudis C, Berendsen RL, Weller DM, Van Wees SC, Bakker PA. Induced systemic resistance by beneficial microbes. Annu Rev Phytopathol 2014; 52: 347-75.
[http://dx.doi.org/10.1146/annurev-phyto-082712-102340] [PMID: 24906124]

[68] Kachroo A, Robin GP. Systemic signaling during plant defense. Curr Opin Plant Biol 2013; 16(4): 527-33.

[http://dx.doi.org/10.1016/j.pbi.2013.06.019] [PMID: 23870750]

[69] Walters DR, Ratsep J, Havis ND. Controlling crop diseases using induced resistance: challenges for the future. J Exp Bot 2013; 64(5): 1263-80.
[http://dx.doi.org/10.1093/jxb/ert026] [PMID: 23386685]

[70] Henry G, Thonart P, Ongena M. Focus on: PAMPs, MAMPs, DAMPs and others: an update on the diversity of plant immunity elicitors. Biotechnol Agron Soc Environ 2012; 16: 257-68.

[71] Luna E, Bruce TJ, Roberts MR, Flors V, Ton J. Next-generation systemic acquired resistance. Plant Physiol 2012; 158(2): 844-53.
[http://dx.doi.org/10.1104/pp.111.187468] [PMID: 22147520]

[72] Slaughter A, Daniel X, Flors V, Luna E, Hohn B, Mauch-Mani B. Descendants of primed Arabidopsis plants exhibit resistance to biotic stress. Plant Physiol 2012; 158(2): 835-43.
[http://dx.doi.org/10.1104/pp.111.191593] [PMID: 22209872]

[73] Vlot AC, Dempsey DA, Klessig DF. Salicylic Acid, a multifaceted hormone to combat disease. Annu Rev Phytopathol 2009; 47: 177-206.
[http://dx.doi.org/10.1146/annurev.phyto.050908.135202] [PMID: 19400653]

[74] Gutjahr C, Paszkowski U. Weights in the balance: jasmonic acid and salicylic acid signaling in root-biotroph interactions. Mol Plant Microbe Interact 2009; 22(7): 763-72.
[http://dx.doi.org/10.1094/MPMI-22-7-0763] [PMID: 19522558]

[75] Pieterse CM, Leon-Reyes A, Van der Ent S, Van Wees SC. Networking by small-molecule hormones in plant immunity. Nat Chem Biol 2009; 5(5): 308-16.
[http://dx.doi.org/10.1038/nchembio.164] [PMID: 19377457]

[76] Ton J, Flors V, Mauch-Mani B. The multifaceted role of ABA in disease resistance. Trends Plant Sci 2009; 14(6): 310-7.
[http://dx.doi.org/10.1016/j.tplants.2009.03.006] [PMID: 19443266]

[77] Thaler JS, Owen B, Higgins VJ. The role of the jasmonate response in plant susceptibility to diverse pathogens with a range of lifestyles. Plant Physiol 2004; 135(1): 530-8.
[http://dx.doi.org/10.1104/pp.104.041566] [PMID: 15133157]

[78] Yasuda M, Ishikawa A, Jikumaru Y, *et al.* Antagonistic interaction between systemic acquired resistance and the abscisic acid-mediated abiotic stress response in Arabidopsis. Plant Cell 2008; 20(6): 1678-92.
[http://dx.doi.org/10.1105/tpc.107.054296] [PMID: 18586869]

[79] Fu ZQ, Dong X. Systemic acquired resistance: turning local infection into global defense. Annu Rev Plant Biol 2013; 64: 839-63.
[http://dx.doi.org/10.1146/annurev-arplant-042811-105606] [PMID: 23373699]

[80] Wang D, Amornsiripanitch N, Dong X. A genomic approach to identify regulatory nodes in the transcriptional network of systemic acquired resistance in plants. PLoS Patholog 2006; 2: e123.
[http://dx.doi.org/10.1371/journal.ppat.0020123] [PMID: PMC1635530]

[81] Spoel SH, Mou Z, Tada Y, Spivey NW, Genschik P, Dong X. Proteasome-mediated turnover of the transcription coactivator NPR1 plays dual roles in regulating plant immunity. Cell 2009; 137(5): 860-72.
[http://dx.doi.org/10.1016/j.cell.2009.03.038] [PMID: 19490895]

[82] Saleh AA, El Komy M, Eranthodi A, *et al.* Variation in a molecular marker for resistance of Saudi date palm germplasm to *Fusarium oxysporum f. sp. albedinis* the causal agent of Bayoud disease. Eur J Plant Pathol 2015; •••
[http://dx.doi.org/10.1007/s10658-015-0703-y]

[83] Conrath U, Beckers GJ, Langenbach CJ, Jaskiewicz MR. Priming for enhanced defense. Annu Rev Phytopathol 2015; 53: 97-119.
[http://dx.doi.org/10.1146/annurev-phyto-080614-120132] [PMID: 26070330]

[84] Po-Wen C, Singh P, Zimmerli L. Priming of the Arabidopsis pattern-triggered immunity response upon infection by necrotrophic *Pectobacterium carotovorum* bacteria. Mol Plant Pathol 2013; 14(1): 58-70.
[http://dx.doi.org/10.1111/j.1364-3703.2012.00827.x] [PMID: 22947164]

[85] Luna E, López A, Kooiman J, Ton J. Role of NPR1 and KYP in long-lasting induced resistance by β-aminobutyric acid. Front Plant Sci 2014; 5: 184.
[http://dx.doi.org/10.3389/fpls.2014.00184] [PMID: 24847342]

[86] Brodersen P, Sakvarelidze-Achard L, Bruun-Rasmussen M, *et al.* Widespread translational inhibition by plant miRNAs and siRNAs. Science 2008; 320(5880): 1185-90.
[http://dx.doi.org/10.1126/science.1159151] [PMID: 18483398]

[87] Gurtan AM, Sharp PA. The role of miRNAs in regulating gene expression networks. J Mol Biol 2013; 425(19): 3582-600. [See comment in PubMed Commons b].
[http://dx.doi.org/10.1016/j.jmb.2013.03.007] [PMID: 23500488]

[88] Xie Z, Qi X. Diverse small RNA-directed silencing pathways in plants. Biochim Biophys Acta 2008; 1779(11): 720-4.
[http://dx.doi.org/10.1016/j.bbagrm.2008.02.009] [PMID: 18358247]

[89] Weiberg A, Wang M, Lin FM, *et al.* Fungal small RNAs suppress plant immunity by hijacking host RNA interference pathways. Science 2013; 342(6154): 118-23.
[http://dx.doi.org/10.1126/science.1239705] [PMID: 24092744]

[90] Song R, Hennig GW, Wu Q, Jose C, Zheng H, Yan W. Male germ cells express abundant endogenous siRNAs. Proc Natl Acad Sci USA 2011; 108(32): 13159-64.
[http://dx.doi.org/10.1073/pnas.1108567108] [PMID: 21788498]

[91] Ruiz-Medrano R, Xoconostle-Cazares B, Kragler F. The plasmodesmatal transport pathway for homeotic proteins, silencing signals and viruses. Curr Opin Plant Biol 2004; 7(6): 641-50.
[http://dx.doi.org/10.1016/j.pbi.2004.09.012] [PMID: 15491912]

[92] Navarro L, Dunoyer P, Jay F, *et al.* A plant miRNA contributes to antibacterial resistance by repressing auxin signaling. Science 2006; 312(5772): 436-9.
[http://dx.doi.org/10.1126/science.1126088] [PMID: 16627744]

[93] Guo W, Liew JY, Yuan YA. Structural insights into the arms race between host and virus along RNA silencing pathways in *Arabidopsis thaliana.* Biol Rev Camb Philos Soc 2013; 89(2): 337-55.
[http://dx.doi.org/10.1111/brv.12057] [PMID: 24034134]

[94] Pollard M, Beisson F, Li Y, Ohlrogge JB. Building lipid barriers: biosynthesis of cutin and suberin. Trends Plant Sci 2008; 13(5): 236-46.
[http://dx.doi.org/10.1016/j.tplants.2008.03.003] [PMID: 18440267]

[95] St Clair DA. Quantitative disease resistance and quantitative resistance Loci in breeding. Annu Rev Phytopathol 2010; 48: 247-68.
[http://dx.doi.org/10.1146/annurev-phyto-080508-081904] [PMID: 19400646]

[96] Michelmore RW, Christopoulou M, Caldwell KS. Impacts of resistance gene genetics, function, and evolution on a durable future. Annu Rev Phytopathol 2013; 51: 291-319.
[http://dx.doi.org/10.1146/annurev-phyto-082712-102334] [PMID: 23682913]

[97] Galvez LC, Banerjee J, Pinar H, Mitra A. Engineered plant virus resistance. Plant Sci 2014; 228: 11-25.
[http://dx.doi.org/10.1016/j.plantsci.2014.07.006] [PMID: 25438782]

[98] Sprague SJ, Balesdent MH, Brun H, *et al.* Major gene resistance in *Brassica napus* (oilseed rape) is overcome by changes in virulence of populations of *Leptosphaeria maculans* in France and Australia. Eur J Plant Pathol 2006; 114: 33-40.
[http://dx.doi.org/10.1007/s10658-005-3683-5]

[99] Jung C, Cai D, Kleine M. Engineering nematode resistance in crop species. Trends Plant Sci 1998; 3: 266-71.
[http://dx.doi.org/10.1016/S1360-1385(98)01247-3]

[100] Hussain B. Modernization in plant breeding approaches for improving biotic stress resistance in crop plants. Turk J Agric For 2015; 39
[http://dx.doi.org/10.3906/tar-1406-176]

[101] Miah G, Rafii MY, Ismail MR, *et al.* Blast resistance in rice: a review of conventional breeding to molecular approaches. Mol Biol Rep 2013; 40(3): 2369-88.
[http://dx.doi.org/10.1007/s11033-012-2318-0] [PMID: 23184051]

[102] McIntosh RA, Wellings CR, Park RF. Wheat rusts: an atlas of resistance genes. Dordrecht: Kluwer Academic Publishers 1995.
[http://dx.doi.org/10.1007/978-94-011-0083-0]

[103] Dangl JL, Jones FGJ, Park RF. Plant pathogens and integrated defence responses to infection. Nature 2001; 411: 826-33.

[104] Vleeshouwers VG, Raffaele S, Vossen JH, *et al.* Understanding and exploiting late blight resistance in the age of effectors. Annu Rev Phytopathol 2011; 49: 507-31.
[http://dx.doi.org/10.1146/annurev-phyto-072910-095326] [PMID: 21663437]

[105] Boyd LA, Ridout C, OSullivan DM, Leach JE, Leung H. Plant-pathogen interactions: disease resistance in modern agriculture. Trends Genet 2013; 29(4): 233-40.
[http://dx.doi.org/10.1016/j.tig.2012.10.011] [PMID: 23153595]

[106] Panstruga R. Serpentine plant MLO proteins as entry portals for powdery mildew fungi. Biochem Soc Trans 2005; 33(Pt 2): 389-92.
[http://dx.doi.org/10.1042/BST0330389] [PMID: 15787613]

[107] Eichmann R, Hückelhoven R. Accommodation of powdery mildew fungi in intact plant cells. J Plant Physiol 2008; 165(1): 5-18.
[http://dx.doi.org/10.1016/j.jplph.2007.05.004] [PMID: 17602788]

[108] Hückelhoven R, Dechert C, Kogel KH. Overexpression of barley BAX inhibitor 1 induces breakdown of mlo-mediated penetration resistance to *Blumeria graminis.* Proc Natl Acad Sci USA 2003; 100(9): 5555-60.
[http://dx.doi.org/10.1073/pnas.0931464100] [PMID: 12704231]

[109] Poland JA, Balint-Kurti PJ, Wisser RJ, Pratt RC, Nelson RJ. Shades of gray: the world of quantitative disease resistance. Trends Plant Sci 2009; 14(1): 21-9.
[http://dx.doi.org/10.1016/j.tplants.2008.10.006] [PMID: 19062327]

[110] Niks RE, Qi X, Marcel TC. Quantitative resistance to biotrophic filamentous plant pathogens: concepts, misconceptions, and mechanisms. Annu Rev Phytopathol 2015; 53: 445-70.
[http://dx.doi.org/10.1146/annurev-phyto-080614-115928] [PMID: 26047563]

[111] Miklas PN, Kelly JD, Beebe SE, Blair MW. Common bean breeding for resistance against biotic and abiotic stresses: from classical to MAS breeding. Euphytica 2006; 147: 105-31.
[http://dx.doi.org/10.1007/s10681-006-4600-5]

[112] Miklas PN, Larsen KM, Terpstra K, *et al.* QTL analysis of ICA Bunsi-derived resistance to white mold in a pinto x navy bean cross. Crop Sci 2007; 47: 174-9.
[http://dx.doi.org/10.2135/cropsci2006.06.0410]

[113] Schweizer P, Stein N. Large-scale data integration reveals colocalization of gene functional groups with meta-QTL for multiple disease resistance in barley. Mol Plant Microbe Interact 2011; 24(12): 1492-501.
[http://dx.doi.org/10.1094/MPMI-05-11-0107] [PMID: 21770767]

[114] Van Inghelandt D, Melchinger AE, Martinant JP, Stich B. Genome-wide association mapping of

flowering time and northern corn leaf blight (*Setosphaeria turcica*) resistance in a vast commercial maize germplasm set. BMC Plant Biol 2012; 12: 56.
[http://dx.doi.org/10.1186/1471-2229-12-56] [PMID: 22545925]

[115] Kump KL, Bradbury PJ, Wisser RJ, *et al.* Genome-wide association study of quantitative resistance to southern leaf blight in the maize nested association mapping population. Nat Genet 2011; 43(2): 163-8.
[http://dx.doi.org/10.1038/ng.747] [PMID: 21217757]

[116] Poland JA, Bradbury PJ, Buckler ES, Nelson RJ. Genome-wide nested association mapping of quantitative resistance to northern leaf blight in maize. Proc Natl Acad Sci USA 2011; 108(17): 6893-8.
[http://dx.doi.org/10.1073/pnas.1010894108] [PMID: 21482771]

[117] Yu LX, Lorenz A, Rutkoski J, *et al.* Association mapping and gene-gene interaction for stem rust resistance in CIMMYT spring wheat germplasm. Theor Appl Genet 2011; 123(8): 1257-68.
[http://dx.doi.org/10.1007/s00122-011-1664-y] [PMID: 21811818]

[118] Gurung S, Mamidi S, Bonman JM, Xiong M, Brown-Guedira G, Adhikari TB. Genome-wide association study reveals novel quantitative trait Loci associated with resistance to multiple leaf spot diseases of spring wheat. PLoS One 2014; 9(9): e108179.
[http://dx.doi.org/10.1371/journal.pone.0108179] [PMID: 25268502]

[119] Yu LX, Lorenz A, Rutkoski J, *et al.* Association mapping and gene-gene interaction for stem rust resistance in CIMMYT spring wheat germplasm. Theor Appl Genet 2011; 123(8): 1257-68.
[http://dx.doi.org/10.1007/s00122-011-1664-y] [PMID: 21811818]

[120] FAO. WFP & iFADThe state of food insecurity in the World Economic growth is necessary but not sufficient to accelerate reduction of hunger and malnutrition. Rome: FAO 2012.

[120a] Schellenberger U, Orai J, Rosen BA, *et al.* A selective insecticidal protein from Pseudomonas for controlling corn rootworms Science 2016.
[http://dx.doi.org/I0.1126/science.aaf6056]

[121] Keisa A, Kanberga-Silina K, Nakurte I, Kunga L, Rostoks N. Differential disease resistance response in the barley necrotic mutant nec1. BMC Plant Biol 2011; 11: 66.
[http://dx.doi.org/10.1186/1471-2229-11-66] [PMID: 21496226]

[122] Rostoks N, Schmierer D, Mudie S, *et al.* Barley necrotic locus nec1 encodes the cyclic nucleotide-gated ion channel 4 homologous to the Arabidopsis HLM1. Mol Genet Genomics 2006; 275(2): 159-68.
[http://dx.doi.org/10.1007/s00438-005-0073-9] [PMID: 16341885]

[123] Farnham G, Baulcombe DC. Artificial evolution extends the spectrum of viruses that are targeted by a disease-resistance gene from potato. Proc Natl Acad Sci USA 2006; 103(49): 18828-33.
[http://dx.doi.org/10.1073/pnas.0605777103] [PMID: 17021014]

[124] Brunner S, Hurni S, Streckeisen P, *et al.* Intragenic allele pyramiding combines different specificities of wheat Pm3 resistance alleles. Plant J 2010; 64(3): 433-45.
[http://dx.doi.org/10.1111/j.1365-313X.2010.04342.x] [PMID: 20804457]

[125] Sikora P, Chavade A, Larsson M, *et al.* Mutagenesis as a Tool In Plant genetics, functional genomics, and breeding. Internat J Plant Genomics 2011; 13 pages.
[http://dx.doi.org/http://dx.doi.org/10.1155/2011/314829] [PMID: 314829]

[126] Desjardins AE, Proctor RH. Molecular biology of Fusarium mycotoxins. Int J Food Microbiol 2007; 119(1-2): 47-50.
[http://dx.doi.org/10.1016/j.ijfoodmicro.2007.07.024] [PMID: 17707105]

[127] Walter S, Nicholson P, Doohan FM. Action and reaction of host and pathogen during Fusarium head blight disease. New Phytol 2010; 185(1): 54-66.
[http://dx.doi.org/10.1111/j.1469-8137.2009.03041.x] [PMID: 19807873]

[128] He X, Skinnes H, Oliver RE, Jackson EW, Bjørnstad A. Linkage mapping and identification of QTL affecting deoxynivalenol (DON) content (Fusarium resistance) in oats (*Avena sativa* L.). Theor Appl Genet 2013; 126(10): 2655-70.
[http://dx.doi.org/10.1007/s00122-013-2163-0] [PMID: 23959525]

[129] Chen X, Shang J, Chen D, *et al.* A B-lectin receptor kinase gene conferring rice blast resistance. Plant J 2006; 46(5): 794-804.
[http://dx.doi.org/10.1111/j.1365-313X.2006.02739.x] [PMID: 16709195]

[130] Atkinson HJ, Lilley CJ, Urwin PE. Strategies for transgenic nematode control in developed and developing world crops. Curr Opin Biotechnol 2012; 23(2): 251-6.
[http://dx.doi.org/10.1016/j.copbio.2011.09.004] [PMID: 21996368]

[130a] Rudolph C, Schreier PH, Uhrig JF. Peptide-mediated broad-spectrum plant resistance to tospoviruses. Proc Natl Acad Sci USA 2003; 100(8): 4429-34.
[http://dx.doi.org/10.1073/pnas.0730832100] [PMID: 12682295]

[131] Fuchs M, Gonsalves D. Safety of virus-resistant transgenic plants two decades after their introduction: lessons from realistic field risk assessment studies. Annu Rev Phytopathol 2007; 45: 173-202.
[http://dx.doi.org/10.1146/annurev.phyto.45.062806.094434] [PMID: 17408355]

[132] Balconi C, Lanzanova C, Motto M. Ribosome-inactivating proteins in cereals In: Lord J, Hartley MR, Eds. Springer series Plant Cell Monographs. Berlin: Springer-Verlag 2010.

[133] Zhang L, Kars I, Essenstam B, *et al.* Fungal endopolygalacturonases are recognized as microbe-associated molecular patterns by the arabidopsis receptor-like protein RESPONSIVENESS TO BOTRYTIS POLYGALACTURONASES1. Plant Physiol 2014; 164(1): 352-64.
[http://dx.doi.org/10.1104/pp.113.230698] [PMID: 24259685]

[134] Lengyel P. Tumor-suppressor genes: news about the interferon connection. Proc Natl Acad Sci USA 1993; 90(13): 5893-5.
[http://dx.doi.org/10.1073/pnas.90.13.5893] [PMID: 7687056]

[135] Mitra A, Higgins DW, Langenberg WG, Nie H, Sengupta DN, Silverman RH. A mammalian 25A system functions as an antiviral pathway in transgenic plants. Proc Natl Acad Sci USA 1996; 93(13): 6780-5.
[http://dx.doi.org/10.1073/pnas.93.13.6780] [PMID: 8692895]

[136] Ogawa K, Kanematsu S, Asada K. Intra- and extracellular localization of "cytosolic" Cu, Zn-superoxide dismutase in spinach leaf and hypocotyl. Plant Cell Physiol 1996; 37: 790-9.
[http://dx.doi.org/10.1093/oxfordjournals.pcp.a029014]

[137] Honda A, Takahashi H, Toguri T, *et al.* Activation of defense-related gene expression and systemic acquired resistance in cucumber mosaic virus-infected tobacco plants expressing the mammalian 25oligoadenylate system. Arch Virol 2003; 148(5): 1017-26.
[http://dx.doi.org/10.1007/s00705-002-0959-3] [PMID: 12721807]

[138] Vleeshouwers VG, Oliver RP. Effectors as tools in disease resistance breeding against biotrophic, hemibiotrophic, and necrotrophic plant pathogens. Mol Plant Microbe Interact 2014; 27(3): 196-206.
[http://dx.doi.org/10.1094/MPMI-10-13-0313-IA] [PMID: 24405032]

[139] Hummel AW, Doyle EL, Bogdanove AJ. Addition of transcription activator-like effector binding sites to a pathogen strain-specific rice bacterial blight resistance gene makes it effective against additional strains and against bacterial leaf streak. New Phytol 2012; 195(4): 883-93.
[http://dx.doi.org/10.1111/j.1469-8137.2012.04216.x] [PMID: 22747776]

[140] Carroll D. Genome engineering with targetable nucleases. Annu Rev Biochem 2014; 83: 409-39.
[http://dx.doi.org/10.1146/annurev-biochem-060713-035418] [PMID: 24606144]

[141] Voytas DF. Plant genome engineering with sequence-specific nucleases. Annu Rev Plant Biol 2013; 64: 327-50.
[http://dx.doi.org/10.1146/annurev-arplant-042811-105552] [PMID: 23451779]

[142] Li T, Liu B, Spalding MH, Weeks DP, Yang B. High-efficiency TALEN-based gene editing produces disease-resistant rice. Nat Biotechnol 2012; 30(5): 390-2. [http://dx.doi.org/10.1038/nbt.2199] [PMID: 22565958]

[143] Pedigo LP, Rice ME. Entomology and pest management. 6th ed. Columbus, OH: Pearson Prentice Hall 2014; p. 784.

[144] Agriotos GN. Plant pathology. 4th ed., San Diego, CA: Academic Press 1997.

[145] Maxwell FG, Jennings PR, Eds. Breeding Plants Resistant To Insects. New York: John Wiley 1980.

[146] Painter RH. Insect resistance in crop plants. Lawrence: University Press of Kansas 1958.

[147] Howe GA, Jander G. Plant immunity to insect herbivores. Annu Rev Plant Biol 2008; 59: 41-66. [http://dx.doi.org/10.1146/annurev.arplant.59.032607.092825] [PMID: 18031220]

[147a] Arimura G, Ozawa R, Maffei ME. Recent advances in plant early signaling in response to herbivory. Int J Mol Sci 2011; 12(6): 3723-39. [http://dx.doi.org/10.3390/ijms12063723] [PMID: 21747702]

[148] War AR, Paulraj MG, Ahmad T, *et al.* Mechanisms of plant defense against insect herbivores. Plant Signal Behav 2012; 7(10): 1306-20. [http://dx.doi.org/10.4161/psb.21663] [PMID: 22895106]

[149] Fürstenberg-Hägg J, Zagrobelny M, Bak S. Plant defense against insect herbivores. Int J Mol Sci 2013; 14(5): 10242-97. [http://dx.doi.org/10.3390/ijms140510242] [PMID: 23681010]

[150] Mithöfer A, Boland W. Plant defense against herbivores: chemical aspects. Annu Rev Plant Biol 2012; 63: 431-50. [http://dx.doi.org/10.1146/annurev-arplant-042110-103854] [PMID: 22404468]

[151] Aharoni A, Jongsma MA, Bouwmeester HJ. Volatile science? Metabolic engineering of terpenoids in plants. Trends Plant Sci 2005; 10(12): 594-602. [http://dx.doi.org/10.1016/j.tplants.2005.10.005] [PMID: 16290212]

[152] Morant AV, Jørgensen K, Jørgensen C, *et al.* beta-Glucosidases as detonators of plant chemical defense. Phytochemistry 2008; 69(9): 1795-813. [http://dx.doi.org/10.1016/j.phytochem.2008.03.006] [PMID: 18472115]

[153] Frey M, Schullehner K, Dick R, Fiesselmann A, Gierl A. Benzoxazinoid biosynthesis, a model for evolution of secondary metabolic pathways in plants. Phytochemistry 2009; 70(15-16): 1645-51. [http://dx.doi.org/10.1016/j.phytochem.2009.05.012] [PMID: 19577780]

[154] Bergvinson DJ, Hamilton RI, Arnason JT. Leaf profile of maize resistance factors to European corn borer,Ostrinia nubilalis. J Chem Ecol 1995; 21(3): 343-54. [http://dx.doi.org/10.1007/BF02036722] [PMID: 24234065]

[155] Birkett MA, Pickett JA. Prospects of genetic engineering for robust insect resistance. Curr Opin Plant Biol 2014; 19: 59-67. [http://dx.doi.org/10.1016/j.pbi.2014.03.009] [PMID: 24747775]

[156] Hansen BG, Kerwin RE, Ober JA, *et al.* A novel 2-oxoacid-dependent dioxygenase involved in the formation of the goiterogenic 2-hydroxybut-3-enyl glucosinolate and generalist insect resistance in Arabidopsis. Plant Physiol 2008; 148(4): 2096-108. [http://dx.doi.org/10.1104/pp.108.129981] [PMID: 18945935]

[157] Mewis I, Appel HM, Hom A, Raina R, Schultz JC. Major signaling pathways modulate Arabidopsis glucosinolate accumulation and response to both phloem-feeding and chewing insects. Plant Physiol 2005; 138(2): 1149-62. [http://dx.doi.org/10.1104/pp.104.053389] [PMID: 15923339]

[158] Barth C, Jander G. Arabidopsis myrosinases TGG1 and TGG2 have redundant function in

glucosinolate breakdown and insect defense. Plant J 2006; 46(4): 549-62. [http://dx.doi.org/10.1111/j.1365-313X.2006.02716.x] [PMID: 16640593]

[159] Kim JH, Jander G. Myzus persicae (green peach aphid) feeding on Arabidopsis induces the formation of a deterrent indole glucosinolate. Plant J 2007; 49(6): 1008-19. [http://dx.doi.org/10.1111/j.1365-313X.2006.03019.x] [PMID: 17257166]

[160] Halkier BA, Gershenzon J. Biology and biochemistry of glucosinolates. Annu Rev Plant Biol 2006; 57: 303-33. [http://dx.doi.org/10.1146/annurev.arplant.57.032905.105228] [PMID: 16669764]

[161] Tattersall DB, Bak S, Jones PR, *et al.* Resistance to an herbivore through engineered cyanogenic glucoside synthesis. Science 2001; 293(5536): 1826-8. [http://dx.doi.org/10.1126/science.1062249] [PMID: 11474068]

[162] Truong DH, Heuskin S, Delaplace P, *et al.* VOC emissions and protein expression mediated by the interactions between herbivorous insects and Arabidopsis plant. A review. Biotechnol Agron Soc Environ 2014; 18: 455-64.

[163] Chen H, Wilkerson CG, Kuchar JA, Phinney BS, Howe GA. Jasmonate-inducible plant enzymes degrade essential amino acids in the herbivore midgut. Proc Natl Acad Sci USA 2005; 102(52): 19237-42. [http://dx.doi.org/10.1073/pnas.0509026102] [PMID: 16357201]

[164] Mithöfer A, Boland W. Plant defense against herbivores: chemical aspects. Annu Rev Plant Biol 2012; 63: 431-50. [http://dx.doi.org/10.1146/annurev-arplant-042110-103854] [PMID: 22404468]

[165] Lawrence PK, Koundal KR. Plant protease inhibitors in control of phytophagous insects. Electron J Biotechnol 2002; 5: 93-109. [http://dx.doi.org/10.2225/vol5-issue1-fulltext-3]

[166] Dunse KM, Stevens JA, Lay FT, Gaspar YM, Heath RL, Anderson MA. Coexpression of potato type I and II proteinase inhibitors gives cotton plants protection against insect damage in the field. Proc Natl Acad Sci USA 2010; 107(34): 15011-5. [http://dx.doi.org/10.1073/pnas.1009241107] [PMID: 20696895]

[167] Azzouz H, Cherqui A, Campan ED, *et al.* Effects of plant protease inhibitors, oryzacystatin I and soybean Bowman-Birk inhibitor, on the aphid Macrosiphum euphorbiae (Homoptera, Aphididae) and its parasitoid Aphelinus abdominalis (Hymenoptera, Aphelinidae). J Insect Physiol 2005; 51(1): 75-86. [http://dx.doi.org/10.1016/j.jinsphys.2004.11.010] [PMID: 15686649]

[168] Hartl M, Giri AP, Kaur H, Baldwin IT. Serine protease inhibitors specifically defend *Solanum nigrum* against generalist herbivores but do not influence plant growth and development. Plant Cell 2010; 22(12): 4158-75. [http://dx.doi.org/10.1105/tpc.109.073395] [PMID: 21177479]

[169] Lisón P, Rodrigo I, Conejero V. A novel function for the cathepsin D inhibitor in tomato. Plant Physiol 2006; 142(3): 1329-39. [http://dx.doi.org/10.1104/pp.106.086587] [PMID: 17012408]

[170] Zavala JA, Patankar AG, Gase K, Hui D, Baldwin IT. Manipulation of endogenous trypsin proteinase inhibitor production in Nicotiana attenuata demonstrates their function as antiherbivore defenses. Plant Physiol 2004; 134(3): 1181-90. [http://dx.doi.org/10.1104/pp.103.035634] [PMID: 14976235]

[171] Bayés A, Comellas-Bigler M, Rodríguez de la Vega M, *et al.* Structural basis of the resistance of an insect carboxypeptidase to plant protease inhibitors. Proc Natl Acad Sci USA 2005; 102(46): 16602-7. [http://dx.doi.org/10.1073/pnas.0505489102] [PMID: 16260742]

[172] Rivard D, Cloutier C, Michaud D. Colorado potato beetles show differential digestive compensatory responses to host plants expressing distinct sets of defense proteins. Arch Insect Biochem Physiol

2004; 55(3): 114-23.
[http://dx.doi.org/10.1002/arch.10136] [PMID: 14981656]

[173] Chakraborti D, Sarkar A, Mondal HA, Das S. Tissue specific expression of potent insecticidal, Allium sativum leaf agglutinin (ASAL) in important pulse crop, chickpea (*Cicer arietinum* L.) to resist the phloem feeding Aphis craccivora. Transgenic Res 2009; 18(4): 529-44.
[http://dx.doi.org/10.1007/s11248-009-9242-7] [PMID: 19184504]

[174] Vandenborre G, Smagghe G, Van Damme EJ. Plant lectins as defense proteins against phytophagous insects. Phytochemistry 2011; 72(13): 1538-50.
[http://dx.doi.org/10.1016/j.phytochem.2011.02.024] [PMID: 21429537]

[175] Konno K, Hirayama C, Nakamura M, *et al.* Papain protects papaya trees from herbivorous insects: role of cysteine proteases in latex. Plant J 2004; 37(3): 370-8.
[http://dx.doi.org/10.1046/j.1365-313X.2003.01968.x] [PMID: 14731257]

[176] Mohan S, Ma PW, Pechan T, Bassford ER, Williams WP, Luthe DS. Degradation of the S. frugiperda peritrophic matrix by an inducible maize cysteine protease. J Insect Physiol 2006; 52(1): 21-8.
[http://dx.doi.org/10.1016/j.jinsphys.2005.08.011] [PMID: 16243350]

[177] Gill RS, Gupta K, Taggar GK, Taggar MS. Role of oxidative enzymes in plant defenses against herbivory. Acta Phytopathol Entomol Hung 2010; 45: 277-90.
[http://dx.doi.org/10.1556/APhyt.45.2010.2.4]

[178] Chen H, Wilkerson CG, Kuchar JA, Phinney BS, Howe GA. Jasmonate-inducible plant enzymes degrade essential amino acids in the herbivore midgut. Proc Natl Acad Sci USA 2005; 102(52): 19237-42.
[http://dx.doi.org/10.1073/pnas.0509026102] [PMID: 16357201]

[179] Truong DH, Delory BM, Vanderplanck M, *et al.* Temperature regimes and aphid density interactions differentially influence VOC emissions in Arabidopsis. Arthropod-Plant Interact 2014; 8: 317-27.

[180] Schuman MC, Barthel K, Baldwin IT. Herbivory-induced volatiles function as defenses increasing fitness of the native plant Nicotiana attenuata in nature. eLife 2012; 1: e00007.
[http://dx.doi.org/10.7554/eLife.00007] [PMID: 23066503]

[181] Degenhardt J, Hiltpold I, Köllner TG, *et al.* Restoring a maize root signal that attracts insect-killing nematodes to control a major pest. Proc Natl Acad Sci USA 2009; 106(32): 13213-8.
[http://dx.doi.org/10.1073/pnas.0906365106] [PMID: 19666594]

[182] Yencho GC, Cohen MB, Byrne PF. Applications of tagging and mapping insect resistance loci in plants. Annu Rev Entomol 2000; 45: 393-422.
[http://dx.doi.org/10.1146/annurev.ento.45.1.393] [PMID: 10761583]

[183] Kaloshian I. Gene-for-gene disease resistance: bridging insect pest and pathogen defense. J Chem Ecol 2004; 30(12): 2419-38.
[http://dx.doi.org/10.1007/s10886-004-7943-1] [PMID: 15724964]

[184] Dogimont C, Bendahmane A, Pitrat M, *et al.* Gene resistant to Aphis gossypii. US Patent Application US 2007/0016977, 2007.

[185] Jung C, Cai D, Kleine M. Engineering nematode resistance in crop species. Trends Plant Sci 1998; 3: 266-71.
[http://dx.doi.org/10.1016/S1360-1385(98)01247-3]

[186] Hilker M, Meiners T. How do plants notice attack by herbivorous arthropods? Biol Rev Camb Philos Soc 2010; 85(2): 267-80.
[http://dx.doi.org/10.1111/j.1469-185X.2009.00100.x] [PMID: 19961475]

[187] Bonaventure G, VanDoorn A, Baldwin IT. Herbivore-associated elicitors: FAC signaling and metabolism. Trends Plant Sci 2011; 16(6): 294-9.
[http://dx.doi.org/10.1016/j.tplants.2011.01.006] [PMID: 21354852]

[188] Wang J, Song WK, Zhang WB, Liu CY, Hu GH, Chen QS. [Meta-analysis of insect-resistance QTLs in soybean]. Yi Chuan 2009; 31(9): 953-61.
[http://dx.doi.org/10.3724/SP.J.1005.2009.00953] [PMID: 19819849]

[189] Hao W, Lin HX. Toward understanding genetic mechanisms of complex traits in rice. J Genet Genomics 2010; 37(10): 653-66.
[http://dx.doi.org/10.1016/S1673-8527(09)60084-9] [PMID: 21035092]

[190] Yencho GC, Cohen MB, Byrne PF. Applications of tagging and mapping insect resistance loci in plants. Annu Rev Entomol 2000; 45: 393-422.
[http://dx.doi.org/10.1146/annurev.ento.45.1.393] [PMID: 10761583]

[191] Ben-Israel I, Yu G, Austin MB, *et al.* Multiple biochemical and morphological factors underlie the production of methylketones in tomato trichomes. Plant Physiol 2009; 151(4): 1952-64.
[http://dx.doi.org/10.1104/pp.109.146415] [PMID: 19801397]

[192] Momotaz A, Scott JW, Schuster DJ. Identification of quantitative trait loci conferring resistance to Bemisiatabaci in an F2 population of Solanum lycopersicum x Solanum habrochaites accession LA1777. J Am Soc Hortic Sci 2010; 135: 134-42.

[193] Kliebenstein DJ. A quantitative genetics and ecological model system: understanding the aliphatic glucosinolate biosynthetic network via QTLs. Phytochem Rev 2009; 8: 243-54.
[http://dx.doi.org/10.1007/s11101-008-9102-8]

[194] Byrne PF, McMullen MD, Wiseman BR, *et al.* Maize silk maysin concentration and corn earworm antibiosis: QTLs and genetic mechanisms. Crop Sci 1998; 38: 461-71.
[http://dx.doi.org/10.2135/cropsci1998.0011183X003800020032x]

[195] Byrne PF, McMullen MD, Snook ME, *et al.* Quantitative trait loci and metabolic pathways: genetic control of the concentration of maysin, a corn earworm resistance factor, in maize silks. Proc Natl Acad Sci USA 1996; 93(17): 8820-5.
[http://dx.doi.org/10.1073/pnas.93.17.8820] [PMID: 11607699]

[196] Walling LL. Avoiding effective defenses: strategies employed by phloem-feeding insects. Plant Physiol 2008; 146(3): 859-66.
[http://dx.doi.org/10.1104/pp.107.113142] [PMID: 18316641]

[197] Green JM, Owen MD, Rusterucci C, *et al.* Herbicide-resistant crops: utilities and limitations for herbicide-resistant weed management. J Agric Food Chem 2011; 59(11): 5819-29.
[http://dx.doi.org/10.1021/jf101286h] [PMID: 20586458]

[198] Pavan S, Jacobsen E, Visser RG, Bai Y. Loss of susceptibility as a novel breeding strategy for durable and broad-spectrum resistance. Mol Breed 2010; 25(1): 1-12.
[http://dx.doi.org/10.1007/s11032-009-9323-6] [PMID: 20234841]

[199] Broekgaarden C, Snoeren TA, Dicke M, Vosman B. Exploiting natural variation to identify insect-resistance genes. Plant Biotechnol J 2011; 9(8): 819-25.
[http://dx.doi.org/10.1111/j.1467-7652.2011.00635.x] [PMID: 21679292]

[200] Kalule T, Wright DJ. Effect of cultivars with varying levels of resistance to aphids on development time, sex ratio, size and longevity of the parasitoid *Aphidius colemani*. BioControl 2005; 50: 235-46.
[http://dx.doi.org/10.1007/s10526-004-3072-6]

[201] Mitchell C, Johnson SN, Gordon SC, *et al.* Combining plant resistance and a natural enemy to control Amphorophora idaei. BioControl 2010; 55: 321-7.
[http://dx.doi.org/10.1007/s10526-009-9257-2]

[202] Allmann S, Baldwin IT. Insects betray themselves in nature to predators by rapid isomerization of green leaf volatiles. Science 2010; 329(5995): 1075-8.
[http://dx.doi.org/10.1126/science.1191634] [PMID: 20798319]

[203] Degenhardt J, Hiltpold I, Köllner TG, *et al.* Restoring a maize root signal that attracts insect-killing

nematodes to control a major pest. Proc Natl Acad Sci USA 2009; 106(32): 13213-8. [http://dx.doi.org/10.1073/pnas.0906365106] [PMID: 19666594]

[204] Kappers IF, Verstappen FW, Luckerhoff LL, Bouwmeester HJ, Dicke M. Genetic variation in jasmonic acid- and spider mite-induced plant volatile emission of cucumber accessions and attraction of the predator Phytoseiulus persimilis. J Chem Ecol 2010; 36(5): 500-12. [http://dx.doi.org/10.1007/s10886-010-9782-6] [PMID: 20383796]

[205] Snoeren TA, Kappers IF, Broekgaarden C, Mumm R, Dicke M, Bouwmeester HJ. Natural variation in herbivore-induced volatiles in *Arabidopsis thaliana*. J Exp Bot 2010; 61(11): 3041-56. [http://dx.doi.org/10.1093/jxb/erq127] [PMID: 20488836]

[206] Lewis WJ, van Lenteren JC, Phatak SC, Tumlinson JH III. A total system approach to sustainable pest management. Proc Natl Acad Sci USA 1997; 94(23): 12243-8. [http://dx.doi.org/10.1073/pnas.94.23.12243] [PMID: 9356432]

[207] Dicke M. Direct and indirect effects of plants on performance of beneficial organisms. In: Ruberson JR, Ed. Handbook of Pest Management. New York: Marcel Dekker 1999; pp. 105-53.

[208] Alam SN, Cohen MB. Detection and analysis of QTLs for resistance to the brown plant hopper, Nilaparvatalugens, in a doubled-haploid rice population. Theor Appl Genet 1998; 97: 1370-9. [http://dx.doi.org/10.1007/s001220051031]

[209] James C. Global status of commercialized biotech/GM crops: 2014. ISAAA Briefs No 49:. Ithaca, NY 2015.

[210] Christou P, Capell T, Kohli A, Gatehouse JA, Gatehouse AM. Recent developments and future prospects in insect pest control in transgenic crops. Trends Plant Sci 2006; 11(6): 302-8. [http://dx.doi.org/10.1016/j.tplants.2006.04.001] [PMID: 16690346]

[211] Mendelsohn M, Kough J, Vaituzis Z, Matthews K. Are Bt crops safe? Nat Biotechnol 2003; 21(9): 1003-9. [http://dx.doi.org/10.1038/nbt0903-1003] [PMID: 12949561]

[212] Crickmore N. Bacillus thuringiensis toxin nomenclature 2011. Available at http://www.lifesci.sussex.ac.uk/home/Neil_Crickmore/Bt/;

[213] Babu RM, Sajeena A, Seetharaman K, Reddy MS. Advances in genetically engineered (transgenic) plants in pest management – An over view. Crop Prot 2003; 22: 1071-86. [http://dx.doi.org/10.1016/S0261-2194(03)00142-X]

[214] Mehlo L, Gahakwa D, Nghia PT, *et al.* An alternative strategy for sustainable pest resistance in genetically enhanced crops. Proc Natl Acad Sci USA 2005; 102(22): 7812-6. [http://dx.doi.org/10.1073/pnas.0502871102] [PMID: 15908504]

[215] Tabashnik BE, Brévault T, Carrière Y. Insect resistance to Bt crops: lessons from the first billion acres. Nat Biotechnol 2013; 31(6): 510-21. [http://dx.doi.org/10.1038/nbt.2597] [PMID: 23752438]

[216] Saha P, Dasgupta I, Das S. A novel approach for developing resistance in rice against phloem limited viruses by antagonizing the phloem feeding hemipteran vectors. Plant Mol Biol 2006; 62(4-5): 735-52. [http://dx.doi.org/10.1007/s11103-006-9054-6] [PMID: 16941213]

[217] Yao J, Pang Y, Qi H, *et al.* Transgenic tobacco expressing Pinellia ternata agglutinin confers enhanced resistance to aphids. Transgenic Res 2003; 12(6): 715-22. [http://dx.doi.org/10.1023/B:TRAG.0000005146.05655.7d] [PMID: 14713200]

[218] Czapla TH, Lang BA. Effect of plant lectins on the larval development of European corn borer (*Lepidoptera: Pyralidae*) and Southern corn rootworm (*Coleoptera Chrysomelidae*). J Econ Entomol 1990; 83: 2480-5. [http://dx.doi.org/10.1093/jee/83.6.2480]

[219] Deka S, Barthakur S. Overview on current status of biotechnological interventions on yellow stem

borer Scirpophaga incertulas (Lepidoptera: Crambidae) resistance in rice. Biotechnol Adv 2010; 28(1): 70-81.
[http://dx.doi.org/10.1016/j.biotechadv.2009.09.003] [PMID: 19811767]

[220] Mao J, Burt AJ, Ramputh AI, *et al.* Diverted secondary metabolism and improved resistance to European corn borer (*Ostrinia nubilalis*) in maize (*Zea mays* L.) transformed with wheat oxalate oxidase. J Agric Food Chem 2007; 55(7): 2582-9.
[http://dx.doi.org/10.1021/jf063030f] [PMID: 17348672]

[220a] Schellenberger U, Orai J, Rosen BA, Wei JZ, *et al.* A selective insecticidal protein from *Pseudomonas* for controlling corn rootworms. Science 2016; 354: 634-37.
[http://dx.doi.org/10.1126/science.aaf6056] [PMID:27708055]

[221] Younis A, Siddique MI, Kim CK, Lim KB. RNA Interference (RNAi) Induced gene silencing: a promising approach of hi-tech plant breeding. Int J Biol Sci 2014; 10(10): 1150-8.
[http://dx.doi.org/10.7150/ijbs.10452] [PMID: 25332689]

[222] Fire A, Xu S, Montgomery MK, Kostas SA, Driver SE, Mello CC. Potent and specific genetic interference by double-stranded RNA in Caenorhabditis elegans. Nature 1998; 391(6669): 806-11.
[http://dx.doi.org/10.1038/35888] [PMID: 9486653]

[223] Zhu JQ, Liu S, Ma Y, *et al.* Improvement of pest resistance in transgenic tobacco plants expressing dsRNA of an insect-associated gene EcR. PLoS One 2012; 7(6): e38572.
[http://dx.doi.org/10.1371/journal.pone.0038572] [PMID: 22685585]

[224] Mao J, Zeng F. Plant-mediated RNAi of a gap gene-enhanced tobacco tolerance against the *Myzus persicae*. Transgenic Res 2014; 23(1): 145-52.
[http://dx.doi.org/10.1007/s11248-013-9739-y] [PMID: 23949691]

[225] Délye C, Jasieniuk M, Le Corre V. Deciphering the evolution of herbicide resistance in weeds. Trends Genet 2013; 29(11): 649-58.
[http://dx.doi.org/10.1016/j.tig.2013.06.001] [PMID: 23830583]

[226] Botterman J, Leemans J. Engineering herbicide resistance in plants. Trends Genet 1988; 4(8): 219-22.
[http://dx.doi.org/10.1016/S0168-9525(88)80004-0] [PMID: 3072722]

[227] Mazur BJ, Falco SC. The development of herbicide resistant crops. Annu Rev Plant Mol Biol 1989; 40: 441-70.
[http://dx.doi.org/10.1146/annurev.pp.40.060189.002301]

[228] Tan S, Evans RR, Dahmer ML, Singh BK, Shaner DL. Imidazolinone-tolerant crops: history, current status and future. Pest Manag Sci 2005; 61(3): 246-57.
[http://dx.doi.org/10.1002/ps.993] [PMID: 15627242]

[229] Kolkman JM, Berry ST, Leon AJ, *et al.* Single nucleotide polymorphisms and linkage disequilibrium in sunflower. Genetics 2007; 177(1): 457-68.
[http://dx.doi.org/10.1534/genetics.107.074054] [PMID: PMC2013689]

[230] Duhoux A, Délye C. Reference genes to study herbicide stress response in Lolium sp.: up-regulation of P450 genes in plants resistant to acetolactate-synthase inhibitors. PLoS One 2013; 8(5): e63576.
[http://dx.doi.org/10.1371/journal.pone.0063576] [PMID: 23696834]

[231] Beversdorf WD, Kott LS. Development of triazine resistance in crops by classical plant breeding. Weed Sci 1987; 35: 9-11.

[232] Nesaki T. Herbicide-tolerant crop creation by homologous recombination-dependent gene targeting. J Biofertil Biopestici 2013; 4: 134.

[233] Zambrano AY, Demey JR, Gonzalez V. In vitro selection of a glyphosate-tolerant sugarcane cellular line. Plant Mol Biol Rep 2003; 21: 365-73.
[http://dx.doi.org/10.1007/BF02772586]

[234] Steinrucken HC, Amrhein N. The herbicide glyphosate id a potent inhibitor of 5-enolpyruvyl-

shikimate-3-phosphate synthase. Biochem Res Commun 1986; 94: 1207-12. [http://dx.doi.org/10.1016/0006-291X(80)90547-1]

[235] James C. Global status of commercialized biotech/GM crops. ISAAA Briefs: 49. Ithaca, NY 2014.

[236] Senseman SA. Herbicide Handbook, Weed Science. 9th ed., Lawrence, KS: Society of America 2007.

[237] Monsanto. Safety Assessment of Roundup Ready Corn Event GA21 2002. URL http://www.monsanto.com /products/Documents/;

[238] Powles SB, Preston C. Evolved glyphosate resistance in plants: biochemical and genetic basis of resistance. Weed Technol 2006; 20: 282-9. [http://dx.doi.org/10.1614/WT-04-142R.1]

[239] Hérouet C, Esdaile DJ, Mallyon BA, *et al.* Safety evaluation of the phosphinothricin acetyltransferase proteins encoded by the pat and bar sequences that confer tolerance to glufosinate-ammonium herbicide in transgenic plants. Regul Toxicol Pharmacol 2005; 41(2): 134-49. [http://dx.doi.org/10.1016/j.yrtph.2004.11.002] [PMID: 15698537]

[240] Matringe M, Sailland A, Pelissier B, Rolland A, Zink O. p-Hydroxyphenylpyruvate dioxygenase inhibitor-resistant plants. Pest Manag Sci 2005; 61(3): 269-76. [http://dx.doi.org/10.1002/ps.997] [PMID: 15633191]

[241] Michel A, Vail G, Elser D, Wichert R. Introducing bicyclopyrone for broad-spectrum weed control in corn and sugarcane. Proc Weed Sci Soc Am 2010; 50: 131.

[242] Owen MD, Zelaya IA. Herbicide-resistant crops and weed resistance to herbicides. Pest Manag Sci 2005; 61(3): 301-11. [http://dx.doi.org/10.1002/ps.1015] [PMID: 15668920]

[243] Green JM, Owen MD. Herbicide-resistant crops: utilities and limitations for herbicide-resistant weed management. J Agric Food Chem 2011; 59(11): 5819-29. [http://dx.doi.org/10.1021/jf101286h] [PMID: 20586458]

[244] Kujur A, Saxena MS, Bajaj D, Laxmi , Parida SK. Integrated genomics and molecular breeding approaches for dissecting the complex quantitative traits in crop plants. J Biosci 2013; 38(5): 971-87. [http://dx.doi.org/10.1007/s12038-013-9388-6] [PMID: 24296899]

[245] Gaj T, Gersbach CA, Barbas CF III. ZFN, TALEN, and CRISPR/Cas-based methods for genome engineering. Trends Biotechnol 2013; 31(7): 397-405.

CHAPTER 12

Harnessing Apomixis to Improve Crops

Emidio Albertini* and **Fabio Veronesi**

Department of Agricultural, Food and Environmental Sciences, Università degli Studi di Perugia, Perugia, Italy

Abstract: Apomixis is a naturally occurring reproduction mode in flowering plants that results in embryo formation in absence of both meiosis and eggcell fertilization. Apomixis results from the combination of two processes: apomeiosis (unreduced embryosac formation due to diplospory or apospory) and parthenogenesis (development of an embryo without fertilization). Seed-derived progenies of an apomictic plant are genetically identical to the maternal parent, *i.e.*, they are clonal in origin. The impact on agriculture of the introgression of apomixis into sexual crops, will be revolutionary. In fact, apomixis will allow clonal seed production and thus enable efficient and consistent yields of high quality seeds, fruits and vegetables at lower costs. The development of apomixis technology will reduce cost and breeding time also avoiding the complications typical of sexual reproduction (*e.g.,* incompatibility barriers) and vegetative propagation (*e.g.,* viral transfer). Progresses in the search for apomixis genes obtained by several groups could allow the manipulation of apomixis and its transfer to crop species where the apomixis system is not present and to revolutionize modern agriculture. Moreover, when coupled with male-sterility systems, apomictic reproduction (with no need for male contribution) could help in addressing issues related to transgene escape from GM crops to organic or conventional crops, and thereby allow for better coexistence systems.

Keywords: Apomeiosis, Apomixis, Asexual reproduction, Fertilization, Hybrid vigor, Inbreeding depression, Meiosis, Parthenogenesis, Plant reproduction, Sexual reproduction.

APOMIXIS IN PLANTS

In modern agriculture the tremendous yield increase achieved by coupling high-yield varieties with high-input agronomic systems (Green revolution) has been one of the major successes. Plant breeders are now trying to extend this by intensifying the selection, develop more hybrids in greater number of crops, increase the range of plant functions through mutagenesis and advanced genetic

* **Corresponding author Emidio Albertini**: Department of Agricultural, Food and Environmental Sciences, Università degli Studi di Perugia, Perugia, Italy; Tel: +39 075 5856206; Fax: +39 075 5856224; E-mail: emidio.albertini@unipg.it

Roberto Pilu and Giuseppe Gavazzi (Eds.)

improvement. Therefore, breeding will continue to play a crucial role in crop improvement because the needs are varied, the techniques are continuously expanding, the possible new genetic combinations are limitless and the successes achieved in the past are a good omen for future.

The problem for cross-pollinated species is that alleles segregate in progenies and, therefore, the optimal genotype is lost together with the phenotypic evidence of the trait of interest (Fig. **1A**). In nature, a trait, termed apomixis, allows the clonal propagation of a plant through seeds by avoiding the process of genetic recombination connected to meiosis and fertilization.

Apomixis could represent a must in modern agriculture but its introgression from wild relatives in crop species and transformation of sexual genotypes in apomictic genotypes were in long persued by plant breeding (Fig. **1**). The impact of apomictic crops in agriculture would be enormous; it is estimated, for example, that the hybrid rice production could lead to economic benefits in excess of 1.8 billion Euros per year [1, 2]. If applied to clonally propagated crops apomixis could allow the production of disease-free material more easily storable and transportable [3]. Apomixis could make the true potato seed an attractive option for plant breeders and farmers with possible economic benefits estimated up to 2.3 billion Euros per year [1]. But, so far, the exploitation of apomixis has been unsuccessfull in all major agricultural crops where it was attemped and only some features of it have been engineered in model species.

Moreover, apomixis could produce extremely rapid breeding results responsive to specific micro-environments, cropping conditions and markets. This would stimulate diverse strategies of breeding aimed at more sustainable crop management with deep implications in modern agricultural research systems. In simple terms the technique could also facilitate the farmers in avoiding the need to buy seed each year thus reducing the production costs. This will also have positive economic impact for consumers due to the decrease of the expenses for the food supply purchase.

The development of apomixis-based technology will require a deep understanding of the genetic control of the trait. Our molecular understanding of apomixis would be greatly increased if it was possible to identify the genes differentially expressed during the development of embryo and embryo sac. Chaudhury and Peacock [4] hypothesized that genes isolated in model species such as *Arabidopsis thaliana* would be important for the study of apomixis. Analysis of mutations in plants belonging to sexually reproducing species that resulted in phenotypes displaying one or more components of apomixis led to the isolation of a number of candidate genes (reviewed in Bicknell and Catanach 2014) [5]. The major result obtained by this approach has been the creation of a sort of synthetic form of apomixis. In fact several mutations were identified and combined into two Arabidopsis lines, which, when crossed, gave rise to a significant proportion of strictly maternal origin [6 - 8].

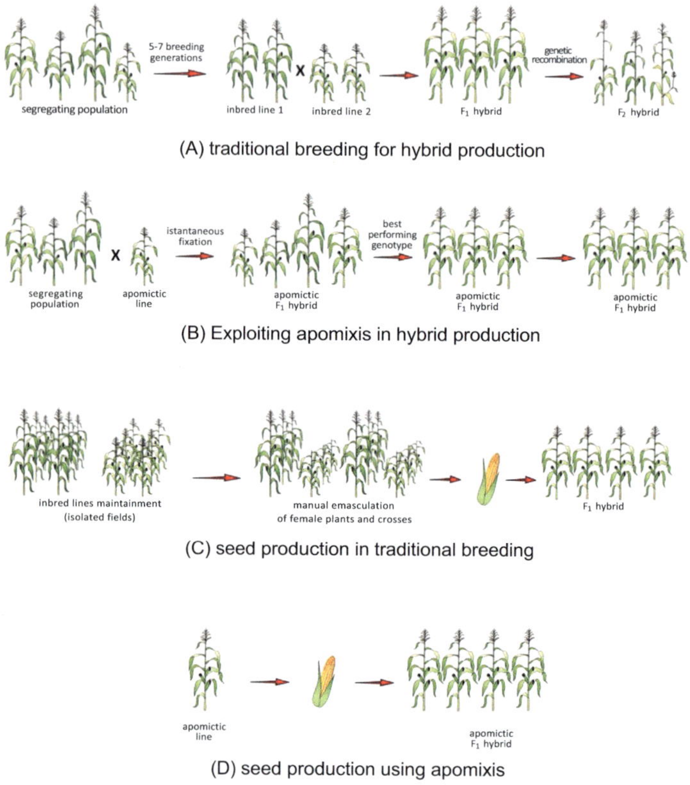

Fig. (1). Conventional *vs* apomixis-mediated breeding as reported in Barcaccia and Albertini [3]. Considering, for example, the conventional maize breeding, the best plants are selected within a segregating population, normally a F_2, and after several generations of selfing and phenotypic selection, their progenies are evaluated for the specific combining ability to identify those which can be used as parental inbred lines in the synthesis of F_1 hybrid varieties able to show heterosis. The best inbred lines are then maintained, multiplied in isolation and crossed each other in pairs in order to give uniform F_1 hybrids vigorous and highly productive. This breeding scheme, however, requires a complex procedure to obtain hybrid seed: the two inbred lines must be kept in isolation in separate fields; then, in order to produce hybrid seed, it is necessary to organize seed production fields where about a quarter of the plants are used as pollinators (*i.e.*, pollen donor inbred) while the F_1 hybrid seed is collected on the seed-bearing plants (*i.e.* maternal inbred) which covers only three quarters of the surface.. Farmers cannot reuse the seeds produced by F_1 hybrids that would give rise to crops characterized by high variability due to recombination and segregation. Using apomictic imbred lines, this problem would not arise. After the best inbred lines to be used as maternal plants are selected, these could be crossed with pollen donor clonal lines carrying the gene for apomixis; this will lead to hybrid seeds characterized by highly heterozygous genotypes. From this moment on, each hybrid F_1 could be maintained as such for many generations in a state of fixed heterosis.

On the other hand Carman *et al.* [9] proposed the existence of an ancient sex/apomixis switch retained, to a greater or lesser extent, in relatively few eukaryote lineages during evolution. Occasionally, this switch (from sex to apomixis or the other way around) may be epigenetically triggered by hybridization and polyploidization but only in materials that have mantained, at the molecular level, a sufficient capacity to express each mechanism.

If the model is correct, then apomixis could be even more complex than it was assumed in the past. It **may be a phenomenon** linked to the life cycle, such as sexual reproduction, including the resetting of the epigenetic clock in every generation. Therefore, apomixis in eukaryotes would share common fundamental pathways, namely the formation of not reduced and epigenetically reset parthenogenetically active cells from the germ line or cell closely associated to it [9]. This hypothesis is strengthen by the evidences that female gametes in plants undergo genome-wide demethylation [10, 11] suggesting that DNA methylation patterns might undergo some degree of reprogramming in plants. Therefore, the so called "epigenetic model", is based on the assumption that genes controlling apomixis could be the target of epigenetic changes such as DNA methylation, histone modification, and changes in chromatin structure. Therefore, if the theory of the genome unbalance [12] is true, the reprogramming of DNA methylation occurring in the wrong cell and/or at the wrong time could cause the cascade of events that lead a cell to change its fate and undergo, for example, to apomixis.

Some confirmation of the role of epigenetics in embryo specification have been obtained in Arabidopsis where a proposed alternative hypothesis to restrict the specification of the embryo sac precursors is linked to the action of *ARGONAUTE* (*AGO*) genes [13, 14]. AGO proteins are known to be involved in post-transcriptional gene silencing mediated by short RNA (both microRNAs and short interfering RNAs) [15].

Mechanisms Of Apomixis

Apomixis can occur sporophytically and gametophytically, based on the tissue involved in the formation of the clonal embryo [5]. The simplest pathway avoids the production of a gametophyte and a maternal embryo originates from one or more somatic cells of the ovule. Such process, known as adventitious embryony, takes the adjective of nucellar or integumental depending on the tissue where the embryogenetic somatic cell differentiates. Adventitious embryony seems to have evolved more frequently in tropical than in temperate flora. Among the agriculturally important species, adventitious embryony has been highlighted in mango (*Mangifera indica*), in various species of *Citrus*, and orchids [16]. In these plants the sexual reproductive system proceeds normally with the fertilization of the egg cell produced by meiosis. The somatic embryo formation takes place in the surrounding ovule tissues, usually coinciding with the initial development of the embryo derived from sexual fertilization [5].

In gametophytic apomixis, the unreduced embryo sac can develop from a somatic nucellar cell acquiring the developmental program of a functional megaspore one (apospory) or from a megaspore mother cell with suppressed or modified meiosis (diplospory) (Fig. **2**).

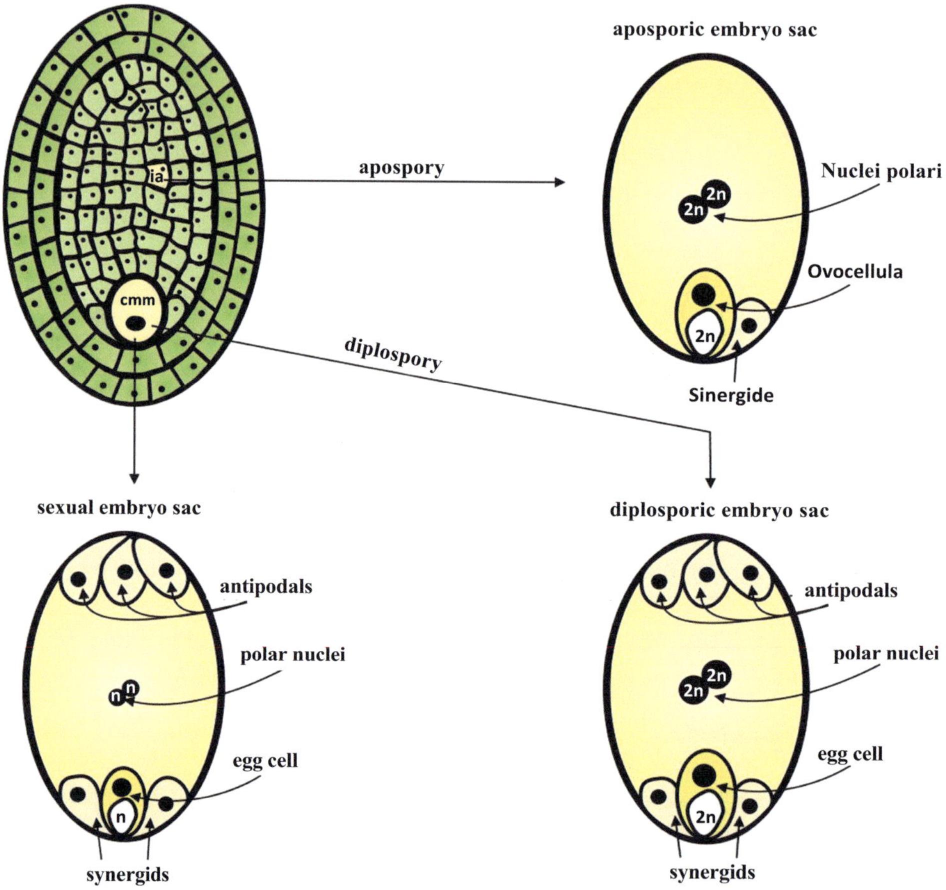

Fig. (2). Sexual, Diplosporic and Aposporic embryo sac developmental. Meiosis of Megaspore Mother Cells (mmc) undergoes meiosis and leads to the formation of functional megaspore that, following a series of mitosis, give rise to the sexual embryo sac. This contains the haploid ovules (n) and two haploid polar nuclei (n). In diplospory, a CMM altered meiosis originates a unreduced megaspore that will develop into an unreduced embryo sac containing an unreduced egg cell (2n) and two unreduced polar nuclei (2n). In aposporic species, a somatic cell from the nucellar tissue change its fate and act as megaspore; its mitotic divisions originates an embryo sac containing a egg cell 2n and two polar nuclei 2n; in this type of embryo sac, some typical cells of the embryo sac sex may miss.

In particular, in diplosporic species, an altered meiosis of the megaspore mother cell gives rise to an unreduced embryo sac. On the other hand, in aposporic apomictic species, the embryo sac forms from somatic cells, called aposporic initials, that differentiate from the nucellus after megaspore mother cell differentiation. Therefore, sexual or aposporic processes can coexist in an ovule, a phenomenon which, by definition, is not possible in diplosporic apomicts. Aposporic embryo sacs develop faster than those of sexual derivation because their development is not slowed by meiotic divisions.

Whatever the mechanism leading to the formation of unreduced embryo sac, the independent development of egg cell in the absence of fertilization (diploid parthenogenesis) constitutes the second component of gametophytic apomixis. Furthermore, a functional endosperm formation is needed for the production of a true apomictic seed and it constitutes the third component of the system. The endosperm development frequently requires a stimulation induced by pollination (pseudogamy) but in several apomictic plants the endosperm is initiated autonomously without any contribution from male gametes. Pseudogamy may be characterized by simple pollen tube growth in the style, cytoplasmic penetration or, most frequently, true fertilization of the polar nuclei in the embryo sac [3].

Apomixis And Plant Breeding

For centuries, plants have been subjected to a strong selection acted both by humans and by the changing environment. Moreover, crop plants have followed a general pattern of introduction, selection, and hybridization proved to be crucial for world agriculture considering that many of the crops important at world level are actually utilized outside their region of domestication. Once introgressed, human intervention has made them adapted to the new environmental conditions allowing through increasingly advanced breeding strategies to produce cultivated varieties with improved agronomical performances. Unfortunately, many of the traits of breeding interest are typically polygenic. The consequence is that only certain unique allelelic combinations are able to generate elite performing genotypes. But, since modern agriculture depends on uniformity of seed and fruit quality, the fixation of a hybrid combination would be desirable. The fixation of specific genotypes occurs naturally in apomictic species. The dream of many plant breeders is to introduce apomixis from wild relatives to crop plants and then transform sexual genotypes in genotypes that are bred apomictically. This would result in revolutionary implications in cross pollinated species such as, the full exploitation of heterosis through the possibility of direct reseeding of seed produced by the best hybrids. Unfortunately apomixis is not a common trait among plants of agronomical interest with few exceptions in certain fodder grasses and some fruit trees. Therefore, the practical use of apomixis in cultivar development is still far. Nevertheless the fixation of hybrid vigor through apomixis is so a desirable objective for breeders and especially for farmers that it is expected to have a revolutionary impact on food and agriculture production. To give an example, the stabilization of heterozygous genotypes *via* apomixis would make breeding programs faster and cheaper (Fig. **1**). In addition, with the introduction of apomixis technology, farmers could produce their own seeds of commercial hybrids which, for the reasons reported above, is currently not possible because of inbreeding effect and decreased performance. Since the genetic composition of the apomictic F_1 hybrid would be maintained, farmers would not be obliged to buy hybrid seed each year [17]. Furthermore, apomictic hybrids will not need the use of cytoplasmic male sterility with restorer genes, which would result in the development of more rapid and more simple strategies for hybrid seed production. Apomixis could also produce benefits in clonally propagated species whose production is severely limited by pathogens (mainly viral and endophytic) able to accumulate their negative effects in subsequent vegetative propagations with serious consequences for the production and also for the exchange of crop genetic resources between different Countries. The exploitation of apomixis in clonally propagated plants would provide the option and benefit of propagation *via* clonal seeds, thus generating disease-free material that could be more easily stored and transported with lower cost.

A deeper knowledge of the mechanisms regulating reproductive development in plants are needed for a real development of apomixis technology in agriculture. Specifically or differentially expressed genes during the formation of the embryo and embryo sac must be identified to increase our molecular understanding of the system. Over the last two decades, many scientists investigated how to isolate genes that control the key points of apomixis and many papers have been published postulating the production of genetically modified plants showing apomictic-like phenotypes. In fact, only a few aspects of apomixis have been genetically engineered in model plants and at today none of the most important crops have been selected for an apomictic behavior. As a consequence even in the era of omics, to reach a good understanding of the genetic control and of the molecular regulation of apomixis is much more complicated than expected.

Candidate genes specifically expressed in either apomictic or sexual ovules were also identified in alfalfa [18, 19]; *Boechera holboellii* [20 - 22], *Brachiaria* [23], *Eragrostis curvula* [24], *Hieracium pilosella* [25], *Hypericum perforatum* [26], *Panicum maximum* [27 - 29], *Paspalum spp* [30 - 35], *Pennisetum ciliare* [36], *Pennisetum squamulatum* [37, 38], *Poa pratensis* [39, 40].

After two decades of studies conducted in many laboratories on model plants, the asexual reproductive strategy defined as "gametophytic apomixis" by Nogler [41] is still an unsolved puzzle. As a result, seed companies seem to have lost interest in this area and in recent years it has been difficult to raise both public and private funds for conducting research on apomixis.

However, we believe that the final steps for the understanding of the genetic system controlling apomixis are within reach and can be achieved in medium term and that future research will be crucial for apomixis to begin to be introduced and successfully exploited in major crops in order to support the agricultural and food interests of the entire world.

In conclusion, apomixis applied to crop plants can contribute greatly to the increase of breeding results, in particular through the exploitation of fixed heterosis able to make hybrid seed production a lot simpler. Once the platform will be feasible for practical use, breeders should be aware that apomixis may result in an evolutionary block so that its exploitation in commercial varieties must be balanced by a careful management of crop genetic resources. Furthermore, the recourse to apomixis in seed production would be economically counterproductive for the seed companies that, as a consequence, would implement such a technology only if regulated by a rigid system of royalties.

CONFLICT OF INTEREST

The authors confirm that they have no conflict of interest to declare for this publication.

ACKNOWLEDGEMENTS

Declared None.

REFERENCES

[1] Spillane C, Curtis MD, Grossniklaus U. Apomixis technology development-virgin births in farmers fields? Nat Biotechnol 2004; 22(6): 687-91. [http://dx.doi.org/10.1038/nbt976] [PMID: 15175691]

[2] Pupilli F, Barcaccia G. Cloning plants by seeds: Inheritance models and candidate genes to increase fundamental knowledge for engineering apomixis in sexual crops. J Biotechnol 2012; 159(4): 291-311. [http://dx.doi.org/10.1016/j.jbiotec.2011.08.028] [PMID: 21906637]

[3] Barcaccia G, Albertini E. Apomixis in plant reproduction: a novel perspective on an old dilemma. Plant Reprod 2013; 26(3): 159-79. [http://dx.doi.org/10.1007/s00497-013-0222-y] [PMID: 23852378]

[4] Chaudhury AM, Peacock JW. Approaches to isolating apomictic mutants in *Arabidopsis thaliana*: prospects and progress. In: Khush GS, Ed. Apomixis: Exploiting hybrid vigour in rice Intl. Manila: Rice Research Institute 1993; pp. 66-71.

[5] Bicknell R, Catanach A. Apomixis: The Asexual Formation of Seed. In: Li X, Donnelly DJ, Jensen TG, Eds. Somatic Genome Manipulation. NY: Springer 2014; pp. 147-67.

[6] Ravi M, Marimuthu MP, Siddiqi I. Gamete formation without meiosis in *Arabidopsis.* Nature 2008; 451(7182): 1121-4.
[http://dx.doi.org/10.1038/nature06557] [PMID: 18272967]

[7] dErfurth I, Jolivet S, Froger N, Catrice O, Novatchkova M, Mercier R. Turning meiosis into mitosis. PLoS Biol 2009; 7(6): e1000124.
[http://dx.doi.org/10.1371/journal.pbio.1000124] [PMID: 19513101]

[8] Ravi M, Chan SW. Haploid plants produced by centromere-mediated genome elimination. Nature 2010; 464(7288): 615-8.
[http://dx.doi.org/10.1038/nature08842] [PMID: 20336146]

[9] Carman JG, Jamison M, Elliott E, Dwivedi KK, Naumova TN. Apospory appears to accelerate onset of meiosis and sexual embryo sac formation in sorghum ovules. BMC Plant Biol 2011; 11: 9.
[http://dx.doi.org/10.1186/1471-2229-11-9] [PMID: 21223576]

[10] Hsieh TF, Ibarra CA, Silva P, *et al.* Genome-wide demethylation of Arabidopsis endosperm. Science 2009; 324(5933): 1451-4.
[http://dx.doi.org/10.1126/science.1172417] [PMID: 19520962]

[11] Gehring M, Bubb KL, Henikoff S. Extensive demethylation of repetitive elements during seed development underlies gene imprinting. Science 2009; 324(5933): 1447-51.
[http://dx.doi.org/10.1126/science.1171609] [PMID: 19520961]

[12] Carman JG. Asynchronous expression of duplicate genes in angiosperms may cause apomixis, bispory, tetraspory, and polyembryony. Biol J Linn Soc Lond 1997; 61: 51-94.
[http://dx.doi.org/10.1111/j.1095-8312.1997.tb01778.x]

[13] Olmedo-Monfil V, Durán-Figueroa N, Arteaga-Vázquez M, *et al.* Control of female gamete formation by a small RNA pathway in *Arabidopsis.* Nature 2010; 464(7288): 628-32.
[http://dx.doi.org/10.1038/nature08828] [PMID: 20208518]

[14] Tucker MR, Okada T, Hu Y, Scholefield A, Taylor JM, Koltunow AM. Somatic small RNA pathways promote the mitotic events of megagametogenesis during female reproductive development in *Arabidopsis.* Development 2012; 139(8): 1399-404.
[http://dx.doi.org/10.1242/dev.075390] [PMID: 22399683]

[15] Baumberger N, Baulcombe DC. Arabidopsis ARGONAUTE1 is an RNA Slicer that selectively recruits microRNAs and short interfering RNAs. Proc Natl Acad Sci USA 2005; 102(33): 11928-33.
[http://dx.doi.org/10.1073/pnas.0505461102] [PMID: 16081530]

[16] Naumova T. Apomixis in Angiosperms: nucellar and integumentary embryony. Boca Raton, FL, USA: CRC Press 1992.

[17] Kandemir N, Saygili I. Apomixis: new horizons in plant breeding. Turk J Agric For 2015.
[http://dx.doi.org/10.3906/tar-1409-74]

[18] Citterio S, Albertini E, Varotto S, *et al.* Alfalfa Mob 1-like genes are expressed in reproductive organs during meiosis and gametogenesis. Plant Mol Biol 2005; 58(6): 789-807.
[http://dx.doi.org/10.1007/s11103-005-8104-9] [PMID: 16240174]

[19] Citterio S, Piatti S, Albertini E, Aina R, Varotto S, Barcaccia G. Alfalfa Mob1-like proteins are involved in cell proliferation and are localized in the cell division plane during cytokinesis. Exp Cell Res 2006; 312(7): 1050-64.
[http://dx.doi.org/10.1016/j.yexcr.2005.12.032] [PMID: 16460730]

[20] Sharbel TF, Voigt ML, Corral JM, *et al.* Molecular signatures of apomictic and sexual ovules in the *Boechera holboellii* complex. Plant J 2009; 58(5): 870-82.
[http://dx.doi.org/10.1111/j.1365-313X.2009.03826.x] [PMID: 19220792]

[21] Sharbel TF, Voigt ML, Corral JM, *et al.* Apomictic and sexual ovules of *Boechera* display heterochronic global gene expression patterns. Plant Cell 2010; 22(3): 655-71.

[http://dx.doi.org/10.1105/tpc.109.072223] [PMID: 20305122]

[22] Corral JM, Vogel H, Aliyu OM, *et al.* APOLLO: A candidate allele linked to apomeiosis in Boechera. Proc Int Plant Anim Gen, San Diego, CA 2013. https://pag.confex.com/pag/xxi/webprogram/Paper5196.html

[23] Leblanc O, Armstead I, Pessino S, *et al.* Nonradioactive mRNA fingerprinting to visualize gene expression in mature ovaries of *Brachiaria* hybrids derived from *B. brizantha*, an apomictic tropical forage. Plant Sci 1997; 126: 49-58. [http://dx.doi.org/10.1016/S0168-9452(97)00067-8]

[24] Cervigni GD, Paniego N, Pessino S, *et al.* Gene expression in diplosporous and sexual *Eragrostis curvula* genotypes with differing ploidy levels. Plant Mol Biol 2008; 67(1-2): 11-23. [http://dx.doi.org/10.1007/s11103-008-9305-9] [PMID: 18311543]

[25] Guerin J, Rossel JB, Robert S, Tsuchiya T, Koltunow A. A DEFICIENS homologue is down-regulated during apomictic initiation in ovules of *Hieracium.* Planta 2000; 210(6): 914-20. [http://dx.doi.org/10.1007/s004250050697] [PMID: 10872222]

[26] Galla G, Volpato M, Sharbel TF, Barcaccia G. Computational identification of conserved microRNAs and their putative targets in the *Hypericum perforatum* L. flower transcriptome. Plant Reprod 2013; 26(3): 209-29. [http://dx.doi.org/10.1007/s00497-013-0227-6] [PMID: 23846415]

[27] Chen L, Miyazaki C, Kojima A, *et al.* Isolation and characterization of a gene expressed during early embryo sac development in apomictic Guinea grass (*Panicum maximum*). J Plant Physiol 1999; 154: 55-62. [http://dx.doi.org/10.1016/S0176-1617(99)80318-6]

[28] Chen L, Guan L, Seo M, Hoffmann F, Adachi T. Developmental expression of ASG- 1 during gametogenesis in apomictic guinea grass (*Panicum maximum*). J Plant Physiol 2005; 162(10): 1141-8. [http://dx.doi.org/10.1016/j.jplph.2005.02.010] [PMID: 16255172]

[29] Yamada-Akiyama H, Akiyama Y, Ebina M, *et al.* Analysis of expressed sequence tags in apomictic guineagrass (*Panicum maximum*). J Plant Physiol 2009; 166(7): 750-61. [http://dx.doi.org/10.1016/j.jplph.2008.10.001] [PMID: 19046615]

[30] Pessino SC, Evans C, Ortiz JP, *et al.* A genetic map of the apospory-region in *Brachiaria* hybrids: identification of two markers closely associated with the trait. Hereditas 1998; 128: 153-8. [http://dx.doi.org/10.1111/j.1601-5223.1998.00153.x]

[31] Pessino SC, Espinoza F, Martínez EJ, Ortiz JP, Valle EM, Quarín CL. Isolation of cDNA clones differentially expressed in flowers of apomictic and sexual *Paspalum notatum.* Hereditas 2001; 134(1): 35-42. [http://dx.doi.org/10.1111/j.1601-5223.2001.00035.x] [PMID: 11525063]

[32] Pupilli F, Martinez EJ, Busti A, Calderini O, Quarin CL, Arcioni S. Comparative mapping reveals partial conservation of synteny at the apomixis locus in *Paspalum* spp. Mol Genet Genomics 2004; 270(6): 539-48. [http://dx.doi.org/10.1007/s00438-003-0949-5] [PMID: 14648202]

[33] Laspina NV, Vega T, Seijo JG, *et al.* Gene expression analysis at the onset of aposporous apomixis in *Paspalum notatum.* Plant Mol Biol 2008; 67(6): 615-28. [http://dx.doi.org/10.1007/s11103-008-9341-5] [PMID: 18481185]

[34] Polegri L, Calderini O, Arcioni S, Pupilli F. Specific expression of apomixis-linked alleles revealed by comparative transcriptomic analysis of sexual and apomictic *Paspalum simplex* Morong flowers. J Exp Bot 2010; 61(6): 1869-83. [http://dx.doi.org/10.1093/jxb/erq054] [PMID: 20231327]

[35] Felitti SA, Seijo JG, González AM, *et al.* Expression of *lorelei*-like genes in aposporous and sexual *Paspalum notatum* plants. Plant Mol Biol 2011; 77(4-5): 337-54.

[http://dx.doi.org/10.1007/s11103-011-9814-9] [PMID: 21826430]

[36] Singh M, Burson BL, Finlayson SA. Isolation of candidate genes for apomictic development in buffelgrass (*Pennisetum ciliare*). Plant Mol Biol 2007; 64(6): 673-82. [http://dx.doi.org/10.1007/s11103-007-9188-1] [PMID: 17541705]

[37] Conner JA, Goel S, Gunawan G, *et al.* Sequence analysis of bacterial artificial chromosome clones from the apospory-specific genomic region of *Pennisetum* and *Cenchrus.* Plant Physiol 2008; 147(3): 1396-411. [http://dx.doi.org/10.1104/pp.108.119081] [PMID: 18508959]

[38] Conner JA, Mookkan M, Huo H, Chae K, Ozias-Akins P. A parthenogenesis gene of apomict origin elicits embryo formation from unfertilized eggs in a sexual plant. Proc Natl Acad Sci USA 2015; 112(36): 11205-10. [http://dx.doi.org/10.1073/pnas.1505856112] [PMID: 26305939]

[39] Albertini E, Marconi G, Barcaccia G, Raggi L, Falcinelli M. Isolation of candidate genes for apomixis in *Poa pratensis* L. Plant Mol Biol 2004; 56(6): 879-94. [http://dx.doi.org/10.1007/s11103-004-5211-y] [PMID: 15821987]

[40] Albertini E, Marconi G, Reale L, *et al.* SERK and APOSTART. Candidate genes for apomixis in Poa pratensis. Plant Physiol 2005; 138(4): 2185-99. [http://dx.doi.org/10.1104/pp.105.062059] [PMID: 16024690]

[41] Nogler GA. Gametophytic apomixis. In: Johri BM, Ed. Embryology of angiosperms. Berlin, Heidelberg, New York: Springer 1984; pp. 475-518. [http://dx.doi.org/10.1007/978-3-642-69302-1_10]

CHAPTER 13

Molecular-Assisted Breeding

Filippo Geuna*

Department of Agricultural and Environmental Sciences – Production, Landscapes, Agroenergy, Università degli Studi di Milano, Milano, Italy

Abstract: The steadily increasing world population and the concurrent reduction in cultivated lands are two major threats to food security, especially in the under-developed countries. Among the many strategies that can mitigate this situation modern biotechnologies play a central role. Many tools are available to scientists to face the challenges of increased production needs and sustainable agriculture. Among these, genetic modification and molecular breeding appear to be the most promising. Both of these approaches require the combined use of technologies such as genome sequencing, genotyping and phenotyping and large-scale data analysis and mining to determine genes and functions amenable to manipulation in the target species. Molecular markers, used in both target gene studies and assisted breeding, represent a powerful tool. This chapter deals with their nature and applications, and also describes some case studies where their use in marker-assisted selection has positively affected some target crop species.

Keywords: Breeding, Cassava, Database, Food security, Genetics, Genome-wide association studies (GWAS), Genomic selection, Genomics, Grapevine, Manihot, Mapping, Marker-assisted selection (MAS), Molecular markers, Nutrition, Pathogen resistance, Polymerase chain reaction (PCR), Quantitative Trait locus (QTL), Tomato, World wide web (WWW).

THE ROLE OF MOLECULAR MARKERS IN MODERN GENETICS AND BREEDING

According to the UN, the world population will be more than 9 billion by 2050 and over 11 billion within the next century. That means that farmers around the world will have to produce increasing amounts of nutrient-rich foods. During a high-level Expert Forum in Rome, held in 2009, the Food and Agriculture Organization of the United Nations (FAO) discussed food needs around the planet. The Forum estimated that to sustain over 9 billion people in 2050 would

* **Corresponding author Filippo Geuna**: Department of Agricultural and Environmental Sciences – Production, Landscapes, Agroenergy, Università degli Studi di Milano, Milano, Italy; Tel: +390250316550; Fax: +390250316553; E-mail: filippo.geuna@unimi.it

Roberto Pilu and Giuseppe Gavazzi (Eds.)

require increasing overall food production by some 70 percent. Production in the developing countries would almost need to double [1 - 5].

Modern biotechnologies are expected to play a steadily growing role in this scenario as the extent of cultivated land will be reduced by the demands of the expanding world population.

In the biotechnological toolbox, molecular markers and marker-assisted breeding (MAB) represent a sector of great potential applied to plants and animals that can improve characters of interest (productivity, resistance to abiotic and biotic stress), through the use of morphological, biochemical and genetic markers and by exploiting the natural potential of recombination. Often these characters are determined by areas of the genome called QTL (quantitative trait loci) whose mapping and characterization represents a major obstacle in the MAB approach. This technique is different from transgenesis, that leads to the production of genetically modified organisms (GMOs). Unlike transgenesis, the final products of marker assisted selection (MAS) are not obtained through genome manipulation and the insertion of foreign pieces of DNA into the species in question, but they derive from a guided selection of naturally occurring recombinations between individuals bearing contrasting characters. Therefore, marker-assisted selection is a technologically-improved version of traditional breeding in both plants and animals.

The availability of increasingly sophisticated and powerful equipment allows us to generate data in increasing quantities and at steadily decreasing costs.

The activity of developing new varieties can last several years. However, today's biotechnological tools have significantly shortened the time to six to ten years for new varieties of crops to get to the market. One of the tools that can make it fast and easy for scientists to select the desired traits is marker-assisted selection or breeding. Some features, such as the color of flowers or kernels, can be determined by a single gene. More complex features, however, such as starch content, crop yield, plant height or drought resistance can be controlled by many genes. Producers have traditionally selected plant species according to their visible or measurable characteristics, known as the phenotype. Such process is difficult, slow to undertake, is affected by the environment and expensive, not only in the development phase, but also uneconomical, as farmers, in the meantime, continue to suffer crop losses.

The molecular markers are identifiable genetic sequences which mark the DNA sequence of the trait under study and are passed on following the laws of inheritance at each generation. As markers are close to genes on the same chromosome, they normally move together, at each new generation (genetic linkage). This connection can be used to predict whether an individual will carry the desired gene. If it is possible to find a marker or markers for the gene, it means that the desired gene itself is probably present. The distance of a marker from its target gene is important to determine its usefulness and to predict the final phenotype as a result of a genetic (marker) state. The ultimate marker, when the physical correlation between marker and genetic trait is total, is represented by the gene itself.

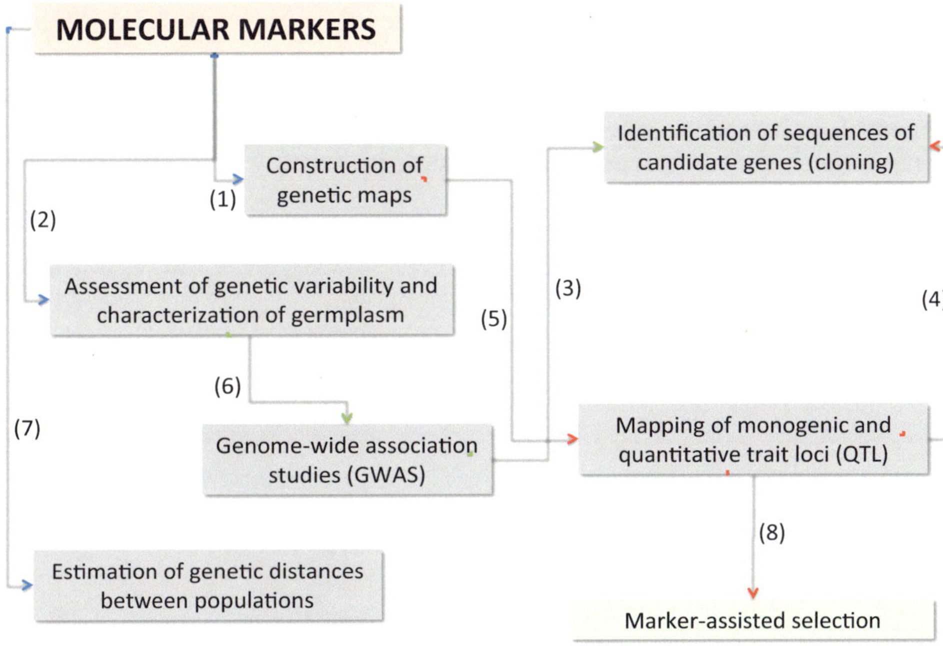

Fig. (1). The main applications of molecular markers in genetic studies and breeding. Blue arrows show the direct applications of molecular markers while red and green arrows mark the downstream logical relations between upper-level applications. For example "Construction of genetic maps" is a prerequisite to "Mapping of monogenic or polygenic (QTL) loci" (link 5). In turn, the latter is a prerequisite to "Identification of sequences of candidate genes (cloning)" (link 4) and to "Marker-assisted selection (link 8). An alternative route to "Identification of sequences of candidate genes (cloning)" passes through "Assessment of genetic variability and characterization of germplasm" and "Genome-wide association studies (GWAS)" (link 6).

Once markers locations on chromosomes and their distance from specific genes are known, it is possible to make a genetic linkage map. In this way it is possible to produce detailed maps in only one generation of plant crossing. The various direct and indirect applications that molecular markers have for genetic studies and breeding are summarized in Fig. (**1**). However, it is important to note that breeding by MAS has some limitations compared to genetic engineering, for the following reasons: 1) its efficacy is restricted to traits already present in a crop; 2) its advantages are limited when breeding species with long generation times (*e.g.*

some fruit and forest species); and 3) it has a biological limit with crops, including some staples, which are propagated clonally because they are sterile or are affected by problems with breeding. Nonetheless, while GMOs may be a more immediately effective way to modify the genetic asset, this latter approach suffers the limitation of the difficulty of transformation and regeneration which affects some important species (*i.e.*, stone fruits).

Beside their use for assisted selection and breeding, molecular markers play a key role in the process of identification of genes by map-based cloning. The construction of high-density molecular maps is the first step towards the creation of high-resolution pictures of a genome that permit us to link traits to markers, that is to say function to position. As in a geographical map, the increased detail of landmark markers, allows us to define with increasing precision the relative positions of target loci (Fig. **2**).

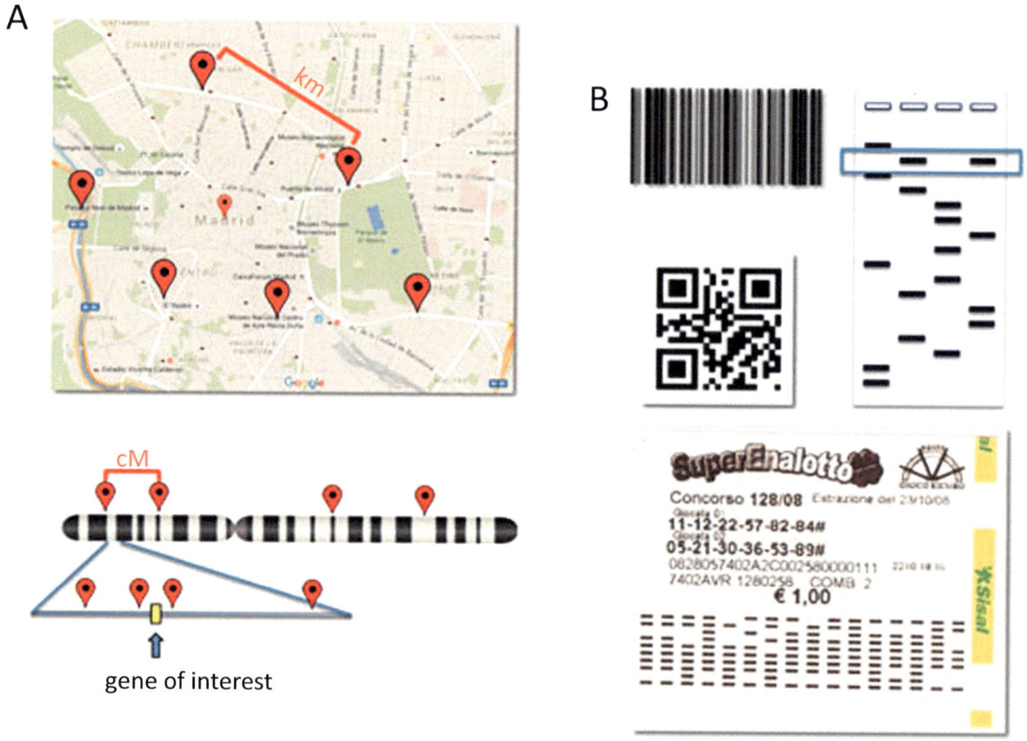

Fig. (2). Molecular markers are used as landmarks and fingerprints in genomes. (**A**) In the genetic maps molecular markers (red symbols) are the equivalent of geographical landmarks. Distance units (*i.e.* "km" in geographic maps), are the equivalent of genetic distance units ("centiMorgans, cM"). (**B**) The same markers act similarly in genetic fingerprinting and in bar coding.

TYPES OF MOLECULAR MARKERS

The analysis of the genome with molecular markers detects diversity which is due to mutations in homologous DNA regions in different individuals of the same species or belonging to different species.

A molecular marker is defined as a locus (region) of the genome detectable by molecular methods and capable of highlighting differences from a single base to several tens or hundreds of bases in different individuals.

In genetics, markers can be of different types:

Morphological: gene regions that determine shapes, colors, male sterility or stress resistance have been mapped in many plant species. In corn, tomato, pea and wheat up to 100 genes have been identified and used as morphological markers (presence/absence of aristae, the color of the leaf sheath, overall dimensions of the plant, seed color, aroma of seed or fruit, *etc.*).

Biochemical: genes for isozymes and storage proteins which are amenable to extraction and analysis, often by enzymatic assays. These markers are affected by the physiological state of the organism and/or tissue specificity. Moreover, they typically require a relatively large sample for protein extraction and lengthy purification steps and biochemical procedures.

Molecular: genes and non-coding DNA regions where mutations can occur and that are detectable and scorable thanks to the techniques of molecular biology.

The first to report an association between a genetic marker to simple Mendelian inheritance and a quantitative trait in plants was Sax [6], who observed in *Phaseolus vulgaris* L. segregation between the character associated to the size of the seed and one for color of the integument.

Rasmusson [7] showed the same between flowering time and the color of the flower.

An ideal marker should have the following characteristics:

- allow the distinction of all possible phenotypes (homo- and heterozygous) by different alleles;
- be present in abundant quantities and be polymorphic;
- it should permit recognition of the phenotype in the early stages of development of the organism;
- the presence of different alleles in the locus of the marker must have no effect on the character of interest (the so-called "neutrality" condition);

- there must be no interaction between markers so as to make possible their massive and simultaneous use in segregating populations.

In order to contribute to the improvement in the creation of novel varieties of vegetable species, markers should ideally be uniformly distributed in the germplasm. Although in reality this is never the case with morphological or biochemical markers, molecular markers address many if not all of the above-mentioned prerequisites. In addition, because of the often difficult political and social implications connected with GMOs, molecular markers are an extra key factor to keep in mind when evaluating strategies of breeding, since they do not imply genome modifications and introgression of heterologous DNA sequences.

Among the many types of molecular markers which are available to scientists and breeders, the following are the most relevant, in terms of popularity and potential:

RFLP

Restriction Fragment Length Polymorphism. They are based on variations that affect the length of fragments produced following DNA cutting by the so called restriction enzymes that precisely recognize and cut sequences of DNA, usually 4-6 base pairs). This is considered a low-throughput method due to the work and amount of DNA necessary to detect and score a sufficiently high number of loci to generate maps. A key and popular technique of the fingerprinting and mapping toolbox, it now continues to keep a certain interest after the introduction of the polymerase chain reaction (PCR) which dramatically reduced the amount of starting genomic DNA required. Indirectly, the principle of RFLP has been inherited and integrated into modern medium- and high-throughput marker systems like AFLPs and RADs [8, 9].

AFLP

Amplified Fragment Length Polymorphism. AFLP marker technology was designed and patented by the Holland-based company Keygene as a high-throughput method to detect polymorphism at the DNA level. The genome is first cleaved by restriction enzymes, and a subset of the resulting DNA fragments is subject to PCR amplification [10]. It is considered a "complexity reduction" system since a subpopulation of the entire genome variability is scored. Technically speaking, it features the concept of DNA restriction typical of RFLP (see above) and the power of DNA amplification made possible by PCR. Following genomic DNA digestion by enzymes, synthetic adaptors are ligated to the sticky ends of the resulting fragments. A subpopulation of the fragments is then selected to be amplified. This selection is operated by primers complementary to the adaptor sequence, the restriction site sequence and a few nucleotides inside the restriction site fragments. The amplified fragments are separated and visualized on denaturing polyacrylamide gels, either by autoradiography or fluorescence methods, or *via* automated capillary sequencing instruments, which increases precision and throughput. Among the most notable features of AFLP are the tiny amount of DNA required and the independence from any previous DNA sequence knowledge. For that reason it is considered an interesting technique in non-model species. Its introduction marked a revolution in the agrobiotech field for both research and commercial applications. Unfortunately, one of its major advantages, the independence from sequence information, is also a major drawback since markers obtained in one individual cannot be exported to different individuals even within the same species. The complexity of AFLP patterns, as multiple loci are screened at the same time, makes its analysis inconvenient for large-scale application, *i.e.* screening hundreds or thousands of individuals [11]. Nevertheless, fine mapping of AFLPs is essential for application in marker-assisted breeding or map-based cloning [12 - 14].

To address this limitation the method was modified [15]. The AFLP method has also been used in combination with next generation sequencing (NGS) for an efficient exploration of polymorphisms [8]. For a technical introduction to AFLP markers many good sources are available on the web [16, 17].

SSR

Simple Sequence Repeat. Also known as Short Tandem Repeats (STR) or microsatellites, these markers represent variation in the number of repetitions of short stretches of DNA sequences (typically 1-5 bp) at a given position [18]. Microsatellite mutations feature the acquisition or loss of one or more repeat unit(s). Mismatches between DNA strands during replication or replication slippage are the main causes of such length changes. Microsatellite alleles can evolve through unequal crossing over during homologous chromosome pairing at meiosis. Microsatellites are integral in the field of forensics and fingerprinting [19]. The microsatellites in use today for genetic analysis, especially finger-printing, are preferably tetra- or penta-nucleotide repeats, as they are less prone to artifacts during the analysis. Microsatellite alleles are identified by the polymerase chain reaction (PCR) process, using the unique sequences of flanking regions as primers and the PCR products are visualized through electrophoretic separation (agarose-, acrylamide- or capillary-based). Many online resources are freely available that collect SSRs for almost any studied species [(20)]. Additionally, genus- or family-specific marker collections are available in the many online databases, some of which are listed in Table **1**.

Table 1. Collections of molecular markers available for major crop species.

Family (Species)	Web site
Gramineae (Maize)	http://www.maizegdb.org/data_center/marker
Gramineae (Rice)	http://archive.gramene.org/markers/
Gramineae (Barley)	http://www.public.iastate.edu/~imagefpc/IBSC%20Webpage/IBSC%20Template-home.html
Solanaceae	https://www.sgn.cornell.edu
Fabaceae (Legumes)	http://comparative-legumes.org
Rosaceae (*e.g.* apple and stone fruits)	http://rosaceae.org
Vitaceae (Grapevine)	http://www.vitaceae.org http://www.vivc.de/docs/dataonbreeding/20150504_Table%20of%20Loci%20within%20VITIS.pdf
Euphorbiaceae (Cassava)	https://cassavabase.org http://cassava.psc.riken.jp/index.pl

SNP

Single Nucleotide Polymorphism. An extremely common, albeit tiny type of variation that occurs at the sequence level of genomes. This kind of genetic polymorphism is widely and routinely used for quantitative, functional and evolutionary studies at the genomic level. The application of modern next generation sequencing (NGS) technologies for whole-genome sequencing or resequencing has the potential to disclose every single genetic variation among individuals. However, with the aim to reduce the cost factor of the process, researchers have developed genome-wide sampling sequencing (GWSS) methods, a set of reduced genome complexity sequencing and selective target genome sequencing. As well as being treated in other chapters of this book, many excellent reviews and tutorials are present on the web dealing with methods and their applications [8, 21, 22]. The NCBI database hosts one of the most important repositories for SNPs along with a set of bioinformatics tools [23, 24]. Extensive and actively growing collections of SNP marker loci are available in the many online databases, some of which are listed in Table **1**.

USE OF MARKERS FOR CONSTRUCTION OF GENETIC MAPS

Prior to the introduction of molecular markers, genetic maps were historically based on morphological characters. A collection of visual traits can be used to map their reciprocal distances based on crossing and segregation. Crosses considering two or three traits taken together can be used to define distance and linear order on chromosome(s). Typically a long set of crossings and field observations are necessary to place morphological traits on a map. Moreover, a sufficient number of traits must be available to be easily scored and to close gaps on chromosomes. The same process applies to molecular markers with the difference that they (i) can be detected in high numbers; (ii) do not require field observations; (iii) are environment-independent; (iv) are easier to obtain based on laboratory tests and relatively inexpensive equipment.

The process of construction of molecular maps falls outside the scope of this review and, once again, many sources are available to the interested reader in the public domain through tutorials, articles and manuals. Nevertheless, a schematic description of the process is necessary to facilitate the comprehension of the following paragraphs [25].

The choice of a DNA marker system in plant breeding is influenced by project objectives, the population structure, the level of genomic diversity of the species under investigation, the availability of genome sequence information and the cost per unit of information in terms of data points (loci per individual).

Marker systems also differ in their efficacy to detect polymorphism across populations, species, and genera. For example, SSRs can be as informative as RFLPs, but primers identified in one species cannot generally be used in another species for the lack of sufficient sequence homology or polymorphism. Similarly, maps based on AFLPs can be constructed in a relatively short period; however, such markers are not useable across populations, because each marker is defined by its length (sequence information might not be sufficient to tag the corresponding locus position).

SSRs are the marker systems of choice whenever the level of polymorphism is low. However, the time and cost necessary to develop such markers can represent limiting factors.

The costs per unit of information are functions of the time required for DNA collection and extraction, the amount of DNA needed for analysis, the possible cloning and sequencing steps required, the amount of potentially useful genetic information achieved, the type of starting genetic information needed, whether the allelic variation can be ascribed to banding patterns (dominant *vs.* codominant), whether the electrophoretic system can be automated, the potential utility of genetic maps, and the proprietary status of the technique (*i.e.* for the AFLP technique). Codominant markers, such as SSRs, are useful for MAS and evolutionary studies, but their use can be time consuming, relatively expensive, and may require instrumentation like capillary electrophoresis-based genetic analyzers and technical expertise in data collection and interpretation (Fig. **3**).

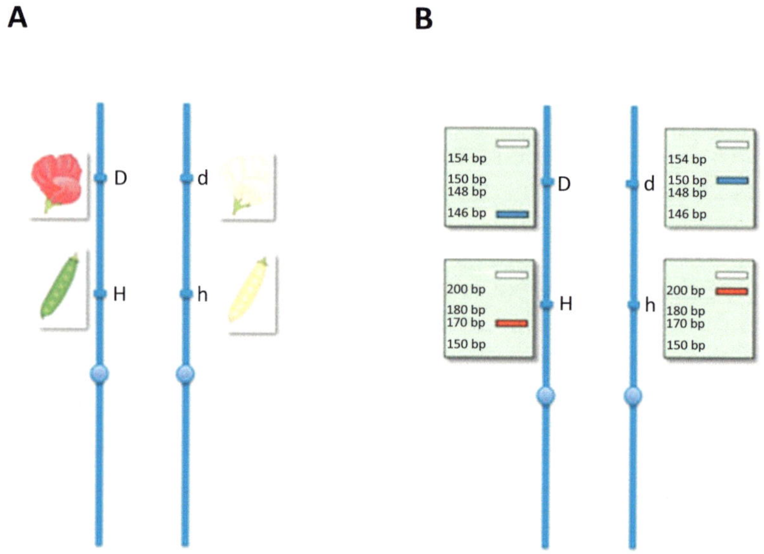

Fig. (3). Similarity between genetic (phenotypic) traits and molecular markers. While different alleles of genetic markers are recognizable by visual inspection of phenotype (**A**), alleles of molecular markers are scored by at the DNA level, in this example by electrophoretic separation of PCR products of two microsatellite loci (**B**). Uppercase and lowercase letters mark the different alleles which are subject to segregation according to Mendelian laws in both cases. The same letter codes and conventions can be used for both genetic and molecular markers. In these picture a dihybrid (DdHh) configuration is shown in both cases.

Map Construction

The development of molecular marker technology and consequent identification of many marker loci has led to a lot of interest in genetic mapping. Genetic map construction is based on the accurate selection of several factors: i) identification of the most appropriate mapping population(s); ii) calculation of pairwise recombination frequencies using these population(s); iii) establishment of linkage groups and estimation of map distances; and iv) determination of markers' order. Map construction relies deeply on software packages to resolve the many possible configurations of locus order and to minimize tensions within markers' groups. Computer packages such as Mapmaker [26, 27], MapManager [28], JoinMap [29] and Carthagene [30] have been developed to aid in the analysis of genetic data for map construction. These programs, starting from genotype data obtained from segregating populations, calculate recombination frequencies that are in turn used to estimate the linear order of markers by keeping the number of recombination events to a minimum.

Mapping Populations

Selection of mapping populations is critical to successful map construction. When planning a map construction, in order to increase its future economic impact, as many qualitatively inherited morphological traits as possible should be included in genetic stocks chosen as parents for its construction. Special consideration must be given to the nature of parents used in the mapping population and their level of domestication or wilderness are key factor to take into account. For example, due to poor or improper chromosome pairing between adapted and exotic parents, recombination frequencies can be interfered with in wide crosses, resulting in reduced linkage distances. Segregating populations made from wide crosses tend to provide relatively large numbers of polymorphisms when compared to progenies segregating in a cross between only adapted parents. To prove valuable in plant improvement programs, a map made from a wide cross must be co-linear (similar order of loci) with maps constructed using adapted parents. The choice of an appropriate mapping population depends on the characters that are under investigation, the ploidy level, the reproductive biology of the species in question and the marker(s) types used.

When two or more genetic maps share a minimum of markers they can be merged to create a more informative map. However, the type of information and precision of recombination frequencies often vary greatly between population sizes and data sets. Hence, any attempt to merge mapping information must consider these types of information to create the most likely map while minimizing the internal tensions. Specific software must be used in such cases [29].

A scheme is provided to show the process of calculation of the distance between a pair of molecular marker (microsatellite) loci starting from a controlled cross (Fig. **4**). In this example, a double-heterozygote for the two markers (mother) is test-crossed by a double homozygote (father). Segregation patterns in the off-spring are shown along with the assignment of the 'parental/recombinant' state for each individual. From the number of recombinants the fraction of recombination and the distance in map units (cM) is calculated. The systematic application of this process, applied to hundreds or thousands of markers, forms the basis of map construction.

Genetic maps in several crop species have been generated using various marker types and population systems. Many resources are available on the web that collect and make available lists of markers (and the corresponding maps) for almost any species of interest to the agrobiotech field. An extract of the most relevant is provided in Table **1**.

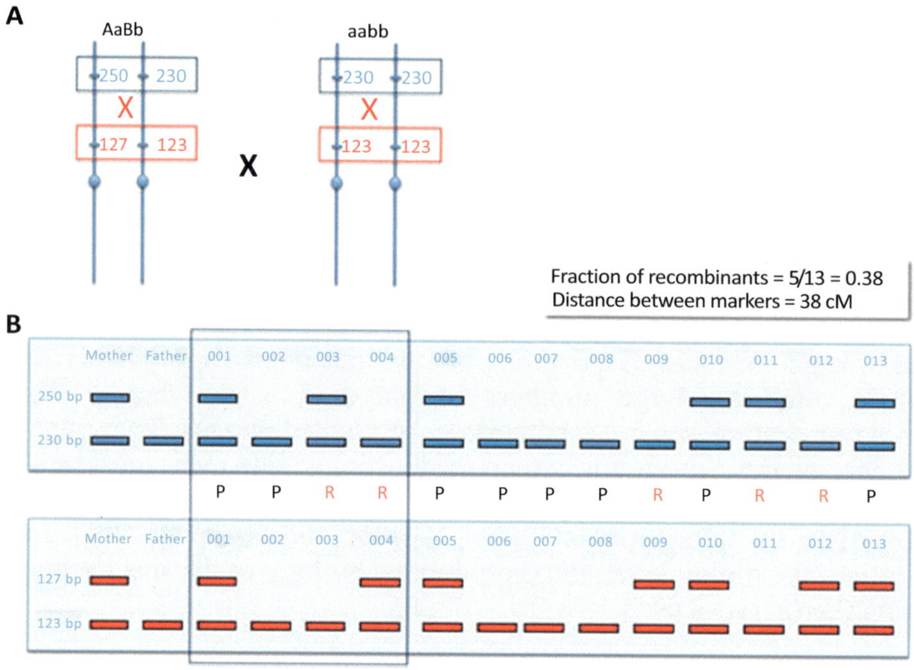

Fig. (4). Calculation of the percentage of recombination between a pair of microsatellite (SSR) markers. Their positions on the chromosome are shown in blue and red boxes, respectively (**A**). A double heterozygous parent "AaBb" (left) is crossed by a double homozygous parent "aabb" (right). Allele sizes (in bp) are different in the double heterozygote and its alleles segregate in the offspring following Mendelian ratios (**B**). Allele sizes are determined by electrophoresis. Here bands are shown corresponding to migration in a hypothetical electrophoretic gel. The marker state of the two parents and 13 random offspring are shown for the corresponding SSR loci with the corresponding colors. The first four offspring with the possible markers configurations are included in a rectangle. Parent- and Recombinant-type configurations are marked by "P" and "R", respectively. The fraction of recombinants and the estimated genetic distance are provided in the grey box.

FROM MAPS TO MARKER-BASED SELECTION

Once a map is constructed it is possible to integrate the genetic traits controlling any character for which polymorphism (variation) and segregation is known. Following the same principle applied to traditional Mendelian genetic analysis, by virtue of crosses and segregation analysis, molecular markers and genetic traits can be combined to form medium- or high-density maps whereby phenotypic traits are assigned to known chromosome positions at a high resolution. The principle is illustrated in Fig. (**5**). Here, an example is provided where a single microsatellite locus is mapped against a phenotypic trait controlling the flower color. The same procedure and calculations outlined for Fig. (**4**) (marker *vs.* marker) are followed. Similarly, a double heterozygote (mother plant) is test-crossed by a double homozygote (father) and the resulting progeny is analyzed for both the molecular marker state and the phenotypic trait (flower color).

Comparison of parent and offspring trait and markers configurations permit the calculation of distance between loci.

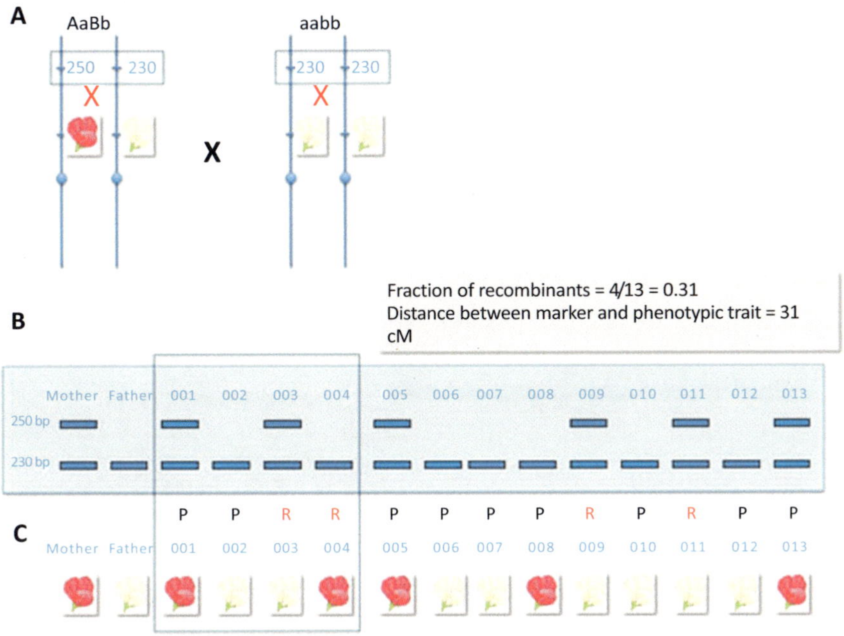

Fig. (5). Calculation of the percentage of recombination between marker and gene. One microsatellite (SSR) marker locus whose position on the chromosome is marked by a blue box, (**A**). A double heterozygous parent "AaBb" (left) is crossed by a double homozygous parent "aabb" (right). The marker state of the two parents and 13 random offspring are shown for the SSR locus as an hypothetical electrophoretic gel (**B**) and the phenotypic trait "flower color" (**C**). The first four offspring with the possible markers' configurations are included in a rectangle. Parent- and Recombinant-type configurations are marked by "P" and "R", respectively. The fraction of recombinants and the estimated genetic distance are provided in the grey box.

By exploiting the information provided by very detailed genetic maps, breeders are able to determine the genetic setup of newly germinated seedling by just analyzing a tiny amount of plant tissue. This very early genotyping is the mean to know whether a given seedling features the appropriate gene(s) and allele(s). If it does not, they can discard it and pass on to another seedling, eventually concentrating the selection activity only to the plants which prove to contain a specific trait. It can be shown that two markers relatively distantly spaced but flanking the target locus are more reliable than a single more closely linked marker in predicting the rate of false positives arising from crossing-over events that disrupt the marker-to-locus linkage (Fig. **6**). The process of marker-assisted selection is normally iterative as new markers, closer to the trait under investigation, can progressively replace the previous ones and advance the resolution of the physical localization of the genomic region bearing the gene(s) involved in its determination. Fig. (**7**) can help to understand this process. In the picture a cross is shown between a parent (mother) bearing, in heterozygosis, an allele for pathogen resistance and a second parent (father) homozygous for the susceptibility allele. Green arrows depict new markers that are found progressively closer to the target locus and that are transmitted with the corresponding phenotype at a higher frequency. The possible marker combinations to assist the selection of the elite offspring are illustrated.

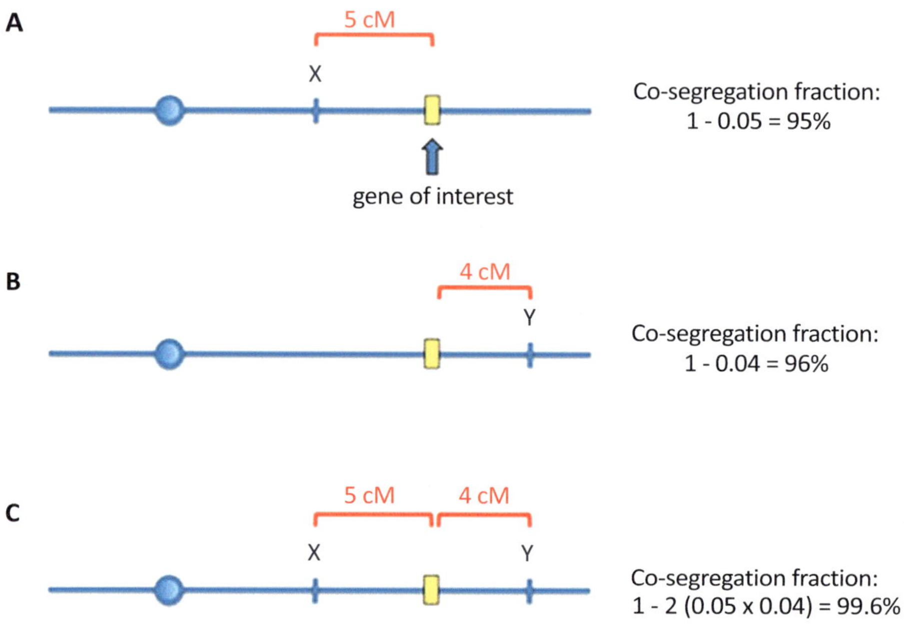

Fig. (6). Predictivity of marker-assisted strategy based on the number of markers flanking the target locus. A hypothetical marker "X" is scored at the genetic distance of 5 cM from the locus of interest (yellow rectangle). A recombination fraction of 0.05 (5% of offspring) is predicted to show recombination between marker and locus (**A**). A similar situation applies to a second marker "Y" placed on the opposite side and at a genetic distance from the locus of 4 cM. In this case a fraction of 0.04 (4% of offspring) will show recombination between marker and locus (**B**). The combined use of two markers flanking the target locus will yield double recombinants at the total frequency of 0.05 x 0.04 = 0.002 (0.2% of offspring) thus providing a more reliable tool to assess that the wanted target allele will be passed on to the progeny. (Adapted from Tanksley, 1983)

FINGERPRINTING AND ASSOCIATION MAPPING

Fingerprinting

Unlike the conventional approach based on making controlled crosses and analyzing trait segregation, fingerprinting and association-based mapping are two applications of molecular markers that permit genomic studies and, ultimately, target locus identification and/or marker-assisted selection. While fingerprinting, *per se*, provides important preliminary data to assess the level of genetic relatedness in germplasm collections used for conventional cross-based mapping, it also poses the foundations for genome-wide correlation studies between genotype and phenotype. This is the prerequisite to Genome-Wide Association Studies (GWAS) and association-based mapping whereby sets of evenly distributed marker loci across the genome are used to genotype at high resolution a panel of individuals not derived by crossing [31].

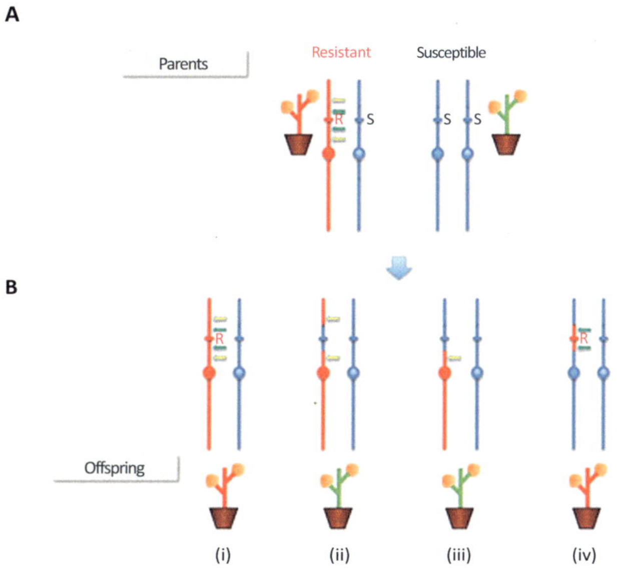

Fig. (7). Scheme of marker-assisted selection. (**A**) A couple of parents with contrasting traits (*i.e.* pathogen resistance) are crossed and their offspring is scored with molecular markers flanking the locus (and the corresponding allele) conferring resistance and carried by the red chromosome. Flanking markers, shown by arrows at varying distance from the locus (deep green being closer and light green being farther away) are used that follow the fate of the chromosome region under selection. (**B**) Various segregations of the trait and markers are possible in the offspring as shown by numbers. (i) the entire locus region with all the flanking markers plus the unwanted extra chromosomal portions are inherited; (ii and iii) based on the distantly-placed markers (light green arrows) a false prediction can be inferred due to double or single recombination events, respectively; (iv) only by following the closer markers (deep green arrows) the inheritance of the chromosome region bearing the target locus can be confirmed with the concurrent exclusion of the extra unwanted red chromosome portions.

A vast array of molecular marker-based fingerprinting techniques and applications has been developed since the earliest works dating back to the 1980s [19]. A literature survey is outside the scope of this chapter and the reader is referred to other chapters in the same book. Additionally, an enormous repertoire of teaching resources can be found online, including technical reviews, books and videos.

Association Mapping

A series of major developments make genome-wide association studies (GWAS) possible. The first is the increasing availability of genome sequences that in turn permits the generation of detailed maps based on single nucleotide polymorphisms (SNPs) and possibly of linkage disequilibrium (LD, statistical dependence of alleles at two genetic loci) profiles for a growing number of plant species. The second is the progress in phenotyping procedures that, for large-scale studies and many species, often represent the true bottleneck. Minimizing labor through automation, the development of remote sensing and improved data integration are factors affecting this actively growing research field [32]. The third major factor to consider for association mapping studies is the availability of statistical models and data computing tools and algorithms that can combine and mine information from the wealth of data generated by genotyping and phenotyping activities.

Genomic Selection

Genomic Selection (GS) is a relatively new plant breeding approach based on statistical modeling to predict, before it is field-tested, the performance of a plant. The increasingly available genomic information, combined with novel statistical models and bioinformatics tools, along with improved phenotyping tools, make the basis to develop breeding methods based on prediction and targeted design.

Genomic Selection is implemented by first genotyping and phenotyping for all the traits of interest to the breeding program a 'training population' comprising plant lines spanning the genetic pool of the relevant material in the breeding program in question. The resulting complex of genetic and phenotypic characteristics for multiple traits make the basis for a full record of the individual makeup in the training population.

Subsequently, the integration of these two data layers allows a prediction model to be created. At this point, whenever proper genotypic information is available for a given line within the breeding population, the model can produce a Genomic Estimated Breeding Value (GEBV) .that is the estimator of the overall predicted quality, or value, of a line if it were used as parent in crosses. The individual plant performance is therefore assessed by the GEBV before its value is tested in the field [33, 34].

Unlike traditional MAS, the GS approach does not rely on a defined subset of significant locus-targeted markers used for selection. Instead, genomic selection exploits all markers in a population together, aiming to explain the total genetic variance with dense genome-wide marker coverage taking into account the sum of marker effects to predict the breeding value of individuals [33, 35].

CASE STUDIES

In the last decade genomic data (*i.e.* genetic and physical maps, ESTs collections and entire genome sequences) have been generated for many plant genomes including model species like Arabidopsis and important crops like rice, maize, tomato and legumes. Four examples, taken from three staple crop species and a fruit species are discussed in this section to give an idea of what can be achieved in plant breeding *via* the application of molecular marker technology.

Tomato

The cultivated tomato, *Lycopersicon esculentum* Mill., is widely grown and constitutes a major agricultural industrial crop. In the world, it is the second most consumed vegetable after potato [36].

A number of pathogens (viruses, bacteria, fungi, and nematodes) are the main cause for severe harvest losses ultimately constraining tomato cultivation and production. The use of resistant lines and the adoption of agrochemicals has partially mitigated the problem. Conventional breeding has significantly impacted resistance of this species but new virulent pathogen strains are continuously appearing which overcome the existing resistances and the long-lasting process of serial crossing and backcrossing, to select for resistant individuals make it difficult to combat effectively the rapidly evolving pathogens.

Modern tomato breeding deeply relies on molecular markers. More than forty genes conferring resistance to major classes of pathogens have been mapped, cloned, and/or sequenced in tomato [37]. Through "pyramiding" several mapped resistance genes into engineered cultivars by MAS it has been possible to cumulate resistances to more pathogens and overcome the possible evolving resistance to single genes. Moreover it is possible to combine resistances to different pathogens through the use of marker-assisted breeding and selection, as in the case of resistance to two major tomato pathogens, namely *Xanthomonas Dowson* the causal agent of bacterial spot, and *Pseudomonas syringae* pv. tomato the aetiological agent of bacterial speck [38]. This example has been the subject of an excellent tutorial for those wishing to deepen the practical aspects of the subject [39].

Excellent reviews describing the application of molecular breeding to tomato are available [40, 41].

Reference genomes, although useful for elucidating many questions, are of limited usefulness to address the interconnection between phenotypes and genotypes, a key question of biology. To this aim a single reference genome is not sufficient but extra data categories must be included such as sequence variation and phenotypic data. The combination of these data is the prerequisite for Genome-Wide Association Studies (GWAS) that can identify genomic regions contributing to a trait of interest if necessary criteria are met. While genotyping has become cheaper to perform compared to one decade ago, it is the phenotyping that presents a bottleneck [32] and its improvement is a new and important challenge for modern breeding.

The next step in the genomics advance is to transform breeding. Novel systems are required that combine traditional genomics databases with breeding applications. The Sol Genomics Network [42] was created to store and organize genetic and genomic data, mainly genetic mapping and expressed sequence tags (ESTs) information for Solanaceae including potato, tomato,, eggplant, pepper, tobacco and petunia. It is the central repository for the tomato genome sequencing project and hosts whole genome sequences of tomato wild accessions, potato, pepper, *Nicotiana benthamiana*, tobacco and eggplant. The SGN platform has recently been supplemented with a breeding management system that is tailored to genomic-based breeding approaches, such as Genomic Selection.

Cassava

Cassava (*Manihot esculenta* Crantz), a perennial woody species native to Latin America, is primarily grown as an annual crop in the humid tropical areas. Cultivated as both a cash and a subsistence crop, cassava's large starchy roots, are a low-cost source of carbohydrates for 500 million people across the globe. It is used for household consumption as fresh roots cooked, boiled, baked or fried, up to food additive in the form of highly processed starch. Africa is the continent that most depends on cassava for food security. It is a staple widely consumed and an important food reservoir that can save lives during food emergencies being stored in the ground as a source of food. Despite its importance in the African continent, cassava has received comparatively little attention in comparison with other staples such as rice, maize and wheat. To exploit the full potential of cassava it is imperative to invest in modern breeding. Resistance to drought, long-term storability of the roots in the ground, ability to grow in marginal soils and low input requirements are characteristics that make cassava a crop of high plasticity and adaptability. Studies indicate this species as the only staple crop with a potential to benefit from climate change. Cassava’s importance as a staple is predicted to steadily grow around the globe as more less land becomes usable due to temperature changes and rainfall patterns.

Breeding of cassava is lengthy as the release of new varieties can take up to a decade to be completed. Small-scale farmers' rapidly changing breeding targets could be greatly addressed by a reduction of the length of the breeding cycle by researchers employing molecular strategies.

One of the limiting factors of cassava breeding is flowering and seed set. Many cassava genotypes are in fact affected to various extent by poor flowering, thus limiting crossing possibilities. For this reason some cassava lines featuring interesting characters cannot be used in breeding programs. Improved flowering and seed set would enable breeders to count on a definitely better exploitation of the genetic resources of cassava in breeding programs.

Since cassava's origin is in Latin America, the vast genetic diversity at the center of origin is an invaluable resource for cassava breeders in Africa. Here, breeding programs and goals could benefit by the access to new genetic stocks. facilitating germplasm exchange and ultimately the process of generating new cassava varieties.

A greater number of projects producing large quantities of phenotypic and genotypic information is the result of increasing research attention to cassava. Improved information exchange has prompted the creation of a platform to facilitate communication and data exchange between cassava researchers and breeders with the aim to harmonize cassava research worldwide. To this end the establishment of information resources [43] like the Cassavabase web portal [44] with connections to major information resources, including the Cassava Online Archive database [45] is an integral tool to promote and assist research on this species.

Legumes

The Fabaceae/Leguminosae or legume family with around 20,000 species rank at the third position as the largest botanical family and second most important after Gramineae or Poaceae for human food, especially as source of proteins. Legumes uniquely feature the property of fixing atmospheric nitrogen *via* symbiosis, resulting in a key component of sustainable agricultural production systems [46, 47].

Pulses are annual legumes yielding between one and twelve grains or seeds of variable size, shape and color in a pod, used both for human and animal nutrition. The term 'pulse' is limited to species harvested for dry grain only, thus excluding species harvested green for food, which fall under the classification of vegetable crops, as well as those crops used principally for the extraction of oil, and pulses that are used solely for sowing, according to the definition of 'vegetables and derived products' of FAO.

Legumes such as lentils, beans, peas and chickpeas are a central part of the food supply. They are a vital source of vegetable protein and amino acids for people all over the world and are recommended as part of a healthy diet to prevent and fight chronic diseases like obesity, diabetes, atherosclerosis and cancer [48]. However they are also an important source of proteins of vegetable origin for animal feeding.

In comparison to other crops, legumes have a low impact on food waste. They are also highly efficient in water use, especially when compared to other sources of protein. For example, a pound of cooked beef meat requires ten times more water than one kg of pulses. Furthermore, they require minimal processing and no refrigeration, limiting the consumption of natural resources in the following stages of the food supply chain. Due to their stability, pulses have the potential for long-term storage during many months and even years without losing nutritional value. This can reduce the likelihood of consumer food waste, and makes pulses a smart choice in case of food insecurity.

As for protein deficiency, it must be emphasized that currently over one billion people globally are suffering from micronutrient and protein malnutrition [2]. In this context, pulses due to their high protein, vitamin and mineral content represent a key factor, especially in the less-developed countries, in countering micronutrient deficiencies, undernourishment or protein caloric malnutrition (PCM) [49].

Most of the agronomic traits studied in pulses are polygenic. For nearly half a century genetics and breeding studies on pulses has been carried out in India and elsewhere in the world, with emphasis on traits like disease resistance, early maturity and growth habit. Initially, selection within and between landraces followed by pedigree, mass-pedigree, and recurrent backcrossing were used [50].

Exhaustive research has been conducted on pulses for improvement through conventional breeding due to their vast agricultural value, resulting in the development and release of several varieties addressing high yield followed by an increase in the global area under pulse cultivation from 64 to more than 77 million ha over the last 50 years [47].

Gene pyramiding is a typical approach to achieve multiple and durable resistance. Genetic markers can be used to identify and select specific genes or combine multiple-resistance genes [51]. It has been successfully demonstrated in soybean for cumulating genes controlling resistance to cyst nematode. Its use has proven to reduce costs associated to traditional greenhouse screening, and has increased the efficiency and speed of developing cultivars resistant to cyst nematode [52]. Also QTLs affecting resistance or tolerance to *Phytophthora* root rot and resistance to soybean mosaic virus (SMV) have been pyramided in this species [53, 54]. Backcrossing and recurrent selection assisted by molecular markers are being utilized for introgressing and/or pyramiding genes for drought-related traits and resistance to diseases (*Ascochyta* blight and *Fusarium* wilt) in elite lines of chickpea at the International Crops Research Institute for the Semi-Arid Tropics (ICRISAT) and Indian Institute of Pulses Research (IIPR) [55, 56].

Grapevine

Among the factors affecting a sustainable agriculture the reduction in the use of agrochemicals is of the utmost importance. Only in Europe, for example, grapevine (*Vitis vinifera* L.) cultivation, while accounting for only 3.5% of the total agricultural area (3.7 million ha out of 105 million in total), is estimated to require more than 30% of the pesticides applied to major crops [57], with a dramatic negative impact and a cause of concern for both consumers' health and the environment.

Most *V. vinifera* cultivars are susceptible to downy and powdery mildew infections and these infections cause large losses in production if not adequately treated, mainly done by spraying with fungicides. Moreover, the appearance of fungicide resistant strains of *Plasmopara viticola* and *Erysiphe necator* (formerly *Uncinula necator*), the causative agents of downy and powdery mildew respectively, emphasize the importance of durable host resistance. Genetic host resistances to both these pathogens are available in various cultivars, but these cultivars are mainly derived from non-vinifera species of North American and Asian origin, with resistance varying from complete to partial. To date more than thirteen *P. viticola* major and minor resistance loci have been identified and mapped with assigned names in the catalogue of mapped resistance loci (http://www.vivc.de).

As with downy mildew resistance, several powdery mildew resistance loci have been identified in different *Vitis* species and named either *Run* (resistance to *Uncinula necator*) or *Ren* (resistance to *Erysiphe necator*) loci. The availability of molecular maps from many grapevine crosses allowed the identification of markers linked to resistance genes/QTLs and gene pyramiding strategies [58, 59]. Soviet breeders first introgressed a downy mildew resistance gene of the Asian species *V. amurensis* into *V. vinifera* and such resistance was then combined by European breeders with other resistance genes from North American grapevines. Several chromosome regions bearing the most studied North American resistance gene for downy mildew (*Rpv3*) have been introgressed since the mid 19th century into many varieties [58]. The *Rpv3* gene has been mapped to chromosome 18 [60] and confers a race-specific hypersensitive response (HR) against pathogen strains that carry the corresponding avirulence factor [61]. As for the resistance to powdery mildew, the *Run1* region on chromosome 12 was defined by a pair of flanking markers (VMC4F3-1 and VMC8G9) [62] while the *REN1* region on chromosome 13 was defined by the markers UDV124 and VMCNG4E10-1 [63]. The combined use of these flanking markers allowed the massive screening of hundreds to thousands of individuals generated by several crosses and permitted the early identification of promising seedlings whereby the two resistance genes were combined [64].

Marker-assisted selection is thus achievable with the DNA information made available by the identification of markers linked to pathogen resistance. The haplotype analysis of the introgressed segment provides markers to track the gene and guide the elimination of linkage drag (unwanted chromosome regions flanking the introgressed gene). Background selection is determinant of the quality of the new grape varieties created by conventional breeding [58]

CONFLICT OF INTEREST

The author confirms that author has no conflict of interest to declare for this publication.

ACKNOWLEDGEMENTS

Declared None.

REFERENCES

[1] Pingali P. Westernization of Asian diets and the transformation of food systems: Implications for research and policy. Food Policy 2007; 32: 281-98. [http://dx.doi.org/10.1016/j.foodpol.2006.08.001]

[2] Godfray HC, Beddington JR, Crute IR, *et al.* Food security: the challenge of feeding 9 billion people. Science 2010; 327(5967): 812-8. [http://dx.doi.org/10.1126/science.1185383] [PMID: 20110467]

[3] Foley JA, Ramankutty N, Brauman KA, *et al.* Solutions for a cultivated planet. Nature 2011; 478(7369): 337-42. [http://dx.doi.org/10.1038/nature10452] [PMID: 21993620]

[4] Tilman D, Balzer C, Hill J, Befort BL. Global food demand and the sustainable intensification of agriculture. Proc Natl Acad Sci USA 2011; 108(50): 20260-4. [http://dx.doi.org/10.1073/pnas.1116437108] [PMID: 22106295]

[5] Ray DK, Mueller ND, West PC, Foley JA. Yield trends are insufficient to double global crop production by 2050. PLoS One 2013; 8(6): e66428. [http://dx.doi.org/10.1371/journal.pone.0066428] [PMID: 23840465]

[6] Sax K. The association of size differences with seed-coat pattern and pigmentation in *Phaseolus vulgaris*. Genetics 1923; 8(6): 552-60. [PMID: 17246026]

[7] Rasmusson J. Studies on the inheritance of quantitative characters in *Pisum*. Hereditas 1935; 20: 161-

80.
[http://dx.doi.org/10.1111/j.1601-5223.1935.tb03184.x]

[8] Davey JW, Hohenlohe PA, Etter PD, Boone JQ, Catchen JM, Blaxter ML. Genome-wide genetic marker discovery and genotyping using next-generation sequencing. Nat Rev Genet 2011; 12(7): 499-510.
[http://dx.doi.org/10.1038/nrg3012] [PMID: 21681211]

[9] Wang S, Meyer E, McKay JK, Matz MV. 2b-RAD: a simple and flexible method for genome-wide genotyping. Nat Methods 2012; 9(8): 808-10.
[http://dx.doi.org/10.1038/nmeth.2023] [PMID: 22609625]

[10] Vos P, Hogers R, Bleeker M, *et al.* AFLP: a new technique for DNA fingerprinting. Nucleic Acids Res 1995; 23(21): 4407-14.
[http://dx.doi.org/10.1093/nar/23.21.4407] [PMID: 7501463]

[11] Pafundo S, Agrimonti C, Maestri E, Marmiroli N. Applicability of SCAR markers to food genomics: olive oil traceability. J Agric Food Chem 2007; 55(15): 6052-9.
[http://dx.doi.org/10.1021/jf0701638] [PMID: 17592855]

[12] Moon H, Nicholson JS. AFLP and SCAR Markers Linked to Resistance in Tobacco. Crop Sci 2007; 47: 1887-94.
[http://dx.doi.org/10.2135/cropsci2007.01.0002]

[13] Brennan R, Jorgensen L, Gordon S, Loades K, Hackett C, Russell J. The development of a PCR-based marker linked to resistance to the blackcurrant gall mite (*Cecidophyopsis ribis Acari*: Eriophyidae). Theor Appl Genet 2009; 118(2): 205-11.
[http://dx.doi.org/10.1007/s00122-008-0889-x] [PMID: 18813905]

[14] Mahanil S, Ramming D, Cadle-Davidson M, *et al.* Development of marker sets useful in the early selection of Ren4 powdery mildew resistance and seedlessness for table and raisin grape breeding. Theor Appl Genet 2012; 124(1): 23-33.
[http://dx.doi.org/10.1007/s00122-011-1684-7] [PMID: 21904846]

[15] Yuste-Lisbona FJ, Capel C, Gómez-Guillamón ML, Capel J, López-Sesé AI, Lozano R. Codominant PCR-based markers and candidate genes for powdery mildew resistance in melon (*Cucumis melo* L.). Theor Appl Genet 2011; 122(4): 747-58.
[http://dx.doi.org/10.1007/s00122-010-1483-6] [PMID: 21243332]

[16] Wikipedia. Available from: https://en.wikipedia.org/wiki/Amplified_fragment_length_polymorphism

[17] Chial H. DNA Fingerprinting using Amplified Fragment Length Polymorphisms (AFLP). Nat Educ 2008; 1: 176.

[18] Wikipedia. Available from: https://en.wikipedia.org/wiki/Microsatellite

[19] Nybom H, Weising K, Rotter B. DNA fingerprinting in botany: past, present, future. Investig Genet 2014; 5(1): 1.
[http://dx.doi.org/10.1186/2041-2223-5-1] [PMID: 24386986]

[20] Molecular Ecology Resources Database. Available from: http://tomato.bio.trinity.edu

[21] Van Tassell CP, Smith TP, Matukumalli LK, *et al.* SNP discovery and allele frequency estimation by deep sequencing of reduced representation libraries. Nat Methods 2008; 5(3): 247-52.
[http://dx.doi.org/10.1038/nmeth.1185] [PMID: 18297082]

[22] Jiang Z, Wang H, Michal JJ, *et al.* Genome wide sampling sequencing for SNP genotyping: Methods, challenges and future development. Int J Biol Sci 2016; 12(1): 100-8.
[http://dx.doi.org/10.7150/ijbs.13498] [PMID: 26722221]

[23] Kitts A, Sherry S. 5 The Single Nucleotide Polymorphism Database (dbSNP) of Nucleotide Sequence Variation. N.C. 2002.Handbook

[24] Bhagwat M. Searching NCBI's dbSNP database. Curr Protoc Bioinforma. 2010; Chapter 1: p.

(SUPP.32)Unit 1.19.

[25] Tanksley SD. Molecular markers in plant breeding. Plant Mol Biol Rep 1983; 1: 3-8. [http://dx.doi.org/10.1007/BF02680255]

[26] Lander ES, Botstein D. Mapping complex genetic traits in humans: new methods using a complete RFLP linkage map. Cold Spring Harb Symp Quant Biol 1986; 51(Pt 1): 49-62. [http://dx.doi.org/10.1101/SQB.1986.051.01.007] [PMID: 2884068]

[27] Lander ES, Green P, Abrahamson J, *et al.* MAPMAKER: an interactive computer package for constructing primary genetic linkage maps of experimental and natural populations. Genomics 1987; 1(2): 174-81. [http://dx.doi.org/10.1016/0888-7543(87)90010-3] [PMID: 3692487]

[28] Manly KF, Cudmore RH Jr, Meer JM. Map Manager QTX, cross-platform software for genetic mapping. Mamm Genome 2001; 12(12): 930-2. [http://dx.doi.org/10.1007/s00335-001-1016-3] [PMID: 11707780]

[29] Stam P. Construction of integrated genetic linkage maps by means of a new computer package: Join Map. Plant J 1993; 3: 739-44. [http://dx.doi.org/10.1111/j.1365-313X.1993.00739.x]

[30] de Givry S, Bouchez M, Chabrier P, Milan D, Schiex T. CARHTA GENE: multipopulation integrated genetic and radiation hybrid mapping. Bioinformatics 2005; 21(8): 1703-4. [http://dx.doi.org/10.1093/bioinformatics/bti222] [PMID: 15598829]

[31] Pritchard JK, Stephens M, Rosenberg NA, Donnelly P. Association mapping in structured populations. Am J Hum Genet 2000; 67(1): 170-81. [http://dx.doi.org/10.1086/302959] [PMID: 10827107]

[32] Cobb JN, Declerck G, Greenberg A, Clark R, McCouch S. Next-generation phenotyping: requirements and strategies for enhancing our understanding of genotype-phenotype relationships and its relevance to crop improvement. Theor Appl Genet 2013; 126(4): 867-87. [http://dx.doi.org/10.1007/s00122-013-2066-0] [PMID: 23471459]

[33] Meuwissen TH, Hayes BJ, Goddard ME. Prediction of total genetic value using genome-wide dense marker maps. Genetics 2001; 157(4): 1819-29. [PMID: 11290733]

[34] Heffner EL, Sorrells ME, Jannink J. Genomic selection for crop improvement. Crop Sci 2009; 49: 1-12. [http://dx.doi.org/10.2135/cropsci2008.08.0512]

[35] Tecle IY, Edwards JD, Menda N, *et al.* solGS: a web-based tool for genomic selection. BMC Bioinformatics 2014; 15: 398. [http://dx.doi.org/10.1186/s12859-014-0398-7] [PMID: 25495537]

[36] FAO statistics. The Food and Agriculture Organization of the United Nations 2014. Available from: http://faostat3.fao.org/faostat-gateway/go/to/download/Q/QC/E\nhttp://faostat3.fao.org/

[37] Grube RC, Radwanski ER, Jahn M. Comparative genetics of disease resistance within the solanaceae. Genetics 2000; 155(2): 873-87. [PMID: 10835406]

[38] Yang W, Francis DM. Marker-assisted Selection for combining resistance to Bacterial Spot and Bacterial Speck in Tomato. J Am Soc Hortic Sci 2005; 130: 716-21.

[39] Francis DM, Merk HL, Namuth-Covert D, Robbins M. Marker-assisted selection for bacterial spot resistance in tomato: A case study. Plant Breeding and Genomics 2013. Available from: http://articles.extension.org/pages/32561/marker-assisted-selection-for-bacterial-spot-resistance-in-tomato:-a-case-study

[40] Foolad MR. Genome mapping and molecular breeding of tomato. Int J Plant Genomics 2007; 2007: 64358.
[http://dx.doi.org/10.1155/2007/64358]

[41] Foolad MR, Panthee DR. Marker-assisted selection in tomato breeding. CRC Crit Rev Plant Sci 2012; 31: 93-123.
[http://dx.doi.org/10.1080/07352689.2011.616057]

[42] The Sol Genomics Network (SGN). Available from: http://solgenomics.net

[43] Ayling S, Ferguson M, Rounsley S, Kulakow P. Information resources for cassava research and breeding. Trop Plant Biol 2012; 5: 140-51.
[http://dx.doi.org/10.1007/s12042-012-9093-x]

[44] Fernandez-Pozo N, Menda N, Edwards JD, *et al.* The Sol Genomics Network (SGN)from genotype to phenotype to breeding. Nucleic Acids Res 2015; 43(Database issue): D1036-41.
[http://dx.doi.org/10.1093/nar/gku1195] [PMID: 25428362]

[45] Sakurai T, Mochida K, Yoshida T, *et al.* Genome-wide discovery and information resource development of DNA polymorphisms in cassava. PLoS One 2013; 8(9): e74056.
[http://dx.doi.org/10.1371/journal.pone.0074056] [PMID: 24040164]

[46] Crews TE, Peoples MB. Legume *versus* fertilizer sources of nitrogen: Ecological tradeoffs and human needs. Agric Ecosyst Environ 2004; 102: 279-97.
[http://dx.doi.org/10.1016/j.agee.2003.09.018]

[47] Bohra A, Pandey MK, Jha UC, *et al.* Genomics-assisted breeding in four major pulse crops of developing countries: present status and prospects. Theor Appl Genet 2014; 127(6): 1263-91.
[http://dx.doi.org/10.1007/s00122-014-2301-3] [PMID: 24710822]

[48] Jati IR, Vadivel V, Biesalski HK. Antioxidant activity of anthocyanins in common legume grains. Bioactive food as dietary interventions for liver and gastrointestinal disease. 2013; pp. 485-97.
[http://dx.doi.org/10.1016/B978-0-12-397154-8.00007-5]

[49] Broughton WJ, Hernández G, Blair M, Beebe S, Gepts P, Vanderleyden J. Beans (*Phaseolus* spp.) - Model food legumes. Plant Soil 2003; 252: 55-128.
[http://dx.doi.org/10.1023/A:1024146710611]

[50] Nadarajan NG. Role of classical breeding in improvement of pulse crops. Electron J Plant Breed 2010; 1: 1099-106.

[51] Gupta PK, Kumar J, Mir RR, Kumar A. Marker-Assisted Selection as a component of conventional plant breeding. Plant Breeding Reviews. 2010; pp. 145-217.
[http://dx.doi.org/10.1002/9780470535486.ch4]

[52] Concibido VC, Diers BW, Arelli PR. A decade of QTL mapping for cyst nematode resistance in soybean. Crop Sci 2004; •••: 1121-31.
[http://dx.doi.org/10.2135/cropsci2004.1121]

[53] Shi A, Chen P, Li D, Zheng C, Zhang B, Hou A. Pyramiding multiple genes for resistance to soybean mosaic virus in soybean using molecular markers. Mol Breed 2009; 23: 113-24.
[http://dx.doi.org/10.1007/s11032-008-9219-x]

[54] Li X, Han Y, Teng W, *et al.* Pyramided QTL underlying tolerance to Phytophthora root rot in mega-environments from soybean cultivars Conrad and Hefeng 25. Theor Appl Genet 2010; 121(4): 651-8.
[http://dx.doi.org/10.1007/s00122-010-1337-2] [PMID: 20390244]

[55] Chamarthi SK, Kumar A, Voung TD, Blair MW, Gaur PM, Nguyen HT. Trait mapping and molecular breeding. In: Pratap A, Kumar J, Eds. Biology and Breeding of Food Legumes. Oxfordshire: CAB International 2011; pp. 296-313.
[http://dx.doi.org/10.1079/9781845937669.0296]

[56] Kumar J, Pratap A, Solanki RK, *et al.* Genomic resources for improving food legume crops. J Agric

Sci 2012; 150: 289-318.
[http://dx.doi.org/10.1017/S0021859611000554]

[57] European Union - Eurostat. The use of plant protection products in the European Union 2007. Available from: http://solgenomics.net

[58] Venuti S, Copetti D, Foria S, *et al.* Historical introgression of the downy mildew resistance gene Rpv12 from the Asian species *Vitis amurensis* into grapevine varieties. PLoS One 2013; 8(4): e61228. [http://dx.doi.org/10.1371/journal.pone.0061228] [PMID: 23593440]

[59] Heerden CJ, Burger P, Vermeulen A, Prins R. Detection of downy and powdery mildew resistance QTL in a `Regent' x `RedGlobe' population. Euphytica 2014; 200: 281-95. [http://dx.doi.org/10.1007/s10681-014-1167-4]

[60] Bellin D, Peressotti E, Merdinoglu D, *et al.* Resistance to Plasmopara viticola in grapevine Bianca is controlled by a major dominant gene causing localised necrosis at the infection site. Theor Appl Genet 2009; 120(1): 163-76.
[http://dx.doi.org/10.1007/s00122-009-1167-2] [PMID: 19821064]

[61] Casagrande K, Falginella L, Castellarin SD, Testolin R, Di Gaspero G. Defence responses in Rpv3-dependent resistance to grapevine downy mildew. Planta 2011; 234(6): 1097-109.
[http://dx.doi.org/10.1007/s00425-011-1461-5] [PMID: 21735199]

[62] Barker CL, Donald T, Pauquet J, *et al.* Genetic and physical mapping of the grapevine powdery mildew resistance gene, Run1, using a bacterial artificial chromosome library. Theor Appl Genet 2005; 111(2): 370-7.
[http://dx.doi.org/10.1007/s00122-005-2030-8] [PMID: 15902396]

[63] Hoffmann S, Di Gaspero G, Kovács L, *et al.* Resistance to Erysiphe necator in the grapevine Kishmish vatkana is controlled by a single locus through restriction of hyphal growth. Theor Appl Genet 2008; 116(3): 427-38.
[http://dx.doi.org/10.1007/s00122-007-0680-4] [PMID: 18064436]

[64] Di Gaspero G, Cattonaro F. Application of genomics to grapevine improvement. Aust J Grape Wine Res 2010; 16: 122-30.
[http://dx.doi.org/10.1111/j.1755-0238.2009.00072.x]

CHAPTER 14

Genetic Engineering for Crop Yield

Daniele Rosellini*

Department of Agricultural, Food and Environmental Sciences, Università degli Studi di Perugia, Perugia, Italy

Abstract: By genetic engineering, one or a few genes are typically introduced into plants, so major direct improvements on complex quantitative traits such as yield are not easily realized. However, yield can be, and has been indirectly increased by reducing the gap between potential and actual crop yield, introducing resistance to diseases, pests, abiotic stresses, or herbicides, alone or combined. New traits are appearing on the market and many others are in the pipeline, which promise to contribute to global food production, such as tolerance to drought, to acid or saline soil, and nutrient use efficiency. The possibility to boost the photosynthetic potential by genetic engineering is also attractive. In this chapter, after an brief presentation of plant engineering techniques, I provide an overview of the actual contribution and potentialities of genetic engineering for enhancing crop yield.

Keywords: Acid soils, Agriculture, Biotechnology, Crop yield, Drought, Genetic engineering, GMO, Herbicide tolerance, Insect resistance, Photosynthesis, Salinity, Transformation, Transgenic plants, Virus resistance.

INTRODUCTION

Genetic engineering (GE) is the modification of the genome of an organism by means of the recombinant DNA techniques. Besides introducing new genes, it is possible to shut off the expression of an endogenous gene (gene silencing), by introducing sequences that interfere with its expression.

Transferring genes between species is regarded by many as potentially dangerous, but routine use of GE microorganisms to produce drugs and vaccines and almost 20 years of large scale cultivation of GE plants (more than 180 million hectares in 2014) [1] have demonstrated that GE does not pose intrinsic risks [2].

Yield is a complex, quantitative trait affected by many genes, each gene exerting a small effect. Since GE manipulates one or a few genes at a time, large direct

* **Corresponding author Daniele Rosellini**: Department of Agricultural, Food and Environmental Sciences, Università degli Studi di Perugia, Perugia, Italy; Tel: +39 0755856211; Fax: +39 0755856224; E-mail: daniele.rosellini@unipg.it

Roberto Pilu and Giuseppe Gavazzi (Eds.)

improvements of yield are difficult to realize. Progress in the dissection of quantitative traits by QTL mapping and cloning will certainly bear fruit (for example [3],) and the molecular dissection of key developmental steps may open ways to manipulate plant architecture and productive potential (for example [4],). In particular, the elucidation of the role of transcription factors in regulating plant growth and development may open the way to direct improvement of yield. For example, the expression of a truncated *Arabidopsis thaliana* HD-Zip transcription factor (ATHB17) in maize results in increased ear biomass at the early reproductive stage, which can brought about yield improvements [5]. Until now, however, the impact of plant GE on yield has been an indirect one.

The *potential yield* of a crop is the yield that can be obtained under optimal agronomic, environmental and management condition. Potential yield can be increased by breeding more efficient genotypes and by improving agricultural practices. *Actual yield* is what a crop actually produces, and can be increased both by increasing potential yield and, often more effectively, by reducing yield losses caused by biotic and abiotic stresses. So far, GE has been successful in making actual yield closer to potential yield, that is, in filling the "yield gap", by reducing the negative effects of weeds, insect pests and viruses; recently, drought tolerance has been tackled by GE with some success (see below). Acid or saline soil stress are major limiting factors for agriculture worldwide because large agricultural land areas are affected by these problems in many parts of the world [6]; GE holds the promise to make these soils capable to sustain crop production. Nutrient use efficiency is another key factor directly affecting yields and sustainability of agriculture [7].

It is important to consider that food production is strictly connected with environmental sustainability: only if agriculture is sustainable it will continue to produce food in the long term. Disease and stress resistance, efficiency in the use of water and nutrients can make crops more sustainable. In this chapter, I will show how GE has already, and will continue to play a positive role to increase food production and agricultural sustainability.

THE PLANT GENETIC ENGINEERING TECHNOLOGY IN BRIEF

When a breeding objective cannot be easily pursued with classic crossing and selection, it is possible to recur to GE, which allows to source genes from any organism, or even to use artificially synthesized genes.

Genetic Constructs

The genes that are to be introduced in a crop plant must be designed for optimal expression in the host organism by choosing suitable control sequences (promoter, terminator, targeting sequences) which must be assembled in the lab into a functional gene construct (Fig. **1**). The elements of the construct are briefly described.

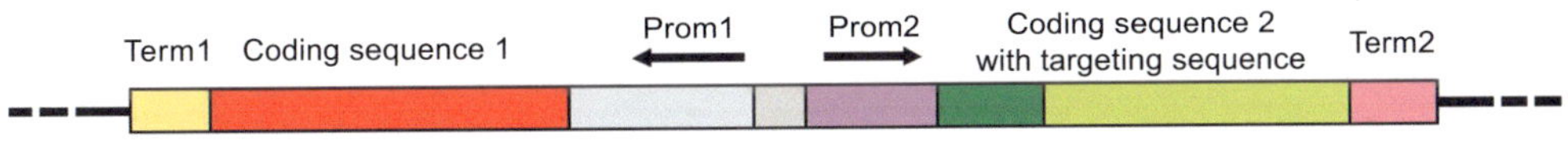

Fig. (1). Schematic representation of a typical genetic construct for plant transformation. Here, two genes were cloned so that their transcription is divergent.

Coding sequence. A gene coding sequence spans from the start to the stop codon. Eucariotic genes are usually cloned from their cDNAs, thus excluding introns from genetic constructs. An important factor for optimal gene expression is codon choice, or codon usage. Up to six different codons specify a single aminoacid but generally one of them is preferred by each species, which can be different in different species. If in the introduced gene the preferred codon for a given aminoacid is scarcely used in the host organism, the expression of the introduced gene could be inefficient, due to scarcity of the tRNA for that codon. The problem can be fixed by replacing the codons of the gene with those preferred by the host. Codon optimization has been shown to increase gene expression in plants in many cases, and is provided by several companies that synthesize genes.

Targeting sequences. These are portions of the proteins, usually the amino terminus, that codes for peptides that allow for the transport of the protein to the correct cell compartment. If, for example, a bacterial protein must be targeted to the plastid, a DNA sequence encoding a chloroplast transit peptide should be added to the 5' of the coding sequence. When the protein reaches its destination, the targeting sequence is removed by proteases.

Promoter and terminator. The promoter is the key component of a genetic construct, because it regulates the timing, the site and the rate of transcription. The promoters is located upstream (5') of the coding sequence and contains the sequences recognized by transcription factors, proteins that control binding and activation of RNA polymerase. Promoters vary widely in length, generally from a few hundred to a few thousands bp. A wealth of information is available on gene sequence and expression, allowing for a fine regulation of transgene expression by choosing a promoter with the desired characteristics. Constitutive promoters are preferred when a high transcription level is desired in all cells. Among these, the promoter of the 35S RNA gene of the cauliflower mosaic virus (*CaMV35S*) is by far the most popular. Other plant-derived constitutive promoters are taken from actin or ubiquitin genes. Promoters of storage protein genes such as glutelin (from cereals) or napin (oilseed rape) or patatin (potato tuber) are useful to accumulate the gene products in harvested organs. In other cases, fine, tissue specific gene expression regulation is pursued; for example, promoters from anther- or ovule-specific genes were used to engineer male sterility or parthenogenesis. Inducible promoters can be used to control transcription so that the introduced gene is expressed only when and where it is necessary: stress- or chemically inducible promoters have been described [8].

The terminator contains the sequences that signal the end of transcription and control mRNA polyadenylation; it affects mRNA stability and translation efficiency. The most popular terminator for plant GE is that of the Nopaline synthase (*Nos*) gene of *Agrobacterium tumefaciens*, but many others are available.

Selectable marker genes. The techniques to introduce DNA in the plant cells are inefficient, and only a few of the treated cells become transgenic. After the transformation treatment, the plant cells are cultured *in vitro* to induce the regeneration of plants from single cells. If all the cells are given the chance to regenerate, only very few of the plants obtained will be transgenic, and finding them would require a lot of work and waste of resources. Therefore, along with the gene of interest, a second gene is introduced, which most often acts by making the transgenic cells resistant to a phytotoxin that is added to the culture medium. Many options are available for the choice of a suitable selectable marker gene for a crop species of interest, including genes derived from plants [9]. Techniques also exist to remove the selectable marker gene, so that the final transgenic plant is *marker-free* [10].

Plant Genetic Transformation

Two methods among the several reported in the literature are by far the most utilized to introduce DNA constructs into the plant cells. One employs the natural ability of *Agrobacterium tumefaciens* to integrate genes in the plant nuclear genome. The other method relies on direct introduction of DNA by micro-projectiles.

A. tumefaciens is a soil bacterium that causes the crown gall, a tumor disease of fruit trees. The virulence of the bacterium is dependent on the *Ti* (tumor inducing) plasmid. The bacterium takes advantage of wounds at the plant crown to transfer a portion of the Ti plasmid, the *transferred DNA* (T-DNA), to the nucleus of plant cells in the wound. The plant cells that integrate and express the bacterial genes contained in the T-DNA proliferate and form a tumor in which the bacterium thrives (Fig. **2**).

Ti-derived plasmids are used as plant transformation vectors. The native T-DNA genes are replaced by the gene to be introduced: the transfer process is not affected, but tumor formation is suppressed. In the lab, after co-culture of *A.*

tumefaciens with the plant tissue (usually leaf portions), transgenic cells are selected *in vitro* and regenerated into plants. Other than *Agrobacterium*, a few symbiotic bacteria were found to be able to transform plan cells with low efficiency (*Sinorhizobium meliloti* 1021, *S. fredii* NGR234, *Mesorhizobium loti* MAFF303099) [12]. Recently, the rhizosphere bacterium *Ensifer adhaerens* OV14 was used for plant transformation, with an efficiency similar to that of *A. tumefaciens* [13].

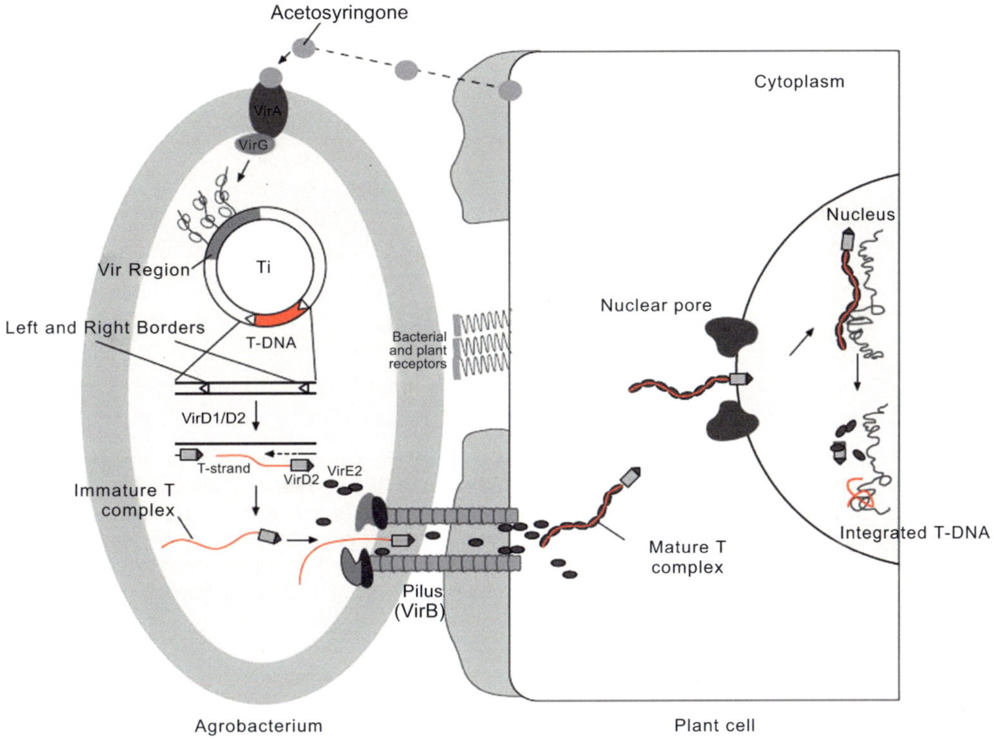

Fig. (2). Scheme of the infection process of *A. tumefaciens*. Phenolic substances present in the wound, like acetosyringone, are perceived by a receptor in the bacterial cell membrane encoded by the *VirA* gene. A signal cascade activates the VirD1 and VirD2 genes encoding nucleases that excise the T-DNA. This is transported to the cell cytoplasm and then enters the nucleus, thanks to the nuclear localization peptide carried by the VirD2 protein, and becomes integrated in a plant chromosome. Thirty five Vir genes participate in the T-DNA integration pathway, and several plant gens are also involved (modified from Lorenzetti *et al.*, 2011).

For plants that are not susceptible to *Agrobacterium* infection, the biolistic (bio-ballistic) method can be used. The DNA is precipitated onto the surface of microscopic gold particles (diameter 0.6-1 μm) that are then literally bombarded through the plant cell walls using the particle delivery system, or gene gun (Fig. **3**).

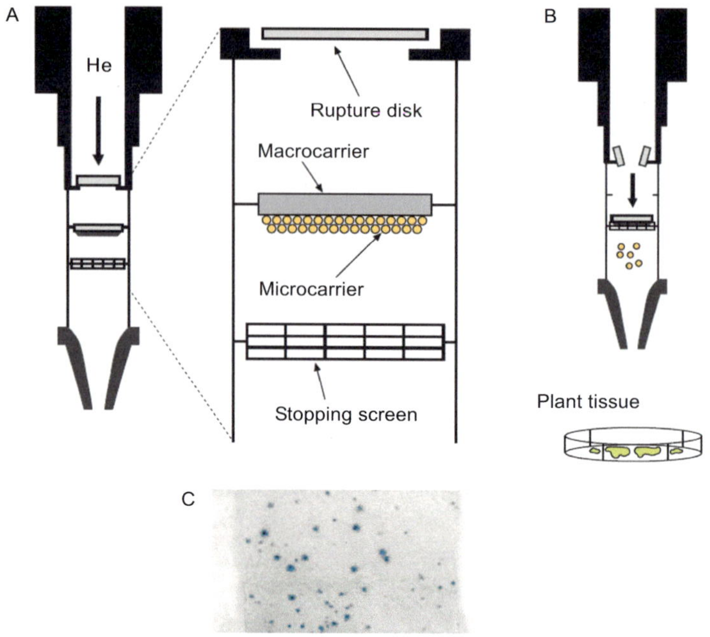

Fig. (3). Schematic diagram of gene delivery to plant cells by the biolistic method. In the upper chamber helium is pressurized (**A**) until a plastic disk (*rupture disk*) breaks (**B**). The sudden release of helium violently projects a second plastic disk (*macrocarrier*) against a metal grid (*stopping screen*) that blocks the macrocarrier but allows the metal particles (*microcarriers*) that had been deposited on the lower surface of the macrocarrier to proceed their course and hit the plant tissues below. The apparatus must be set according to the type of plant species and tissue. The expression of a reporter gene in the cells hit by the metal particles (black spots in C, blue in reality) can be used to demonstrate the functionality of the introduced DNA (modified from Lorenzetti *et al.*, 2011).

Plastid Genome Transformation

Introducing genes into the plastid genome has interesting peculiarities [14]. First of all, the integration occurs in a pre-determined site thanks to homologous recombination, which occurs spontaneously in the plastid, as in bacteria; this is realized by flanking the sequences to be introduced with sequences from the insertion site. Secondly, the high number of copies of the plastid genome in each cell normally results in very high gene expression and protein accumulation. Thirdly, the plastids are generally inherited maternally, and this facilitates transgene containment. Gene delivery is accomplished with the biolistic method, and is not readily attained in all plant species, being relatively easy in some crop species (tobacco, tomato, potato, lettuce) but still difficult in many others. Plastid transformation is particularly useful to produce high value proteins, such as drugs and vaccines, in plants.

Regeneration And Characterization Of Transgenic Plants

Tissues subjected to a transformation treatment are placed on a selective growth medium *in vitro*, allowing only transgenic cells to regenerate. However, due to incomplete selective pressure, non transgenic plants may also be obtained. Therefore, the plants must undergo molecular analyses to demonstrate their transgenic nature and the structure of the "transformation event".

PCR amplification of gene construct-specific sequences is the first step. Southern hybridization then serves to confirm genomic integration and to determine how many genomic sites contain the construct. *Northern* blot or RT-PCR are performed to demonstrate transcription and Western blot is used to prove the formation of the protein. The phenotype resulting from transgene expression is finally searched for. Subsequent essential analyses regard the insertion site(s) in the genome, the stability of transgene the expression and of the phenotype in the following generations and in field conditions.

Once a transgenic event has passed this long screening phase, cultivated varieties can be endowed with the transgene by crossing with the donor transgenic plant. Recurrent backcrossing with the cultivated variety allows to recover the genotype of the variety with the added transgene. Such backcrossing programs have been carried out to obtain many different varieties, adapted to diverse agricultural environments, all carrying the same transgenic event.

New Trends Of Plant Genetic Engineering

Plant GE techniques are constantly improving tanks to better knowledge of the molecular mechanism involved. Importantly, it is becoming easier to overcome one limitation of plant transformation, the randomness of integration in the nuclear genome, which could cause unwanted mutations of plant genes and variability of gene expression. Today, gene targeting, site-specific mutagenesis, and precise genome editing are actively pursued for crop improvement. Some of the plant products of these new breeding techniques may not even be considered transgenic, because they do not differ from genotypes produced by spontaneous mutations.

Among the so-called "new biotechnologies for plant breeding" [15], cisgenesis and intragenesis are receiving much attention [16]. Cisgenesis can be defined as a "restricted" use of GE in which only unmodified genomic fragments cloned from the same species or from sexually compatible species can be introduced in plants. Addition of useful alleles can be realized this way. Intragenesis allows more freedom to operate, because DNA sequences from sexually compatible species can be reassembled so that, for example, promoters can be changed, introns removed, and gene silencing constructs realized. The rationale behind these concepts it that regulatory requirement of the varieties obtained with these techniques may be less difficult and costly to meet [15].

YIELD INCREASES FROM CURRENT GE APPLICATIONS

According to Oerke (2006) [17] the potential loss of crop productions due to weeds, pathogens and animal pests in 2001-2003 was 26–29% for soybean, wheat and cotton, 31% for maize, 37% for rice and 40% for potato. Weeds are responsible for 34% of the losses; remarkably, hand weeding in poor African countries takes about 2000 hours per hectare, mostly of women work [18]. Animal pests cause 18% losses and pathogens 16%. It is evident that crop protection from these biotic stress factors is key to much higher food production.

The current applications of GE to crops are resistance to total herbicides (herbicide resistance, HR) and insect pests (IR), with still limited use of virus resistance. Stacking of HR and IR in the same varieties is more and more practiced to combine the advantages of different transgenes [1].

After almost 20 years from the beginning of large scale cultivation in 1996, the potential of GE for yield improvement has been convincingly demonstrated. In the 1996-2004 period GM crops brought about yield increases that varied among crop species, transgenic trait, and country. HT delivered clear yield gains only in canola (+6% USA, +11% Canada). IR GM crops provided 5-63% yield increases, the highest figures recorded for IR cotton in India. Considering only food crops, the maximum reported yield increases were 25% for IR maize and 50% for virus-resistant papaya in Hawaii. Notably, HR soybeans allowed many farmers to plant a second crop, usually following wheat, thanks to facilitated soil management; in fact, minimum- or no-tillage techniques can be adopted if weed control is done by means of total herbicides to which the crop plants are resistant [19]; it was estimated that, in Argentina, second crop soybean produced 14 M tons in 2007 [20].

Brookes and Barfoot (2009) [20] estimated that in 2007, at the global level, the yields of soybeans, corn and canola were 6.5%, 1.9% and 1.1% higher, respectively, than those that would have been obtained if biotech traits had not been used by farmers. In Europe, 4-12.5% yield gains were reported from adoption of IR corn in 2007; in Romania the average yield of HT soybean improved by over 30%, and HT corn in Argentina and the Philippines delivered yield improvements of +9% and +15% respectively.

The yield gains of corn hybrids in the USA appear to have improved since the introduction of GM varieties [21]. This positive effect of the GM technology is indirectly confirmed by stagnation of corn yield in Europe, where GM corn is grown in very limited areas [22]. This stagnation may be partly due to the impossibility to cultivate recent, more productive GM hybrids, which adds to the yield penalty of not deploying the IR and HR traits.

Stacking of multiple transgenic traits in single varieties is now common practice, especially in corn and cotton, in which IR and HR traits are often combined. A comparison of yields of corn hybrids combining IR transgenes for resistance to European corn borer and corn root worm showed that additional 8.4 million tons of corn were produced from the 16.5 million ha of stacked hybrids harvested in the United States in 2010, equivalent to cultivating about 875,000 ha of additional land. Interestingly, regional data showed that the positive effect of IR traits increased as non transgenic yield decreased, indicating that yield benefits from insect-resistance traits are higher when insect infestation is higher, or insect control is less efficient; this may be the case of less developed countries [23].

Reduced risk of cultivation related to heavy insect damage or weed infestation played a role in attracting small farmers toward GM varieties; this is reflected by the lower yield gains recorded in developed countries (IR corn +4%, HT soybean +7%) with respect to those recorded in developing countries (IR corn +16%; IR white corn +22%; HT corn +85%, HT soybean +21%) [24].

More recently, Brookes and Barfoot [25, 26] reported yield increases up to 31% for HT soybean, 15% for HT maize, 12% for HT canola, 13% for HT sugar beet, 24% for corn borer IR maize, and 5% for rootworm IR maize.

A recent meta-analysis [27] of 147 publications on GE crop impact worldwide, gives an updated and clear picture of the effects of GE on farm economy (Fig. **4**). Average yield increases for IR and HT crops were 25 and 9%, respectively, with a mean of 21.6%; profits were raised much more than yields (68.2%), thanks to reduced costs of pesticide applications. Interestingly, the profit gains of GM crops were 60% higher in developing than in developed countries, due to higher GM yield gains and stronger pesticide cost savings in developing countries.

The general picture emerging from the data presented above (recently reinforced by a study of the US National Academies of Sciences, Engineering, and Medicine [28] is that GE is increasing yields in all the crops in which it has been deployed. Unfortunately, the most impressive results have been obtained in cotton, a non-food crop, and in two commodity crops, soybean and corn, that are largely used for animal feed and energy production rather than for direct human consumption.

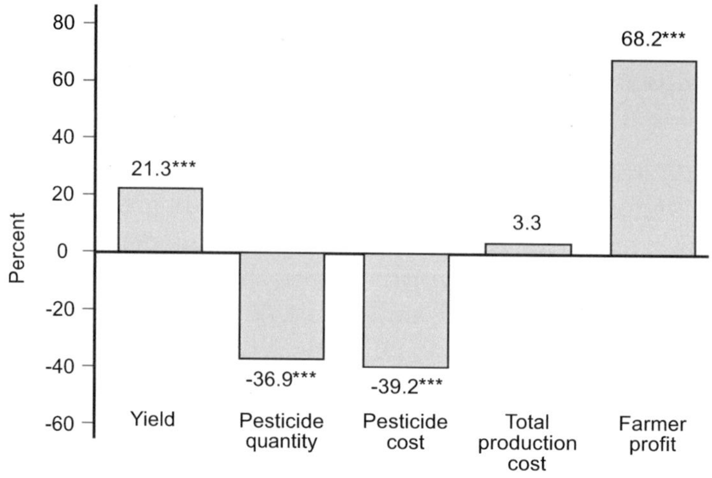

Fig. (4). Impacts of GM crop adoption. Average percentage differences between GM and non-GM crops are shown. Results refer to all GM crops, including HT and IR traits. *** indicates statistical significance at the 1% level (from Klümper *et al.*, 2014).

APPLYING CURRENT GE TECHNOLOGIES TO OTHER CROP SPECIES

The full potential of GE for food production will be realized when it is applied to important crops for direct human consumption, especially staple crops of poor countries [29]. Fortunately, this is starting to happen. Bt rice is being developed in China [1]. Cassava, a major staple crop for 700 million people worldwide is being improved by GE. The cassava brown streak disease and cassava mosaic disease are devastating pathogens of this crop. The virus resistant cassava for Africa (VIRCA) project, funded by several US foundations, aims to develop virus resistant cassava varieties especially designed for East Africa [30].

Cowpea, a staple leguminous crop in Africa, is heavily attacked by the lepidopteran *Maruca vitrata*, which causes 300 million USD in annual losses to small-scale farmers in Africa. This insect pest is controlled by expensive pesticides, which cost Nigeria about 500 million USD a year. IR resistant cowpea expressing the *Bt Cry1Ab* protein was developed at Ahmadu Bello University in Zaria, Nigeria and is being tested in field trials [18].

Xanthomonas wilt resistant banana is in an advanced field trial phase in Uganda [31]. It can be an effective solution to tackle a disease that is threatening banana production, the livelihoods of smallholder growers and food security in East and Central Africa, where banana is consumed daily in large quantities by the local population.

Eggplant, a staple food in the Indian subcontinent, is another example: it is heavily attacked by insect pests that cause up to 70% yield losses and require numerous and polluting pesticide applications. IR eggplant was authorized for cultivation in Bangladesh in 2014, and holds the promise to raise yields and reducing insecticide sprays: it was estimated that Bt varieties could increase yields by 33% while reducing pesticide use by up to 80% [29]. The same is expected for India, where Bt eggplant authorization is pending.

New IR strategies are being developed [32]; this is useful to reduce the risk of development of populations of insect pests resistant to Bt toxins.

Nichterlein *et al.* (2013) [33] reported about field trials of HR sugar beet in Europe: they found that yields were 4-18% higher than with conventional weed control systems, and were obtained with less herbicide treatments and lower dosage of active ingredient (glyphosate). This indicated that both yield and the environmental impact of sugar beet cultivation may be improved in Europe by adopting HR varieties.

It appears that GE for virus resistance has not yet been fully exploited. Papaya and squash are the only crops for which virus resistant transgenic varieties are presently on the market, but the powerful RNA silencing technology (for example [34],), as well as new strategies employing plant genes [35] have not yet reached the farmers' field. Transgenic common bean resistant to golden mosaic virus (BGMV) obtained by public research was approved for commercialization in 2011 in Brasil.

It is necessary to point out that when introducing resistance to pathogens and pests, any solution, conventional or biotechnological, is never the definitive one, due to the evolution of resistance, which can be delayed by a smart use of the technologies, but not prevented. Therefore, the biotech approach should not be regarded as a panacea, but as powerful tool in the ever continuing struggle to protect crops.

In a sense, nutritional enhancement can be regarded as a way to increase yield, if a lower quantity of a food is necessary to reach the minimum daily intake of an essential nutrient. Golden rice [36, 37] will soon be released for cultivation, after many years in the pipeline, and promises to bring significant health improvement in South East Asia by combating vitamin A deficiency. Enrichment in mineral content (iron and zinc) will further add nutritional value to golden rice. Enrichment of cassava with beta carotene and iron is underway to address nutritional deficiencies, especially in East African countries [38].

NEW GE TRAITS

Traits capable to contribute significantly to raising crop yield in the coming decades are in the pipeline. Some of them will probably be in the field soon; others are still in the phase of proof of concept. I will briefly mention the main new GE traits under development, which could significantly increase yields. Other chapters in this book deal with some of these traits more thoroughly.

Tolerance To Problem Soils: Acid And Saline Soils

Acid soils are estimated to add up to 35% of arable land, and are especially widespread in sub-Saharan Africa and Asia [6]. Most crop plants do not grow and produce well on acid soils due to Al^{3+} toxicity at the root tip, and many research efforts are made to understand the molecular basis of acid soil toxicity and tolerance [39]. Based on the observation that many acid-tolerant plants respond to low pH by overproduction and/or increased efflux of citrate and malate at the root tip, the first attempts to engineer acid soil tolerance in crop plants were by overexpressing a bacterial citrate synthase gene or an alfalfa malate dehydrogenase gene (reviewed by [6]). These experiments allowed to obtain appreciable Al tolerance, but not in all cases. More significant progress is expected from the study and manipulation of plant genes involved in transport of organic acids, namely aluminium-activated malate transporter (ALMT) genes and the multidrug and toxic compound exudation (MATE) genes. If successful in the field, GE acid-tolerant crops can boost global food production.

Soil salinity is a major global constraint for agriculture, affecting about 20% of irrigated land [40]. The increased salinity of arable land is expected to result in up to 50% land loss due to irreversible soil degradation and desertification by the middle of the twenty-first century [41, 42]. Different strategies to engineer salt tolerance have been adopted (partly overlapping with those addressing drought, an often concomitant stress, see below): limiting the uptake of salt ions by roots, increasing the active extrusion of salt ions, compartmentalizing of salt ions in the vacuole, modifying salt stress signaling pathways by engineering transcription factors. Genes encoding osmoprotectants have also been the targets of GE experiments. The role of rhizosphere and symbiotic microorganisms has been implicated, adding another level to the complexity of the trait. Despite much effort, justified by the huge potential positive impact, salt-tolerant crops have not yet been obtained, due to the complexity of the salt stress response mechanism [42].

Drought Tolerance And Water Use Efficiency

GE approaches to improve drought stress tolerance in plants follow two strategies: the first is overexpressing structural genes involved in cellular stress response, such as enzymes or chaperones; the second is manipulating the expression of regulatory genes that activate multiple mechanisms of stress tolerance [43, 44].

Several examples of the former approach can be made. The first GE drought tolerant technology on the market is the Monsanto *DroughtGard*™ maize, with event MON 87460, planted for the first time in the US on 50,000 hectares in 2013 and 275,000 hectares in 2014. It appears to give the best results in the presence of moderate levels of drought. This transgenic corn expresses the cold shock protein gene CspB from *Bacillus subtilis*, encoding a molecular chaperon [45, 46]. This technology is used in the not-for-profit project Water Efficient Maize for Africa (WEMA), a public-private partnership designed to deliver the first GE drought tolerant maize to African countries (South Africa, Kenya, Uganda, Tanzania and Mozambique) starting 2017 [1].

Drought tolerant sugarcane expressing the *betA* gene from *Rhizobium meliloti* is expected to be commercially released in 2015 in Indonesia [1]. This gene encodes choline dehydrogenase, an enzyme that converts choline to betaine aldehyde, an osmoprotectant.

Canola has been engineered to downregulate the α or β subunits of farnesyl transferase, an enzyme involved in the regulation of the signaling of abscisic acid (ABA), a growth regulator that plays a fundamental role in drought response through stomata opening control. Down-regulation of farnesyl transferase by means of RNA interference and the *AtHPR1* shoot specific promoter increased ABA sensitivity and drought tolerance [47]. This approach is patented as "Yield Protection Technology".

It has been argued that constitutive expression of stress resistance genes could be a reason for their scarce success. In fact, in the absence of stress, the constitutive expression of the transgene can be a considerable metabolic cost. On the contrary, targeting transcription to the developmental phases and tissue or organ where stress exerts its negative effects can be key to success [44]. The previous case of farnesyl transferase serves as an example of the advantages of using of a tissue-specific promoter.

The latter approach based on transcription factors is more recent. Many transcriptional regulators are known to be involved in plant responses to drought or high salinity stress, and their up- or down-regulation has been shown to affect abiotic stress tolerance. Particularly DREB (Dehydration Responsive Element Binding) proteins, belonging to AP2/ERF family, have been studied for their role in response to abiotic stresses (reviewed by [44]). In Argentina, drought stress tolerant soybean is under evaluation toward market release. It expresses a stress inducible, modified Hahb4 homeodomain-leucine zipper transcription factor from sunflower, which decreases sensitivity to ethylene. Transgenic plants maintain metabolic activity longer under stress, with 7-14% reported yield increases in field trials [48].

A new approach to make crops more drought tolerant could be engineering a receptor of abscisic acid, a regulator of drought stress responses. In an experiment conducted in the model plant *Arabidopsis thaliana*, the PYR1 ABA receptor was engineered to make it responsive to an available agrochemical [49]. Transgenic plants overexpressing the engineered ABA receptors showed much higher survival rate after water deprivation. Even though if and how this translates into higher crop yield under drought stress remains to be demonstrated, this research shows that it could become possible to modulate stress resistance responses in the field by genetic engineering plant receptors so that they can be activated with agrochemicals.

Nutrient Use Efficiency

Nitrogen (N) and phosphorous (P) are the main mineral elements that limit agricultural productivity. They are also polluting elements that contribute to the environmental footprint of agriculture: improving the efficiency of crop plants in N and P utilization can improve both productivity and sustainability [7].

Promising results from transformation experiments of model and crop plants with several candidate genes for nitrogen use efficiency have been reported [7]. These include genes involved in transport, translocation, and remobilization of nitrogen, genes encoding nitrate reductase, nitrite reductase, glutamine-2-oxoglutarate aminotransferase, glutamine synthetase, and genes encoding transcription factors and other regulatory genes.

Possibly closer to field use are GE plants capable to better utilize P. P fertilization is usually necessary to obtain high yields. However, P sources are becoming limited and the cost of P fertilization is increasing [50, 51]. On the other hand, a large amount of the P administered with fertilization is not used by the plants and contributes to water pollution.

One strategy for better P utilization is solubilization of calcium phosphates in alkaline soils by means of organic acid secretion from roots [6, 7], the same mechanism that may help to obtain tolerance to acid soils. Another strategy is overexpression of phytase and phosphatase genes in roots to release P from its organic forms that plants cannot uptake. Phytase expression was realized in several plant species that were capable of growing in media containing phytate as the only P source [52, 53]. More recently, alfalfa, a major forage crop, has been engineered with phytase and phosphatase genes from the related species

Medicago truncatula: the roots of these transgenic plants secreted the enzymes and were more efficient in P utilization, especially when grown in P-poor soils [54].

A very interesting approach to the P nutrition problem can stem from the finding that phosphite, a P form that plants cannot metabolize, can be used as a P source by overexpressing a bacterial *ptxD* gene encoding a phosphite-specific oxidoreductase that oxidizes phosphite to orthophosphate [7]. Since phosphite inhibits plant growth, it may be used as both a P source and a broad spectrum herbicide for *ptxD*-transgenic crops.

Another way to improve P utilization is addressing the poor assimilation of P in feed by animals. In fact, P is mostly present in animal feed in the form of phytate, which accounts for 60%-80% of the total P in seeds (corn, soybean). Phytate not metabolized by monogastric animals, and P supplementation of feed is required for optimal animal growth. Consequently, large amounts of P are wasted in manure, a major cause of water pollution. Expression of phytase in seeds is a promising strategy to improve animal nutrition while reducing P pollution. Canola and soybean plants expressing phytases showed phytate-hydrolysing capacity in seeds [52]. Fungal phytase was also expressed in alfalfa [55] and maize grains [56].

Photosynthesis

The idea of boosting photosynthesis in plants to enhance carbon fixation and crop yield has long been pursued as a way to increase crop yields (reviewed by [57]). Plant photosynthesis converts only 2% of the intercepted solar radiation into biomass, corresponding to one-fifth of the theoretical efficiency of 10% for C3 crops (wheat, rice) or 13% for C4 crops (maize, sorghum). Unfortunately, there is no genetic variation to be exploited for breeding due to high conservation of the systems across higher plants.

However, there has been great progress in dissecting the photosynthetic mechanism from light capture by pigment molecules to production of storage carbohydrates, including the knowledge of the genes coding for the key components and the structures of the key proteins. In addition, the use of *in silico* modeling of simulation and optimization has indicated which are the candidate molecules to target by genetic engineering. The promise of this approach was demonstrated by the identification of the enzyme sedoheptulose-1:7-bisphosphatase as a key player: indeed, its overexpression in tobacco increased photosynthesis and yield in the field [58].

Long *et al.* (2015) [57] list ten possible, non alternative ways to improve photosynthetic efficiency by manipulating ten key elements of the process. Models estimate that successful manipulation of a single of these ten elements could improve efficiency from 10 to 60%. Recently, one of these elements was manipulated, by replacing the large subunit of the Rubisco enzyme in tobacco chloroplast with the cyanobacterial one, which can be much more efficient [59].

CONCLUDING REMARKS

Sustainable intensification is an imperative if agriculture is to meet future food demand: it means increasing food production from existing farmland, while reducing the environmental footprint, so that it will be possible to continue producing food in the future [60]. Also, agriculture needs to become "climate smart". This comes down to three objectives: increasing productivity, farm incomes, food security and development; increasing adaptive capacity to climate change; decreasing greenhouse gas emissions and increasing carbon sinks [61, 62]. All these challenges can only be met by innovation in all agricultural sectors, especially by building better and better gene packages in the seeds. Plant science, plant breeding and plant GE, both public and private, are providing, and will continue to provide tools and solutions to cope with the challenges facing agriculture in the next decades.

CONFLICT OF INTEREST

The author confirms that author has no conflict of interest to declare for this publication.

ACKNOWLEDGEMENTS

I wish to acknowledge Fabio Veronesi for critical reading of this manuscript.

LIST OF ABBREVIATIONS

GE	=	Genetic engineering/engineered
IR	=	Insect resistance/resistant
HT	=	Herbicide tolerance/tolerant

REFERENCES

[1] James C. Global Status of Commercialized Biotech/GM Crops: 2014 ISAAA Brief No 49. Ithaca, NY: ISAAA 2014.

[2] Nicolia A, Manzo A, Veronesi F, Rosellini D. An overview of the last 10 years of genetically engineered crop safety research. Crit Rev Biotechnol 2014; 34(1): 77-88. [http://dx.doi.org/10.3109/07388551.2013.823595] [PMID: 24041244]

[3] Xing Y, Zhang Q. Genetic and molecular bases of rice yield. Annu Rev Plant Biol 2010; 61: 421-42.

[http://dx.doi.org/10.1146/annurev-arplant-042809-112209] [PMID: 20192739]

[4] Bommert P, Nagasawa NS, Jackson D. Quantitative variation in maize kernel row number is controlled by the FASCIATED EAR2 locus. Nat Genet 2013; 45(3): 334-7. [http://dx.doi.org/10.1038/ng.2534] [PMID: 23377180]

[5] Monsanto. "Petition for the Determination of Nonregulated Status for Increased Ear Biomass MON 87403 Maize." Submitted by Cordts, John M, Registration Manager Monsanto Company St Louis, MO 2014.

[6] Ryan PR, Tyerman SD, Sasaki T, *et al.* The identification of aluminium-resistance genes provides opportunities for enhancing crop production on acid soils. J Exp Bot 2011; 62(1): 9-20. [http://dx.doi.org/10.1093/jxb/erq272] [PMID: 20847099]

[7] López-Arredondo DL, Leyva-González MA, Alatorre-Cobos F, Herrera-Estrella L. Biotechnology of nutrient uptake and assimilation in plants. Int J Dev Biol 2013; 57(6-8): 595-610. [http://dx.doi.org/10.1387/ijdb.130268lh] [PMID: 24166442]

[8] Corrado G, Karali M. Inducible gene expression systems and plant biotechnology. Biotechnol Adv 2009; 27(6): 733-43. [http://dx.doi.org/10.1016/j.biotechadv.2009.05.006] [PMID: 19460424]

[9] Rosellini D. Selectable Markers and Reporter Genes: A Well Furnished Toolbox for Plant Science and Genetic Engineering. Crit Rev Plant Sci 2012; 31: 401-53. [http://dx.doi.org/10.1080/07352689.2012.683373]

[10] Breyer D, Kopertekh L, Reheul D. Alternatives to Antibiotic Resistance Marker Genes for *In Vitro* Selection of Genetically Modified Plants – Scientific Developments, Current Use, Operational Access and Biosafety Considerations. Crit Rev Plant Sci 2014; 33: 286-330. [http://dx.doi.org/10.1080/07352689.2013.870422]

[11] Lorenzetti F, Ceccarelli S, Rosellini D, Veronesi F. Genetica Agraria. Bologna: Patron 2011.

[12] Broothaerts W, Mitchell HJ, Weir B, *et al.* Gene transfer to plants by diverse species of bacteria. Nature 2005; 433(7026): 629-33. [http://dx.doi.org/10.1038/nature03309] [PMID: 15703747]

[13] Wendt T, Doohan F, Mullins E. Production of Phytophthora infestans-resistant potato (*Solanum tuberosum*) utilising Ensifer adhaerens OV14. Transgenic Res 2012; 21(3): 567-78. [http://dx.doi.org/10.1007/s11248-011-9553-3] [PMID: 21912851]

[14] Bock R. Engineering plastid genomes: methods, tools, and applications in basic research and biotechnology. Annu Rev Plant Biol 2015; 66: 211-41. [http://dx.doi.org/10.1146/annurev-arplant-050213-040212] [PMID: 25494465]

[15] Lusser M, Parisi C, Plan D, Rodríguez-Cerezo E. Deployment of new biotechnologies in plant breeding. Nat Biotechnol 2012; 30(3): 231-9. [http://dx.doi.org/10.1038/nbt.2142] [PMID: 22398616]

[16] Schouten HJ, Jacobsen E. Cisgenesis and intragenesis, sisters in innovative plant breeding. Trends Plant Sci 2008; 13(6): 260-1. [http://dx.doi.org/10.1016/j.tplants.2008.04.005] [PMID: 18486525]

[17] Oerke EC. Crop losses to pests. J Agric Sci 2006; 144: 31-43. [http://dx.doi.org/10.1017/S0021859605005708]

[18] Juma C. Preventing hunger: biotechnology is key. Nature 2011; 479(7374): 471-2. [http://dx.doi.org/10.1038/479471a] [PMID: 22113675]

[19] Brookes G, Barfoot P. GM Crops: The Global Economic and Environmental Impact: The First Nine Years. Ag Bio Forum. 8: 187-96.

[20] Brookes G, Barfoot P. GM crops: global socio-economic and environmental impacts 1996-2007. Dorchester, UK: PG economics 2009.

[21] Troyer AF. Adaptedness and heterosis in corn and mule hybrids. Crop Sci 2006; 46: 528-43. [http://dx.doi.org/10.2135/cropsci2005.0065]

[22] Parrott WA. Response to "Failure to yield" by Doug Gurian-Sherman. Union of Concerned Scientists 2010.

[23] Edgerton MD, Fridgen J, Anderson JR Jr, *et al.* Transgenic insect resistance traits increase corn yield and yield stability. Nat Biotechnol 2012; 30(6): 493-6. [http://dx.doi.org/10.1038/nbt.2259] [PMID: 22678382]

[24] Carpenter JE. Peer-reviewed surveys indicate positive impact of commercialized GM crops. Nat Biotechnol 2010; 28(4): 319-21. [http://dx.doi.org/10.1038/nbt0410-319] [PMID: 20379171]

[25] Brookes G, Barfoot P. Economic impact of GM crops: the global income and production effects 19962012. GM Crops Food 2014; 5(1): 65-75. [http://dx.doi.org/10.4161/gmcr.28098] [PMID: 24637520]

[26] Brookes G, Barfoot P. GM crops: global socio-economic and environmental impacts 1996-2013. Dorchester, UK: PG economics 2015.

[27] Klümper W, Qaim M. A meta-analysis of the impacts of genetically modified crops. PLoS One 2014; 9(11): e111629. [http://dx.doi.org/10.1371/journal.pone.0111629] [PMID: 25365303]

[28] National Academies of Sciences, Engineering, and Medicine Genetically Engineered Crops: Experiences and Prospects. Washington, D.C.: National Academies Press 2016.

[29] Yuan D, Bassie L, Sabalza M, *et al.* The potential impact of plant biotechnology on the Millennium Development Goals. Plant Cell Rep 2011; 30(3): 249-65. [http://dx.doi.org/10.1007/s00299-010-0987-5] [PMID: 21249369]

[30] Taylor NJ, Halsey M, Gaitán-Solís E, *et al.* The VIRCA Project: virus resistant cassava for Africa. GM Crops Food 2012; 3(2): 93-103. [http://dx.doi.org/10.4161/gmcr.19144] [PMID: 22572842]

[31] Tripathi L, Tripathi JN, Kiggundu A, Korie S, Shotkoski F, Tushemereirwe WK. Field trial of Xanthomonas wilt disease-resistant bananas in East Africa. Nat Biotechnol 2014; 32(9): 868-70. [http://dx.doi.org/10.1038/nbt.3007] [PMID: 25203031]

[32] Christou P, Capell T, Kohli A, Gatehouse JA, Gatehouse AM. Recent developments and future prospects in insect pest control in transgenic crops. Trends Plant Sci 2006; 11(6): 302-8. [http://dx.doi.org/10.1016/j.tplants.2006.04.001] [PMID: 16690346]

[33] Nichterlein H, Matzk A, Kordas L, Kraus J, Stibbe C. Yield of glyphosate-resistant sugar beets and efficiency of weed management systems with glyphosate and conventional herbicides under German and Polish crop production. Transgenic Res 2013; 22(4): 725-36. [http://dx.doi.org/10.1007/s11248-012-9678-z] [PMID: 23271374]

[34] Aragão FJ, Faria JC. First transgenic geminivirus-resistant plant in the field. Nat Biotechnol 2009; 27(12): 1086-8. [http://dx.doi.org/10.1038/nbt1209-1086] [PMID: 20010580]

[35] Cavatorta J, Perez KW, Gray SM, Van Eck J, Yeam I, Jahn M. Engineering virus resistance using a modified potato gene. Plant Biotechnol J 2011; 9(9): 1014-21. [http://dx.doi.org/10.1111/j.1467-7652.2011.00622.x] [PMID: 21668622]

[36] Ye X, Al-Babili S, Klöti A, *et al.* Engineering the provitamin A (β-carotene) biosynthetic pathway into (carotenoid-free) rice endosperm. Science 2000; 287(5451): 303-5. [http://dx.doi.org/10.1126/science.287.5451.303] [PMID: 10634784]

[37] Paine JA, Shipton CA, Chaggar S, *et al.* Improving the nutritional value of Golden Rice through increased pro-vitamin A content. Nat Biotechnol 2005; 23(4): 482-7.

[http://dx.doi.org/10.1038/nbt1082] [PMID: 15793573]

[38] Welsch R, Arango J, Bär C, *et al.* Provitamin A accumulation in cassava (Manihot esculenta) roots driven by a single nucleotide polymorphism in a phytoene synthase gene. Plant Cell 2010; 22(10): 3348-56.
[http://dx.doi.org/10.1105/tpc.110.077560] [PMID: 20889914]

[39] Kochian LV, Piñeros MA, Liu J, Magalhaes JV. Plant Adaptation to Acid Soils: The Molecular Basis for Crop Aluminum Resistance. Annu Rev Plant Biol 2015; 66: 571-98.
[http://dx.doi.org/10.1146/annurev-arplant-043014-114822] [PMID: 25621514]

[40] Tuteja N. Mechanisms of High Salinity Tolerance in Plants. In: Sies DH, , Eds. Methods in Enzymology. Academic Press 2007.
[http://dx.doi.org/10.1016/S0076-6879(07)28024-3]

[41] Mahajan S, Tuteja N. Cold, salinity and drought stresses: an overview. Arch Biochem Biophys 2005; 444(2): 139-58.
[http://dx.doi.org/10.1016/j.abb.2005.10.018] [PMID: 16309626]

[42] Cabot C, Sibole JV, Barceló J, Poschenrieder C. Lessons from crop plants struggling with salinity. Plant Sci 2014; 226: 2-13.
[http://dx.doi.org/10.1016/j.plantsci.2014.04.013] [PMID: 25113445]

[43] Deikman J, Petracek M, Heard JE. Drought tolerance through biotechnology: improving translation from the laboratory to farmers fields. Curr Opin Biotechnol 2012; 23(2): 243-50.
[http://dx.doi.org/10.1016/j.copbio.2011.11.003] [PMID: 22154468]

[44] Cominelli E, Conti L, Tonelli C, Galbiati M. Challenges and perspectives to improve crop drought and salinity tolerance. N Biotechnol 2013; 30(4): 355-61.
[http://dx.doi.org/10.1016/j.nbt.2012.11.001] [PMID: 23165101]

[45] Nelson DE, Repetti PP, Adams TR, *et al.* Plant nuclear factor Y (NF-Y) B subunits confer drought tolerance and lead to improved corn yields on water-limited acres. Proc Natl Acad Sci USA 2007; 104(42): 16450-5.
[http://dx.doi.org/10.1073/pnas.0707193104] [PMID: 17923671]

[46] Castiglioni P, Warner D, Bensen RJ, *et al.* Bacterial RNA chaperones confer abiotic stress tolerance in plants and improved grain yield in maize under water-limited conditions. Plant Physiol 2008; 147(2): 446-55.
[http://dx.doi.org/10.1104/pp.108.118828] [PMID: 18524876]

[47] Wang Y, Beaith M, Chalifoux M, *et al.* Shoot-specific down-regulation of protein farnesyltransferase (alpha-subunit) for yield protection against drought in canola. Mol Plant 2009; 2(1): 191-200.
[http://dx.doi.org/10.1093/mp/ssn088] [PMID: 19529821]

[48] Waltz E. First stress-tolerant soybean gets go-ahead in Argentina. Nat Biotechnol 2015; 33(7): 682.
[http://dx.doi.org/10.1038/nbt0715-682] [PMID: 26153997]

[49] Park SY, Peterson FC, Mosquna A, Yao J, Volkman BF, Cutler SR. Agrochemical control of plant water use using engineered abscisic acid receptors. Nature 2015; 520(7548): 545-8.
[http://dx.doi.org/10.1038/nature14123] [PMID: 25652827]

[50] Cordell D, Drangert JO, White S. The story of phosphorus: global food security and food for thought. Glob Environ Change 2009; 19: 292-305.
[http://dx.doi.org/10.1016/j.gloenvcha.2008.10.009]

[51] Gilbert N. Environment: The disappearing nutrient. Nature 2009; 461(7265): 716-8.
[http://dx.doi.org/10.1038/461716a] [PMID: 19812648]

[52] Brinch-Pedersen H, Sørensen LD, Holm PB. Engineering crop plants: getting a handle on phosphate. Trends Plant Sci 2002; 7(3): 118-25.
[http://dx.doi.org/10.1016/S1360-1385(01)02222-1] [PMID: 11906835]

[53] George TS, Simpson RJ, Hadobas PA, Richardson AE. Expression of a fungal phytase gene in *Nicotiana tabacum* improves phosphorus nutrition of plants grown in amended soils. Plant Biotechnol J 2005; 3(1): 129-40. [http://dx.doi.org/10.1111/j.1467-7652.2004.00116.x] [PMID: 17168905]

[54] Ma XF, Tudor S, Butler T, *et al.* Transgenic expression of phytase and acid phosphatase genes in alfalfa (Medicagosativa) leads to improved phosphate uptake in natural soils. Mol Breed 2012; 30(1): 377-91. [http://dx.doi.org/10.1007/s11032-011-9628-0] [PMID: 22707914]

[55] Austin-Phillips S, Ziegelhoffer T. The Production of Value-Added Proteins in Transgenic Alfalfa. In: Spangenberg G, Ed. Molecular Breeding of Forage Crops. Springer Netherlands 2001; pp. 285-301. [http://dx.doi.org/10.1007/978-94-015-9700-5_18]

[56] Chen R, Xue G, Chen P, *et al.* Transgenic maize plants expressing a fungal phytase gene. Transgenic Res 2008; 17(4): 633-43. [http://dx.doi.org/10.1007/s11248-007-9138-3] [PMID: 17932782]

[57] Long SP, Marshall-Colon A, Zhu XG. Meeting the global food demand of the future by engineering crop photosynthesis and yield potential. Cell 2015; 161(1): 56-66. [http://dx.doi.org/10.1016/j.cell.2015.03.019] [PMID: 25815985]

[58] Rosenthal DM, Locke AM, Khozaei M, Raines CA, Long SP, Ort DR. Over-expressing the C(3) photosynthesis cycle enzyme Sedoheptulose-17 Bisphosphatase improves photosynthetic carbon gain and yield under fully open air CO(2) fumigation (FACE). BMC Plant Biol 2011; 11: 123. [http://dx.doi.org/10.1186/1471-2229-11-123] [PMID: 21884586]

[59] Lin MT, Occhialini A, Andralojc PJ, Parry MA, Hanson MR. A faster Rubisco with potential to increase photosynthesis in crops. Nature 2014; 513(7519): 547-50. [http://dx.doi.org/10.1038/nature13776] [PMID: 25231869]

[60] Garnett T, Appleby MC, Balmford A, *et al.* Agriculture. Sustainable intensification in agriculture: premises and policies. Science 2013; 341(6141): 33-4. [http://dx.doi.org/10.1126/science.1234485] [PMID: 23828927]

[61] Food and Agriculture Organization of the United Nations. Climate Smart Agriculture Sourcebook. FAO 2013.

[62] Vermeulen SJ, Campbell BM, Ingram JS. Climate change and food systems. Annu Rev Environ Resour 2012; 37: 195. [http://dx.doi.org/10.1146/annurev-environ-020411-130608]

CHAPTER 15

Plant Breeding and Next Generation Sequencing (NGS)

Carlo Pozzi*

Department of Agricultural and Environmental Sciences – Production, Landscapes, Agroenergy, Università degli Studi di Milano, Milano, Italy

Abstract: Next generation sequencing (NGS) refers to a set of technologies based on massive parallel sequencing. NGS technologies can be included in three main groups: sequencing by synthesis, by ligation, and single-molecule sequencing. NGS technologies are improving at a fast pace and the cost of sequencing per base is decreasing, thus allowing to use NGS to approach a variety of biological questions as impossible before. In plants, tools based on genome decoding, building on NGS, contribute to increase the rate of genetic gain during selection and the precision when choosing superior genotypes. The tools used to facilitate the process leading to genomes drafts is described. The chapter also presents new methods, such as genomic selection or genome-wide association mapping, which are based on NGS to disclose an unprecedented amount of genetic variability made available to plant breeding. In addition, the use of NGS to decode the epigenome and the transcriptome is reported, as well as its power, when combined with appropriate genetic designs, to map and clone quantitative trait loci. It is out of scope of this review to provide a comprehensive repertory of sequencing efforts and techniques that are rapidly evolving and quickly becoming obsolete. The paper, moreover, does not provide a comprehensive list of all the too numerous experiments conducted in the field, but rather it describes rationales and examples of possible applications of plant genome sequencing.

Keywords: Breeding, Breeding by genotyping, Crop improvement, Genome sequencing, Genome wide association studies, Genomic selection, Next generation sequencing, Plant breeding, Pyrosequencing, QTL-seq, RNA-seq.

SEQUENCING TECHNIQUES

In the past few years, sequencing techniques witnessed an impressive increase of technical advancement. Improvement in sequencing contributed to the so-called "next generation sequencing" (NGS) protocols based in most cases on massive

* **Corresponding author Carlo Pozzi**: Department of Agricultural and Environmental Sciences – Production, Landscapes, Agroenergy, Università degli Studi di Milano, Milano, Italy; Tel: 02 503 16568; Fax: +39.02.50315057; E-mail: carlo.pozzi@unimi.it

Roberto Pilu and Giuseppe Gavazzi (Eds.)

parallel sequencing of single isolated DNA molecules. Technological advances are mirrored by a diminishing cost per base sequenced: the first human genome was fully sequenced in 2003 at the cost of USD 300 million, and involving dozens of corporations and institutions. In 2007, the genome of two persons (C. Venter and J. Watson) was sequenced at USD 1 million each. A year later, the genome of a human female was sequenced in 1 month at roughly USD 60,000. The target, also valid for plant genomes, is to cut every 1.9 years [1] half of the cost of sequencing of a single DNA base pair.

The output of the sequencing efforts are DNA reads, which are available in different formats:

1. Twenty-250 bp sequences, ready to be attributed to a single location in the genome;
2. Sequences originating from ends of DNA templates that have been circularized so that distant ends are physically ligated and read together ("mate-paired" fragments);
3. Paired-end, when the sequences originate from each end of a DNA template.

One of the current limits of NGS is the reading length. This is hindering the possibility of an accurate assembling the highly repetitive plant genomes. Sanger sequencing takes care of small repeats, at least of those shorter than the 700 bp. Moreover, sequencing of clones ranging in size from two to 150 kb covers even long repeats, also common in plants. Methods to capture more structural information useful in genome assembling using NGS technologies are being developed. In the following section, we will shortly review sequencing techniques. For more detailed reviews the readers may consult [2 - 4].

Sanger Sequencing

The Sanger technique has been used for more than 30 years, including its adoption in the sequencing of the first decoded plant genomes. The protocol is based on the cloning of DNA fragments in bacterial vectors (such as bacterial artificial chromosomes, BAC) and, following an amplification step, on direct sequencing of shorter fragments. Dideoxynucleotides (ddNTPs) are mixed with DNA resident dNTPs. ddNTPs contain a hydrogen group on the 3'-carbon, instead of a hydroxyl group. The incorporated ddNTP prevents the addition of further nucleotides and the DNA chain terminates. ddNTP can be labeled with different fluorescent dyes, thus allowing the detection of all terminated sequence products through electrophoresis based on size separation. The reactions need to be separated by size of products using electrophoresis, which can be done on capillary sequencers.

This protocol produces high quality read length of more than 1 kb [5]. The drawbacks of the technique include the laborious separation step and the need to prepare clonal populations of DNA in *E. coli*, a procedure that is costly and not adequate for large-scale operations.

The Arabidopsis [6] and the rice [7] genomes, completed respectively in 2000 and 2004, were produced using the Sanger sequencing approach.

Next Generation Sequencing (NGS)

The term next generation sequencing (NGS) relates to technologies other than the Sanger one. NGS used at first a massive parallel sequencing where hundreds of millions of reactions are detected per instrument run. The protocol eliminates the need for the bacterial cloning step while introducing the amplification of single DNA molecules and their subsequent massive and parallel analysis. In a second generation of NGS, millions of single-stranded DNA molecules are immobilized on a solid surface such as a glass slide or beads. In the third generation, the length of DNA fragments to be sequenced is much larger and the running time and costs are reduced due to multiplexing. More in details, the second generation requires the amplification of the template molecules prior to sequencing, while the third generation relies on sequencing directly molecules of DNA extracted from plant tissues.

Currently, six NGS technologies are in use and a seventh is in development. Most platforms require a short template DNA (<1000 bp) and each template contains forward and reverse primer binding sites (*i.e.* a library of templates is needed) [8]. Steps in common to all NGS platforms are DNA shearing, ends polishing, and custom DNA linkers (adapters) ligation. The fragments of the library are then amplified on a solid support that provides DNA sequences hybridizing to the library adapters. The amplification step can occur either by emulsion PCR, *in situ* polonies or bridge PCR [2]. The PCR steps introduce several potential risks of inaccuracy: PCR fragments can be preferentially amplified, or false positive artifacts derive from polymerase errors. Artifacts due to unbalanced G/C content are particularly pronounced in the case of bridge amplification.

The detection of the nucleotide incorporated by each amplified library fragment set occurs step-by-step and is mediated by a polymerase or a ligase. The reaction is monitored by image acquisition, typically of fluorescence signals.

Based on the type of molecular representation, NGS technologies can be divided in two groups:

1. Amplification-dependent. Needs the amplification of the target by PCR;

2. Amplification–independent. In this type of NGS, sequencing occurs directly on single DNA molecules, following methods referred to as "third generation sequencing".

The following describes the most used commercial options. Fig. (**1**) summarizes some of the most relevant aspects of the technologies described.

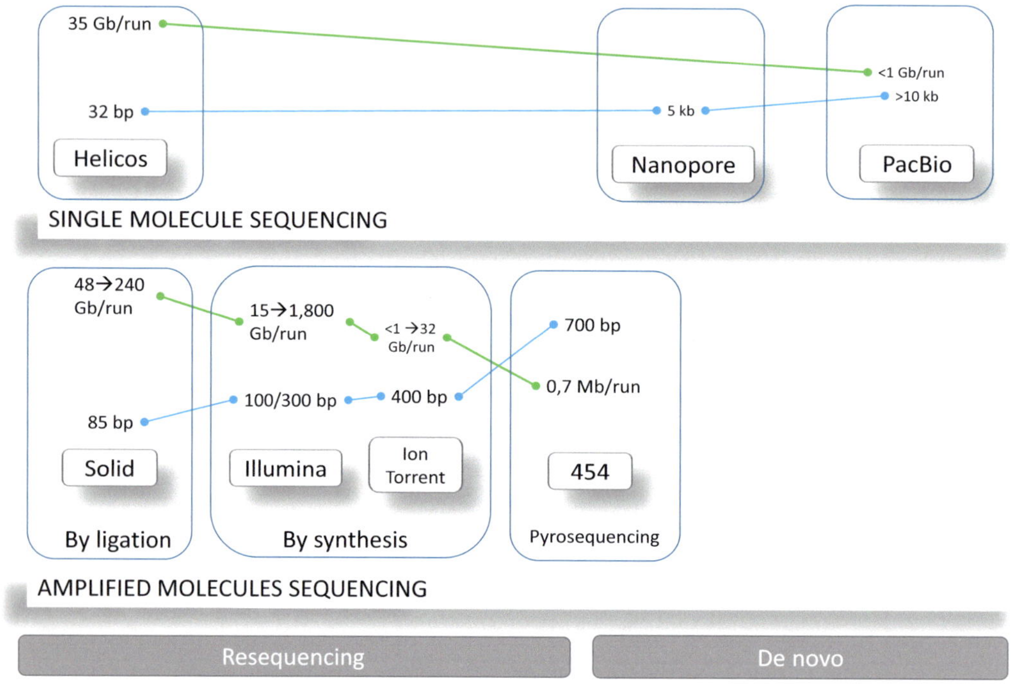

Fig. (1). NGS technologies can be grouped based on the scope of the experiment (resequencing and de novo), or based on the target of the sequencing reactions (single *vs.* amplified molecules). The scheme depicts several techniques, distributed based on output (bp/run) and read length (bp).

454

454 sequencing is a member of the first group [9]. It is based on polony (for polymerase colony) amplification and pyrosequencing. Polony reactions take place on DNA molecules adhering to micro-beads, where each bead captures only one DNA molecule. In pyrosequencing, each cycle consists of the introduction of a single nucleotide species, followed by addition of substrates (luciferin and adenosine 5′-phosphosulphate) to drive light production where polymerase-driven incorporation of the specific nucleotide occurs. The clusters are excited by a light source and a characteristic signal is emitted for each added nucleotide [10]. An apyrase wash follows to remove unincorporated nucleotides. The company 454, now part of Roche, commercialized the first commercial platform based on this principle. The hardware is composed of a picotitre plate of millions wells, where

each microwell holds only a single DNA-coated bead. Parallel sequencing of every bead occurs by flowing pyrosequencing reagents across the plate.

Illumina

The Illumina platform was formerly known as Solexa [11]. The protocol includes a step of DNA shredding and immobilization on a solid glass surface. A solid phase amplification (bridge PCR) follows which produces small clusters of identical molecules. Sequencing begins with the extension of each molecule in the cluster starting with sequencing primers [12]. In each sequencing cycle, four uniquely tagged nucleotides are added to the solution and one is incorporated based on the sequence of the template cluster. Each dNTP is a terminator, and further extension (a process called "sequencing-by-synthesis") entails the cleavage of the dye and of the terminator to allow incorporation of the next base. The platform is able to overcome the problems connected with short DNA tandem repeats affecting the 454 technology, but the reads are short (up to 300 bp on a MiSeq platform, with an average of 150 bp). This method is currently optimized for re-sequencing approaches. The strong point of this platform is the low cost/base.

SOLiD

In this group-1 platform, owned by Applied Biosystems/Life Technologies, sequencing occurs by oligonucleotide ligation and detection. The samples derive from libraries of fragments or mate-paired libraries. In both cases, adapters are added to fragments prior to emulsion PCR on beads. Bar coding allows reaching high paralleled sequencing on the same physical support. Sequencing occurs by measuring dye-labeled oligonucleotides serially ligated to a universal sequencing primer by a ligase. The offset binding of successive universal primers allows for a dual measurement of each base. Reduced read lengths, difficulties in emulsion PCR, and resolution of palindromic sequences are the main issues with this technique [13].

Ion Torrent

The Ion Torrent [14] platform is produced by Life Technologies and uses a strategy similar to the one of 454. The fundamental differences are: 1. the detection of H^+ ions (instead of the products of the pyrophosphatase reactions) which eliminates the need for laser, cameras or dyes; 2. the solid surface is equivalent to a commercial microchip (thus lowering the manufacturing cost). The protocol starts with the fragmentation of the DNA followed by its attachment to beads. Each bead is deposited in one of the millions of wells of a microchip. Each of the well is a phMeter and measures (even quantitatively) the voltage change occurring after the incorporation of every nucleotide (which causes the release of H^+ ions).

Helicos

This technology belongs to the second group. The HeliScope instrument is the first commercial single-molecule sequencer [15]. The peculiarity of the system is that no clonal amplification is required. The libraries of fragments are hybridized to the surface of a support and produce an array of primed single-molecule sequencing templates. Each DNA molecule hybridized to the sequencing surface generates its own sequencing reaction and this allows reaching the density of billions of molecule sequenced per run. The protocol is based on sequencing the single molecule by adding fluorescently labelled nucleotides, which recognize the sequence of the template fragment and emit fluorescence in the process. Washing and addition of further nucleotides allow the reconstruction of the template's sequence [2]. The strong point of this technology is the number of bases sequenced per hour.

PacBIo

This platform uses the real-time sequencing of individual DNA molecules, following nucleotide incorporation by DNA polymerases linked to the surface of microscopic slides. Each dNTP (actually a proprietary phospholinked nucleotide) has a unique fluorescent label that is detected prior to being cleaved off during synthesis. The camera creates a movie of the reaction occurring at each polymerase in each well. It is possible to reach the length of 5 kbp, even if the error rate can be, arguably, as high as 15% [16]. Fragments are circularized and sequenced multiple times to distinguish similar repeats or paralogues. The technology is particularly suitable to connect small bits of the genome [17] and has been applied in the *de novo* sequencing of the desiccation-tolerant grass *Oropetium thomaeum* [18].

Nanopore

Oxford Nanopore [19] is developing a platform combining the abilities of pore-forming proteins with molecule-based sequencing. A exonuclease let the nucleotide fall into a pore and single molecules entering the nanopore cause characteristic disruptions in the current existing between the two surfaces of the support. Every nucleotide causes a specific disruption that allows its identification. As a DNA strand enters the pore one nucleotide at the time, the sequence of the filament is read in real time. No reagent is needed but a high error rate is still hampering the technology. The production of *de novo* genome sequencing and assembly using only nanopore data is still a challenge. Algorithms are being developed that could facilitate the identification and correction of errors and improve the coverage depth and accuracy of this type of sequencing [20]. The technology is showing the first results in portable genome sequencer technology (MiniION) [21 - 23], *i.e.* a palm-sized gene sequencer, able to plug into the USB port of a laptop.

PLANT GENOMES COMPLEXITY AND SEQUENCING ISSUES

Genomic sequences can be classified under six quality levels: standard draft; high-quality draft; improved high-quality draft; annotation directed improvement; noncontiguous finished, and finished [24]. The quality of the sequence, and the effort needed, depend from the purpose of the sequencing project [25], and from the type of biological information and possibility of use of the sequence in further applications. Inherent errors in the sequencing technology and/or lack of software that address these errors do affect the sequence quality. The annotation software provides closure approaches for almost all gaps, misassemblies, and low-quality regions (noncontiguous finished). Draft genome sequences are used to define gene catalogs and repeat content, as well as in comparative genomics and syntenic studies [26]. The annotation directed sequences still leave out repeat-rich regions, while are useful for gene comparisons, alternative splicing analysis and pathway reconstruction. Finished sequences are credited to have less than one error per 100,000 bp and each replicon is assembled into a single contiguous sequence. Low coverage sequences cannot decipher the repetitive fraction of the genome that contains key elements for the analysis of evolutionary process. Inferences about lineage-specific features (*i.e.* taxonomically restricted sequences that share no similarity with sequences of other lineages in one taxonomic group) can only be achieved with high quality complete sequences.

The average haploid genome size in plants is around 6,000 Mbp [27]. Plant genomes pose specific challenges to sequencers, due to their high average size and complexity [28]. Size is variable depending particularly from the amount of repetitive DNA and transposable elements The Arabidopsis genome has only 10% of this type of genomic DNA, while wheat has more than 80%. This repeated DNA accounts for 82 Mbp in *Utricularia gibba* [29], 120 Mbp in Arabidopsis, and 16 Gbp in wheat. A second factor influencing genome size is the level of ploidy. Strategies used to overcome the hurdles of sequencing complex genomes are exemplified in maize, a paleopolyploid [30], where a BAC-by-BAC sequencing approach was adopted: about 160 hundreds of BACs were shotgun-sequenced, physically anchored and assembled [31]. The positioning of BACs on chromosomal macromolecules was possible thanks to the anchoring of the reference genome to a high-resolution gene map. A drawback of this approach is that the order and orientation of many contigs[1] within a given BAC remain unknown.

In plant genomes, frequently polyploidy creates the need to differentiate between homoeologous gene copies and to analyze chromosomal rearrangements and amplified repetitive DNA [32]. Consequently, the list of polyploid species whose genome is completely sequenced is limited. For example, sequencing in a doubled haploid accession allowed the assembly only of a draft sequence of the genome of *Musa acuminata* (banana) [33, 34]. Another example of polyploid sequencing is the *Brassica napus* genome (AACC genome), which is privately held by Bayer CropScience. The brassica genome was sequenced by combining the *B. rapa* (AA genome) and the *B. oleracea* (CC genome) with *B. napus*. Based on these data, it was further investigated the allopoliploidization process that originated *B. napus* [35].

In the polyploid potato (*Solanum tuberosum*) the problems connected with genome complexity were reduced by NGS sequencing an homozygous doubled-monoploid clone. Subsequent genome analysis revealed a palaeoploydization event that led to potato genome duplication [36]. A similar approach could not be adopted for wheat (*Triticum aestivum*), because the ABD genome does not correspond to the sum of the current, still present in the wild, forms ancestral to genome A (*T. Urartu*), B (probably an *Aegilops speltoides*-like genome) and D (from *A. tauschii*). The approach used by the International Wheat Sequencing Consortium was rather to sequence isolated chromosome arms [37]. A further approach was to study genome assemblies of bread wheat and five diploid relatives to estimate evolutionary relatedness and their divergence times [38].

NGS was successfully applied to decipher the paleopolyploid genome of soybean (*Glycine max*) [39]. The soybean genome derives from two ancient whole genome duplication events followed by gene diversification, gene loss, and chromosome rearrangements. The tetraploidization event was estimated to have occurred around 10-15 million years ago. Homologous regions have diverged sufficiently to allow their assembling apart from one another using a shotgun strategy [40].

In cases, sequencing led to the discovery of hidden genome duplications. For example, in poplar the assembled genome revealed a whole-genome duplication event leading to about 8,000 pairs of duplicated genes. A second older duplication event was coincident with the divergence of the *Populus* and Arabidopsis lineages [41].

A combination of Sanger and NGS sequencing led to the assembly of the coffee (*Coffea canephora*) genome, starting from a doubled-haploid accession. This achievement is considered to be preliminary to the sequencing of the tetraploid

Coffea arabica. The *C. canephora* genome does not have signs of polyploidization, but contains three paralogous regions for each of the seven ancestral chromosomes predating the genome triplication reported for the core eudicots [42]. The sequencing of the cacao (*Theobroma cacao*) genome allowed to hypothesize that eleven fusions had a role in shaping the present chromosome status of the species [43].

Most of crops species are heterozygous. The main difficulty in heterozygotes sequencing is the misleading alignment of paralogous and heterozygous regions. In these cases, the sequencing results in the assembly of haplotypes are difficult to order. For this reason, in most cases homozygous derivatives were sequenced, thus improving the mapping of the variation reported for the heterozygotes. This is the rationale followed in sequencing the highly homozygous grape (*Vitis vinifera*) cultivar of Pinot Noir [44], instead of producing the heterozygous genome, as chosen by another consortium [45]. Also the black cottonwood tree (*Populus trichocarpa*) has been sequenced starting from a highly heterozygous genotype, exploiting the simplification granted by its small genome size (485 Mb). It was sequenced using WGS integrated with a BAC-based physical map to guide the sequence assembly. The first draft still contained a significant portion of the reads unassembled [41]. A BAC-by-BAC approach was adopted for pear [46]. In other cases, the sequence was based on diploid progenitor species (*T. Urartu* for wheat [47]), or on the haploid cultivars/lines (in peach [48] and citrus [49, 50]).

NGS provides data on the distribution of mutation rates in the genome and on its association with the heterozygous state, when parents and offspring (from F1 to F2s) are sequenced and compared. This approach was used to provide evidence that mutation rates are higher in heterozygotes and occur in proximity to crossover events [51].

PLANT GENOMES SEQUENCING

The contribution of NGS platforms can be summarized as follows: 1. Whole genome sequencing, where the entire genome (including non-coding regions) is the target; 2. Exome sequencing, which is specifically addressed to the coding regions; 3. Amplicon-based sequencing, where only a sub-fraction of the genome is sequenced after PCR amplification; 4. Transcriptome sequencing, which evaluates the expressed RNA and provides a measure of gene expression.

BAC-By-BAC Sequencing

The sequence of the entire genome can be obtained following the BAC-by-BAC approach, known also as hierarchical shotgun sequencing. A physical map of the target genome is produced by using BAC clones, and a minimal tiling path (MTP) of overlapping clones is ordered along the chromosomes. Each of the MTP clones is fragmented, subcloned, and shotgun sequenced using the dideoxynucleotide chemistry. Alternatively, sequencing can occur *via* NGS. The alignment of the sequences produces contigs, which are assembled into a finished sequence. A pooling strategy of BAC clones allows the reduction of costs and is often associated with a bar-coding strategy to assign the contigs to specific BACs. A major issue in the construction of a BAC physical map is sequence redundancy typical of large and polyploid genomes [52]. On the other hand, a physical map is the tool that allows NGS sequences to be assembled in groups ("bins"). This facilitate order confirmation and provides a strong structure in ordering the data.

In the case of Arabidopsis, a BAC-by-BAC approach was adopted in the first genome-sequencing project which produced an incomplete genome sequence [6]. A later finishing of the sequence was based on WGS [53]. The case stresses the importance of developing combined sequencing strategies.

Maize has been the most important crop sequenced using a BAC-by-BAC approach, starting with the inbred line B73 [31]. In this case, after shotgun sequencing, BACs were anchored genetically and physically. The sequences were further refined by resequencing MTP BACs with Illumina technologies. Notably, a second inbred line, Mo17, was aligned to B73 using 454 sequencing, as reported in the Phytozome website.

The Barley Sequencing Consortium released a whole-genome shotgun sequence integrated within a physical and genetic framework. The MTP served as a template for hierarchical clone-by-clone map-based shotgun sequencing. The fine structure was refined and about 2/3 of the genome is represented in a tiling path. The physical map is anchored genetically using WGS from individuals of two mapping populations and BAC sequences [54, 55].

WHOLE GENOME SHOTGUN (WGS) SEQUENCING

The WGS protocol entails an initial step of genome fragmentation. The subcloning of fragments into plasmid vectors, or their direct NGS sequencing, eliminates the need of the physical map. Vectors used can also be fosmids, or BACs. The sequencing of several subclones guarantees a high coverage of the genome. The power of NGS lies in its capacity to generate a huge volume of paired-end reads from cloned DNA fragments. Its weakness is that reads are rather short. The access to various computational methods allows the assembling of the short reads into a consensus sequence. Most of recently sequenced genomes (with the relevant exception of maize) were processed using a WGS approach, because of its ease, cost-efficiency, and speed. Problems remain concerning large repeat-rich plant genomes, like those of the grasses. WGS genome sequences are usually highly fragmented, but the coverage can go up to 95%. The coverage of a WGS-sequenced genome, or at least of the "gene space"[2], is often determined by comparison with a set of EST assemblies (from the same or a different genotype) that have homology to proteins of model species, *i.e. Arabidopsis*.

Cassava is a recent example of WGS sequencing. The genome was obtained by combining a one-fold coverage of Sanger reads with 454 GS-FLX Titanium reads. Additionally, in a second approach, single-end reads[3] were assembled alone and then the paired-end reads were added. Additionally, longer reads contributed to the closing of many gaps, as reported by the Phytozome sequence repository [56].

The soybean genome was mostly (85%) assembled from WGS sequence data [40]. A step of genetic mapping was necessary to orient 95% of the scaffolds, eventually assembled into the 20 pseudomolecules corresponding to soybean chromosomes. The data reported in Phytozome [56] are correcting cases of pseudomolecule reconstruction. Soybean is currently the most complex plant genome sequenced using the WGS strategy. The genome is largely complete with respect to "gene space" and even large parts of repetitive sequences are assembled [40].

WGS draft assemblies are available for the bread wheat progenitors, *Triticum Urartu* [47] and *Aegilops tauschii* [57].

The rice genome initiated using a clone-by-clone strategy [58], while the finished sequences, produced by three groups (Monsanto, Beijing Genome Institute, and Syngenta), used a BAC-by-BAC integrated with WGS approaches and considering *Oryza indica* and *O. japonica*. The information from the draft sequences was integrated as the result of the International Rice Genome Sequencing Project and contributed to the gold standard map-based sequence [7].

A recent development that could reveal useful in overcoming the WGS problems is the Moleculo (TruSeq) process. The protocol first creates DNA fragments about 10 kb in length. Shearing and amplifying of the fragments is followed by tagging them with unique barcodes that helps in assigning them to the larger fragments from which they originated. The assembly process is thus facilitated and can take as short as two days. The protocol was tested on rice, although at present the owner, Illumina, has not released any data.

FINISHING

The first phase of large-scale sequencing projects ends with the production of the contigs. The final steps of sequencing include "cleaning" the sequence to reach the accuracy expected by the standards described in [24, 59], and contigs and scaffolds assembling. This step is known as "finishing" and entails the ordering of contigs and their linking with no ambiguities left. The finishing step is expensive, also in terms of time, and requires the right combination of sequence coverage, read length, and read quality. A recent example of finishing is provided by the genome of sorghum (*Sorghum bicolor*) [60]. The groups involved divided the gene space in 1 Mb-overlapping pieces. Each piece was manually inspected and finished using several methods: primer walks on subclones and fosmid templates, transposon sequencing on subcloned templates, shotgun sequencing and finishing of complete fosmid and BAC clones, and small insert shatter libraries (obtained by 454 and Illumina sequencing). A final step of the visual examination of subcloned paired-ends and repeat structures and validation of any remaining lower quality regions, including regions with high quality base pair discrepancies, allowed to "close" the sequences and reduce the error rate [60].

NGS contributes to an increasing number of draft sequences, and the genome produced have, frequently, several gaps. Even the Arabidopsis genome, probably the currently best assembled among plants, still has a total gap length of approximately 185,000 bp [24]. In cases, this is acceptable, as the effort required to finish the sequence is not justified by the relatively marginal increment of information that could be obtained.

Finished sequences are particularly important to understand lineage-specific features and in *de novo* sequencing studies. When resequencing is the approach, a finished sequence is necessary to map precisely the variation underlying the studied phenotypes [61]. A study used expressed sequence tags (ESTs) to evaluate the qualities of the genomes of 32 plant species: the results indicate that quality for most of the analyzed genomes is lower than previously hypothesized [62]. Very large genomes are particularly prone to be largely defective in their finishing assembly. For example, only 25% of the Norway spruce genome (approx. 20 Gb) has been assembled into scaffolds longer more than 10 Kb [63].

Various experimental designs that can facilitate a finer sequence assembly include longer reads, paired-end reads, mate-pair libraries (large insert libraries), RNA-seq data, reduced representation libraries, optical mapping and genetic mapping [64]. Radiation hybrid, in complex genomes such as *Aegilops*, has also proven to be efficient in developing dense maps useful to anchor and orient BAC contigs [65].

DE NOVO SEQUENCING

De novo sequencing refers to sequencing and assembling a plant genome without any reference genome sequence available. The powerful high-throughput of NGS allows moving beyond model species and gaining an understanding of the genomes of crops and specialty species. The complexity of plants genomes can create highly fragmented assemblies, with isolated gene islands among the background of high copy repeats, a picture further complicated by aligning properly the members of gene families. Some of the issues in aligning *de novo* sequenced genomes can be addressed by using data from closely related model or major crop species. The alignment of more than two genotypes from a single species should allow to take care of short sequences and to detect sequence variants. Alternatively, it would be possible to use whole transcriptome or reduced-representation genome sequences [66].

A combination of different strategies is adopted to compensate for problems inherent to NGS. The rise of third-generation sequencing technologies (such as PacBio) provides a solution for some of the *de novo* sequencing uncertainties. A complementary approach is the development of methods for localizing sequence position, for example using the BAC-by-BAC sequencing strategy. Optical sequencing can also be used during *de novo* genome assembly, because it produces genome-wide maps that can serve as scaffold for sequence orientation [67]. For the largest plant genomes, a hybrid approach combining short and long reads and long-range mapping technologies seems the best option.

At first, the tendency was to develop reference genomes and then resequence other cultivars based on such templates, to identify the genetic variation present in specific genotypes. It was subsequently demonstrated that cultivars may share only a fraction of the genome of the species (named the "core-genome") being each characterized by a variety-specific dispensable genome. The most prominent examples are maize and rice [68 - 70]. The pan-genome is the sum of the core and variably distributed genome, and can be assembled by comparing multiple *de novo* sequenced genomes. Thus, the paradigm shifted towards the *de novo* sequencing of several accessions, in particular for species with repetitive-rich genomes, with the purpose of capturing the whole diversity of a species. For this target, *de novo* sequencing is cost effective and of high quality, two parameters which are yet to be matched by current protocols, unlcss complexity reduction approaches are introduced.

RESEQUENCING

In resequencing approaches, reads are aligned to an available complete genome sequence, whose quality determines the precision of the resequenced genome. The resequencing of different genotypes is instrumental in discovering variants of interest to breeders, particularly when genome or EST sequences are available. Resequencing is particularly suited for identification of DNA structural variations, mutations, copy number variations, and rearrangements. Variants are identified on a genome-wide scale or by comparison to the reference genotype. Resequencing projects have the potential to identify alleles responsible for phenotypic diversity across the entire genome and the entire species. In general, resequencing projects are devoted to the discovery and analysis of germplasm genetic diversity and variation (both at DNA and RNA level), to crop evolution and to monogenic and QTL identification.

Resequencing Projects

As the "1000 genomes" is producing more and more data in humans [71], similar approaches are being conducted in plants. One example of major resequencing-based project is the "1001 genome Arabidopsis project" [72], in which sequencing of germplasm isolated throughout the Northern hemisphere is being undertaken using Solexa, 454, and SOLiD technologies [53]. The aim of the project is to discover the whole genome sequence variation of 1001 inbred lines of Arabidopsis [73, 74]. Plants are grown, phenotyped in different environments, and sequenced. The information is used in association studies at biochemical, metabolic, physiological, morphological, and whole plant fitness level. As of September 2014, over 1100 lines have been resequenced.

Rice is among the most considered crop in resequencing projects, in virtue of the availability of at least three reference genomes [75]. In one of the most recent studies, the resequencing of 104 rice varieties of *O. sativa* identified repetitive elements and recurrent copy number variation. The exercise led to the identification of over 6 million high-quality SNPs. Genotyping made possible the reconstruction of the individual haplotypes of elite varieties of United States and Latin America, along with 10 accessions of their wild progenitors (*Oryza rufipogon* and *Oryza nivara*) [76]. These data complemented other experiments [77], identified genes with lower sequence diversity in cultivated compared to wild rice: such regions are candidates for having being selected during domestication. Some variants are associated with important biological features, whereas others have yet to be functionally characterized. The International Rice Research Institute announced that the 3,000-rice genome project has successfully sequenced 3,024 rice varieties grown in 89 countries. This improves our understanding of the natural genetic variation of rice and should lead the discovery of novel genes at the basis of economically important traits [78].

Resequencing efforts are extensively conducted in maize. The scope of one of these projects is to identify the genes underlying maize evolution through the resequencing of 75 wild, landraces, and improved lines. On average, more genomic regions have been under a strong selection during domestication compared to the effects of maize breeding, suggesting that in modern breeding much space is left for the use of diversity at previously untapped loci [79]. In a second experiment, researchers resequenced 4,100 lines of maize and teosinte [80]. A substantial variation was found at the level of SNPs and of even bigger variations.

Resequencing approaches are instrumental for mapping mutations. In rice and in *Arabis alpina* a reference-free algorithm allows for the precise discovery of homozygous mutations, based on comparison of short subsequences (K-mers). Short sequences combined with syntenic information present in public databases and with *de novo* gene annotation result in a very efficient functional annotation [81]. This method of forward genetics is exportable to any species amenable to mutagenesis.

SEQUENCING BY REDUCING THE GENOMIC COMPLEXITY

A complete sequence information on the whole genome may exceeds the experimental needs. A number of methods (known as "genome partitioning", "enrichment" or "capture") are used to reduce the complexity of the sought after genome, pointing to the sequencing only of parts of the genome. When the candidate genomic region has been identified, they are sequenced and genes and mutations underlying a trait are identified. The effectiveness of methods "concentrating" on specific genomic regions is measured by several parameters: sensitivity (how many bases are represented by one or more sequence reads?); specificity (how many sequences are mapping to the target?); reproducibility (to which extent are the results reproducible in different experiments?); feasibility (cost and ease of use). Most of the target-enrichment strategies have already been described in details [82], and we will here concentrate on some examples only.

Chromosome Sorting

The reduction of genomic complexity contributes to address the complexity of polyploid genomes. This may be accomplished *via* the reduction of the full nuclear genome to its individual chromosome macromolecules [83]. The flow-sorting approach uses flow-cytometry to isolate single chromosomes [84]. The combination of NGS technology and chromosome sorting represents the most affordable approach to obtain the sequence of single chromosomes, and offers opportunities for the study of genome organization at the single-chromosome level. The DNA acquired by the application of this method has proven to be suitable for many applications, including PCR-based physical mapping, *in situ* hybridization, DNA arrays production, development of DNA markers, construction of BAC libraries, and positional cloning. Moreover, the approach is useful in comparative analyses between related species or subgenomes of the same species [85], and for the validation of whole genome assemblies.

In the case of wheat [86, 87] a full set of chromosome-specific BAC libraries are available, and a MTP has been assembled for each of them, prior to NGS sequencing. A recent assembly of shotgun sequencing of flow-sorted wheat chromosome 5D using 454 technology [88] provides a complementary dataset to the WGS sequences [37]. In wheat and other species it is possible to develop a "whole chromosome shotgun" (WCS) sequencing approach. An example of WCS is the sequencing of barley chromosome 1H [84].

Exome Sequencing

The exome is the genome including all the exons of genes. To avoid the hurdles connected with the sequencing of large and repetitive genomes, it is possible to sequence only the exome, which provides a good deal of information about coding, and even non-coding (miRNA, for example) subgenomes. The methods used for exome sequencing entail the selection of fragments from a whole genome library, by hybridizing with probes that correspond to most of the exons (known as "exome capture kit") [89]. The probe DNAs are biotinylated to allow for the streptavidin-based selection. Eventually the samples are PCR-amplified prior to sequencing. Array-based methods are similar, but the probes are deposited on high-density microarray. It is even possible to synthesize probes to target specific loci that may be of agronomical or breeding interest [89]. In fact, the exome kit can be used to identify induced mutations in the genome [90], or to spot the genetic variation potentially important for plant's phenotype. This approach was undertaken in poplar where it was proven that exome sequencing could be useful for single nucleotide polymorphisms (SNP) detection. Markers identified in this way are further used in assisted breeding of the species [91].

An exome capture kit has been designed for wheat based on transcriptome data [92]. A similar approach in barley led to the development of a kit based on the gene space assembly and it identified mutations involved in early maturation [93]. In barley, exome capture has been useful to discriminate between marker alleles derived from cultivated varieties and those from accessions of *H. bulbosum*. The target is to avoid the introgression of unwanted wild genomic fragment along with the selected agronomically relevant traits, such as pathogen tolerance [94].

Probe/hybridization/capture strategies have also been devised to sequence specific genomic regions, for example those surrounding miRNA genes [95]. In switchgrass (*Panicum vigatum*) the exome sequencing of 66 diverse populations allowed to distinguish ecotypes and ecotype-restricted SNPs predicted to be relevant in controlling flowering time and biomass production [96].

GENOMIC PROXIES AND NGS

The decision to select model species to be sequenced has considered two rationales: the genome dimension (small first) and the agricultural relevance (based on key nodes of the plant phylogenetic trees leading to agriculturally important families or tribes). The alignment of sequences of related species could be used to facilitate the identification of the order of sequence reads in species where the complete reference genome is missing [56]. If there is sufficient synteny, the gene order along chromosomal regions in a model species is a proxy of the gene order of a related specie. Extensive synteny is present in several model and crop species and in families. Significant examples are Arabidopsis and *brassicaceae* [97]; *Brachypodium*, rice, and *triticeae* [98, 99]; *Medicago truncatula* and *fabaceae*, and tomato and *solanaceae* [100]. When the rice genome was completed, further understanding of rice biology was accelerated by comparative analyses with the Arabidopsis genome; this approach is proving valid with the progressive access to more and more crop genome sequences [101]. The drawback is a bias of the assembly towards the reference genome, and the un-detection of major genome rearrangements.

In wheat, 5,400 genes were identified comparing available genomic sequences with genes from rice and sorghum, and with ESTs from barley and wheat. The "genome zipper" approach produces a sequence-based gene map of a partially sequenced species [99, 102]. The method relies on chromosome flow-sorting, repeats masking, and alignment of reads to model syntenous genomes (for example *Brachypodium*, rice, and sorghum, in the case of barley). Genes are aligned using as a scaffold an EST-based map. The result, in barley, was the localization of six out of seven centromeres, and the establishment of gene order in proximal chromosome regions [103]. A similar approach led, in ryegrass (*Lolium perenne*), to a synteny-based gene order of the genome [104]. The International Wheat Genome Sequencing Consortium compared the bread wheat [38] with five diploid relatives (*Triticum boeoticum*, *T. polonicum*, *T. macha*, *T. dicoccoides*), and found that the A and B genomes have diverged from the common ancestor about 7 million years ago. It was also clarified that the ancestral lineages of the basic genomes are the result of ancestral reciprocal hybridizations and that the current bread wheat is the result of multiple rounds of hybridization and speciation [38, 105].

Banana (*Musa spp.*) belong to a taxonomic group sister of the *Poales*. The sequencing of its genome represents a crucial step for comparative genome analysis in plants [34]. The banana genome, in fact, is the first monocotyledonous whole-genome sequence reported outside the *Poales*. One of the most striking finding deriving from the work is the discovery of non-coding sequences older than the divergence between monocots and dicots.

The complex genome of the octoploid strawberry (*Fragaria* x *ananassa*) was dissected through comparison with its wild relatives, *F. iinumae, F. nipponica, F. nubicola*, and *F. orientalis*, using Illumina and Roche 454 sequencing. The strategy led to the construction of a virtual reference genome where the heterozygous regions are eliminated during assembly [106], and which allows for the elucidation of the genome structures of the *Fragaria* species. The complex genomes of allotetraploid cotton were studied using NGS mapping methods, and this led to hypotheses concerning the relationships between the A and D subgenomes [107].

The application of NGS to wild relatives of crop species is helping in characterizing and understanding complex domesticated species, and in shedding additional light on plant evolution and adaptation. For example, *Setaria viridis* is being used as a comparative model for C4 plants [108]. *Amborella* genome is helping reconstructing the ancestral gene arrangement in eudicots [109, 110]. The shotgun sequencing of wild relatives generates reads that can be aligned against the domesticated genome, revealing regions that suffered variation bottlenecks, and eventually allowing to reintroduce some (targeted) lost diversity [111].

NGS And Phylogenetic Studies

In plants, DNA elements exist which are species-specific and which in several cases originated from horizontal DNA transfer from organelles. This complicates their usage in phylogenetic studies. The possibility to combine high-throughput sequencing platforms with multiplexing approaches makes it possible to analyze more loci and more taxa at once. Genome skimming refers to the shallow sequencing of genomic DNA, that leads to a deep sequencing of the high-copy fraction of the genome (for example the pastidial DNA) [112]. The starting material can be fresh, frozen or even herbarium specimens [113]. The approach is thus particularly suitable for taxonomy, DNA barcoding and extensive phylogenic analyses [114]. The genome skimming method is combined with target enrichment in the Hyb-Seq application [115]. The procedure is based on the sequencing of an enriched portion of a genome following the integration of genomic and transcriptomic data.

NGS: Beyond the Genome

Transcriptome Sequencing

The genome sequence *per se* may not be sufficient for breeding applications, but its informativeness increases when combined with transcriptomics data. The sequencing of the transcriptome relies on a powerful technique to reveal gene expression on a large scale, including low-abundance transcripts, and to refine the structure of the transcripts. These digital expression techniques are replacing microarrays. The high-throughput sequencing of RNA (RNA-seq) is primarily used to measure levels of genes expression in different cells or tissues, both for coding and non-coding sequences. Other RNA-seq goals are to determine start and end and splicing patterns of genes. A method based on paired-end RNA-seq was used to identify alternative splicing sites [116, 117]. Protocols are available for RNA-seq, such as cDNA-seq, strand specific RNA-seq, polyA RNA-seq, ribosomal RNA depletion RNA-seq, and small RNA-seq. This variety, while useful and target-specific, makes it difficult to compare data generated from different projects and laboratories [117, 118].

In wheat, RNA-seq was used to identify SNPs associated with resistance to rust, being diagnostic for R-genes (such as *Yr15*) to be used in marker assisted selection (MAS) [119]. SNPs, in this case, were identified in F_2s derived from nearly isogenic lines (differing for the resistance locus).

NGS platforms are instrumental for the sequencing of small RNAs and, therein, micro RNAs (miRNA) [118, 120], important components of post-transcriptional gene regulation. The goal is to obtain novel insights about genome organization and function by measuring small RNAs distribution at the chromosome level. Micro-RNA genes were studied in the context of domestication in rice, by sequencing them in three different tissues in *Oryza rufipogon.* miRNA specific for cultivated rice were found, suggesting their role in domestication [121].

RNA-seq is a valuable alternative to genome sequencing in polyploid species, where association genetics is hindered by genomic complexity. Associative transcriptomics [122] uses transcript sequences to develop breeding markers. Following this approach, an SNP map of *Brassica napus* allowed to orient sequences of *B. rapa* and *B. oleracea* (the ancestors of one of the genomes of *B. napus*) and to create a pseudomolecular representation of *B. napus*. Pseudo-molecules were then used to align unigenes assembled from mRNA-seq dataset. Eventually also transcripts sequenced from a diversity panel were aligned to the unigenes and more than 100,000 SNPs were detected. The gene sequences markers data were instrumental for association mapping experiments, in combination with gene expression markers (*i.e.* markers based on the quantification of transcript abundance).

RNA-seq has been of help in characterizing the extent of the variation in genome content between individuals within a species, along with genome resequencing exercises. For example, the maize pan-genome: 503 maize inbred lines were analyzed with RNA-seq to demonstrate that only a minor subset of transcripts were common to all lines [123].

Variation of gene expression can itself be a quantitative trait. When its inheritance is studied, loci are discovered called "expression QTL" (eQTL) [124]. An eQTL is detected when genetic variation within the genome results in changes in transcript abundance. The combination of eQTLs with traditional QTLs shed light on the role of the allelic variation revealed by genome sequencing. RNA-seq can map expression QTL (eQTL) in an experiment conducted on the progeny of two parents also analyzed. The results support the existence of genes with canonical (*i.e.* mendelian) and non-canonical (*i.e.* paramutation-like) expression patterns [125].

Epigenomics

When it comes to translate the sequence diversity between individuals into functional causal variants of quantitative variation, approaches able to consider the epigenetic regulation have to be considered. The epigenetic decorations and gene expression control need to be addressed and studied in a cell- and time-specific way. DNA methylation, RNAs and small RNAs, binding of transcription factors, nucleosome positioning and post-translational modification of histones are some of the mechanisms known to be important in gene expression regulation of plants.

In cases, epigenomic maps facilitate a more rapid identification of causal variants by providing an additional layer of information, particularly in species with a large genome. Mutants are a vital resource in this respect, and there are already numerous projects aimed at creating large mutant populations of diverse plant species using T-DNA, transposon tagging, or TILLING mutagenesis. The mutant populations, in combination with results from the Plant Encyclopedia of DNA Elements (pENCODE), are the base to identify trait-associated sequence variants [126]. ENCODE projects aim at clarifying the function of the plant epigenome [126, 127], with the goal of identifying large numbers of trait-associated sequence variants localized to regulatory DNA elements. The protocols are those devised by the plants epigenomics Consortium [127], mostly based on DNA-proteins interaction. Thus, chromatin immunoprecipitation is used in combination with NGS to identify at the genomic level the DNA- bound factors and histone tail modifications (ChiP-seq [128]), both indicative of gene silencing. The cross-linked chromatin is sequenced to identify genomic regions associated with the protein or histone tail modifications of interest. The ChIP-seq approach was used in tomato to identify loci regulated by a transcription factor involved in alleviating the stress caused by drought [129].

NGS proved to be instrumental in decoding the multiple layers of epigenetic regulation, including cytosine methylation, and small RNA transcripts. This approach was conducted in wild-type and DNA methyltransferase or demethylase defective mutants of Arabidopsis, shedding some light on the complex network of relationships between layers [130]. The protocol MeDIP-Seq used specific antibodies to immunoprecipitate methylated cytosines [131]. Further accuracy was reached with a treatment with sodium bisulfite under denaturing conditions, to convert only unmethylated cytosines into uracil. Subsequently, the converted DNA (where only unmethylated cytosines appear as thymines) is sequenced by NGS methods [132, 133] and compared with a reference genome, reaching a resolution of single-base in determining the methylation status.

Other techniques allow to sequence regulatory elements (FAIRE-Seq [134]), or regions sensitive to DNAse I (DNAase-seq [135]): these are able to identify putative bound sites in nucleosome-depleted regions at the genome scale.

NGS is also instrumental in comparative epigenomics where epigenomic maps identify similarities and differences in epigenomes within and between species. Between-species comparisons provide insight into the evolution of the epigenome and on how the same epigenomic tool-kit is used by different species [136].

APPLICATIONS OF NGS TO PLANT BREEDING

The improvements in sequencing technologies are changing the way plant scientists look at genomes, in the same time paving the way for the next wave of agricultural discoveries and methodologies to become part of breeding activities.

Sequencing of plant genomes was credited from the very first to generate a high-throughput battery of molecular markers to be used in genetic analysis, gene discovery, and breeding programs. By using an ever-growing number of molecular markers, it is now feasible to establish and utilize relationships between genotype and phenotype in crop improvement. The identification of SNPs in plants has been hampered by genome size and ploidy level. In addition, the ease by which in plants large populations can be obtained, persuaded molecular breeders to increase the number of individuals analyzed rather than the number of molecular markers. SNP discovery (and subsequent scoring) is now becoming an essential pre-breeding tool for crops improvement programs, because it allows the exploitation of naturally occurring variation at genes presumed important in the control of agronomic traits. Sequencing-based polymorphisms can help to trace allelic evolution, to identify new useful haplotypes and to guide the development of allele-specific markers basic for marker-assisted selection. NGS and high-throughput genotyping lead now to the discovery of massive amounts of SNPs further used in genomics-assisted breeding, thus speeding the integration of genomic tools in breeding practices. Fig. (**2**) provides a scheme of the dependencies between.

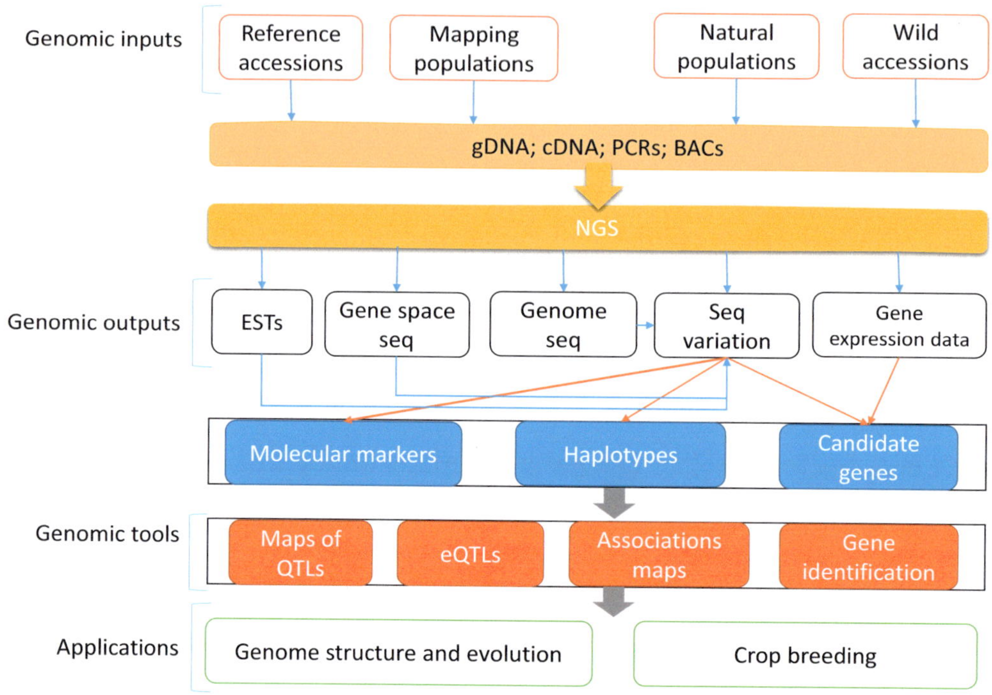

Fig. (2). Overview of NGS applications in crop genetics and breeding. NGS is critically positioned in genomics-assisted breeding, as it allows the production of molecular tools to be sued in deciphering the genomes. Molecular markers, haplotypes, and genes enter in molecular maps that allow to understand how the genomes are structured and how they evolved. NGS is also instrumental in the ability to use markers in breeding based on association genetics, and in marker assisted selection.

NGS As A Source Of Molecular Markers

SNPs resulting from a single-base pair variation are the most abundant and evenly distributed DNA markers on the whole genome scale. In the past decade SNP array-based genotyping platforms were considered as the favorite options for gene/QTL mapping and marker-assisted crop breeding. These markers (especially when functional [137]) are essential in comparative mapping, evolutionary studies, and in MAS. NGS is instrumental in developing molecular markers and effectively accelerates breeding, particularly in species for which genome and/or transcriptome sequence data already exist [138 - 140].

The most comprehensive source for SNP discovery is the comparison of sequenced genomes from individuals of a given species. This can be achieved when a complete reference genome is available, or when SNP are identified from the comparison of single-copy sequences between individuals [141]. The necessity of first discovering SNP and then scoring them is today simplified into a single process: bioinformatics allows analyzing the NBS sequence data, discovering, and genotyping SNPs at once.

The number of SNPs in genes is limited but complementary strategies are available, in many instances, based on the combinations of complexity reduction technologies, involving the construction of libraries prior to sequencing. Genome representation libraries are either (i) complexity reduced representation libraries constructed by using restriction enzymes, or (ii) sequence capture libraries not involving restriction digestion [142]. Examples of the first published methods (restriction site associated genomic DNA or RAD [143]) are still labor intensive, involving restriction enzyme digestion, DNA shearing and repair, ligation of adaptors and gel purification steps. Whole genome resequencing [144] is simpler than RAD, but still involves DNA shearing and repair prior to adaptor ligation. Additional methods have been developed for SNP discovery and scoring using NGS, more amenable to large mapping efforts [145 - 147].

In rice, 3.6 million SNPs were identified in 500 landraces, resulting in a high-density haplotype map. This experiment was crucial for further genome-wide association studies [148]. In another set of experiments, resequencing of 50 accessions of cultivated and wild rice was carried out to detect genetic diversity and the markers associated with agronomically important genes and 6.5 million SNPs were identified [77]. Similarly, the alignment of 4.6 billion of reads from 950 rice varieties allowed the identification of more than 4 million SNPs, 44% of which were low frequency [149]. In this experiment, the identification of SNPs was preliminary to GWAS. A database of 20 million SNPs is now available for this species, derived from the alignment of the reads from 3,000 rice genomes with the Nipponbare genome [150]. In maize, 3.3 million SNPs were used to produce the first haplotype map (HapMap) and to define the distribution of recombination along the chromosomes. The markers were identified in the low-copy genome, based on a panel of inbred lines [151]. The SNP were identified based on the knowledge of the B73 genome sequence [152], and were instrumental in comparative genomics experiments leading to the identification of regions associated with the domestication and geographic differentiation of maize [153]. The HapMap2 [80] is further addressing the observation that, in maize, the reference genome is capturing only about 70% of the gene space of inbred lines. HapMap2 characterized 103 inbred lines and produced 55 million SNPs. The authors hypothesize the existence of a genome "in flux", where structural variations are pervasive, especially at loci controlling important traits. The agronomically important heterosis phenomenon could also in part depend from the expression of haplotypes in a line-specific manner [68].

All SNPs identified *via* NGS are not 100% reliable due to sequencing errors. To overcome the problem a "sliding window approach" is adopted. Contigs of different haplotypic groups are aligned to the reference genome. In a rice experiment, RILs developed from a cross between *indica* and *japonica* cultivars were compared to the resequenced parents and SNPs identified. Illumina resequencing of the RILs allows for sorting and alignment of SNPs with the pseudomolecules of parental genome sequences [144]. The detection of SNPs was carried out in small chromosomal segments (windows) moving along the macromolecule one base at a time. For each window, the ratio of the number of SNPs derived from either parent was calculated. The sliding of the window allows for genotype calling and recombination breakpoints determination, eventually resulting in the construction of a recombination map [154].

A number of approaches are designed to reduce the cost of NGS so that it can be used for routine genotyping. One of the currently most adopted approach is genotyping-by-sequencing (GBS) [155]. GBS is a highly multiplexed genotyping system for constructing libraries for NGS. GBS involves DNA digestion with different enzymes and the construction of a reduced representation library sequenced using an NGS platform after barcode tagging. The complexity reduction step is based on methylation-sensitive restriction enzymes, and other strategies have also been developed [156]. GBS does not require prior information on SNPs position (even if it could benefit from the knowledge of a reference genome to which anchor the markers). Key features of the GBS technology are the low cost, the reduced number of PCRs and purifications needed, and the absence of size fractionation [157]. GBS enables the detection of thousands of SNPs in large populations or collections of lines that can be used for mapping, genetic diversity and for evolutionary studies [158]. Currently the low cost of GBS makes it the technology of choice to produce high-density maps [159, 160].

GBS needs to be fine-tuned in species with complex genomes. In such cases, experiments include variations on the barcoding theme reducing costs and experimental complexity [158]. The method has been applied to the genome of maize (*Zea mays*), and the first haplotype map was produced [152]. GBS was adopted to genotype 2,815 maize inbreds. The data obtained (among them more than 680,000 SNPs) should be instrumental to study QTLs architecture [161]. Again in maize, GBS of 14,000 inbred lines facilitated the assembling of pan-genome, whose anchors are instrumental in *de novo* assembling and characterizing the line-specific parts of the genome [162]. In barley (*Hordeum vulgare*), GBS was implemented targeting regions flanking restriction endonuclease sites [163]. Mapping confirmed the position of a barley semi-dwarfing gene [164]. In potato the method allowed to map more than 100,000 variants and identify alleles strongly associated with maturity and flesh color

[165]. The GBS method, with modifications, was the base to localize crossovers locations and to carry out fine-scale genetic mapping in Arabidopsis [166]. GBS is also useful in highly polyploid species, such as oat [167], and as a valid tool for genomic diversity studies in mustard [168]. In common bean (*Phaseoulus vulgaris*) this technology made it possible to discover SNPs in tight linkage with resistance to the bean yellow mosaic virus [169].

NGS, Mutants, And Candidate Genes

NGS is used to map and clone candidate genes of interest [170, 171]. In sequence-based mapping, two DNA pools derived from individuals from the phenotypic extremes of a segregating population are deep-sequenced, to identify candidate genes associated with the phenotype of interest [172].This is important in species of agronomic and economical value where genomic resources are not fully established, such as barley. In this species, the many noded dwarf phenotype (a shorter plant with narrow leaves) was associated with a candidate gene (HvMND) using a mapping-by-sequencing approach: the mutant and wild type alleles were mapped on the integrated physical and genetic map. A fine-mapping approach allowed the candidate gene to be identified. Mutant and wild type genomes were exome sequenced [173].

An efficient method to isolate and map mutants was at first designed in Arabidopsis (SHOREMap), where a group of recessive EMS-induced mutants could be associated to specific genes and DNA lesions. SHOREMap could also map quantitative trait loci, large deletions and recessive lethal or dominant mutations [174, 175]. The SHOREMap approach requires a sequenced reference genome; to overcome this shortcoming it was proposed to use the genome of related organisms as "bridge" between the mutant and wild-type genomes [174]. The implementation of refined algorithms allows developing reference-free discovery of homozygous mutations in WGS data [81]. The GBS protocol is also useful in recovering mutants following mutagenesis. It is applied, for example, in the MutMap approach in rice [176, 177]. The strategy was used to identify a causal SNP in a gene for the *pale green leaf* mutant, and for a second major gene responsible for blast resistance [177, 178].

NGS may be combined with reverse genetics approaches, such as the Targeted Induced Local Lesion in Genomes. This approach (including a step of exon-capture for genome enrichment) has been adopted for wheat and rice [90, 179, 180].

In species where genomic information is almost complete, NGS-based methods are even more fast and accurate [181]. WGS in rice [144] concerned 150 recombinant inbred lines giving rise to an ultrahigh-density SNP linkage map.

The work was instrumental for mapping 49 QTLs for 14 agronomic traits [182]. The method can assist in mapping the recombination breakpoints for each RILs. Haplotypes are then established to assemble a bin-map for mapping exercises.

An additional case of ultra-high SNP map in rice is reported in [183]. The quality of this map was assessed based on known positions of cloned genes, including *GS3* and *GW5/qSW5* which control grain length and width, respectively, and *OsC1*, a locus regulating pigmentation. The SNP map applied to QTL analysis for yield and yield-component traits, proved to agree with QTL positions based on RFLPs/SSRs. In the same species whole-genome re-sequencing identified QTLs influencing plant shape [184], up to the identification of candidate genes [185]. A combination of bulked segregant analysis and NGS was capable to identify candidate genome regions conferring resistance to two diseases in *Cajanus cajan* (pigeonpea) [186].

The increasing density of genetic maps obtained with NGS is determinant in narrowing down to the nucleotide level QTL identification and mapping. For this purpose NGS is particularly useful when combined with SNP bin-mapping. Indeed, the release of the high-quality rice genome sequence has accelerated the cloning of QTL for flowering time, disease resistance, plant architecture, and abiotic stress [187].

QTL-seq is an additional method for the rapid identification of QTLs. It relies on whole-genome resequencing of RILs or F2 populations composed of 20–50 individuals per population, selecting extreme opposites in trait values for a given phenotype. The strategy was adopted, for example, to identify QTLs for seedling vigor and resistance to blast disease in rice [187].

NGS is also instrumental in the design of novel approaches to molecular breeding. For example, QTL pyramiding was successfully carried out with NGS in rice [188] for eight QTLs related to increased panicle and spikelet size.

NGS And Genome-Wide Association Studies (GWAS)

Mapping of QTLs has been ongoing for many years, based on biparental populations phenotyped and genotyped. Methods relying on bi-parental populations are detecting only a fraction of the possible recombination events, and QTL are generally localized to large chromosomal regions. Moreover, the QTL can be identified, in a controlled cross, only when the phenotype is present and segregating in the experimental population. To overcome the limits of family-based mapping, several methods are currently interrogating the genetic diversity that controls quantitative traits within a species, at the population level and based on linkage disequilibrium. In plants, GWAS approaches are now particularly amenable to the creation of lines with nearly identical background, "Immortalized" as inbred, doubled haploid or recombinant inbred lines [155, 176]. NGS is instrumental both in producing high-density molecular linkage maps from controlled crosses, and in linkage disequilibrium approaches to QTL mapping. The extent of LD in a given species and the availability of molecular markers influences the choice between candidate-gene approach and GWAS. In the latter, which strongly relies on NGS, enough markers are genotyped across the genome so that the locus of interest will be in LD with at least one of them [189]. GWAS requires the high-density genotyping of large numbers of individuals from a diversity panel. The use of NGS in the context of GWAS makes it possible to genotype larger populations of plants with a higher density of markers than was previously possible, and this contributes to increased mapping resolution [190, 191].

GWAS was used, for example, in maize to identify SNPs in association with kernel color and sweet/starchy corn. The same approach was also adopted for a complex trait, such as the growing degree-days to silking [161]. In the same species, NGS and GWAS made possible to apply the nested association mapping (NAM) [153]. NAM creates detailed genotypes of the founder lines and then projects these genotypes onto progeny based on genetic markers. In this way, genome-wide association studies can be built on ancient recombination histories, and associations between traits and loci are detected at a very fine scale. A recent success of the NAM strategy is the dissection of the leaf architecture trait [192] and of stalk strength in maize [193]. A similar application is being developed in Arabidopsis with the Multiparent Advanced Generation Inter-Cross (MAGIC) population [194].

In rice, GWAS was the basis to analyze 950 worldwide rice varieties, including the *Oryza sativa indica* and *Oryza sativa japonica* subspecies [195], and a total of 32 new loci associated with flowering time and with ten grain-related traits were identified. This study showed an increased power to detect trait-associated variants using GWAS. This approach facilitates the detection of potential causal variants in follow-up analyses of GWAS, which have typically included PCR amplification of target regions, followed by resequencing [149].

In rice, in a different study, an ultra-high density haplotype maps was obtained and GWAS for 14 agronomic traits was performed. The loci identified through GWAS explained, on average, approximately 36% of the phenotypic variance. The peak signals at six loci were close to previously identified genes [148].

Conventional GWAS, as described, requires high-density genotyping of large number of individuals from a diversity panel. A further refinement of the GWAS protocols relies on genotyping pools of individuals from a diversity panel characterized by extremely different phenotypes, as demonstrated in maize [196]. This approach can be used when extensive genotyping resources are not available.

NGS And Genomic Selection (GS)

Genomic selection (GS) combines molecular markers with phenotypic and pedigree data to increase breeding accuracy [197]. Unlike MAS (and derivatives, such as MARS), GS assists in the production of superior lines *via* genome-wide marker profile data [198]. The use of GS considers complex quantitative traits using whole-genome markers to predict the genomic estimated breeding value (GEBV) in the unknown phenotypes of a target species. GS uses a training population to create a statistical model and assign breeding values to all markers across the genome using both phenotypes and genotypes. The training population is often comprised of breeding lines, and considers phenotyping data for overall performance. The model is then used to assign GEBV to new individuals derived from a "candidate population", based solely on their genotype [199]. The candidate population is typically the one used by breeders [200]. GS identifies the highest GEBVs instead of novel gene(s), in other words it aims to estimate GEBV without mapping or identifying QTL [201]. In GS, phenotyping is used only to verify the accuracy of the statistical models, and the entire subsequent process is based solely on the predictive capacity of molecular markers [202].

GS has been applied in maize [203], sugar beet [204], wheat [205], and in general when it is important to predict hybrid performance, in particular when established heterotic pools are not available [201]. GBS provided, in soybean, the set of SNPs that allowed to conduct a GS study on grain yield [206].

LISTS OF ONGOING AND COMPLETED GENOME SEQUENCING PROJECTS AND WEB RESOURCES

The number of genomes being drafted or finished, and whose sequencing is being started, is growing at a pace that it would not fit into a static picture [207]. As of May 2015 [138], the genomes of 84 species were sequenced, to various degrees of accuracy. Thirty-four of these are crops, 67 of which published in the last five years. The quality of the published sequences is not homogeneous: it is often assumed that available assemblies are highly polished, but in most cases they are still at the draft level, with varying degrees of quality [61]. The plant genomes published so far fall in the listed categories.

1. Crops for food, animal feed, energy.
2. Model organism for plant biology, and crop genomics.
3. Relatives or ancestors of crops or of model plants and thus used in comparative

genomics and syntenic analyses.
4. Accessions used in breeding programs.

Some online resources to retrieve updated information about sequenced genomes and ongoing sequencing efforts are also listed as follows.

WIKIPEDIA provides an updated list at http://en.wikipedia.org/wiki/List_of_sequenced_plant_genomes.

A list of published, and some of the unpublished/unreleased/incomplete genomes is provided and maintained at the CoGe Pedia website (https://genom-evolution.org/wiki/index.php/Sequenced_plant_genomes.)

A number of websites are offering a set of integrated tools to retrieve and analyze sequencing data from plant genomes. For example, the PlantGDB [208] provides tools specifically designed for comparative genomics (updated until July 2015): http://www.plantgdb.org. Another resource for comparative genomics maintained at GreenPhyl (http://www.greenphyl.org), while http://plantgenie.org/ [209] is offering genomics tools particularly for poplar and Norway spruce.

The Ensemble Genomes website provides navigable comparative analyses of plant genomes, at the website http://ensemblgenomes.org/.

Further examples are the Gramene database (http://www.gramene.org [210]), designed for grasses; the TAIR dataset, which is dedicated to Arabidopsis but also contains useful information on other plant species (www.arabidopsis.org); the MaizeGDB website with data concerning maize, including a genome browser (http://www.maizegdb.org/).

Many plant genomic sequencing projects are still ongoing and at a various degrees of advancement. For example, the progresses obtained in the *Solanaceae*, and more specifically in tomato, are monitored at https://solgenomics.net/organism/sol100/view and http://www.tomatogenome.net/ [211], respectively.

NOTES

[1] A contig is a set of overlapping DNA fragments representing a consensus region of DNA. It could consist of sequence reads of overlapping clones aligned in a physical map.

[2] "Gene space" refers to the part of the genome that includes protein-coding genes. In plants with large genomes, it refers also to gene-rich regions, surrounded by gene-poor regions.

[3] Single-end reads. A genomic fragment is generated by shearing whole genomic DNA, and differential adaptors are ligated to a genomic DNA fragment. An adaptor-specific primer is used to sequence a single end of the fragment. In Paired-end reads sequencing, both end of the fragment are sequenced in two separate reactions. In mate-pair reads sequencing a biotin-mediated subtraction step allows to sequence only fragments containing the biotin tag.

CONFLICT OF INTEREST

The author confirms that he has no conflict of interest to declare for this publication.

ACKNOWLEDGEMENTS

The author wishes to thank Prof. Francesco Salamini for critical reading of the manuscript.

LIST OF ABBREVIATIONS

BAC	=	bacterial artificial chromosome
bp/Kbp/Mbp/Gbp	=	base pair / kilo base pair /mega base pair / Giga base pair
dNTP / ddNTP	=	deoxynucleotyide; dideoxynucleotide
EMS	=	ethyl-methanesulfonate
EST	=	expressed sequence tags
MAS	=	marker assisted selection
miRNA	=	micro RNA
MTP	=	minimum tiling path
NGS	=	next generation sequencing
PCR	=	polymerase chain reaction
QTL	=	quantitative trait loci
QTN	=	quantitative trait nucleotide
SNP	=	single nucleotide polymorphism
T-DNA	=	transfer DNA of the Ti plasmid of Agrobacterium
TILLING	=	targeted induced local lesions in genomes
WCS	=	whole chromosome sequencing
WGS	=	whole genome sequencing

REFERENCES

[1] Bolger ME, Weisshaar B, Scholz U, Stein N, Usadel B, Mayer KF. Plant genome sequencing - applications for crop improvement. Curr Opin Biotechnol 2014; 26: 31-7. [http://dx.doi.org/10.1016/j.copbio.2013.08.019] [PMID: 24679255]

[2] Shendure J, Ji H. Next-generation DNA sequencing. Nat Biotechnol 2008; 26(10): 1135-45.

[http://dx.doi.org/10.1038/nbt1486] [PMID: 18846087]

[3] Metzker ML. Sequencing technologies - the next generation. Nat Rev Genet 2010; 11(1): 31-46. [http://dx.doi.org/10.1038/nrg2626] [PMID: 19997069]

[4] Mardis ER. The impact of next-generation sequencing technology on genetics. Trends Genet 2008; 24(3): 133-41. [http://dx.doi.org/10.1016/j.tig.2007.12.007] [PMID: 18262675]

[5] Smith LM, Sanders JZ, Kaiser RJ, *et al.* Fluorescence detection in automated DNA sequence analysis. Nature 1986; 321(6071): 674-9. [http://dx.doi.org/10.1038/321674a0] [PMID: 3713851]

[6] Arabidopsis Genome I. Arabidopsis Genome Initiative. Analysis of the genome sequence of the flowering plant *Arabidopsis thaliana*. Nature 2000; 408(6814): 796-815. [http://dx.doi.org/10.1038/35048692] [PMID: 11130711]

[7] International Rice Genome Sequencing Project. The map-based sequence of the rice genome. Nature 2005; 436(7052): 793-800. [http://dx.doi.org/10.1038/nature03895] [PMID: 16100779]

[8] Glenn TC. Field guide to next-generation DNA sequencers. Mol Ecol Resour 2011; 11(5): 759-69. [http://dx.doi.org/10.1111/j.1755-0998.2011.03024.x] [PMID: 21592312]

[9] Margulies M, Egholm M, Altman WE, *et al.* Genome sequencing in microfabricated high-density picolitre reactors. Nature 2005; 437(7057): 376-80. [PMID: 16056220]

[10] Rothberg JM, Leamon JH. The development and impact of 454 sequencing. Nat Biotechnol 2008; 26(10): 1117-24. [http://dx.doi.org/10.1038/nbt1485] [PMID: 18846085]

[11] Bennett S. Solexa Ltd. Pharmacogenomics 2004; 5(4): 433-8. [http://dx.doi.org/10.1517/14622416.5.4.433] [PMID: 15165179]

[12] Quail MA, Smith M, Coupland P, *et al.* A tale of three next generation sequencing platforms: comparison of Ion Torrent, Pacific Biosciences and Illumina MiSeq sequencers. BMC Genomics 2012; 13: 341. [http://dx.doi.org/10.1186/1471-2164-13-341] [PMID: 22827831]

[13] Shendure J, Porreca GJ, Reppas NB, *et al.* Accurate multiplex polony sequencing of an evolved bacterial genome. Science 2005; 309(5741): 1728-32. [http://dx.doi.org/10.1126/science.1117389] [PMID: 16081699]

[14] Rothberg JM, Hinz W, Rearick TM, *et al.* An integrated semiconductor device enabling non-optical genome sequencing. Nature 2011; 475(7356): 348-52. [http://dx.doi.org/10.1038/nature10242] [PMID: 21776081]

[15] Harris TD, Buzby PR, Babcock H, *et al.* Single-molecule DNA sequencing of a viral genome. Science 2008; 320(5872): 106-9. [http://dx.doi.org/10.1126/science.1150427] [PMID: 18388294]

[16] Roberts RJ, Carneiro MO, Schatz MC. The advantages of SMRT sequencing. Genome Biol 2013; 14(7): 405. [http://dx.doi.org/10.1186/gb-2013-14-6-405] [PMID: 23822731]

[17] Eid J, Fehr A, Gray J, *et al.* Real-time DNA sequencing from single polymerase molecules. Science 2009; 323(5910): 133-8. [http://dx.doi.org/10.1126/science.1162986] [PMID: 19023044]

[18] VanBuren R, Bryant D, Edger PP, *et al.* Single-molecule sequencing of the desiccation-tolerant grass Oropetium thomaeum. Nature 2015; 527(7579): 508-11. [http://dx.doi.org/10.1038/nature15714] [PMID: 26560029]

[19] Clarke J, Wu HC, Jayasinghe L, Patel A, Reid S, Bayley H. Continuous base identification for single-molecule nanopore DNA sequencing. Nat Nanotechnol 2009; 4(4): 265-70. [http://dx.doi.org/10.1038/nnano.2009.12] [PMID: 19350039]

[20] Szalay T, Golovchenko JA. *De novo* sequencing and variant calling with nanopores using PoreSeq. Nat Biotechnol 2015; 33(10): 1087-91. [http://dx.doi.org/10.1038/nbt.3360] [PMID: 26352647]

[21] Check Hayden E. Data from pocket-sized genome sequencer unveiled. Nature 2014.

[22] Check Hayden E. Pint-sized DNA sequencer impresses first users. Nature 2015; 521(7550): 15-6. [http://dx.doi.org/10.1038/521015a] [PMID: 25951262]

[23] Laver T, Harrison J, ONeill PA, *et al.* Assessing the performance of the Oxford Nanopore Technologies MinION. Biomol Detect Quantif 2015; 3: 1-8. [http://dx.doi.org/10.1016/j.bdq.2015.02.001] [PMID: 26753127]

[24] Chain PS, Grafham DV, Fulton RS, *et al.* Genomic Standards Consortium Human Microbiome Project Jumpstart Consortium. Genomics. Genome project standards in a new era of sequencing. Science 2009; 326(5950): 236-7. [http://dx.doi.org/10.1126/science.1180614] [PMID: 19815760]

[25] Jain M, Misra G, Patel RK, *et al.* A draft genome sequence of the pulse crop chickpea (*Cicer arietinum* L.). Plant J 2013; 74(5): 715-29. [http://dx.doi.org/10.1111/tpj.12173] [PMID: 23489434]

[26] Paterson AH. Leafing through the genomes of our major crop plants: strategies for capturing unique information. Nat Rev Genet 2006; 7(3): 174-84. [http://dx.doi.org/10.1038/nrg1806] [PMID: 16485017]

[27] Gregory TR, Nicol JA, Tamm H, *et al.* Eukaryotic genome size databases. Nucleic Acids Res 2007; 35(Database issue): D332-8. [http://dx.doi.org/10.1093/nar/gkl828] [PMID: 17090588]

[28] Dodsworth S, Leitch AR, Leitch IJ. Genome size diversity in angiosperms and its influence on gene space. Curr Opin Genet Dev 2015; 35: 73-8. [http://dx.doi.org/10.1016/j.gde.2015.10.006] [PMID: 26605684]

[29] Ibarra-Laclette E, Lyons E, Hernández-Guzmán G, *et al.* Architecture and evolution of a minute plant genome. Nature 2013; 498(7452): 94-8. [http://dx.doi.org/10.1038/nature12132] [PMID: 23665961]

[30] Schnable JC, Springer NM, Freeling M. Differentiation of the maize subgenomes by genome dominance and both ancient and ongoing gene loss. Proc Natl Acad Sci USA 2011; 108(10): 4069-74. [http://dx.doi.org/10.1073/pnas.1101368108] [PMID: 21368132]

[31] Schnable PS, Ware D, Fulton RS, *et al.* The B73 maize genome: complexity, diversity, and dynamics. Science 2009; 326(5956): 1112-5. [http://dx.doi.org/10.1126/science.1178534] [PMID: 19965430]

[32] Edwards D, Batley J, Snowdon RJ. Accessing complex crop genomes with next-generation sequencing. Theor Appl Genet 2013; 126(1): 1-11. [http://dx.doi.org/10.1007/s00122-012-1964-x] [PMID: 22948437]

[33] Droc G, Lariviere D, Guignon V, *et al.* The banana genome hub. Database 2013; 2013: bat035. [http://dx.doi.org/10.1093/database/bat035]

[34] DHont A, Denoeud F, Aury JM, *et al.* The banana (*Musa acuminata*) genome and the evolution of monocotyledonous plants. Nature 2012; 488(7410): 213-7. [http://dx.doi.org/10.1038/nature11241] [PMID: 22801500]

[35] Chalhoub B, Denoeud F, Liu S, *et al.* Plant genetics. Early allopolyploid evolution in the post-

Neolithic Brassica napus oilseed genome. Science 2014; 345(6199): 950-3.
[http://dx.doi.org/10.1126/science.1253435] [PMID: 25146293]

[36] Xu X, Pan S, Cheng S, *et al.* Potato Genome Sequencing Consortium. Genome sequence and analysis of the tuber crop potato. Nature 2011; 475(7355): 189-95.
[http://dx.doi.org/10.1038/nature10158] [PMID: 21743474]

[37] International Wheat Genome Sequencing Consortium (IWGSC). A chromosome-based draft sequence of the hexaploid bread wheat (*Triticum aestivum*) genome. Science 2014; 345(6194): 1251788.
[http://dx.doi.org/10.1126/science.1251788] [PMID: 25035500]

[38] Marcussen T, Sandve SR, Heier L, *et al.* International Wheat Genome Sequencing Consortium. Ancient hybridizations among the ancestral genomes of bread wheat. Science 2014; 345(6194): 1250092.
[http://dx.doi.org/10.1126/science.1250092] [PMID: 25035499]

[39] Xu X, Zeng L, Tao Y, *et al.* Pinpointing genes underlying the quantitative trait loci for root-knot nematode resistance in palaeopolyploid soybean by whole genome resequencing. Proc Natl Acad Sci USA 2013; 110(33): 13469-74.
[http://dx.doi.org/10.1073/pnas.1222368110] [PMID: 23898176]

[40] Schmutz J, Cannon SB, Schlueter J, *et al.* Genome sequence of the palaeopolyploid soybean. Nature 2010; 463(7278): 178-83.
[http://dx.doi.org/10.1038/nature08670] [PMID: 20075913]

[41] Tuskan GA, Difazio S, Jansson S, *et al.* The genome of black cottonwood, Populus trichocarpa (Torr. & Gray). Science 2006; 313(5793): 1596-604.
[http://dx.doi.org/10.1126/science.1128691] [PMID: 16973872]

[42] Denoeud F, Carretero-Paulet L, Dereeper A, *et al.* The coffee genome provides insight into the convergent evolution of caffeine biosynthesis. Science 2014; 345(6201): 1181-4.
[http://dx.doi.org/10.1126/science.1255274] [PMID: 25190796]

[43] Argout X, Salse J, Aury JM, *et al.* The genome of Theobroma cacao. Nat Genet 2011; 43(2): 101-8.
[http://dx.doi.org/10.1038/ng.736] [PMID: 21186351]

[44] Jaillon O, Aury JM, Noel B, *et al.* French-Italian Public Consortium for Grapevine Genome Characterization. The grapevine genome sequence suggests ancestral hexaploidization in major angiosperm phyla. Nature 2007; 449(7161): 463-7.
[http://dx.doi.org/10.1038/nature06148] [PMID: 17721507]

[45] Velasco R, Zharkikh A, Troggio M, *et al.* A high quality draft consensus sequence of the genome of a heterozygous grapevine variety. PLoS One 2007; 2(12): e1326.
[http://dx.doi.org/10.1371/journal.pone.0001326] [PMID: 18094749]

[46] Wu J, Wang Z, Shi Z, *et al.* The genome of the pear (Pyrus bretschneideri Rehd.). Genome Res 2013; 23(2): 396-408.
[http://dx.doi.org/10.1101/gr.144311.112] [PMID: 23149293]

[47] Ling HQ, Zhao S, Liu D, *et al.* Draft genome of the wheat A-genome progenitor Triticum urartu. Nature 2013; 496(7443): 87-90.
[http://dx.doi.org/10.1038/nature11997] [PMID: 23535596]

[48] Verde I, Abbott AG, Scalabrin S, *et al.* International Peach Genome Initiative. The high-quality draft genome of peach (*Prunus persica*) identifies unique patterns of genetic diversity, domestication and genome evolution. Nat Genet 2013; 45(5): 487-94.
[http://dx.doi.org/10.1038/ng.2586] [PMID: 23525075]

[49] Xu Q, Chen LL, Ruan X, *et al.* The draft genome of sweet orange (*Citrus sinensis*). Nat Genet 2013; 45(1): 59-66.
[http://dx.doi.org/10.1038/ng.2472] [PMID: 23179022]

[50] Wu GA, Prochnik S, Jenkins J, *et al.* Sequencing of diverse mandarin, pummelo and orange genomes

reveals complex history of admixture during citrus domestication. Nat Biotechnol 2014; 32(7): 656-62. [http://dx.doi.org/10.1038/nbt.2906] [PMID: 24908277]

[51] Yang S, Wang L, Huang J, *et al.* Parent-progeny sequencing indicates higher mutation rates in heterozygotes. Nature 2015; 523(7561): 463-7. [http://dx.doi.org/10.1038/nature14649] [PMID: 26176923]

[52] Marx V. Next-generation sequencing: The genome jigsaw. Nature 2013; 501(7466): 263-8. [http://dx.doi.org/10.1038/501261a] [PMID: 24025842]

[53] Cao J, Schneeberger K, Ossowski S, *et al.* Whole-genome sequencing of multiple *Arabidopsis thaliana* populations. Nat Genet 2011; 43(10): 956-63. [http://dx.doi.org/10.1038/ng.911] [PMID: 21874002]

[54] Ariyadasa R, Mascher M, Nussbaumer T, *et al.* A sequence-ready physical map of barley anchored genetically by two million single-nucleotide polymorphisms. Plant Physiol 2014; 164(1): 412-23. [http://dx.doi.org/10.1104/pp.113.228213] [PMID: 24243933]

[55] Muñoz-Amatriaín M, Lonardi S, Luo M, *et al.* Sequencing of 15 622 gene-bearing BACs clarifies the gene-dense regions of the barley genome. Plant J 2015; 84(1): 216-27. [http://dx.doi.org/10.1111/tpj.12959] [PMID: 26252423]

[56] Goodstein DM, Shu S, Howson R, *et al.* Phytozome: a comparative platform for green plant genomics. Nucleic Acids Res 2012; 40(Database issue): D1178-86. [http://dx.doi.org/10.1093/nar/gkr944] [PMID: 22110026]

[57] Jia J, Zhao S, Kong X, *et al.* International Wheat Genome Sequencing Consortium. Aegilops tauschii draft genome sequence reveals a gene repertoire for wheat adaptation. Nature 2013; 496(7443): 91-5. [http://dx.doi.org/10.1038/nature12028] [PMID: 23535592]

[58] Ouyang S, Zhu W, Hamilton J, *et al.* The TIGR Rice Genome Annotation Resource: improvements and new features. Nucleic Acids Res 2007; 35(Database issue): D883-7. [http://dx.doi.org/10.1093/nar/gkl976] [PMID: 17145706]

[59] Mardis E, McPherson J, Martienssen R, Wilson RK, McCombie WR. What is finished, and why does it matter. Genome Res 2002; 12(5): 669-71. [http://dx.doi.org/10.1101/gr.032102] [PMID: 11997333]

[60] Paterson AH, Bowers JE, Bruggmann R, *et al.* The Sorghum bicolor genome and the diversification of grasses. Nature 2009; 457(7229): 551-6. [http://dx.doi.org/10.1038/nature07723] [PMID: 19189423]

[61] Feuillet C, Leach JE, Rogers J, Schnable PS, Eversole K. Crop genome sequencing: lessons and rationales. Trends Plant Sci 2011; 16(2): 77-88. [http://dx.doi.org/10.1016/j.tplants.2010.10.005] [PMID: 21081278]

[62] Shangguan L, Han J, Kayesh E, *et al.* Evaluation of genome sequencing quality in selected plant species using expressed sequence tags. PLoS One 2013; 8(7): e69890. [http://dx.doi.org/10.1371/journal.pone.0069890] [PMID: 23922843]

[63] Nystedt B, Street NR, Wetterbom A, *et al.* The Norway spruce genome sequence and conifer genome evolution. Nature 2013; 497(7451): 579-84. [http://dx.doi.org/10.1038/nature12211] [PMID: 23698360]

[64] Birney E. Assemblies: the good, the bad, the ugly. Nat Methods 2011; 8(1): 59-60. [http://dx.doi.org/10.1038/nmeth0111-59] [PMID: 21191376]

[65] Kumar A, Seetan R, Mergoum M, *et al.* Radiation hybrid maps of the D-genome of Aegilops tauschii and their application in sequence assembly of large and complex plant genomes. BMC Genomics 2015; 16: 800. [http://dx.doi.org/10.1186/s12864-015-2030-2] [PMID: 26475137]

[66] Schatz MC, Witkowski J, McCombie WR. Current challenges in de novo plant genome sequencing

and assembly. Genome Biol 2012; 13(4): 243.
[http://dx.doi.org/10.1186/gb-2012-13-4-243] [PMID: 22546054]

[67] Tang H, Lyons E, Town CD. Optical mapping in plant comparative genomics. Gigascience 2015; 4: 3. [http://dx.doi.org/10.1186/s13742-015-0044-y] [PMID: 25699175]

[68] Springer NM, Ying K, Fu Y, *et al.* Maize inbreds exhibit high levels of copy number variation (CNV) and presence/absence variation (PAV) in genome content. PLoS Genet 2009; 5(11): e1000734. [http://dx.doi.org/10.1371/journal.pgen.1000734] [PMID: 19956538]

[69] Marroni F, Pinosio S, Morgante M. Structural variation and genome complexity: is dispensable really dispensable? Curr Opin Plant Biol 2014; 18: 31-6. [http://dx.doi.org/10.1016/j.pbi.2014.01.003] [PMID: 24548794]

[70] Schatz MC, Maron LG, Stein JC, *et al.* Whole genome *de novo* assemblies of three divergent strains of rice, *Oryza sativa*, document novel gene space of aus and indica. Genome Biol 2014; 15(11): 506. [PMID: 25468217]

[71] Auton A, Brooks LD, Durbin RM, *et al.* 1000 Genomes Project Consortium. A global reference for human genetic variation. Nature 2015; 526(7571): 68-74. [http://dx.doi.org/10.1038/nature15393] [PMID: 26432245]

[72] Weigel D, Mott R. The 1001 genomes project for *Arabidopsis thaliana*. Genome Biol 2009; 10(5): 107. [http://dx.doi.org/10.1186/gb-2009-10-5-107] [PMID: 19519932]

[73] Ossowski S, Schneeberger K, Clark RM, Lanz C, Warthmann N, Weigel D. Sequencing of natural strains of *Arabidopsis thaliana* with short reads. Genome Res 2008; 18(12): 2024-33. [http://dx.doi.org/10.1101/gr.080200.108] [PMID: 18818371]

[74] 1001 genomes: a catalog of Arabidopsis thaliana genetic variation [Internet] 2015.

[75] Guo L, Gao Z, Qian Q. Application of resequencing to rice genomics, functional genomics and evolutionary analysis. Rice (N Y) 2014; 7(1): 4. [http://dx.doi.org/10.1186/s12284-014-0004-7] [PMID: 25006357]

[76] Duitama J, Silva A, Sanabria Y, *et al.* Whole genome sequencing of elite rice cultivars as a comprehensive information resource for marker assisted selection. PLoS One 2015; 10(4): e0124617. [http://dx.doi.org/10.1371/journal.pone.0124617] [PMID: 25923345]

[77] Xu X, Liu X, Ge S, *et al.* Resequencing 50 accessions of cultivated and wild rice yields markers for identifying agronomically important genes. Nat Biotechnol 2011; 30(1): 105-11. [http://dx.doi.org/10.1038/nbt.2050] [PMID: 22158310]

[78] 3,000 rice genomes project. The 3,000 rice genomes project. Gigascience 2014; 3: 7. [http://dx.doi.org/10.1186/2047-217X-3-7] [PMID: 24872877]

[79] Hufford MB, Xu X, van Heerwaarden J, *et al.* Comparative population genomics of maize domestication and improvement. Nat Genet 2012; 44(7): 808-11. [http://dx.doi.org/10.1038/ng.2309] [PMID: 22660546]

[80] Chia JM, Song C, Bradbury PJ, *et al.* Maize HapMap2 identifies extant variation from a genome in flux. Nat Genet 2012; 44(7): 803-7. [http://dx.doi.org/10.1038/ng.2313] [PMID: 22660545]

[81] Nordström KJ, Albani MC, James GV, *et al.* Mutation identification by direct comparison of whole-genome sequencing data from mutant and wild-type individuals using k-mers. Nat Biotechnol 2013; 31(4): 325-30. [http://dx.doi.org/10.1038/nbt.2515] [PMID: 23475072]

[82] Mamanova L, Coffey AJ, Scott CE, *et al.* Target-enrichment strategies for next-generation sequencing. Nat Methods 2010; 7(2): 111-8. [http://dx.doi.org/10.1038/nmeth.1419] [PMID: 20111037]

[83] Doležel J, Vrána J, Cápal P, Kubaláková M, Burešová V, Simková H. Advances in plant chromosome genomics. Biotechnol Adv 2014; 32(1): 122-36. [http://dx.doi.org/10.1016/j.biotechadv.2013.12.011] [PMID: 24406816]

[84] Mayer KF, Taudien S, Martis M, *et al.* Gene content and virtual gene order of barley chromosome 1H. Plant Physiol 2009; 151(2): 496-505. [http://dx.doi.org/10.1104/pp.109.142612] [PMID: 19692534]

[85] Wicker T, Mayer KF, Gundlach H, *et al.* Frequent gene movement and pseudogene evolution is common to the large and complex genomes of wheat, barley, and their relatives. Plant Cell 2011; 23(5): 1706-18. [http://dx.doi.org/10.1105/tpc.111.086629] [PMID: 21622801]

[86] The International Wheat Genome Sequencing Consortium 2013.

[87] Brenchley R, Spannagl M, Pfeifer M, *et al.* Analysis of the bread wheat genome using whole-genome shotgun sequencing. Nature 2012; 491(7426): 705-10. [http://dx.doi.org/10.1038/nature11650] [PMID: 23192148]

[88] Lucas SJ, Akpınar BA, Šimková H, Kubaláková M, Doležel J, Budak H. Next-generation sequencing of flow-sorted wheat chromosome 5D reveals lineage-specific translocations and widespread gene duplications. BMC Genomics 2014; 15: 1080. [http://dx.doi.org/10.1186/1471-2164-15-1080] [PMID: 25487001]

[89] Warr A, Robert C, Hume D, Archibald A, Deeb N, Watson M. Exome Sequencing: Current and Future Perspectives. G3 (Bethesda) 2015; 5(8): 1543-50. [http://dx.doi.org/10.1534/g3.115.018564] [PMID: 26139844]

[90] Henry IM, Nagalakshmi U, Lieberman MC, *et al.* Efficient Genome-Wide Detection and Cataloging of EMS-Induced Mutations Using Exome Capture and Next-Generation Sequencing. Plant Cell 2014; 26(4): 1382-97. [http://dx.doi.org/10.1105/tpc.113.121590] [PMID: 24728647]

[91] Zhou L, Holliday JA. Targeted enrichment of the black cottonwood (*Populus trichocarpa*) gene space using sequence capture. BMC Genomics 2012; 13: 703. [http://dx.doi.org/10.1186/1471-2164-13-703] [PMID: 23241106]

[92] Winfield MO, Wilkinson PA, Allen AM, *et al.* Targeted re-sequencing of the allohexaploid wheat exome. Plant Biotechnol J 2012; 10(6): 733-42. [http://dx.doi.org/10.1111/j.1467-7652.2012.00713.x] [PMID: 22703335]

[93] Mascher M, Richmond TA, Gerhardt DJ, *et al.* Barley whole exome capture: a tool for genomic research in the genus Hordeum and beyond. Plant J 2013; 76(3): 494-505. [http://dx.doi.org/10.1111/tpj.12294] [PMID: 23889683]

[94] Wendler N, Mascher M, Nöh C, *et al.* Unlocking the secondary gene-pool of barley with next-generation sequencing. Plant Biotechnol J 2014; 12(8): 1122-31. [http://dx.doi.org/10.1111/pbi.12219] [PMID: 25040223]

[95] Ma Z, Axtell MJ. Long-range genomic enrichment, sequencing, and assembly to determine unknown sequences flanking a known microRNA. PLoS One 2013; 8(12): e83721. [http://dx.doi.org/10.1371/journal.pone.0083721] [PMID: 24376739]

[96] Evans J, Crisovan E, Barry K, *et al.* Diversity and population structure of northern switchgrass as revealed through exome capture sequencing. Plant J 2015; 84(4): 800-15. [http://dx.doi.org/10.1111/tpj.13041] [PMID: 26426343]

[97] Cheng F, Wu J, Fang L, Wang X. Syntenic gene analysis between Brassica rapa and other Brassicaceae species. Front Plant Sci 2012; 3: 198. [http://dx.doi.org/10.3389/fpls.2012.00198] [PMID: 22969786]

[98] Huo N, Vogel JP, Lazo GR, *et al.* Structural characterization of Brachypodium genome and its

syntenic relationship with rice and wheat. Plant Mol Biol 2009; 70(1-2): 47-61. [http://dx.doi.org/10.1007/s11103-009-9456-3] [PMID: 19184460]

[99] Pfeifer M, Martis M, Asp T, *et al.* The perennial ryegrass GenomeZipper: targeted use of genome resources for comparative grass genomics. Plant Physiol 2013; 161(2): 571-82. [http://dx.doi.org/10.1104/pp.112.207282] [PMID: 23184232]

[100] Sato S, Tabata S, Hirakawa H, *et al.* Tomato Genome Consortium. The tomato genome sequence provides insights into fleshy fruit evolution. Nature 2012; 485(7400): 635-41. [http://dx.doi.org/10.1038/nature11119] [PMID: 22660326]

[101] Tang H, Bowers JE, Wang X, Ming R, Alam M, Paterson AH. Synteny and collinearity in plant genomes. Science 2008; 320(5875): 486-8. [http://dx.doi.org/10.1126/science.1153917] [PMID: 18436778]

[102] Mayer KF, Martis M, Hedley PE, *et al.* Unlocking the barley genome by chromosomal and comparative genomics. Plant Cell 2011; 23(4): 1249-63. [http://dx.doi.org/10.1105/tpc.110.082537] [PMID: 21467582]

[103] Mayer KF, Waugh R, Brown JW, *et al.* International Barley Genome Sequencing Consortium. A physical, genetic and functional sequence assembly of the barley genome. Nature 2012; 491(7426): 711-6.
[PMID: 23075845]

[104] Byrne SL, Nagy I, Pfeifer M, *et al.* A synteny-based draft genome sequence of the forage grass Lolium perenne. Plant J 2015; 84(4): 816-26. [http://dx.doi.org/10.1111/tpj.13037] [PMID: 26408275]

[105] Eversole K, Feuillet C, Mayer KF, Rogers J. Slicing the wheat genome. Introduction. Science 2014; 345(6194): 285-7. [http://dx.doi.org/10.1126/science.1257983] [PMID: 25035484]

[106] Hirakawa H, Shirasawa K, Kosugi S, *et al.* Dissection of the octoploid strawberry genome by deep sequencing of the genomes of Fragaria species. DNA Res 2014; 21(2): 169-81. [http://dx.doi.org/10.1093/dnares/dst049] [PMID: 24282021]

[107] Wang S, Chen J, Zhang W, *et al.* Sequence-based ultra-dense genetic and physical maps reveal structural variations of allopolyploid cotton genomes. Genome Biol 2015; 16: 108. [http://dx.doi.org/10.1186/s13059-015-0678-1] [PMID: 26003111]

[108] Bennetzen JL, Schmutz J, Wang H, *et al.* Reference genome sequence of the model plant Setaria. Nat Biotechnol 2012; 30(6): 555-61. [http://dx.doi.org/10.1038/nbt.2196] [PMID: 22580951]

[109] Amborella Genome P. Amborella Genome Project. The Amborella genome and the evolution of flowering plants. Science 2013; 342(6165): 1241089. [http://dx.doi.org/10.1126/science.1241089] [PMID: 24357323]

[110] Tiley GP, Burleigh JG. The relationship of recombination rate, genome structure, and patterns of molecular evolution across angiosperms. BMC Evol Biol 2015; 15: 194. [http://dx.doi.org/10.1186/s12862-015-0473-3] [PMID: 26377000]

[111] Zhou Z, Jiang Y, Wang Z, *et al.* Resequencing 302 wild and cultivated accessions identifies genes related to domestication and improvement in soybean. Nat Biotechnol 2015; 33(4): 408-14. [http://dx.doi.org/10.1038/nbt.3096] [PMID: 25643055]

[112] Dodsworth S. Genome skimming for next-generation biodiversity analysis. Trends Plant Sci 2015; 20(9): 525-7. [http://dx.doi.org/10.1016/j.tplants.2015.06.012] [PMID: 26205170]

[113] Straub SC, Parks M, Weitemier K, Fishbein M, Cronn RC, Liston A. Navigating the tip of the genomic iceberg: Next-generation sequencing for plant systematics. Am J Bot 2012; 99(2): 349-64. [http://dx.doi.org/10.3732/ajb.1100335] [PMID: 22174336]

[114] Malé PJ, Bardon L, Besnard G, *et al.* Genome skimming by shotgun sequencing helps resolve the phylogeny of a pantropical tree family. Mol Ecol Resour 2014; 14(5): 966-75. [PMID: 24606032]

[115] Weitemier K, Straub SC, Cronn RC, *et al.* Hyb-Seq: Combining target enrichment and genome skimming for plant phylogenomics. App Plant Sci 2014; p. 2.

[116] Mortazavi A, Williams BA, McCue K, Schaeffer L, Wold B. Mapping and quantifying mammalian transcriptomes by RNA-Seq. Nat Methods 2008; 5(7): 621-8. [http://dx.doi.org/10.1038/nmeth.1226] [PMID: 18516045]

[117] Wang Z, Gerstein M, Snyder M. RNA-Seq: a revolutionary tool for transcriptomics. Nat Rev Genet 2009; 10(1): 57-63. [http://dx.doi.org/10.1038/nrg2484] [PMID: 19015660]

[118] Ma X, Tang Z, Qin J, Meng Y. The use of high-throughput sequencing methods for plant microRNA research. RNA Biol 2015; 12(7): 709-19. [http://dx.doi.org/10.1080/15476286.2015.1053686] [PMID: 26016494]

[119] Ramirez-Gonzalez RH, Segovia V, Bird N, *et al.* RNA-Seq bulked segregant analysis enables the identification of high-resolution genetic markers for breeding in hexaploid wheat. Plant Biotechnol J 2015; 13(5): 613-24. [http://dx.doi.org/10.1111/pbi.12281] [PMID: 25382230]

[120] Donovan WP, Zhang Y, Howell MD. Large-scale sequencing of plant small RNAs. Methods Mol Biol 2011; 744: 159-73. [http://dx.doi.org/10.1007/978-1-61779-123-9_11] [PMID: 21533692]

[121] Wang Y, Bai X, Yan C, *et al.* Genomic dissection of small RNAs in wild rice (*Oryza rufipogon*): lessons for rice domestication. New Phytol 2012; 196(3): 914-25. [http://dx.doi.org/10.1111/j.1469-8137.2012.04304.x] [PMID: 22994911]

[122] Harper AL, Trick M, Higgins J, *et al.* Associative transcriptomics of traits in the polyploid crop species Brassica napus. Nat Biotechnol 2012; 30(8): 798-802. [http://dx.doi.org/10.1038/nbt.2302] [PMID: 22820317]

[123] Hirsch CN, Foerster JM, Johnson JM, *et al.* Insights into the maize pan-genome and pan-transcriptome. Plant Cell 2014; 26(1): 121-35. [http://dx.doi.org/10.1105/tpc.113.119982] [PMID: 24488960]

[124] Holloway B, Li B. Expression QTLs: applications for crop improvement. Mol Breed 2010; 26: 381-91. [http://dx.doi.org/10.1007/s11032-010-9396-2]

[125] Li L, Petsch K, Shimizu R, *et al.* Mendelian and non-Mendelian regulation of gene expression in maize. PLoS Genet 2013; 9(1): e1003202. [http://dx.doi.org/10.1371/journal.pgen.1003202] [PMID: 23341782]

[126] Lane AK, Niederhuth CE, Ji L, Schmitz RJ. pENCODE: a plant encyclopedia of DNA elements. Annu Rev Genet 2014; 48: 49-70. [http://dx.doi.org/10.1146/annurev-genet-120213-092443] [PMID: 25149370]

[127] Committee EP. EPIC Planning Committee. Reading the second code: mapping epigenomes to understand plant growth, development, and adaptation to the environment. Plant Cell 2012; 24(6): 2257-61. [http://dx.doi.org/10.1105/tpc.112.100636] [PMID: 22751210]

[128] Kaufmann K, Muiño JM, Østerås M, Farinelli L, Krajewski P, Angenent GC. Chromatin immunoprecipitation (ChIP) of plant transcription factors followed by sequencing (ChIP-SEQ) or hybridization to whole genome arrays (ChIP-CHIP). Nat Protoc 2010; 5(3): 457-72. [http://dx.doi.org/10.1038/nprot.2009.244] [PMID: 20203663]

[129] Ricardi MM, González RM, Zhong S, *et al.* Genome-wide data (ChIP-seq) enabled identification of

cell wall-related and aquaporin genes as targets of tomato ASR1, a drought stress-responsive transcription factor. BMC Plant Biol 2014; 14: 29. [http://dx.doi.org/10.1186/1471-2229-14-29] [PMID: 24423251]

[130] Lister R, OMalley RC, Tonti-Filippini J, *et al.* Highly integrated single-base resolution maps of the epigenome in Arabidopsis. Cell 2008; 133(3): 523-36. [http://dx.doi.org/10.1016/j.cell.2008.03.029] [PMID: 18423832]

[131] Zhang X, Yazaki J, Sundaresan A, *et al.* Genome-wide high-resolution mapping and functional analysis of DNA methylation in arabidopsis. Cell 2006; 126(6): 1189-201. [http://dx.doi.org/10.1016/j.cell.2006.08.003] [PMID: 16949657]

[132] Cokus SJ, Feng S, Zhang X, *et al.* Shotgun bisulphite sequencing of the Arabidopsis genome reveals DNA methylation patterning. Nature 2008; 452(7184): 215-9. [http://dx.doi.org/10.1038/nature06745] [PMID: 18278030]

[133] Hardcastle TJ. High-throughput sequencing of cytosine methylation in plant DNA. Plant Methods 2013; 9(1): 16. [http://dx.doi.org/10.1186/1746-4811-9-16] [PMID: 23758782]

[134] Giresi PG, Kim J, McDaniell RM, Iyer VR, Lieb JD. FAIRE (Formaldehyde-Assisted Isolation of Regulatory Elements) isolates active regulatory elements from human chromatin. Genome Res 2007; 17(6): 877-85. [http://dx.doi.org/10.1101/gr.5533506] [PMID: 17179217]

[135] Crawford GE, Holt IE, Whittle J, *et al.* Genome-wide mapping of DNase hypersensitive sites using massively parallel signature sequencing (MPSS). Genome Res 2006; 16(1): 123-31. [http://dx.doi.org/10.1101/gr.4074106] [PMID: 16344561]

[136] Ji L, Neumann DA, Schmitz RJ. Crop Epigenomics: Identifying, Unlocking, and Harnessing Cryptic Variation in Crop Genomes. Mol Plant 2015; 8(6): 860-70. [http://dx.doi.org/10.1016/j.molp.2015.01.021] [PMID: 25638564]

[137] Andersen JR, Lübberstedt T. Functional markers in plants. Trends Plant Sci 2003; 8(11): 554-60. [http://dx.doi.org/10.1016/j.tplants.2003.09.010] [PMID: 14607101]

[138] Michael TP, VanBuren R. Progress, challenges and the future of crop genomes. Curr Opin Plant Biol 2015; 24: 71-81. [http://dx.doi.org/10.1016/j.pbi.2015.02.002] [PMID: 25703261]

[139] Gao Q, Yue G, Li W, Wang J, Xu J, Yin Y. Recent progress using high-throughput sequencing technologies in plant molecular breeding. J Integr Plant Biol 2012; 54(4): 215-27. [http://dx.doi.org/10.1111/j.1744-7909.2012.01115.x] [PMID: 22409591]

[140] Bräutigam A, Gowik U. What can next generation sequencing do for you? Next generation sequencing as a valuable tool in plant research. Plant Biol (Stuttg) 2010; 12(6): 831-41. [http://dx.doi.org/10.1111/j.1438-8677.2010.00373.x] [PMID: 21040298]

[141] Schneeberger K, Weigel D. Fast-forward genetics enabled by new sequencing technologies. Trends Plant Sci 2011; 16(5): 282-8. [http://dx.doi.org/10.1016/j.tplants.2011.02.006] [PMID: 21439889]

[142] Thomson MJ. High-Throughput SNP Genotyping to Accelerate Crop Improvement. Plant Breeding and Biotechnology 2014; 2: 195-212. [http://dx.doi.org/10.9787/PBB.2014.2.3.195]

[143] Baird NA, Etter PD, Atwood TS, *et al.* Rapid SNP discovery and genetic mapping using sequenced RAD markers. PLoS One 2008; 3(10): e3376. [http://dx.doi.org/10.1371/journal.pone.0003376] [PMID: 18852878]

[144] Huang X, Feng Q, Qian Q, *et al.* High-throughput genotyping by whole-genome resequencing. Genome Res 2009; 19(6): 1068-76. [http://dx.doi.org/10.1101/gr.089516.108] [PMID: 19420380]

[145] Andolfatto P, Davison D, Erezyilmaz D, *et al.* Multiplexed shotgun genotyping for rapid and efficient genetic mapping. Genome Res 2011; 21(4): 610-7. [http://dx.doi.org/10.1101/gr.115402.110] [PMID: 21233398]

[146] Kim JE, Oh SK, Lee JH, Lee BM, Jo SH. Genome-wide SNP calling using next generation sequencing data in tomato. Mol Cells 2014; 37(1): 36-42. [http://dx.doi.org/10.14348/molcells.2014.2241] [PMID: 24552708]

[147] Ray S, Satya P. Next generation sequencing technologies for next generation plant breeding. Front Plant Sci 2014; 5: 367. [http://dx.doi.org/10.3389/fpls.2014.00367] [PMID: 25126091]

[148] Huang X, Wei X, Sang T, *et al.* Genome-wide association studies of 14 agronomic traits in rice landraces. Nat Genet 2010; 42(11): 961-7. [http://dx.doi.org/10.1038/ng.695] [PMID: 20972439]

[149] Huang X, Zhao Y, Wei X, *et al.* Genome-wide association study of flowering time and grain yield traits in a worldwide collection of rice germplasm. Nat Genet 2011; 44(1): 32-9. [http://dx.doi.org/10.1038/ng.1018] [PMID: 22138690]

[150] Alexandrov N, Tai S, Wang W, *et al.* SNP-Seek database of SNPs derived from 3000 rice genomes. Nucleic Acids Res 2015; 43(Database issue): D1023-7. [http://dx.doi.org/10.1093/nar/gku1039] [PMID: 25429973]

[151] Yu J, Holland JB, McMullen MD, Buckler ES. Genetic design and statistical power of nested association mapping in maize. Genetics 2008; 178(1): 539-51. [http://dx.doi.org/10.1534/genetics.107.074245] [PMID: 18202393]

[152] Gore MA, Chia JM, Elshire RJ, *et al.* A first-generation haplotype map of maize. Science 2009; 326(5956): 1115-7. [http://dx.doi.org/10.1126/science.1177837] [PMID: 19965431]

[153] McMullen MD, Kresovich S, Villeda HS, *et al.* Genetic properties of the maize nested association mapping population. Science 2009; 325(5941): 737-40. [http://dx.doi.org/10.1126/science.1174320] [PMID: 19661427]

[154] Chen H, He H, Zhou F, Yu H, Deng XW. Development of genomics-based genotyping platforms and their applications in rice breeding. Curr Opin Plant Biol 2013; 16(2): 247-54. [http://dx.doi.org/10.1016/j.pbi.2013.04.002] [PMID: 23706659]

[155] Deschamps S, Llaca V, May GD. Genotyping-by-Sequencing in Plants. Biology (Basel) 2012; 1(3): 460-83. [http://dx.doi.org/10.3390/biology1030460] [PMID: 24832503]

[156] Sonah H, Bastien M, Iquira E, *et al.* An improved genotyping by sequencing (GBS) approach offering increased versatility and efficiency of SNP discovery and genotyping. PLoS One 2013; 8(1): e54603. [http://dx.doi.org/10.1371/journal.pone.0054603] [PMID: 23372741]

[157] Narum SR, Buerkle CA, Davey JW, Miller MR, Hohenlohe PA. Genotyping-by-sequencing in ecological and conservation genomics. Mol Ecol 2013; 22(11): 2841-7. [http://dx.doi.org/10.1111/mec.12350] [PMID: 23711105]

[158] Elshire RJ, Glaubitz JC, Sun Q, *et al.* A robust, simple genotyping-by-sequencing (GBS) approach for high diversity species. PLoS One 2011; 6(5): e19379. [http://dx.doi.org/10.1371/journal.pone.0019379] [PMID: 21573248]

[159] He J, Zhao X, Laroche A, Lu ZX, Liu H, Li Z. Genotyping-by-sequencing (GBS), an ultimate marker-assisted selection (MAS) tool to accelerate plant breeding. Front Plant Sci 2014; 5: 484. [http://dx.doi.org/10.3389/fpls.2014.00484] [PMID: 25324846]

[160] Poland JA, Rife TW. Genotyping-by-Sequencing for Plant Breeding and Genetics. Plant Genome 2012; 5: 92.

[http://dx.doi.org/10.3835/plantgenome2012.05.0005]

[161] Romay MC, Millard MJ, Glaubitz JC, *et al.* Comprehensive genotyping of the USA national maize inbred seed bank. Genome Biol 2013; 14(6): R55. [http://dx.doi.org/10.1186/gb-2013-14-6-r55] [PMID: 23759205]

[162] Lu F, Romay MC, Glaubitz JC, *et al.* High-resolution genetic mapping of maize pan-genome sequence anchors. Nat Commun 2015; 6: 6914. [http://dx.doi.org/10.1038/ncomms7914] [PMID: 25881062]

[163] Poland JA, Brown PJ, Sorrells ME, Jannink JL. Development of high-density genetic maps for barley and wheat using a novel two-enzyme genotyping-by-sequencing approach. PLoS One 2012; 7(2): e32253. [http://dx.doi.org/10.1371/journal.pone.0032253] [PMID: 22389690]

[164] Liu H, Bayer M, Druka A, *et al.* An evaluation of genotyping by sequencing (GBS) to map the Breviaristatum-e (ari-e) locus in cultivated barley. BMC Genomics 2014; 15: 104. [http://dx.doi.org/10.1186/1471-2164-15-104] [PMID: 24498911]

[165] Uitdewilligen JG, Wolters AM, Dhoop BB, Borm TJ, Visser RG, van Eck HJ. A next-generation sequencing method for genotyping-by-sequencing of highly heterozygous autotetraploid potato. PLoS One 2013; 8(5): e62355. [http://dx.doi.org/10.1371/journal.pone.0062355] [PMID: 23667470]

[166] Rowan BA, Patel V, Weigel D, Schneeberger K. Rapid and inexpensive whole-genome genotyping-by-sequencing for crossover localization and fine-scale genetic mapping. G3 (Bethesda) 2015; 5(3): 385-98. [http://dx.doi.org/10.1534/g3.114.016501] [PMID: 25585881]

[167] Huang YF, Poland JA, Wight CP, Jackson EW, Tinker NA. Using genotyping-by-sequencing (GBS) for genomic discovery in cultivated oat. PLoS One 2014; 9(7): e102448. [http://dx.doi.org/10.1371/journal.pone.0102448] [PMID: 25047601]

[168] Fu Y-B, Cheng B, Peterson GW. Genetic diversity analysis of yellow mustard (*Sinapis alba* L.) germplasm based on genotyping by sequencing. Genet Resour Crop Evol 2013; 61: 579-94. [http://dx.doi.org/10.1007/s10722-013-0058-1]

[169] Hart JP, Griffiths PD. Genotyping-by-Sequencing Enabled Mapping and Marker Development for the Potyvirus Resistance Allele in Common Bean. Plant Genome 2015; 8: 0. [http://dx.doi.org/10.3835/plantgenome2014.09.0058]

[170] Mokry M, Nijman IJ, van Dijken A, *et al.* Identification of factors required for meristem function in Arabidopsis using a novel next generation sequencing fast forward genetics approach. BMC Genomics 2011; 12: 256. [http://dx.doi.org/10.1186/1471-2164-12-256] [PMID: 21599977]

[171] Schneeberger K. Using next-generation sequencing to isolate mutant genes from forward genetic screens. Nat Rev Genet 2014; 15(10): 662-76. [http://dx.doi.org/10.1038/nrg3745] [PMID: 25139187]

[172] Spindel J, Wright M, Chen C, *et al.* Bridging the genotyping gap: using genotyping by sequencing (GBS) to add high-density SNP markers and new value to traditional bi-parental mapping and breeding populations. Theor Appl Genet 2013; 126(11): 2699-716. [http://dx.doi.org/10.1007/s00122-013-2166-x] [PMID: 23918062]

[173] Mascher M, Jost M, Kuon JE, *et al.* Mapping-by-sequencing accelerates forward genetics in barley. Genome Biol 2014; 15(6): R78. [http://dx.doi.org/10.1186/gb-2014-15-6-r78] [PMID: 24917130]

[174] Galvão VC, Nordström KJ, Lanz C, *et al.* Synteny-based mapping-by-sequencing enabled by targeted enrichment. Plant J 2012; 71(3): 517-26. [PMID: 22409706]

[175] Schneeberger K, Ossowski S, Lanz C, *et al.* SHOREmap: simultaneous mapping and mutation identification by deep sequencing. Nat Methods 2009; 6(8): 550-1. [http://dx.doi.org/10.1038/nmeth0809-550] [PMID: 19644454]

[176] Takagi H, Tamiru M, Abe A, *et al.* MutMap accelerates breeding of a salt-tolerant rice cultivar. Nat Biotechnol 2015; 33(5): 445-9. [PMID: 25798936]

[177] Abe A, Kosugi S, Yoshida K, *et al.* Genome sequencing reveals agronomically important loci in rice using MutMap. Nat Biotechnol 2012; 30(2): 174-8. [http://dx.doi.org/10.1038/nbt.2095] [PMID: 22267009]

[178] Fekih R, Takagi H, Tamiru M, *et al.* MutMap+: genetic mapping and mutant identification without crossing in rice. PLoS One 2013; 8(7): e68529. [http://dx.doi.org/10.1371/journal.pone.0068529] [PMID: 23874658]

[179] Tsai H, Howell T, Nitcher R, *et al.* Discovery of rare mutations in populations: TILLING by sequencing. Plant Physiol 2011; 156(3): 1257-68. [http://dx.doi.org/10.1104/pp.110.169748] [PMID: 21531898]

[180] King R, Bird N, Ramirez-Gonzalez R, *et al.* Mutation Scanning in Wheat by Exon Capture and Next-Generation Sequencing. PLoS One 2015; 10(9): e0137549. [http://dx.doi.org/10.1371/journal.pone.0137549] [PMID: 26335335]

[181] Austin RS, Vidaurre D, Stamatiou G, *et al.* Next-generation mapping of Arabidopsis genes. Plant J 2011; 67(4): 715-25. [http://dx.doi.org/10.1111/j.1365-313X.2011.04619.x] [PMID: 21518053]

[182] Wang L, Wang A, Huang X, *et al.* Mapping 49 quantitative trait loci at high resolution through sequencing-based genotyping of rice recombinant inbred lines. Theor Appl Genet 2011; 122(2): 327-40. [http://dx.doi.org/10.1007/s00122-010-1449-8] [PMID: 20878143]

[183] Yu H, Xie W, Wang J, *et al.* Gains in QTL detection using an ultra-high density SNP map based on population sequencing relative to traditional RFLP/SSR markers. PLoS One 2011; 6(3): e17595. [http://dx.doi.org/10.1371/journal.pone.0017595] [PMID: 21390234]

[184] Xie W, Feng Q, Yu H, *et al.* Parent-independent genotyping for constructing an ultrahigh-density linkage map based on population sequencing. Proc Natl Acad Sci USA 2010; 107(23): 10578-83. [http://dx.doi.org/10.1073/pnas.1005931107] [PMID: 20498060]

[185] Lim JH, Yang HJ, Jung KH, Yoo SC, Paek NC. Quantitative trait locus mapping and candidate gene analysis for plant architecture traits using whole genome re-sequencing in rice. Mol Cells 2014; 37(2): 149-60. [http://dx.doi.org/10.14348/molcells.2014.2336] [PMID: 24599000]

[186] Singh VK, Khan AW, Saxena RK, *et al.* Next-generation sequencing for identification of candidate genes for Fusarium wilt and sterility mosaic disease in pigeonpea (*Cajanus cajan*). Plant Biotechnol J 2015. [PMID: 26397045]

[187] Takagi H, Abe A, Yoshida K, *et al.* QTL-seq: rapid mapping of quantitative trait loci in rice by whole genome resequencing of DNA from two bulked populations. Plant J 2013; 74(1): 174-83. [http://dx.doi.org/10.1111/tpj.12105] [PMID: 23289725]

[188] Zong G, Wang A, Wang L, *et al.* A pyramid breeding of eight grain-yield related quantitative trait loci based on marker-assistant and phenotype selection in rice (*Oryza sativa* L.). J Genet Genom 2012; 39: 335-50.

[189] Myles S, Peiffer J, Brown PJ, *et al.* Association mapping: critical considerations shift from genotyping to experimental design. Plant Cell 2009; 21(8): 2194-202. [http://dx.doi.org/10.1105/tpc.109.068437] [PMID: 19654263]

[190] Nordborg M, Weigel D. Next-generation genetics in plants. Nature 2008; 456(7223): 720-3. [http://dx.doi.org/10.1038/nature07629] [PMID: 19079047]

[191] Varshney RK, Nayak SN, May GD, Jackson SA. Next-generation sequencing technologies and their implications for crop genetics and breeding. Trends Biotechnol 2009; 27(9): 522-30. [http://dx.doi.org/10.1016/j.tibtech.2009.05.006] [PMID: 19679362]

[192] Tian F, Bradbury PJ, Brown PJ, *et al.* Genome-wide association study of leaf architecture in the maize nested association mapping population. Nat Genet 2011; 43(2): 159-62. [http://dx.doi.org/10.1038/ng.746] [PMID: 21217756]

[193] Peiffer JA, Flint-Garcia SA, De Leon N, McMullen MD, Kaeppler SM, Buckler ES. The genetic architecture of maize stalk strength. PLoS One 2013; 8(6): e67066. [http://dx.doi.org/10.1371/journal.pone.0067066] [PMID: 23840585]

[194] Kover PX, Valdar W, Trakalo J, *et al.* A Multiparent Advanced Generation Inter-Cross to fine-map quantitative traits in *Arabidopsis thaliana.* PLoS Genet 2009; 5(7): e1000551. [http://dx.doi.org/10.1371/journal.pgen.1000551] [PMID: 19593375]

[195] Han B, Huang X. Sequencing-based genome-wide association study in rice. Curr Opin Plant Biol 2013; 16(2): 133-8. [http://dx.doi.org/10.1016/j.pbi.2013.03.006] [PMID: 23562592]

[196] Yang J, Jiang H, Yeh CT, *et al.* Extreme Phenotype-Genome-Wide Association Study (XP-GWAS): a method for identifying trait-associated variants by sequencing pools of individuals selected from a diversity panel. Plant J 2015. [http://dx.doi.org/10.1111/tpj.13029]

[197] Jannink JL, Lorenz AJ, Iwata H. Genomic selection in plant breeding: from theory to practice. Brief Funct Genomics 2010; 9(2): 166-77. [http://dx.doi.org/10.1093/bfgp/elq001] [PMID: 20156985]

[198] Desta ZA, Ortiz R. Genomic selection: genome-wide prediction in plant improvement. Trends Plant Sci 2014; 19(9): 592-601. [http://dx.doi.org/10.1016/j.tplants.2014.05.006] [PMID: 24970707]

[199] Wallace JG, Larsson SJ, Buckler ES. Entering the second century of maize quantitative genetics. Heredity (Edinb) 2014; 112(1): 30-8. [http://dx.doi.org/10.1038/hdy.2013.6] [PMID: 23462502]

[200] Varshney RK, Mohan SM, Gaur PM, *et al.* Achievements and prospects of genomics-assisted breeding in three legume crops of the semi-arid tropics. Biotechnol Adv 2013; 31(8): 1120-34. [http://dx.doi.org/10.1016/j.biotechadv.2013.01.001] [PMID: 23313999]

[201] Zhao Y, Mette MF, Reif JC, Ordon F. Genomic selection in hybrid breeding. Plant Breed 2015; 134: 1-10. [http://dx.doi.org/10.1111/pbr.12231]

[202] Heffner EL, Sorrells ME, Jannink J-L. Genomic Selection for Crop Improvement. Crop Sci 2009; 49: 1. [http://dx.doi.org/10.2135/cropsci2008.08.0512]

[203] Crossa J, Beyene Y, Kassa S, *et al.* Genomic prediction in maize breeding populations with genotyping-by-sequencing. G3 (Bethesda) 2013; 3(11): 1903-26. [http://dx.doi.org/10.1534/g3.113.008227] [PMID: 24022750]

[204] Würschum T, Reif JC, Kraft T, Janssen G, Zhao Y. Genomic selection in sugar beet breeding populations. BMC Genet 2013; 14: 85. [http://dx.doi.org/10.1186/1471-2156-14-85] [PMID: 24047500]

[205] Longin CF, Mi X, Wurschum T. Genomic selection in wheat: optimum allocation of test resources and comparison of breeding strategies for line and hybrid breeding. TAG 2015.

[206] Jarquín D, Kocak K, Posadas L, *et al.* Genotyping by sequencing for genomic prediction in a soybean breeding population. BMC Genomics 2014; 15: 740. [http://dx.doi.org/10.1186/1471-2164-15-740] [PMID: 25174348]

[207] Michael TP, Jackson S. The First 50 Plant Genomes. Plant Genome 2013; 6: 0.

[208] Dong Q, Schlueter SD, Brendel V. PlantGDB, plant genome database and analysis tools. Nucleic Acids Res 2004; 32(Database issue): D354-9. [http://dx.doi.org/10.1093/nar/gkh046] [PMID: 14681433]

[209] Sundell D, Mannapperuma C, Netotea S, *et al.* The Plant Genome Integrative Explorer Resource: PlantGenIE.org. New Phytol 2015; 208(4): 1149-56. [http://dx.doi.org/10.1111/nph.13557] [PMID: 26192091]

[210] Jaiswal P. Gramene database: a hub for comparative plant genomics. Methods Mol Biol 2011; 678: 247-75. [http://dx.doi.org/10.1007/978-1-60761-682-5_18] [PMID: 20931385]

[211] Aflitos S, Schijlen E, de Jong H, *et al.* 100 Tomato Genome Sequencing Consortium. Exploring genetic variation in the tomato (Solanum section Lycopersicon) clade by whole-genome sequencing. Plant J 2014; 80(1): 136-48. [http://dx.doi.org/10.1111/tpj.12616] [PMID: 25039268]

CHAPTER 16

Genome Editing in Crop Species

Francesca Galbiati, Damiano Martignago[#], Martina Landini[#], Jorge Gomez-Ariza[#], Vittoria Brambilla[#] and **Fabio Fornara[*]**

Department of Biosciences, Università degli Studi di Milano, Milano, Italy

Abstract: The possibility to feed an increasing world population will largely depend on the capacity to increase yield and the nutritional value of crops. Breeding has played a pivotal role so far but more is to be done to meet the challenging objective of feeding 9 billion people by 2050. Applied plant genetics is facing the dawn of a new era, in which novel genome editing technologies are opening unexpected horizons in basic and applied research. The study of DNA nucleases that can be engineered to land on specific loci of the DNA and alter its sequence is providing incredible tools for plant geneticists. This Chapter will describe genome editing technologies and the molecular bases that govern their function. Application to plant species is recent but advancing fast. Several traits of interest have already been successfully introduced or modified in crops and the first applications are starting to leave the lab and enter the path leading to commercial approval. This process is raising issues related to the regulation of genome-edited crops that governments all over the world will soon be called to rule.

Keywords: CRISPR/Cas9, Crop genetics, DNA repair, Double strand break, Genome editing, GMO regulation, Meganuclease, Oligonucleotide mediated mutagenesis, TALEN, Zinc finger nuclease.

INTRODUCTION

The world population has tremendously increased during the last decades, going from 3 to 7 billions in less than 50 years. This trend is predicted to be maintained for the near future and the growth rates will be higher in developing countries, putting pressure on communities and governments for access to resources and energy [1]. Feeding an increasing world population becomes a necessity and represents a multi-faceted problem that can be addressed in different ways.

From an agricultural perspective, increased productions of food and feed would be desirable but will have to be obtained using fewer inputs (land, fertilizers and

[*] **Corresponding author Fabio Fornara**: Department of Biosciences, Università degli Studi di Milano, Milano, Italy; Tel/Fax: +390250314817; E-mail: fabio.fornara@unimi.it

[#] These authors contributed equally to this work.

Roberto Pilu and Giuseppe Gavazzi (Eds.)

pesticides), as intensive farming has limitations for sustainable production increases [2].

During the past century, genetics has greatly helped to address some of the challenges mentioned above, increasing food production and securing yields, also thanks to the introduction of novel wheat, barley and rice varieties during the Green Revolution [3]. Classical breeding can still provide solutions to many problems but the scientific community is calling for a Green Revolution 2.0, in which the contribution of biotechnology has to be thoroughly considered [4]. Most crops are being improved to be more resistant to pathogens and abiotic stresses (*e.g.* water scarcity is an issue in many areas of the globe), to be more resilient to climate changes (often hitting locally and unpredictably) and to be prone to cultivation for biomass rather than for a specific edible part of the plant. Time is a major issue that classical breeding has to deal with, because problems can arise abruptly and require a rapid response, often incompatible with the time required to produce and validate novel crop varieties [5].

The success of most varieties has depended upon the modification of the genetic information of a species, in which selection has guided the choice of a plethora of desired traits. Variation at the DNA level is a prerequisite to be able to direct such modification in any crop species. Genetic variation is often present in nature, for example in the form of rare alleles that can be exploited by breeders when identified, passed onto superior strains and used to confer specific characteristics. For example, a crucial step during crop domestication has been the selection of mutations that reduced seed shattering, an essential trait for offspring propagation in nature, but the major cause of yield loss for ancient farmers [6]. In several polymorphism and not polymorphisms species, a single gene modification, often a single nucleotide polymorphisms (SNP), allowed to pass from shattering wild plants to domesticated, non-shattering crops [7, 8].

When genetic variation is low or absent, it can be induced in a genome using chemical or physical treatments to introduce heritable changes in the DNA such as SNPs, insertions and deletions (Indels), whole genome rearrangements, including translocations, duplications or even polyploidizations [9]. The effects of these treatments on the genome of a plant are severe and in most cases unpredictable, because they occur randomly on the DNA sequence. However, the ultimate effect is on the underlying genetic information of the organism.

The use of transgenesis (the transfer of genetic information from an organism to another by means of molecular technologies) has paved the way to introduction of novel genetic information, not previously encoded in the DNA of the species. However, genetically modified organisms (GMOs) did not meet wide acceptance and their cultivation is banned in many countries because of ethical or political issues.

The discovery of novel technologies for targeted genome editing opens novel perspectives and incredible potential for plant geneticists and crop improvement [10]. Traits of interest can be introduced or modified by acting directly at specific loci of the genome and can quickly lead to yield and agricultural productivity increases. Hundreds of plant varieties have been developed using conventional mutagenesis, often leading to random mutations in the genome and producing plants with multiple and unspecific genetic changes [11, 12]. On the contrary, genome editing technologies can be used to produce site-specific modifications in an efficient and precise manner, with little disturbance on the rest of the genome. The time required to engineer a plant genome to meet the needs of a breeding program can be strongly reduced, because the time-consuming cycles of crosses and selection could be shortened. The purity of the genetic background would be safeguarded by the targeted nature of the approach, while deleterious mutations would not be an issue. Finally, the type of genetic modification introduced in most instances would be indistinguishable from that induced by conventional mutagenesis. This issue is important from a regulatory perspective as it might allow the scientific community to move forward from a stalled debate on the use of GM plants in agriculture, while calling for a thorough reconsideration of national and international regulations governing registration, safety assessments and cultivation of such novel plant types.

A Primer On Genome Editing Technologies

Genome editing technologies collectively comprise a diverse set of molecular tools that allow the targeted modification of a DNA sequence within a genome. Precise genome modification starts by inducing double strand breaks (DSBs) in the DNA, exploiting engineered site-specific nucleases that cleave the target locus at the desired position and trigger the activation of repair pathways.

Plant cells can use two types of endogenous repair systems: non-homologous end joining (NHEJ), which involves rejoining of the broken ends, and homology-directed repair (HDR), which uses homologous DNA sequences as template for reparation [13] (Fig. **1A**). The term "non-homologous" indicates a feature of this reparation pathway in which ligation of free chromosome ends takes place without the need for a homologous template [14]. Two distinct types of NHEJ mechanisms have been described, including a classical one (cNHEJ) and an alternative one (aNHEJ) [15, 16] (Fig. **1A**). In cNHEJ the double-stranded ends are bound by ku70-ku80 heterodimers, which prevent DNA damage and recruit DNA ligase and its cofactors at the broken ends [17 - 19]. This repair pathway can generate nucleotide insertions or deletions at the DNA broken site. The aNHEJ is a ku-independent pathway facilitated by the presence of few complementary nucleotides (microhomologies) on single strand chromosomal ends. Micro-homology regions anneal with one another, while the most distal nucleotides are resected.

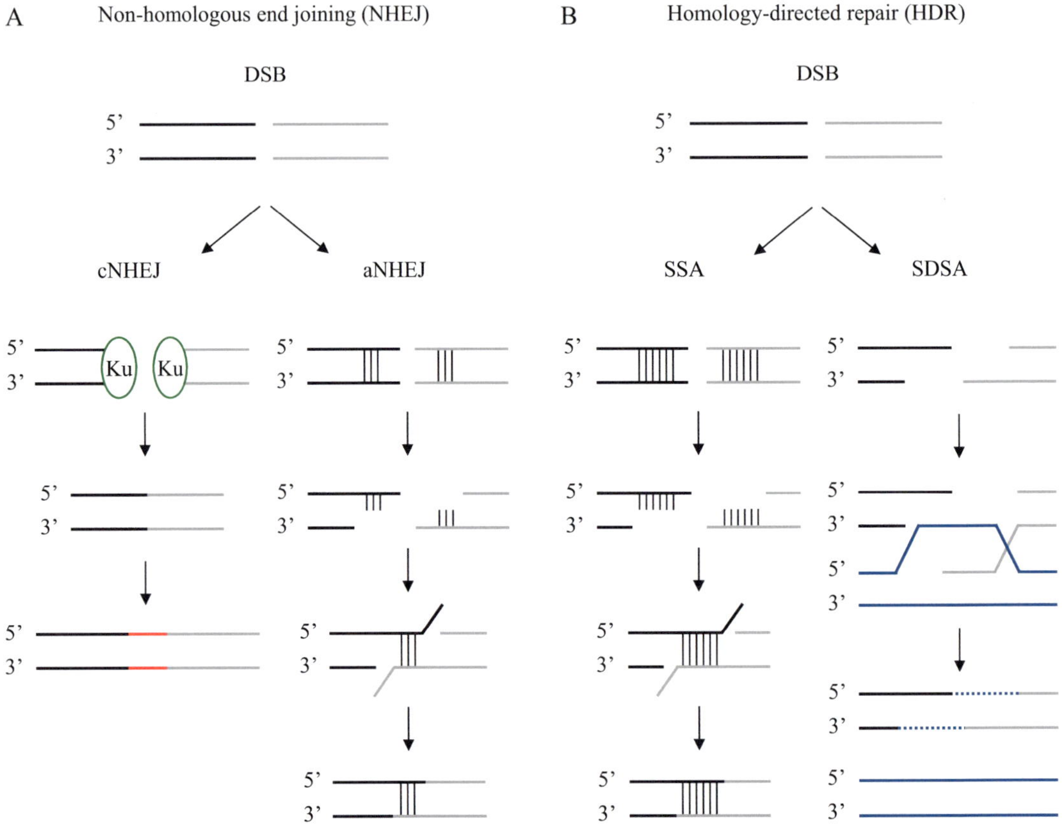

Fig. (1). DNA repair mechanisms in plant cells. (**A**) In the classical form of NHEJ, the ends of the DSB are protected by ku70-ku80 heterodimers (green circles). DNA ligation may result in deletions or insertions of nucleotides at the junction site. In aNHEJ, DNA sequences between microhomology regions are resected, leading to DNA re-ligation at the microhomology sites. Thus, aNHEJ produces larger nucleotide deletions compared to cNHEJ. (**B**) In the SSA and SDSA pathways DNA reparation starts by production of free 3' ends. SSA is a non-conservative mechanism in which homology regions present in close proximity of the DSB anneal and delete the region between the homologous repeats. In SDSA, a 3' single strand invades a double strand DNA template molecule with homology to the break site, forming a "D-loop" used as template for repair. Genetic information is maintained by the SDSA pathway (conservative mechanism).

This produces the deletion of some bp at the junction where re-ligation occurs [20] (Fig. **1A**). Both classical and alternative NHEJ repair mechanisms are considered error-prone, causing indels at the repair site and changes in the DNA sequence [21]. Rearrangements in the surroundings of the DSB could also take place. If sequence modifications cause disruption of the reading frame in a coding region, or alteration of key amino acid residues, a loss of function mutation is created [22 - 25].

As opposed to NHEJ, HDR is usually an error-free DSB repair mechanism, since it normally uses the sister chromatid as homologous template to repair DNA. However, if a specific exogenous template with a high level of homology to the sequences surrounding the break is delivered along with the site-specific nuclease, HDR mechanisms could recombine it at the target locus to repair the DSB. Any DNA sequence contained between the homology arms would be introduced, opening the possibility of manipulation of the locus, including the targeted introduction of transgenes [26].

HR appears to be an efficient and conserved repair mechanism during meiosis in eukaryotes, however it is not frequently used in somatic plant cells [27, 28]. Two distinct HDR repair mechanisms can be triggered upon DNA damage: single-strand annealing (SSA) and synthesis-dependent strand annealing (SDSA) (Fig. **1B**). In both pathways, once a DSB occurs, double strand DNA is resected and 3' single stranded overhangs are produced. If two homologous sequences are present in close proximity and the DSB takes place between them, SSA is activated. The DNA break is repaired by losing the sequence information between the homologous repeats [29]. By contrast, SDSA is a conservative repair mechanism in which a homologous template sequence is copied into the break site. No genetic information is lost during the process (Fig. **1B**) [16, 29]. A 3' single strand invades a double-strand sequence, with homology to the break site that acts as donor template forming a so called "D-loop". After the copying process, the double strand DNA is restored. Usually, SDSA uses sequences either on the same chromosome or on the sister chromatid, but rarely ectopic homology sequences can also be used as template [30, 31].

Homologous recombination occurs in plants with a very low frequency, ranging from 10^{-4} to 10^{-6} events/genome [25] and cvcn if DNA is delivered to the cells to induce homologous recombination at a specific locus, it is likely that its integration will occur *via* NHEJ [12, 32, 33] or *via* a combination of both HDR and NHEJ [30].

Both NHEJ and HDR repair mechanisms have been exploited for genome editing to produce site directed mutagenesis leading to gene correction or gene silencing. Moreover, if two DSBs simultaneously cleave the DNA at two different sites, deletions of larger portions of DNA, inversions or translocations might occur [34, 35].

During the last years, oligonucleotide-based and nuclease-based technologies have been developed to edit plant genomes [36 - 40] (Table **1**). Four types of nucleases have been engineered to produce DSBs, including engineered meaganucleases (EMNs), zinc finger nucleases (ZFNs), transcription activator-like effector nucleases (TALENs), and clustered regularly interspaced short palindromic repeats (CRISPR) coupled to CRISPR associated protein 9 (Cas9) (reviewed in [36 - 39]). Genome editing can also be directed by oligonucleotides integrating at the target locus by sequence-based homology [40]. Despite not based on the use of engineered enzymes, this system currently provides the first example of a GE strategy producing a variety ready to reach the market [40, 41].

Table 1. Examples of genome editing techniques applied to economically important traits in different crops species

species	technique	effect	Reference
Rice	ODM	herbicide tolerance	[45]
	TALEN	bacterial blight resistance	[75]
	TALEN	rice fragrance	[77]
	TALEN	multiplex gene editing	[99]
	CRISPR	multiplex gene editing	[101]
	CRISPR	bacterial blight resistance	[97]
	CRISPR	blast resistance	[97]
Maize	ODM	herbicide tolerance	[44]
	EMN	male sterility	[48]
	ZFN	herbicide tolerance	[55]
	ZFN	herbicide tolerance	[55]
Tobacco	ODM	herbicide tolerance	[56]
	ODM	herbicide tolerance	[56]
	ZFN	herbicide tolerance	[56]
	TALEN	herbicide tolerance	[89]
Wheat	TALEN	powdery mildew resistance	[76]
Canola	ODM	herbicide tolerance	[41]
Cotton	EMN	herbicide tolerance	[49]
Soybean	ZFN	multiplex gene editing	[57]
	TALEN	high oleic acid oil content	[57]
	CRISPR	herbicide tolerance	[103]
	CRISPR	multiplex gene editing	[104]
Barley	TALEN	herbicide tolerance	[104]

(Table 1) contd.....

species	technique	effect	Reference
Tomato	TALEN	organ dimension	[104]
Potato	TALEN	low reducing sugars and acrylamide	[78]
	CRISPR	herbicide tolerance	[107]

Oligonucleotide-Directed Mutagenesis

Oligonucleotide-directed mutagenesis (ODM) includes a range of vector-free oligo-based technologies that do not require the use of chemical mutagens or proteins. It consists of the introduction of a specific oligonucleotide in plant cells that drives DNA modification of the host genome. The oligonucleotide is designed to be fully homologous to the target DNA, with the exception of the nucleotide(s) to be modified. Oligonucleotides can range in size between 20 and 100 base-pairs, and can be single-stranded DNA molecules, chimeric RNA/DNA or double stranded DNA, depending on the type of ODM method. The oligonucleotide hybridizes with the targeted genomic location to create a mismatched base-pair(s) which acts as triggering signal for gene repair mechanisms [42, 43]. Reparation of the DNA using the oligonucleotide as template causes the introduction of the mutation. If germ-line cells are modified, the mutation is stabilized and transmitted to the progeny of the individual cell that received the oligonucleotide. Oligonucleotides can be delivered through proto-plast transfection, electroporation or by biolistic methods. The main advantage of this technique is that it does not make use of vectors and it does not introduce transgenic DNA in the host cell. Major drawbacks include the low frequency of gene modification, the difficulty to select and regenerate plants bearing mutations (due to the absence of a selectable marker), as well as the spontaneous occurrence of somatic mutations and off-target mutations.

The first example of ODM was presented in 1999 by Beetham and colleagues, showing that a site specific nucleotide substitution in the *ACETOLACTATE SYNTHASE* (*ALS*) gene conferred herbicide resistance in tobacco cells [40]. Since those early years, the strategy has been applied to several plant species including maize, rice and wheat, to confer herbicide resistance [44 - 46]. However, ODM is not the elite choice for plant genome editing due to its low efficiency compared to other methods and the lack of a selection marker that makes it difficult to screen for the modified plant amid non-modified ones. Nevertheless, absence of selection markers and the possibility to avoid the use of agro infiltration and introduction of foreign DNA, makes ODM not subject to the specific regulations that apply to other methods. For these reasons, crops produced by ODM can reach the market bypassing the regulatory steps that GM crops need to satisfy. Recently, a canola variety tolerant to sulphunylurea and produced by ODM, has received approval for commercialization by the Canadian Food Inspection Agency [40, 41]. The commercial release of the first genome-edited crop is thus expected in 2016 (http://www.producer.com/2014/03/growers-to-see-new-ht-canola-in-2016/).

Engineered Mega Nucleases

Engineered Mega Nucleases (EMN), also referred to as homing nucleases, include specific types of nucleases that recognize target DNA sequences of 14 to 40 nucleotides and cause DSBs upon binding (Fig. **2A**). These enzymes have been found in archea, eubacteria, unicellular eukaryotes and also in the organelles of higher organisms as plants and fungi [47]. The long DNA target sequence makes it generally unique in any given genome, reducing the incidence of off-target effects. However, the use of nucleases for genome editing is limited because the DNA recognition and cleavage functions are encoded by a single domain of the protein and modifying specificity of target recognition without affecting nuclease activity is challenging. For this reason, other more recently developed nuclease-based techniques having distinct DNA recognition strategies are nowadays often preferred for plant genome editing.

The I-CreI protein found in the chloroplasts of *Chlamydomonas reinhardti* has been the only enzyme of this class used for genome editing to date. The first successful experiments led to modification of the *LIGULELESS1* (*LG1*) locus in maize, mostly causing deletions and occasionally small insertions of nucleotides. Mutations could frequently be recovered in a bi-allelic state, demonstrating the potential of this technique to plant engineering [23]. The same approach was applied to target the *MS26* maize gene, encoding a cytochrome P450 mono-oxygenase. Mutations in *MS26* cause altered tapetum development resulting in male sterility [48]. DNA editing of the locus caused small deletions and insertions and homozygous siblings or bi-allelic T_0 mutant plants were male-sterile. D'Halluin and colleagues reported the use of a meganuclease to induce DSB and HR-mediated gene insertion in cotton [49]. In this study, the authors used transgenic Bt cotton plants resistant to glufosinate to further introduce the *epsps* and *hppd* genes (conferring resistance to glyphosate and to tembotrione, respectively) in the vicinity of the existing transgenes. This work provided the first example of stacking of multiple genes by targeted genome engineering, in which the gene stack behaved as single genetic unit.

Zinc Finger Nucleases

As opposed to meganucleases in which DNA recognition and cleavage are mediated by the same domain, the zinc-finger nucleases (ZFNs) are synthetic restriction enzymes with separate DNA-binding and DNA-cleavage domains [37, 50, 51] (Fig. **2B**). In a ZFN monomer, the non-specific DNA-cleavage domain of the *Fok*I restriction enzyme is fused with an artificial Cys_2-His_2 zinc finger (ZF) domain. Zinc fingers are the most common DNA binding motifs found in eukaryotes. Each unit consists of ~30 amino acids in a conserved ββα configuration and is bound to an atom of zinc [51, 52]. Due to their modular structure, a typical ZF domain is composed of 3-4 individual fingers, each able to recognize a specific 3-bp sequence upon binding to the major groove of the DNA.

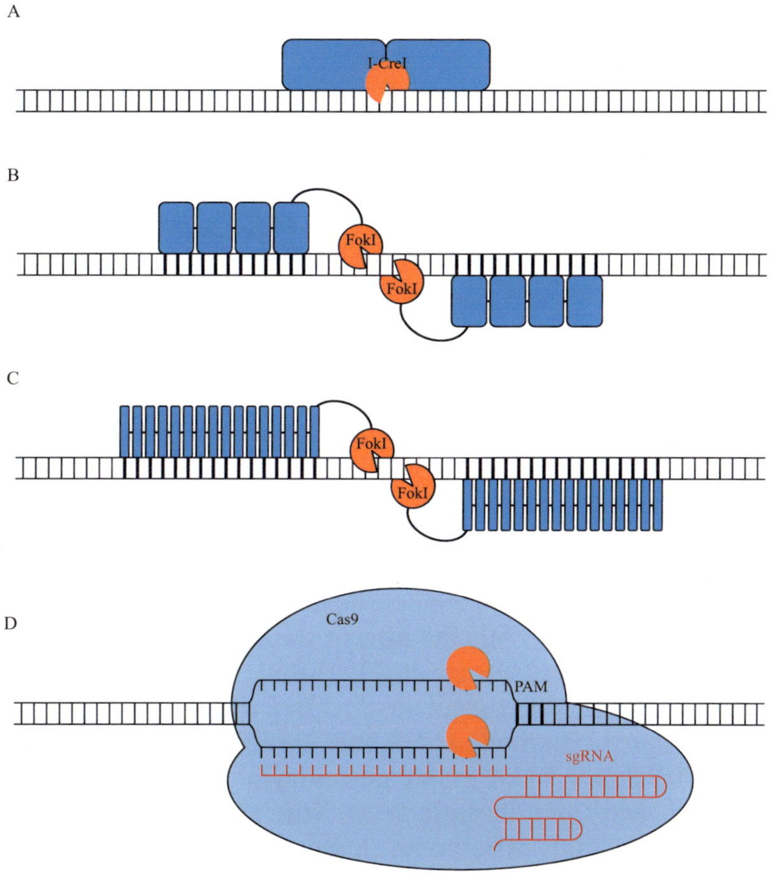

Fig. (2). Overview of engineered nucleases used to induce genome editing. (**A**) The dimeric meganuclease I–CreI catalytic domain also determines target specificity. The DNA binding domain can be modified to recognize specific target loci to induce DSB. (**B**) Zinc finger nucleases are dimeric enzymes composed of a DNA binding domain and a nuclease domain. The DNA binding domain is composed of three to four zinc fingers, each of them recognizing three nucleotides. The nuclease domain of the FokI endonuclease is responsible for DNA cleavage. (**C**) Transcription-Activator-Like Effector Nucleases also act as dimers to induce DSB *via* FokI. Each DNA binding domain consists of repeated modules that recognize specific single nucleotides. (**D**) The Cas9 nuclease has two catalytic domains, responsible for the cleavage of double strand DNA. This enzyme is guided to recognize the cleavage site by a single-guide RNA (sgRNA), which confers binding specificity to the target. The position of the protospacer-adjacent motiv (PAM) is highlighted. The nuclease domain of each enzyme is indicated in blue. The catalytic domain, cutting 3 nucleotides downstream of the PAM is indicated in orange.

The *Fok*I cleavage domain cuts DNA as a dimer [37, 53]. For editing applications, two ZFN monomers are designed to recognize specific sequences flanking a 5-6 bp spacer at the target locus. Such a mechanism, combined with the modular structure of zinc finger proteins, has made ZFNs an attractive tool for designing artificial DNA-binding proteins [37, 50, 51]. A heterodimer composed of two DNA binding domains specifically recognizing 9 bp is able to bind a target sequence of approximately 24 bp, including the spacer region, providing enough specificity for single-target selection even in a complex genome, while minimizing off-target effects [50].

Genome targeting mediated by ZFN has been applied to several crop species [24, 54 - 59]. This technology was successfully used to produce maize varieties with reduced phytate, an anti-nutritional component of seeds [55]. The final step of phytate-biosynthesis is mainly catalyzed by the *IPK1* gene, encoding for inositol-1,3,4,5,6-pentakisphosphate 2-kinase. Four ZFN pairs were designed to cleave the *IPK1* locus at different positions and delivered to maize cells, together with donor plasmids carrying a fully functional *PAT*-expression cassette, conferring resistance to Glufosinate. Multiple independent transgenic plants were recovered that exhibited herbicide resistance and disruption of the *IPK1* gene, leading to low phytate and inorganic phosphate accumulation [55]. This work provides an excellent example of rapid trait stacking in a complex genome with the aim of producing a commercial variety. Low phytate lines are currently undergoing regulatory assessment [60].

ZFN site-specific mutagenesis in the acetolactate synthase genes ALS, SuRA and SuRB was induced in maize, with the aim to confer resistance to imidazolinone and sulphonylurea. Mutations conferring herbicide resistance were achieved by using sets of ZFNs, designed to target several loci of both SuR genes. The ability of ZFNs to stimulate HR-mediated gene targeting in SuR loci was also investigated by using donor DNA molecules carrying missense or silent mutations in SuR genes [56]. Multiple mutations in SuR genes were introduced in the regenerated plants with high efficiency, demonstrating the possibility of higher order gene targeting in a single transformation [56].

The possibility of parallel editing of more than one gene was also tested in soybean using engineered ZFN, targeting nine different genes involved in RNA silencing. The efficiency of ZFN mutagenesis was tested at each target locus indicating that seven out of nine genes were mutated [24]. Moreover, this technology was also used to create mutations in multiple gene family members [57], demonstrating how ZFN can be an efficient method for studying redundant genes or for enabling genetic engineering in polyploid or paleopolyploid genomes. Traditional breeding is still the most important method used for developing new cultivars of fruit trees, even if several years are required to achieve the desired results. Very recently, ZFNs were efficiently used to generate targeted mutations in apples and in fig trees, paving the way for accelerating breeding of these species [59].

Despite the potentiality and the versatility of the technology, further research is required to extend it to a variety of model and crop species. This challenge relies not only on the development of novel ZFNs but also on enhancing methods for ZFN delivery, expression and validation in plant cells. Obtaining functionally active ZFNs is the key step for enhancing efficiency but the main obstacles will be improving both affinity and specificity of a particular ZF set [50, 61]. At least three fingers in each ZFN are required for ensuring affinity, but not all fingers make equal contributions to DNA binding. Moreover, depending on the genomic regions, specific targets might not be accessible to ZFNs due to their compact chromatin structure, making DNA cleavage difficult. Specificity of ZF binding is also crucial [37]. Adding fingers can enhance it, but the extent of binding to off-target sites still needs to be thoroughly evaluated. Finally, delivering ZFNs into a target cell can be challenging, particularly in plant species where the transformation efficiency is low or regeneration protocols are unavailable. Therefore, improved methods for delivering, expressing and selecting ZFNs are very much awaited.

Transcription Activator-Like Effector Nucleases

Transcription-Activator-Like Effector (TALE) proteins are a large family of Hrp-type III proteins also known as AvrBs3/Pth proteins, first identified in *Xanthomonas spp.* and studied as plant pathogenesis factors [62, 63]. TALE bacterial proteins carry an N-terminal secretion and translocation signal peptide that allows them to be secreted from the bacteria during infection. Subsequently, they are imported into the plant cell where they are targeted to the nucleus thanks to their C-terminal nuclear localization signals (NLS). TALE proteins are able to bind the DNA with their DNA-binding domain, and to influence transcription thanks to their C-terminal activation domain [64]. *Xanthomonas spp.* encode for a plethora of different TALE proteins that act on many pathways, allowing the bacteria to suppress plant defenses and exploit the host transcriptome to support their own growth [65].

The DNA binding domain of TALE proteins presents a series of tandem repeats, each of them comprising 33-35 amino acids and recognizing a single nucleotide (Fig. **2C**). DNA recognition specificity is conferred by the highly variable amino acids at positions 12 and 13 of each repeat (repeat-variable diresidues, RVDs). By acting on these two amino acid residues and then assembling a series of custom-designed tandem repeats, sequence-specific DNA binding domains were built [66, 67]. Chimeric Transcription Activator-Like Effector Nucleases (TALENs) were then constructed by combining the DNA binding domain of a TALE and the FokI nuclease [68, 69]. Subsequently, arrays of highly specific and efficient RDVs were identified that minimized off-target cleavage and toxicity. Assembly strategies were also improved, allowing versatile construction of plasmids for targeted gene editing [70].

As in ZFNs, FokI has to dimerize to become active and two TALENs need to be designed flanking the target site to bring FoKI subunits in contact. Homodimeric TALENs may produce off-target DBS, affecting specificity and increasing toxicity. To overcome this problem, hetrodimeric TALENs need to be generated, increasing specificity but complicating the cloning steps [71]. An interesting development that bypasses this problem has been the cloning of TALENs fused with two FokI subunits spaced by a polypeptide linker. Such single chain TALENs require only one binding site [72]. By inactivating one of the two FokI subunits, TALENs will introduce a single strand DNA break instead of a DSB. The resulting active/inactive FokI heterodimer is called nickase, and its activity tends to trigger DNA reparation by HR rather than by NHEJ [73]. Finally, instead of fusing TALEs with endonucleases, the TALE DNA-binding domain could be exploited to create transcription activators or repressors targeting specific gene promoters [69, 74].

Genome editing enabled by TALENs has been applied to several crops. The technology was first used in rice, targeting the blight susceptibility *OsSWEET14* gene [75]. The OsSWEET14 protein encodes a sucrose-efflux transporter that the plant pathogen *Xanthomonas oryzae* activates upon infection by binding to specific promoter elements, in order to exploit plant sugars produced by the host. TALENs were designed to target the motifs recognized by *X. Oryzae* in the promoter of *OsSWEET14*. The resulting plants displayed increased resistance to infection [75]. Following the same approach, bread wheat (*Triticum aestivum*) was engineered to induce mutations affecting the MILDEW-RESISTANCE LOCUS (TaMLO) proteins [76]. Mutations in the cognate genes increase resistance to powdery mildew in several species. However, *Triticum aestivum* is hexaploid (2n=42, genomes AABBDD) and obtaining simultaneous mutations of three homeoalleles by mutagenesis or classical breeding has never been possible. Artificial TALENs were designed to simultaneously target a conserved region of three *MLO* genes, resulting in plants where all six alleles were mutated. The *tamlo-aabbdd* showed strongly increased resistance to a virulent race of powdery mildew, indicating the flexibility of the technology to target multiple alleles in a polyploidy genome.

Nutritional traits have also been targeted using TALENs, in both monocots and dicots. TALEN-induced disruption of the *OsBADH2* gene of rice caused accumulation of 2-acetyl-1-pyrroline, a major fragrance compound of rice [77]. In potato tubers, cold storage causes the accumulation of reducing sugars that react with free amino acids upon cooking, producing a bitter taste and increasing the levels of acrylamide, a carcinogenic compound. TALEN enzymes were used to silence the *VACUOLAR INVERTASE* (*VInv*) gene, encoding a protein breaking sucrose into glucose and fructose. Mutagenized plants produced tubers with undetectable levels of reducing sugars and low levels of acrylamide once processed at high temperature [78].

The TALENs technology is relatively simple and straightforward to apply and an increasing number of online design tools facilitate the work of researchers. However, the choice of the correct protein design and the subsequent assembly of individual DNA-binding proteins are still labor- and time-consuming. Simpler technologies, including CRISPR/CAS9 could limit TALENs to a marginal role in genome editing.

The CRISPR/Cas9 System

The Clustered Regularly Interspaced Short Palindromic Repeats/CRISPR-Associated Protein 9 (CRISPR/Cas9) system is a genome editing technology that exploits the bacterial Cas9 protein and a synthetic guide RNA to introduce double strand breaks at target loci within the genome (Fig. **2D**). A fundamental difference between the editing methods described above and CRISPR/Cas9 is that the editing machinery is not directed to the DNA by the binding domain of an engineered protein but by a small RNA oligonucleotide that binds specific sequences by complementarity.

The CRISPR/Cas9 system derives from an adaptive immune system present in bacteria and archaea that protects against plasmid or viral DNA invasion. Ishino and colleagues first observed that in the genome of *Escherichia coli*, a set of short repeats were separated by spacer sequences [79]. Such genomic structures (the CRISPR arrays) were later observed also in archaeal genomes [80]. In 2005, several studies suggested that the spacer DNA between the repeats (or protospacers) was of foreign origin [81 - 83]. It is now understood that CRISPR arrays develop by acquiring exogenous short DNA fragments integrated between palindromic DNA repeats in the genome of the host. Foreign DNA in the array provides a specific signature (a form of memory) that can be later recognized and attacked upon invasion.

Most CRISPR arrays are transcribed as single RNAs and processed into shorter CRISPR RNAs (crRNAs), each composed of a spacer portion and a protospacer repeat. The crRNAs hybridize with a transactivating CRISPR RNA (tracrRNA). The Cas genes associated to CRISPR arrays are translated into proteins containing nuclease and helicase domains and assemble with the crRNA:tracrRNA to produce a fully functional complex that is directed against invading DNA, complementary to the spacer [81, 83 - 85].

Currently, numerous CRISPR/Cas systems with diverse genomic architectures and high rate of evolution have been described, and the most recent classification has divided them into three types (I,II and III) [86]. The most interesting system for genome editing applications is the type II system, which requires a single Cas protein for RNA-guided DNA recognition and cleavage, instead of a large complex of Cas proteins typical of type I and III systems [87]. The family of Cas9 proteins is exclusively associated with type II systems and is characterized by two distinct nuclease domains (RuvC and HNH) each cleaving a strand of DNA, producing a DSB.

The work of Jinek and colleagues represented a turning point towards the biotechnological exploitation of the system [88]. The dual crRNA:tracrRNA molecule could be engineered into a single guide RNA (sgRNA), directing Cas9 to target sites cleaved by the RuvC and HNH domains. Upon modification of the crRNA spacer sequence the complex could be directed to genes of interest to induce DSBs, paving the way for using the technology for genome editing outside its original host.

For the CRISPR/Cas9 system to properly land on target loci, the DNA complementary to the sgRNA must be adjacent to a 3-nt sequence known as protospacer-adjacent motif (PAM). The canonical PAM is 5'-NGG-3', even if other non-canonical sequences have been shown to drive efficient editing in human cells [89]. Due to the fact that the range of sequences that the sgRNA can recognize is constrained by the need for a specific PAM, Cas9 enzymes are being engineered to recognize different PAMs, to increase the range of target loci [90].

The most frequent application of the CRISPR technology is aimed at creating knock-out lines due to insertion or deletion of nucleotides within a coding sequence, upon erroneous DSB repair. However, the technology can also be exploited to control gene expression. Key to this second application is the introduction of mutations in essential residues of the RuvC and HNH nuclease domains that convert Cas9 into an RNA-guided DNA-binding protein [91]. This type of Cas9, also called dead Cas9 (dCas9), can be directed to specific loci without breaking the DNA. If the target locus lies within regulatory regions, transcription can be affected, *e.g.* initiation can be blocked by preventing the binding of RNA polymerase to the promoter [92]. Alternatively, transcription could be enhanced, upon fusing dCas9 with transcriptional activation domains [93].

The list of programmable DNA nucleases amenable to genome editing is destined to quickly expand. A putative class II CRISPR system has been isolated from *Francisella* and *Prevotella* strains and shown to mediate efficient plasmid interference in bacteria [94]. The CRISPR from *Prevotella* and *Francisella* 1 (Cpf1) protein contains no HNH domain and a RuvC-like domain with low similarity to the corresponding domain of Cas9. Features of Cpf1 that distinguish it from Cas9 include the requirement for a single crRNA for DNA targeting, preference for a different PAM sequence and, most interestingly, a cleavage resulting in a 5-nt overhang, as opposed to the blunt cleavage of Cas9. Staggered cleavage sites might open the possibility of directional introduction of DNA and possibly increase the efficiency of integration also in plants, where the efficiency of homologous recombination is low. The finding that Cpf1-family proteins are widespread among bacteria indicates that Cas9 might not be the best enzyme for every editing application, and that the toolbox of engineered nucleases might soon be richer.

The first crop species in which the CRISPR technology has been applied include tobacco, rice, sorghum and wheat [95 - 99]. The Cas9 nuclease has been codon-optimized for expression in rice [95], protocols have been improved to increase mutation frequencies [100] and finally, simultaneous mutagenesis of up to eight genes has been proven possible [101]. The number of crops in which the system has been reported to be functional is rapidly increasing and besides the species already mentioned, the genomes of maize [102], soybean [103, 104] tomato [105, 106], potato [107] as well as tree species including sweet orange and poplar [108, 109] have been edited using Cas9 and sgRNAs directed against diverse targets. Table **1** provides a non-exhaustive list of examples and agronomic traits which are destined to increase very quickly.

Regulation Of Genome-Edited Crops

The cultivation of crops with genomes modified by addition of a transgene, especially if coming from a distantly related species, is prohibited or limited in several countries and has often been accompanied by the hostility of the public opinion. The precautionary approach introduced by the Cartagena Protocol on Biosafety has been the answer to the supposed threat of genetically modified organisms to human health or the environment [110]. In the simplest form of genome editing, the DNA of a plant species can be modified by introducing one or more targeted mutations in genes of interest [111]. The current status of the technology potentially allows this to happen without the need of stable integration of the transgene in the host genome, *e.g.* by transient expression of endonucleases and guide RNAs in protoplasts or by delivery of Cas9 and sgRNA as ribonucleoprotein, followed by plant regeneration [112 - 114]. Even in case of stable integration of endonucleases and sgRNAs in the genome, mutations introduced upon DSB could be selected and segregated from the transgene during subsequent generations, resulting in a mutant plant indistinguishable from that produced by conventional chemical or physical mutagenesis. Such mutant would actually bear a genetically "cleaner" DNA, unaffected by the hundreds of random mutations that characterize a genome treated with chemical or physical mutagens [11, 12].

For all that said above, political decisions are urgently needed to rule the release of genome-edited crops in the environment and their use in agriculture. If deprived of a transgene, genome-edited plant varieties will bear the same DNA changes as conventionally bred varieties and no available technology could distinguish them. The boundaries separating genetic modifications caused by stably integrated transgenes or by genome editing are indistinct and will be even more so as technology develops. In this context, a revision of the rules currently limiting free cultivation of GMOs would be desirable, while considering the longstanding experience of countries that adopted them, and the requirements of a global and competitive international market.

In the EU, deliberate release of GMOs in the environment is ruled by Directive 2001/18/EC that defines GMOs as "organisms, with the exception of human beings, in which the genetic material has been altered in a way that does not occur naturally by mating and/or natural recombination". Mutagenesis and cell fusion of cross-compatible species or varieties fall outside the scope of this definition, as specified by Annex 1B of the same Directive. This definition emphasizes the process through which a GMO is generated and that ultimately decides which regulatory path has to be followed for varietal approval. Besides the EU, also Brazil, Australia and New Zealand have adopted a process-based regulation [12]. In the USA, a product-based regulatory framework is present; however, a process-based regulation is triggered if elements of a plant pathogen (*e.g. Agrobacterium tumefaciens*) are used to develop a GM crop variety [114]. Only Canadian regulatory frameworks assess the potential risks of the products resulting from the use of biotechnology, the focus being only on the phenotypes arising from DNA manipulation rather than the biotechnological process used to obtain them.

A progressive shift from process-based towards product-based regulation would bring several advantages. If transgenic, cisgenic and genome-edited crops were regulated similarly to varieties bred by conventional methods, potential frauds would be avoided and the competition between companies from different countries (where regulation follows either a process-based or a product-based system) would be more equal.

Universities and Public Research Centers would have the possibility to innovate and propose novel products and protocols. According to a survey of 2011 that gathered data from six very large multinational companies active in the crop biotech industry, the costs associated to discovery, development and authorization of a novel transgenic crop variety were extremely high, reaching 136M$. The costs required to satisfy regulatory requirements amounted to ~35M$ [115]. Small companies and public institutions can establish partnerships with such large enterprises but cannot lead innovation themselves. Interestingly, small companies and universities are already finding their way out of the regulatory jungle. A recent report indicates that requests of safety assessment receiving "non-regulated" status by the Animal and Plant Health Inspection Service (APHIS) of the USA have been increasing during the last years [60]. Notably, this group included some instances of varieties produced through genome editing [55, 75]. These figures are destined to increase as technology is applied to more crops and traits. Finally, the interest towards genetic improvement of minor crops that do not have high profit margins would increase.

Current opinions on the regulatory approaches that govern production and use of GMOs and genome edited plant varieties have suggested the introduction of several degrees of scrutiny [12, 114]. This approach might reduce the overall regulatory load while recognizing a diversity of instances where regulation is calibrated depending on the effects of the technology on the genetic background. When recombinant DNA encoding the endonuclease is transiently introduced in the host cell, causing a DSB repaired by NHEJ without insertion of DNA, regulatory attention should be low, as the resulting modified genome could be generated by conventional methods. An intermediate case would arise upon stable introduction of DNA in the host genome followed by selection of null segregants that do not bear the elements encoding for the editing machinery. The final result would be identical to that obtained in the previous example but involves production of a transgenic plant, at least for one generation. Finally, a high level of regulatory attention could still be maintained for stable transgene insertions, where the presence of the transgene is necessary to express the phenotypic trait.

These three broad regulatory levels could be further divided taking into consideration the type of DSB repair system [114], the length of the DNA fragment stably introduced [12] or its origin (distinguishing between transgenes, cisgenes or intragenes). Without discussing the details of these categories, it is important to stress the fact that a case-by-case evaluation would be recommendable and that technological advances will make these differences more blurred in the future.

CONCLUDING REMARKS

Genome editing technologies are advancing fast and new discoveries keep improving the methods. New endonucleases are being cloned, methods for delivering the editing machinery to the host cell are becoming more efficient and new strategies of genetic modification will make the system more flexible and adaptable to different needs. For the scientific community and the end users of the technology, it will be important to explore novel aspects of the methods, increase the efficiency of DNA modification without stable introduction of transgenes and perform quantitative, thorough evaluations of the off-target effects in crops. Of utmost importance will be to improve our understanding of the cellular processes that have been exploited to derive the technology. Basic science will still play a fundamental role. Nobody could have anticipated how the study of the bacterial immune system could have lead to the CRISPR system in such a short time. Almost unexpectedly, the plant scientific community has more powerful yet simple tools to use and can look more positively to the challenges of a second Green Revolution.

CONFLICT OF INTEREST

The authors declare no conflict of interest. This work was supported by an ERC Starting Grant to FF.

ACKNOWLEDGEMENTS

All individuals listed as authors have contributed to the work. FG and FF organized the Chapter, wrote and revised it. FG prepared figures, figure legends and formatted the text before submission. DM, ML, JGA and VB contributed writing the manuscript and revised it.

REFERENCES

[1] Prospects WP. The 2008 Revision [Internet] 2008; 2008. Available from: http://www.un.org/esa/population/publications/wpp2008/wpp2008_highlights.pdf.

[2] Tilman D, Cassman KG, Matson PA, Naylor R, Polasky S. Agricultural sustainability and intensive production practices. 2002; 418: 671-7. [http://dx.doi.org/10.1038/nature01014]

[3] Ram M. Plant breeding methods. Easter Eco 2014.

[4] Pingali PL. Green revolution: impacts, limits, and the path ahead. Proc Natl Acad Sci USA 2012; 109(31): 12302-8. [http://dx.doi.org/10.1073/pnas.0912953109] [PMID: 22826253]

[5] De La Fuente GN, Frei UK, Lübberstedt T. Accelerating plant breeding. Trends Plant Sci 2013;

18(12): 667-72.
[http://dx.doi.org/10.1016/j.tplants.2013.09.001] [PMID: 24080381]

[6] Dong Y, Wang Y-Z. Seed shattering: from models to crops. Front Plant Sci 2015; 6: 476. [http://dx.doi.org/10.3389/fpls.2015.00476] [PMID: 26157453]

[7] Konishi S, Izawa T, Lin SY, *et al.* An SNP caused loss of seed shattering during rice domestication. Science 2006; 312(5778): 1392-6. [http://dx.doi.org/10.1126/science.1126410] [PMID: 16614172]

[8] Lin Z, Li X, Shannon LM, *et al.* Parallel domestication of the Shattering1 genes in cereals. Nat Genet 2012; 44(6): 720-4. [http://dx.doi.org/10.1038/ng.2281] [PMID: 22581231]

[9] Lee LS, Till BJ, Hill H, Huynh OA, Jankowicz-Cieslak J. Mutation and mutation screening. Methods Mol Biol 2014; 1099: 77-95. [http://dx.doi.org/10.1007/978-1-62703-715-0_8] [PMID: 24243197]

[10] Hartung F, Schiemann J. Precise plant breeding using new genome editing techniques: opportunities, safety and regulation in the EU. Plant J 2014; 78(5): 742-52. [http://dx.doi.org/10.1111/tpj.12413] [PMID: 24330272]

[11] Schouten HJ, Jacobsen E. Are mutations in genetically modified plants dangerous? J Biomed Biotechnol 2007; 2007: 82612. [http://dx.doi.org/10.1111/tpj.12413] [PMID: 24330272]

[12] Araki M, Ishii T. Towards social acceptance of plant breeding by genome editing. Trends Plant Sci 2015; 20(3): 145-9. [http://dx.doi.org/10.1016/j.tplants.2015.01.010] [PMID: 25726138]

[13] Lieberman-Lazarovich M, Levy AA. Homologous recombination in plants: an antireview. Methods Mol Biol 2011; 701: 51-65. [http://dx.doi.org/10.1007/978-1-61737-957-4_3] [PMID: 21181524]

[14] Moore JK, Haber JE. Cell cycle and genetic requirements of two pathways of nonhomologous end-joining repair of double-strand breaks in *Saccharomyces cerevisiae*. Mol Cell Biol 1996; 16(5): 2164-73. [http://dx.doi.org/10.1128/MCB.16.5.2164] [PMID: 8628283]

[15] Mladenov E, Iliakis G. Induction and repair of DNA double strand breaks: the increasing spectrum of non-homologous end joining pathways. Mutat Res 2011; 711(1-2): 61-72. [http://dx.doi.org/10.1016/j.mrfmmm.2011.02.005] [PMID: 21329706]

[16] Knoll A, Fauser F, Puchta H. DNA recombination in somatic plant cells: mechanisms and evolutionary consequences. Chromosome Res 2014; 22(2): 191-201. [http://dx.doi.org/10.1007/s10577-014-9415-y] [PMID: 24788060]

[17] Critchlow SE, Jackson SP. DNA end-joining: from yeast to man. Trends Biochem Sci 1998; 23(10): 394-8. [http://dx.doi.org/10.1016/S0968-0004(98)01284-5] [PMID: 9810228]

[18] Tamura K, Adachi Y, Chiba K, Oguchi K, Takahashi H. Identification of Ku70 and Ku80 homologues in Arabidopsis thaliana: evidence for a role in the repair of DNA double-strand breaks. Plant J 2002; 29(6): 771-81. [http://dx.doi.org/10.1046/j.1365-313X.2002.01258.x] [PMID: 12148535]

[19] Deriano L, Roth DB. Modernizing the nonhomologous end-joining repertoire: alternative and classical NHEJ share the stage. Annu Rev Genet 2013; 47: 433-55. [http://dx.doi.org/10.1146/annurev-genet-110711-155540] [PMID: 24050180]

[20] Crespan E, Czabany T, Maga G, Hübscher U. Microhomology-mediated DNA strand annealing and elongation by human DNA polymerases λ and β on normal and repetitive DNA sequences. Nucleic Acids Res 2012; 40(12): 5577-90.

[http://dx.doi.org/10.1093/nar/gks186] [PMID: 22373917]

[21] Gorbunova V, Levy AA. Non-homologous DNA end joining in plant cells is associated with deletions and filler DNA insertions. Nucleic Acids Res 1997; 25(22): 4650-7. [http://dx.doi.org/10.1093/nar/25.22.4650] [PMID: 9358178]

[22] Zhang F, Maeder ML, Unger-Wallace E, *et al.* High frequency targeted mutagenesis in *Arabidopsis thaliana* using zinc finger nucleases. Proc Natl Acad Sci USA 2010; 107(26): 12028-33. [http://dx.doi.org/10.1073/pnas.0914991107] [PMID: 20508152]

[23] Gao H, Smith J, Yang M, *et al.* Heritable targeted mutagenesis in maize using a designed endonuclease. Plant J 2010; 61(1): 176-87. [http://dx.doi.org/10.1111/j.1365-313X.2009.04041.x] [PMID: 19811621]

[24] Curtin SJ, Zhang F, Sander JD, *et al.* Targeted mutagenesis of duplicated genes in soybean with zinc-finger nucleases. Plant Physiol 2011; 156(2): 466-73. [http://dx.doi.org/10.1104/pp.111.172981] [PMID: 21464476]

[25] Voytas DF. Plant genome engineering with sequence-specific nucleases. Annu Rev Plant Biol 2013; 64: 327-50. [http://dx.doi.org/10.1146/annurev-arplant-042811-105552] [PMID: 23451779]

[26] Schiml S, Fauser F, Puchta H. The CRISPR/Cas system can be used as nuclease for in planta gene targeting and as paired nickases for directed mutagenesis in Arabidopsis resulting in heritable progeny. Plant J 2014; 80(6): 1139-50. [n/a–n/a.]. [http://dx.doi.org/10.1111/tpj.12704] [PMID: 25327456]

[27] Osman K, Higgins JD, Sanchez-Moran E, Armstrong SJ, Franklin FC. Pathways to meiotic recombination in *Arabidopsis thaliana.* New Phytol 2011; 190(3): 523-44. [http://dx.doi.org/10.1111/j.1469-8137.2011.03665.x] [PMID: 21366595]

[28] Puchta H. The repair of double-strand breaks in plants: mechanisms and consequences for genome evolution. J Exp Bot 2005; 56(409): 1-14. [PMID: 15557293]

[29] Puchta H, Fauser F. Synthetic nucleases for genome engineering in plants: prospects for a bright future. Plant J 2014; 78(5): 727-41. [http://dx.doi.org/10.1111/tpj.12338] [PMID: 24112784]

[30] Puchta H. Double-strand break-induced recombination between ectopic homologous sequences in somatic plant cells. Genetics 1999; 152(3): 1173-81. [PMID: 10388832]

[31] Gisler B, Salomon S, Puchta H. The role of double-strand break-induced allelic homologous recombination in somatic plant cells. Plant J 2002; 32(3): 277-84. [http://dx.doi.org/10.1046/j.1365-313X.2002.01421.x] [PMID: 12410807]

[32] Ray A, Langer M. Homologous recombination: ends as the means. Trends Plant Sci 2002; 7(10): 435-40. [http://dx.doi.org/10.1016/S1360-1385(02)02327-0] [PMID: 12399177]

[33] Britt AB, May GD. Re-engineering plant gene targeting. Trends Plant Sci 2003; 8(2): 90-5. [http://dx.doi.org/10.1016/S1360-1385(03)00002-5] [PMID: 12597876]

[34] Siebert R, Puchta H. Efficient repair of genomic double-strand breaks by homologous recombination between directly repeated sequences in the plant genome. Plant Cell 2002; 14(5): 1121-31. [http://dx.doi.org/10.1105/tpc.001727] [PMID: 12034901]

[35] Pacher M, Schmidt-Puchta W, Puchta H. Two unlinked double-strand breaks can induce reciprocal exchanges in plant genomes *via* homologous recombination and nonhomologous end joining. Genetics 2007; 175(1): 21-9. [http://dx.doi.org/10.1534/genetics.106.065185] [PMID: 17057227]

[36] Bogdanove AJ, Voytas DF. TAL effectors: customizable proteins for DNA targeting. Science 2011; 333(6051): 1843-6.
[http://dx.doi.org/10.1126/science.1204094] [PMID: 21960622]

[37] Carroll D. Genome engineering with zinc-finger nucleases. Genetics 2011; 188(4): 773-82.
[http://dx.doi.org/10.1534/genetics.111.131433] [PMID: 21828278]

[38] Stoddard BL. Homing endonucleases: from microbial genetic invaders to reagents for targeted DNA modification. Structure (London, England : 1993). London, England 2011; 19: pp. (1)7-15.

[39] Belhaj K, Chaparro-Garcia A, Kamoun S, Patron NJ, Nekrasov V. Editing plant genomes with CRISPR/Cas9. Curr Opin Biotechnol 2015; 32: 76-84.
[http://dx.doi.org/10.1016/j.copbio.2014.11.007] [PMID: 25437637]

[40] Beetham PR, Kipp PB, Sawycky XL, Arntzen CJ, May GD. A tool for functional plant genomics: chimeric RNA/DNA oligonucleotides cause *in vivo* gene-specific mutations. Proc Natl Acad Sci USA 1999; 96(15): 8774-8.
[http://dx.doi.org/10.1073/pnas.96.15.8774] [PMID: 10411951]

[41] Pratt S. Growers to see new HT canola in 2016. the western producer [Internet] 2014. Available from: http://www.producer.com/2014/03/growers-to-see-new-ht-canola-in-2016/

[42] Oh TJ, May GD. Oligonucleotide-directed plant gene targeting. Curr Opin Biotechnol 2001; 12(2): 169-72.
[http://dx.doi.org/10.1016/S0958-1669(00)00194-4] [PMID: 11287232]

[43] Breyer D, Herman P, Brandenburger A, *et al.* Genetic modification through oligonucleotide-mediated mutagenesis. A GMO regulatory challenge? Environ Biosafety Res 2009; 8(2): 57-64.
[http://dx.doi.org/10.1051/ebr/2009007] [PMID: 19833073]

[44] Zhu T, Mettenburg K, Peterson DJ, Tagliani L, Baszczynski CL. Engineering herbicide-resistant maize using chimeric RNA/DNA oligonucleotides. Nat Biotechnol 2000; 18(5): 555-8.
[http://dx.doi.org/10.1038/75435] [PMID: 10802626]

[45] Okuzaki A, Toriyama K. Chimeric RNA/DNA oligonucleotide-directed gene targeting in rice. Plant Cell Rep 2004; 22(7): 509-12.
[http://dx.doi.org/10.1007/s00299-003-0698-2] [PMID: 14634786]

[46] Dong C, Beetham P, Vincent K, Sharp P. Oligonucleotide-directed gene repair in wheat using a transient plasmid gene repair assay system. Plant Cell Rep 2006; 25(5): 457-65.
[http://dx.doi.org/10.1007/s00299-005-0098-x] [PMID: 16404599]

[47] Hafez M, Hausner G. Homing endonucleases: DNA scissors on a mission. Genome 2012; 55(8): 553-69.
[http://dx.doi.org/10.1139/g2012-049] [PMID: 22891613]

[48] Djukanovic V, Smith J, Lowe K, *et al.* Male-sterile maize plants produced by targeted mutagenesis of the cytochrome P450-like gene (MS26) using a re-designed I-CreI homing endonuclease. Plant J 2013; 76(5): 888-99.
[http://dx.doi.org/10.1111/tpj.12335] [PMID: 24112765]

[49] DHalluin K, Vanderstraeten C, Van Hulle J, *et al.* Targeted molecular trait stacking in cotton through targeted double-strand break induction. Plant Biotechnol J 2013; 11(8): 933-41.
[http://dx.doi.org/10.1111/pbi.12085] [PMID: 23777410]

[50] Weinthal D, Tovkach A, Zeevi V, Tzfira T. Genome editing in plant cells by zinc finger nucleases. Trends Plant Sci 2010; 15(6): 308-21.
[http://dx.doi.org/10.1016/j.tplants.2010.03.001] [PMID: 20347379]

[51] Gaj T, Gersbach CA, Barbas CF III. ZFN, TALEN, and CRISPR/Cas-based methods for genome engineering. Trends Biotechnol 2013; 31(7): 397-405.
[http://dx.doi.org/10.1016/j.tibtech.2013.04.004] [PMID: 23664777]

[52] Beerli RR, Barbas CF III. Engineering polydactyl zinc-finger transcription factors. Nat Biotechnol 2002; 20(2): 135-41.
[http://dx.doi.org/10.1038/nbt0202-135] [PMID: 11821858]

[53] Urnov FD, Rebar EJ, Holmes MC, Zhang HS, Gregory PD. Genome editing with engineered zinc finger nucleases. Nat Rev Genet 2010; 11(9): 636-46.
[http://dx.doi.org/10.1038/nrg2842] [PMID: 20717154]

[54] Cai CQ, Doyon Y, Ainley WM, *et al.* Targeted transgene integration in plant cells using designed zinc finger nucleases. Plant Mol Biol 2009; 69(6): 699-709.
[http://dx.doi.org/10.1007/s11103-008-9449-7] [PMID: 19112554]

[55] Shukla VK, Doyon Y, Miller JC, *et al.* Precise genome modification in the crop species *Zea mays* using zinc-finger nucleases. Nature 2009; 459(7245): 437-41.
[http://dx.doi.org/10.1038/nature07992] [PMID: 19404259]

[56] Townsend JA, Wright DA, Winfrey RJ, *et al.* High-frequency modification of plant genes using engineered zinc-finger nucleases. Nature 2009; 459(7245): 442-5.
[http://dx.doi.org/10.1038/nature07845] [PMID: 19404258]

[57] Curtin SJ, Anderson JE, Starker CG, *et al.* Targeted mutagenesis for functional analysis of gene duplication in legumes. Methods Mol Biol 2013; 1069: 25-42.
[http://dx.doi.org/10.1007/978-1-62703-613-9_3] [PMID: 23996306]

[58] Ainley WM, Sastry-Dent L, Welter ME, *et al.*

[59] Peer R, Rivlin G, Golobovitch S, *et al.* Targeted mutagenesis using zinc-finger nucleases in perennial fruit trees. Planta 2015; 241(4): 941-51.
[http://dx.doi.org/10.1007/s00425-014-2224-x] [PMID: 25528147]

[60] Camacho A, Van Deynze A, Chi-Ham C, Bennett AB. Genetically engineered crops that fly under the US regulatory radar. Nat Biotechnol 2014; 32(11): 1087-91.
[http://dx.doi.org/10.1038/nbt.3057] [PMID: 25380439]

[61] Porteus MH. Plant biotechnology: Zinc fingers on target. Nature 2009; 459(7245): 337-8.
[http://dx.doi.org/10.1038/459337a] [PMID: 19458706]

[62] Minsavage GV. Gene-For-Gene Relationships Specifying Disease Resistance in *Xanthomonas campestris* pv. *vesicatoria* - Pepper Interactions. Mol Plant Microbe Interact 1990; 3: 41.
[http://dx.doi.org/10.1094/MPMI-3-041]

[63] Bonas U, Conrads-Strauch J, Balbo I. Resistance in tomato to *Xanthomonas campestris* pv vesicatoria is determined by alleles of the pepper-specific avirulence gene avrBs3. Mol Gen Genet 1993; 238(1-2): 261-9.
[PMID: 8479432]

[64] Boch J, Bonas U. Xanthomonas AvrBs3 family-type III effectors: discovery and function. Annu Rev Phytopathol 2010; 48: 419-36.
[http://dx.doi.org/10.1146/annurev-phyto-080508-081936] [PMID: 19400638]

[65] Hogenhout SA, Van der Hoorn RA, Terauchi R, Kamoun S. Emerging concepts in effector biology of plant-associated organisms. Mol Plant Microbe Interact 2009; 22(2): 115-22.
[http://dx.doi.org/10.1094/MPMI-22-2-0115] [PMID: 19132864]

[66] Boch J, Scholze H, Schornack S, *et al.* Breaking the code of DNA binding specificity of TAL-type III effectors. Science 2009; 326(5959): 1509-12.
[http://dx.doi.org/10.1126/science.1178811] [PMID: 19933107]

[67] Moscou MJ, Bogdanove AJ. A simple cipher governs DNA recognition by TAL effectors. Science 2009 Dec 11;326(5959):1501
[PMID: 19933106]

[68] Christian M, Cermak T, Doyle EL, *et al.* Targeting DNA double-strand breaks with TAL effector

nucleases. Genetics 2010; 186(2): 757-61.
[http://dx.doi.org/10.1534/genetics.110.120717] [PMID: 20660643]

[69] Morbitzer R, Römer P, Boch J, Lahaye T. Regulation of selected genome loci using de *novo*-engineered transcription activator-like effector (TALE)-type transcription factors. Proc Natl Acad Sci USA 2010; 107(50): 21617-22.
[http://dx.doi.org/10.1073/pnas.1013133107] [PMID: 21106758]

[70] Sun N, Zhao H. Transcription activator-like effector nucleases (TALENs): a highly efficient and versatile tool for genome editing. Biotechnol Bioeng 2013; 110(7): 1811-21.
[http://dx.doi.org/10.1002/bit.24890] [PMID: 23508559]

[71] Cade L, Reyon D, Hwang WY, *et al.* Highly efficient generation of heritable zebrafish gene mutations using homo- and heterodimeric TALENs. Nucleic Acids Res 2012; 40(16): 8001-10.
[http://dx.doi.org/10.1093/nar/gks518] [PMID: 22684503]

[72] Sun N, Zhao H. A single-chain TALEN architecture for genome engineering. Mol Biosyst 2014; 10(3): 446-53.
[http://dx.doi.org/10.1039/C3MB70412B] [PMID: 24336919]

[73] Wu Y, Gao T, Wang X, *et al.* TALE nickase mediates high efficient targeted transgene integration at the human multi-copy ribosomal DNA locus. Biochem Biophys Res Commun 2014; 446(1): 261-6.
[http://dx.doi.org/10.1016/j.bbrc.2014.02.099] [PMID: 24589733]

[74] Mahfouz MM, Li L, Piatek M, *et al.* Targeted transcriptional repression using a chimeric TALE-SRDX repressor protein. Plant Mol Biol 2012; 78(3): 311-21.
[http://dx.doi.org/10.1007/s11103-011-9866-x] [PMID: 22167390]

[75] Li T, Liu B, Spalding MH, Weeks DP, Yang B. High-efficiency TALEN-based gene editing produces disease-resistant rice. Nat Biotechnol 2012; 30(5): 390-2.
[http://dx.doi.org/10.1038/nbt.2199] [PMID: 22565958]

[76] Wang Y, Cheng X, Shan Q, *et al.* Simultaneous editing of three homoeoalleles in hexaploid bread wheat confers heritable resistance to powdery mildew. Nat Biotechnol 2014; 32(9): 947-51.
[http://dx.doi.org/10.1038/nbt.2969] [PMID: 25038773]

[77] Shan Q, Zhang Y, Chen K, Zhang K, Gao C. Creation of fragrant rice by targeted knockout of the OsBADH2 gene using TALEN technology. Plant Biotechnol J 2015; 13(6): 791-800.
[http://dx.doi.org/10.1111/pbi.12312] [PMID: 25599829]

[78] Clasen BM, Stoddard TJ, Luo S, *et al.* Improving cold storage and processing traits in potato through targeted gene knockout. Plant Biotechnol J 2015.
[PMID: 25846201]

[79] Ishino Y, Shinagawa H, Makino K, Amemura M, Nakata A. Nucleotide sequence of the iap gene, responsible for alkaline phosphatase isozyme conversion in *Escherichia coli*, and identification of the gene product. J Bacteriol 1987; 169(12): 5429-33.
[http://dx.doi.org/10.1128/jb.169.12.5429-5433.1987] [PMID: 3316184]

[80] Mojica FJ, Díez-Villaseñor C, Soria E, Juez G. Biological significance of a family of regularly spaced repeats in the genomes of Archaea, Bacteria and mitochondria. Mol Microbiol 2000; 36(1): 244-6.
[http://dx.doi.org/10.1046/j.1365-2958.2000.01838.x] [PMID: 10760181]

[81] Bolotin A, Quinquis B, Sorokin A, Ehrlich SD. Clustered regularly interspaced short palindrome repeats (CRISPRs) have spacers of extrachromosomal origin. Microbiology 2005; 151(Pt 8): 2551-61.
[http://dx.doi.org/10.1099/mic.0.28048-0] [PMID: 16079334]

[82] Mojica FJ, Díez-Villaseñor C, García-Martínez J, Soria E. Intervening sequences of regularly spaced prokaryotic repeats derive from foreign genetic elements. J Mol Evol 2005; 60(2): 174-82.
[http://dx.doi.org/10.1007/s00239-004-0046-3] [PMID: 15791728]

[83] Pourcel C, Salvignol G, Vergnaud G. CRISPR elements in Yersinia pestis acquire new repeats by preferential uptake of bacteriophage DNA, and provide additional tools for evolutionary studies.

Microbiology 2005; 151(Pt 3): 653-63.
[http://dx.doi.org/10.1099/mic.0.27437-0] [PMID: 15758212]

[84] Jansen R, Embden JD, Gaastra W, Schouls LM. Identification of genes that are associated with DNA repeats in prokaryotes. Mol Microbiol 2002; 43(6): 1565-75.
[http://dx.doi.org/10.1046/j.1365-2958.2002.02839.x] [PMID: 11952905]

[85] Haft DH, Selengut J, Mongodin EF, Nelson KE. A guild of 45 CRISPR-associated (Cas) protein families and multiple CRISPR/Cas subtypes exist in prokaryotic genomes. PLOS Comput Biol 2005; 1(6): e60.
[http://dx.doi.org/10.1371/journal.pcbi.0010060] [PMID: 16292354]

[86] Makarova KS, Haft DH, Barrangou R, *et al.* Evolution and classification of the CRISPR-Cas systems. Nat Rev Microbiol 2011; 9(6): 467-77.
[http://dx.doi.org/10.1038/nrmicro2577] [PMID: 21552286]

[87] Doudna JA, Charpentier E. Genome editing. The new frontier of genome engineering with CRISPR-Cas9. Science 2014; 346(6213): 1258096-6.
[http://dx.doi.org/10.1126/science.1258096] [PMID: 25430774]

[88] Jinek M, Chylinski K, Fonfara I, Hauer M, Doudna JA, Charpentier E. A programmable dual-RN--guided DNA endonuclease in adaptive bacterial immunity. Science 2012; 337(6096): 816-21.
[http://dx.doi.org/10.1126/science.1225829] [PMID: 22745249]

[89] Zhang Y, Ge X, Yang F, *et al.* Comparison of non-canonical PAMs for CRISPR/Cas9-mediated DNA cleavage in human cells. Sci Rep 2014; 4: 5405.
[PMID: 24956376]

[90] Kleinstiver BP, Prew MS, Tsai SQ, *et al.* Engineered CRISPR-Cas9 nucleases with altered PAM specificities. Nature 2015; 523(7561): 481-5.
[http://dx.doi.org/10.1038/nature14592] [PMID: 26098369]

[91] Bikard D, Marraffini LA. Control of gene expression by CRISPR-Cas systems. F1000Prime Rep 2013; 5: 47.
[http://dx.doi.org/10.12703/P5-47] [PMID: 24273648]

[92] Qi LS, Larson MH, Gilbert LA, *et al.* Repurposing CRISPR as an RNA-guided platform for sequence-specific control of gene expression. Cell 2013; 152(5): 1173-83.
[http://dx.doi.org/10.1016/j.cell.2013.02.022] [PMID: 23452860]

[93] Gilbert LA, Larson MH, Morsut L, *et al.* CRISPR-mediated modular RNA-guided regulation of transcription in eukaryotes. Cell 2013; 154(2): 442-51.
[http://dx.doi.org/10.1016/j.cell.2013.06.044] [PMID: 23849981]

[94] Zetsche B, Gootenberg JS, Abudayyeh OO, *et al.* Cpf1 is a single RNA-guided endonuclease of a class 2 CRISPR-Cas system. Cell 2015; 163(3): 759-71.
[http://dx.doi.org/10.1016/j.cell.2015.09.038] [PMID: 26422227]

[95] Miao J, Guo D, Zhang J, *et al.* Targeted mutagenesis in rice using CRISPR-Cas system. Cell Res 2013; 23(10): 1233-6.
[http://dx.doi.org/10.1038/cr.2013.123] [PMID: 23999856]

[96] Nekrasov V, Staskawicz B, Weigel D, Jones JD, Kamoun S. Targeted mutagenesis in the model plant Nicotiana benthamiana using Cas9 RNA-guided endonuclease. Nat Biotechnol 2013; 31(8): 691-3.
[http://dx.doi.org/10.1038/nbt.2655] [PMID: 23929340]

[97] Jiang W, Zhou H, Bi H, Fromm M, Yang B, Weeks DP. Demonstration of CRISPR/Cas9/sgRNA-mediated targeted gene modification in Arabidopsis, tobacco, sorghum and rice. Nucleic Acids Res 2013; 41(20): e188.
[http://dx.doi.org/10.1093/nar/gkt780] [PMID: 23999092]

[98] Feng Z, Zhang B, Ding W, *et al.* Efficient genome editing in plants using a CRISPR/Cas system. Cell Res 2013; 23(10): 1229-32.

[http://dx.doi.org/10.1038/cr.2013.114] [PMID: 23958582]

[99] Shan Q, Wang Y, Li J, *et al.* Targeted genome modification of crop plants using a CRISPR-Cas system. Nat Biotechnol 2013; 31(8): 686-8. [http://dx.doi.org/10.1038/nbt.2650] [PMID: 23929338]

[100] Mikami M, Toki S, Endo M. Parameters affecting frequency of CRISPR/Cas9 mediated targeted mutagenesis in rice. Plant Cell Rep 2015; 34(10): 1807-15. [http://dx.doi.org/10.1007/s00299-015-1826-5] [PMID: 26134856]

[101] Ma X, Zhang Q, Zhu Q, *et al.* A robustCRISPR/Cas9 system for convenient, high-efficiency multiplex genome editing in monocot and dicotplants. Mol Plant 2015; 8(8): 1274-84. [http://dx.doi.org/10.1016/j.molp.2015.04.007] [PMID: 25917172]

[102] Liang Z, Zhang K, Chen K, Gao C. Targeted mutagenesis in Zea mays using TALENs and the CRISPR/Cas system. J Genet Genomics 2014; 41(2): 63-8. [http://dx.doi.org/10.1016/j.jgg.2013.12.001] [PMID: 24576457]

[103] Li Z, Liu Z-B, Xing A, *et al.* Cas9-guide RNA Directed Genome Editing in Soybean. Plant Physiol 2015; 169: 00783.2015. [http://dx.doi.org/10.1104/pp.15.00783]

[104] Cai Y, Chen L, Liu X, *et al.* CRISPR/Cas9-Mediated Genome Editing in Soybean Hairy Roots. PLoS One 2015; 10(8): e0136064. [http://dx.doi.org/10.1371/journal.pone.0136064] [PMID: 26284791]

[105] Ron M, Kajala K, Pauluzzi G, *et al.* Hairy root transformation using Agrobacterium rhizogenes as a tool for exploring cell type-specific gene expression and function using tomato as a model. Plant Physiol 2014; 166(2): 455-69. [http://dx.doi.org/10.1104/pp.114.239392] [PMID: 24868032]

[106] Brooks C, Nekrasov V, Lippman Z, Van Eck J. Efficient gene editing in tomato in the first generation using the CRISPR/Cas9 system. Plant Physiol 2014; 166: 1292-7. [http://dx.doi.org/10.1104/pp.114.247577] [PMID: 25225186]

[107] Wang S, Zhang S, Wang W, Xiong X, Meng F, Cui X. Efficient targeted mutagenesis in potato by the CRISPR/Cas9 system. Plant Cell Rep 2015; 34(9): 1473-6. [http://dx.doi.org/10.1007/s00299-015-1816-7] [PMID: 26082432]

[108] Jia H, Wang N. Targeted genome editing of sweet orange using Cas9/sgRNA. PLoS One 2014; 9(4): e93806. [http://dx.doi.org/10.1371/journal.pone.0093806] [PMID: 24710347]

[109] Fan D, Liu T, Li C, *et al.* Efficient CRISPR/Cas9-mediated Targeted Mutagenesis in Populus in the First Generation. Sci Rep 2015; 5: 12217. [http://dx.doi.org/10.1038/srep12217] [PMID: 26193631]

[110] Cartagena protocol on biosafety to the convention on biological diversity. Secretariat of the Convention on Biological Diversity 2000. Available from: http://bch.cbd.int/protocol/text/

[111] Osakabe Y, Osakabe K. Genome editing with engineered nucleases in plants. Plant Cell Physiol 2015; 56(3): 389-400. [http://dx.doi.org/10.1093/pcp/pcu170] [PMID: 25416289]

[112] Shan Q, Wang Y, Li J, Gao C. Genome editing in rice and wheat using the CRISPR/Cas system. Nat Protoc 2014; 9(10): 2395-410. [http://dx.doi.org/10.1038/nprot.2014.157] [PMID: 25232936]

[113] Woo JW, Kim J, Kwon S. DNA-free genome editing in plants with preassembled CRISPR-Cas9 ribonucleoproteins. Nat Biotechnol 33: 1162-4. Available from: http://www.ncbi.nlm.nih.gov/pubmed/26479191

[114] Wolt JD, Wang K, Yang B. The Regulatory Status of Genome-edited Crops. Plant Biotechnol J 2015

Aug; 7n/a-a.

[115] McDougall P. The cost and time involved in the discovery, development and authorisation of a new plant biotechnology derived trait 2011. Available from: https://croplife.org/wp-content/uploads/2014/04/Getting-a-Biotech-Crop-to-Market-Phillips-McDougall-Study.pdf

CHAPTER 17

Progress in Small Grain Cereals: A Case Study

A. Michele Stanca[1,*], Enrico Francia[1], Alessandro Tondelli[2], Franz W. Badeck[2] and **Valeria Terzi[2]**

[1] *Department of Life Sciences, Università degli Studi di Modena e Reggio Emilia, Modena, Italy*

[2] *Consiglio per la ricerca in agricoltura e l'analisi dell'economia agraria (CREA), Genomics Research Centre, Fiorenzuola d'Arda, Italy*

Abstract: Small grain cereals (here including durum wheat, bread wheat, barley and oats) have been crucial to the development of mankind providing a regular staple source of food compounds – carbohydrates, proteins, fat and secondary metabolites – since their domestication 10,000 years ago. Historically, genetic studies have their foundations in Mendelian mutants, characterized by altered physiology and/or morphology. In this regard there are examples of morphological mutations described in the past for which the gene/genes responsible have been recently cloned, characterized and used. An example is the Rht-B1b gene that controls plant height in wheat, which induces semidwarf plants due to the effect of a single nucleotide mutation capable of converting the majority of sugar into grain starch. With this model the source-sink relationship has been studied in depth and new varieties based on the concept of "Improved Harvest Index" have been released with an impressive grain yield enhancement in a wide range of environments. The question is: "Can we produce and supply sufficient food in the next 40 years without consuming more land?" On the basis of modern plant science, the answer is positive. Selection is specifically directed to create highly tolerant and/or resistant genotypes to increase the "High Yield Potential and Stability of Yield" and to reduce the gap between high yield potential and the actual yield also in very poor small farms (low or zero input). In fact the interaction between private and public pre-breeding-/breeding programmes, allowing the introduction of modern varieties which are very well adapted in fertile as well as in severe stress conditions, represents the modern vision to improve not only grain yield but even the quality of life of all farmers.

Keywords: Barley, Breeding progress, Genomic revolution, Grain Yield, Oats, Plant ideotype, Wheats.

HISTORICAL CEREAL BREEDING PROGRESS

Cereals have been and are important because they provide two-thirds of the calories and half of the protein in the diets of humans and they are major feed-

[*] **Corresponding author A. Michele Stanca**: Department of Life Sciences, Università degli Studi di Modena e Reggio Emilia, Modena, Italy; Tel: 0039 340 3707674; E-mail: michele@stanca.it

Roberto Pilu and Giuseppe Gavazzi (Eds.)

stuffs for both monogastric and ruminant livestock. In 2015, world cereal production was estimated at 2,525 Million tonnes (FAOSTAT). Small grain cereals (here including durum wheat, bread wheat, barley and oats) have been crucial to the development of mankind, providing a regular staple source of food compounds – carbohydrates, proteins, fat, fibre, minerals, and secondary metabolites. Their history as crops began in the 'Fertile Crescent', a region including parts of Iran, Iraq, Turkey, Syria, Jordan, Lebanon and Israel. The domestication of the wild ancestors, *Hordeum spontaneum* (L.), diploid wheat and later oats, which are still to be found in the Middle East, dates to the Neolithic. Human intervention has been a decisive factor in the breeding of these species down through the millennia, the cultivars currently available manifesting completely different traits from those found in either the original parents or in those populations that have evolved in the wild. The first, and perhaps most outstanding development, was domestication itself – that moment in which man stood before an immense number of different species and chose cereals. In these species the new farmer saw the genetic variants most responsive to his/her needs [1]. He invented by copying from mother nature the earliest rudimentary agronomic techniques: seed and soil selection, ploughing, sowing, manual weed control to protect the crops, seed gathering and storage. With experience, the best plants were likely earmarked for the following year's seed supply rather than for direct consumption, incidentally giving rise to 'mass selection'.

Yet why, we might well wonder, among the thousands of species available, were cereals the first crops bred and continuously cultivated down through the ages? At the dawn of agricultural civilization humans realized that the grain cereals could ensure their survival, an intuition which initiated an instinctive selection, generation after generation, of the genotypes best suited to the farmer's needs. One example of this is the discovery of plants with non-fragile spikes. This was indubitably a spark in the impulse towards cereals cultivation, being in all likelihood the first criterion of selection in the history of plant breeding. Such alterations in the rachis were the pre-requisite for the harvest of the whole ear. Brittle rachis genes are present in all wild species of small grain cereals and very well described. The trait itself, *e.g.* in barley, is actually controlled by the two complementary genes *Btr1* and *Btr2* [2]. The change may seem a simple one to us, yet for Neolithic or late Palaeolithic man it very likely improved his chances of survival. The discovery too of naked seed represented another important milestone on the road to the spread of cereals as a food staple. This selection process was necessarily focused on phenotypes distinguishable for simple traits controlled by one or not more than a few genes - a method that was empirically pursued right up to the end of the nineteenth century.

Another step forward in newborn agriculture was the conservation for the subsequent year's sowing (rather than for direct consumption) of the seeds produced by those plants deemed the best performing. The progress this practice engendered evidently had to be extraordinary in terms of quantitative and qualitative increase (large kernels), given the subsequent Middle Eastern civilizations that sprang up and developed around these plants. This Neolithic model of agricultural technology was exported from the Middle East at a rate of one kilometre per year and it is from here that the species set out first to colonize the Mediterranean Basin and then, following the great river and sea migration routes of the time, fanned out to central and northern Europe and the Far East. The perfecting of crop management techniques (*e.g.* the advent of rotation and the practice of fallowing), in particular during the period of Roman dominion, contributed to the spread of cereals in the fertile as well as in the poorer soils of the Empire. As is well-known, cereal breeding has been developed during three revolutions, beginning with the Neolithic, during which, unconsciously a good gene – non- fragile spike - was discovered and used. The humans learned to sow and to harvest and they worked hard to develop new equipment to improve quantity and quality of food production. It was the birth of agriculture. The second revolution was the Mendelian, during which theoretical and practical goals have been reached. The work of G. Mendel was re-discovered at the beginning of 1900 when the modern age of plant breeding began. Until that time, the increase in food production was obtained simply by bringing new land into cultivation. In fact, the yield per unit of land remained stable from the Roman period to the beginning of the 20th Century. The application of Mendel's laws, the development of modern varieties and the introduction of new agricultural practices gave strong contributions towards improving grain yield around the world. However, more rapid increases in grain production were attained when semi-dwarf varieties were introduced into cultivation. In fact the new varieties increased grain yield by reducing height at the expense of straw biomass, and as a consequence became more resistant to lodging. The increase of the grain-to-straw ratio resulted in improving the Harvest Index (HI). In 1964 CIMMYT varieties were introduced in Asia, the Near East, Latin America, Australia and Europe. Grain production in many cases doubled and countries from importers became exporters. In 1970 the so-called "Green Revolution" was realised. Farmers rapidly adopted the new varieties and new agronomic techniques were developed. The technique of ideotype breeding was born. How the new wheat or barley genotypes make the better use of the environment has been considered as a model to be exported to many other crops. For the first time knowledge of genetics, statistics, biochemistry, pathology, physiology, and more recently molecular biology are synthesized by the breeder. The advances linked to genetics in all countries around the world have been remarkable, and breeding practices (pedigree selection and its modifications) have gradually widened their compass and developed up to the present through the research efforts of both the public and private sectors. Successful winter and spring cultivars marked by excellent quality, more flexible straw and resistance to biotic and abiotic stresses have been released. Modern varieties, when properly grown, during the last century have been demonstrated to highly increase grain yield and grain quality everywhere in the world. The contribution of high yielding varieties to yield increases can be estimated at 2%/*ha*/year from 1920 to the end of that century. Even if barley and wheat are classified self pollinated species, and more specifically cleistogamous, recently F1 wheat and barley hybrids, with a 20% heterotic effect, are under cultivation. Parallel to this activity the oldest and obsolete varieties, landraces, mutants and in general genetic stocks have been collected and stored in germplasm banks – 800,000; 480,000 and 136,000 accessions for wheat, barley and oats, respectively - to be available as reserves of genes for future generations. On the basis of the above considerations, the target of "Yield Potential and Stability of Yield" represents the strategy to design the plant for the future, based on pre-breeding/breeding work, with the final aim to increase the production of numbers of seed per unit of land and to produce more grains for "Feeding Ten Billions". The question is "How can we produce and supply sufficient food in the next 40 years without consuming more land?" Intense pre-breeding/breeding work is necessary to accumulate good alleles in elite germplasm.

SPECIES EVOLUTION

Barley (*Hordeum vulgare* L.)

Barley belongs to the *Triticeae* and has true diploid nature (2n=2x=14). The species possesses seven pairs of distinct chromosomes, five of them without satellites – 1H, 2H, 3H, 4H, 7H, and two with satellites – 5H and 6H. Cultivated barley is inbreeding, with cleistogamous flowers, even if under stress conditions it can be more open flowering. The normal outcrossing rate has been determined to be a maximum of 1.7%-1.8%. Despite a very high inbreeding rate, a high degree of diversity is maintained, which is sufficient for rapid adaptation to varying environmental conditions. Its seven chromosomes represent the basic genome of all *Triticeae*. The first complete barley genome sequence was publicly released in 2012, overcoming the difficulties linked to its size (about 5,500 MB) and the presence of repetitive DNA regions [3]. The springboard for the sequencing of the barley genome has been offered by key molecular tools developed in the recent past, like barley ESTs (Expressed Sequence Tags), BACs (Bacterial Artificial Chromosomes) and physical maps. Historically, barley genetic studies have their foundations in Mendelian mutants, characterized by altered physiology and/or morphology and still preserved as Barley Genetic Stocks [4]. In this regard there are examples of morphological mutations described in the past for which the gene/genes responsible have been recently cloned and characterized. The high diversity of barley phenotypes, easiness of hybridization and cultivation has made barley a model genetic organism since the rediscovery of Mendel's laws.

Barley is considered one of the earliest plants domesticated from its wild ancestor, the two row and brittle rachis (easy shattering seeds) *Hordeum spontaneum*, which is still found in the Middle East. The domestication dates to the Neolithic (10,000 years BC). The Mediterranean lands contained two barley types. The first was the six-row type, which grew rapidly in the mild Mediterranean climate, characterized by low and erratic rainfall. The second, called two-row barley, features high-quality grain and was sown in spring for direct human consumption [5]. Genotypes that had a naked kernel were often resorted to as foodstuffs because of their superior nutritional value. Gladiators, whose diet consisted almost entirely of food made from barley flour, were even called '*hordearii*': "Barley Men". It was during the Roman era too that all the criteria for a rational use of grain and straw were defined and refined to perfection, Columella providing a detailed description in his *Res rustica*. The yield per hectare in the fertile lands was about ten-fold the number of seeds sown. Barley further expanded its horizons following Columbus's second voyage, when it was introduced to the New World. Later Dutch, English, French and Danish colonists promoted its spread throughout North America, as did their Spanish and Portuguese counterparts in Central and South America. On the Upland of the Andean nations of Peru and Bolivia, barley is grown for grain at high elevations – over 4,550 m [6].

Since the turn of the 20th century our insights into plant growth, reproduction and trait inheritance have been of indispensable support to barley field management and the improvement of cultivars. Several morphological forms have evolved, including winter, spring, two row, six row, awned, awnless, hooded, naked and covered grain, malting, feed (grain and forage), and food types (functional foods). Barley today ranks fourth behind wheat, rice and maize among the world's cereals for the importance of its contribution, whether direct or indirect, to the production of food. The global production is estimated in 2015 in 135 Millions of metric tons (Mt) from a harvested area of 50 Millions of hectares with an average grain yield of 2.7 t/ha. Grown in almost every country throughout the globe, it is generally found as a crop in areas that are less conducive to the survival of other cereals. It is very widespread in the drought-prone lands of the Near East, northern Africa and the highlands of India, in the saline soils of Australia, in Nordic countries, and in areas that, while fertile, are marked by a very short growing season (less than 100 days) [7]. This flexibility is doubtless due to the wide range of barley's genetic variants. The plant appears capable of producing high yields in fertile areas but with less energy input than other cereals. Barley as a winter crop can register yields as high as 10 t/ha of grain at lower costs than the other cereals, mainly due to its high tillering capacity, Nitrogen Use Efficiency (NUE) and Water Use Efficiency (WUE) (Fig. **1**). Barley's capacity to produce biomass even under restrictive conditions implies that it will continue to play a strategic role in food production for the foreseeable future in many areas of the world where drought and other adverse climatic phenomena are inimical to plant growth and development [8, 9].

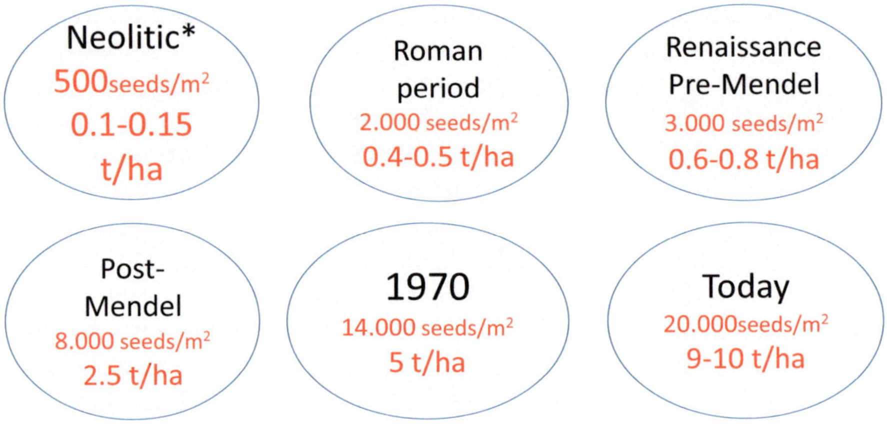

Fig. (1). Estimated "Yield Potential" of barley from domestication (10,000 BC) to the beginning of twenty-first century.**Hordeum spontaneum in situ* conservation (from Stanca, 2016).

Wheats (*Triticum* spp. L.)

For the agronomic and economic value of modern tetraploid durum wheat (2n=4x=28 chromosome; genomic code AABB) and hexaploid bread wheat (2n=6x=42 chromosomes; genomic code AABBDD) the history of their evolution and domestication represents one of the research areas which has been pursued with the greatest intensity. A considerable amount of research has been conducted to elucidate the phylogenetic relationships among cultivated wheats, primitive species, wild species and ancestors of modern wheat. The current bread wheat genome is defined as originally derived from three diploid (2n=2x=14) species: *Triticum urartu* (AA), an unknown close relative of *Aegilops speltoides* (BB), and *Ae. tauschii* (DD) [10]. Likely, the tetraploid *T. turgidum* (AABB) resulted from a first allopolyploidization step involving the A and B genome donors. A subsequent hybridization event between these tetraploid wheat and the wheat genome D donor gave the allohexaploid bread wheat (AABBDD). The noticeable lack of hexaploid wheats in wild populations, and with only archaeological support to estimate the time of origin of bread wheat, raised for a long time questions about the evolution and domestication of the species [11 - 13]. The results were often contradictory due to the different experimental approaches, although currently the emerging picture largely benefits from the results obtained with Next Generation Sequencing technologies. In particular, two recent complementary studies based on overall sequence comparison, phylogenetic analysis and dating of divergence times, shed new light on the complex path undertaken by wheat and the Triticeae genomes during their evolution. In both studies the diploid progenitors and relatives of durum and bread wheats such as *Aegilops speltoides* (BB), *Ae. tauschii* (DD), *Triticum urartu* (AA) and *T. monococcum* (A^mA^m) were included in the analyses and compared to previously published *Brachypodium distachyon* (L.) and *Oryza sativa* (L.) genome sequences as outgroups. Marcussen *et al.* [14], compared the nuclear genome sequences of hexaploid bread wheat subgenomes A, B and D originated from the International Wheat Genome Sequencing Consortium (IWGSC) to estimate the phylogenetic history of the three homologous genome lineages. Topology-based species phylogeny based on a data set of ca. 2,300 gene trees suggested that lineages A and B are more closely related to lineage D individually than to each other. This is in agreement with the hypothesis of reticulate evolution (*i.e.* hybridization of closely related species) and incomplete lineage sorting in large populations that was proposed by Escobar *et al.* [15]. Moreover, further support to the model of a homoploid hybrid origin of the D genome obtained by hybridization between the A and B lineages can be drawn from independent genome sequence analyses that indicated a greater similarity of the A and B genomes to the D lineage, both at the base-pair level as well as in gene content [16]. Middleton *et al.* [17] used chloroplast sequence assembly information for twelve *Triticeae* taxa including bread wheat, barley and rye to provide detailed insights into the evolution of the species. A particularly informative 37 kb segment of chloroplast sequence containing both protein-coding and non-coding regions allowed them to obtain more precise estimates for the divergence times of Triticeae species. Although the same authors emphasized that the results only provide the relationships of the chloroplast lineages, a precise phylogeny and dating of the age of the wheat genome donors was derived and indicated that the donors (or their close relatives) of the A, B and D genomes diverged within the past 3 Million years. Notably, the chloroplast sequences of *T. aestivum* (AABBDD) and *Ae. speltoides* (BB) were found to be 99.87% identical while "the other chloroplast genome sequences of the wheat genome donors *T. urartu* (AA) and *Ae. tauschii* (DD) showed lower sequence identity to *T. aestivum*, ranging from 99.64%–99.69% and *T. urartu* was found to have the greatest similarity to *Ae. tauschii*, with 99.76%" [17]. Therefore, the authors concluded that the donor of the *T. aestivum* chloroplast was closely related, yet not identical, to the current *Ae. speltoides* chloroplast. Today we have available an ordered sequence of 17 Gigabase bread wheat genome produced by sequencing isolated chromosome arms. 124,201 gene loci have been annotated. Comparative gene analysis of wheat subgenomes and diploid and tetraploid wheat relatives show that high sequence similarity and structural conservation are retained, with limited gene losses, after polyploidization. These insights into the genome biology of a polyploid species can provide a faster gene isolation, rapid genetic marker development, and precise prebreeding-breeding programs. Bread wheat (*Triticum aestivum* L.) is the most cultivated cereal crops worldwide with 216,638,762 hectares (10% dedicated to durum wheat) sown in 2015. Due to improved varieties, grain yields have been greatly increased in almost all regions of the world, passing in the last 50 years from less of 2.0 t/ha to 3.1 t/ha registered in 2015 with a yield production per unit of land in many European countries of 7-8 t/ha and with a yield potential of 12 t/ha. The more rapid increases in production began when semi-dwarf varieties were introduced into cultivation [18]. By using semi-dwarf Japanese varieties, N. Strampelli (first) in Italy and N. Borlaug (later) developed high yielding varieties which represent the new era of bread and durum wheat production in the world. Strampelli's success was largely based on two genotypes that introduced novel variation into the European breeding programmes: he used Akagomugi (Japanese) and Wilhelmina (Dutch) varieties carrying *Rht8* (short-straw) and *Ppd* (photoperiod insensitivity) genes, respectively. In 1920-1932 the “Battaglia del grano” was realized and the production doubled [19]. In 1955, N. Borlaug made the first crosses by using Norin 10 (6x) carrying a GA insensitive dwarfing gene Reduced height (*Rht-B1b*). The *Rht-B1b* mutation is a nucleotide substitution that creates a stop codon. T for C substitution converts the CGA codon to a stop codon TGA. The translational termination of the mutant stop codon in *Rht-B1b* permits translational re-initiation at one of several methionines (AUG) that closely follows this stop codon. *Rht-B1b* has been a very lucky gene as all *Rht* genes have pleiotropic effect on plant development and yield potential. In 1970 the so called “Green Revolution” was realised. The farmers rapidly adopted the new varieties and new agronomic techniques were developed. The ideotype breeding was born. The concept how the new wheat or barley genotypes make the better use of the environment has been considered as model to be exported to many other crops. For the first time advances in genetics, statistics, biochemistry, pathology, physiology and more recently molecular biology are synthesized by the breeder. In the last decade the total bread wheat production has not always met the worldwide demand, due to the growing world population, the increased costs for agronomic practices, the negative effects of biotic and abiotic stresses, the competition between food and non-food uses. Therefore, to ensure food security and quality in an environmentally sustainable manner, wheat breeding is now facing big challenges.

Oat (*Avena sativa* L.)

Oat is an autogamous hexaploid (2n = 6x = 42) crop that arose from ancient interspecific hybridization and polyploidization. Oat was first domesticated in Europe ca. 2,000 to 3,000 BC [20] and considerable genetic resources exist with large collections of landraces, wild relatives and cultivars. Significant sources of genetic variation for oat breeding are available in gene banks and breeding programmes [21]. These include adapted cultivars, old landraces, marginally cultivated oats such as *Avena strigosa*, *A. abyssinica*, *A. brevis* and *A. nuda*, crop wild relatives, and genetic stocks bearing defined traits. In the last century, new varieties replaced the traditional landraces. However, the conservation of landraces is important for they potential use in improving crop adaptation and end-use quality. Transfer of useful alleles from primary gene pool (*e.g.* from landraces) can be easily done because of the absence of reproductive barriers. In addition, molecular marker assisted breeding can facilitate the transfer of beneficial traits from landraces to modern cultivars while minimizing linkage drag of non-beneficial alleles. There is therefore a pressing need to radically improve the relevance and quality of the characterization of genetic resources and to combine this with recent advances in genomics to ensure sustainability of agricultural practices [21 - 23]. Despite its excellent nutritional and agronomic properties, many current oat cultivars possess limited yield potential. In the last 50 years the grain yield increase has been very small, passing from 1.4 t/ha in 1960 to 1.8 t/ha in 2015. In parallel with this, the oat area harvested in the same period has been reduced from 35 Million hectares to 15 Million hectares. Opportunities now exist to significantly enhance genetic gain in crop breeding by combining phenotypic selection with precise molecular breeding approaches and exploiting the wider germplasm available; *i.e.*, enhancing the use of oat genetic resources through their use in improving the crops for various farming systems. To expand oat cultivation and increase its use, future oat production will need to meet the challenges posed by the changing climate and the demands of various end-users by identifying cultivars with higher and more stable yield, enhanced grain quality and reduced environmental impacts. In recent years there has been an increase in the demand for oats due to the unique properties of the grain. Oat grains contribute to healthy diets due to the high levels of β-glucan, which is the major endosperm cell wall polysaccharide. Oat β-glucan has approved health claims from EFSA and the US FDA for their cholesterol-lowering effects. Several other health claims, specific to oats and related to reducing the impact of chronic diseases (such as type 2 diabetes, obesity, hypertension, immune-related diseases) are under investigation. Oats also have the highest oil (6-12%) and protein (12-20%) contents in de-hulled grains among cereals and unique types of antioxidants (avenanthramides). As a result of this grain composition, there has been an increase in the use of oats, especially in Europe and North America. The increasing knowledge of grain composition and the potential for various end-uses continues to lead to demand for more oat-derived foods. Traditional oat genetic studies involving for example bi-parental populations have a limited ability to identify markers linked to genes determining key traits. Other novel populations that show wider genetic and phenotypic variation than bi-parental populations, including multi-parent advanced generation inter-cross (MAGIC) and nested association mapping (NAM) populations are currently being developed. Advances in oat genomics will play a key role in the dissection of complex traits and accelerate their selection in oat breeding.

THE GENOMIC REVOLUTION

The extent of genetic and genomic information accumulated during the past decades for several species of the Triticeae tribe has enabled some of the most comprehensive comparative genomics studies in plants. The *Triticeae* belongs to the *Poaceae*, a family which displays great variation in genome, chromosome and gene number and size [24]. Structural genomic variation is reflected by variability in SNPs, mini- and microsatellites, presence/absence of transposable elements, and different forms of chromosomal inversions, translocations and copy number variation. Until recent years, apart from timing and costs, the complexity of plant genomes and the short read length obtained from DNA sequencing platforms posed big challenges to the structural and functional characterization of complex genomes. While Sanger sequences are about 1 kb long with nearly perfect accuracy and can be produced only at very low throughput and high costs, Second and Third Generation Technologies have revolutionized our ability to explore the complexity of the genomes of higher organisms and have demonstrated the capacity to sequence DNA at unprecedented speed. However, the massive amounts of data produced by NGS also present a significant challenge for data storage, analysis, and management solutions. An array of available methods for accessing genome variation in terms of structure and function from NGS data is however available [25, 26].

Using improved sequence alignment criteria and statistical analysis, genomic and expressed sequence data have allowed comparisons of the *Poaceae* genomes and revealed that all species of the family "have evolved from a common ancestor with a basic number of five chromosomes, which underwent a series of whole-genome and segmental duplications, chromosome fusions, and translocations to produce an intermediate ancestral cereal genome of 12 chromosomes" [27]. Sequence-based analysis of translocations and inversions in bread wheat also improved our understanding on the origins of DNA sequences in polyploid species [28].

The impact of next-generation sequencing on genomics of wheat, barley, rye and oats will pave the way for future plant breeding and genetic engineering in crops. Novel strategies to overcome the problems associated to cereals genome complexity have been proposed such as the Genome Zipper concept, reference-guided exome assembly, and "chromosome genomics" based on flow cytometry sorted chromosomes [29].

The vast amount of information deriving from next generation sequencing efforts now needs to be translated into next generation breeding, *i.e.* a suite of genome-driven breeding tools offering new opportunities to deploy genetic diversity in modern agriculture [30]. In small grain cereals the implementation of these tools in pre-breeding and breeding schemes is being achieved, both in the private and the public sectors, with the following major objectives.

i. Characterization of wide genetic resources. Sequence information and comparison of the genomes of different cultivars, landraces and wild accessions has led to the development of high-density genotyping platforms, such as the SNP arrays today available for barley [31], wheat [32] and oat [33]. These tools provide a cheap, easy and robust assay for scoring of thousands of markers across many genotypes and allow us to accurately describe the genetic diversity of different populations [34, 35], define diverse gene pools representing the basis of heterotic potential, and to reveal domestication events and genomic regions under selection [36]. Since genetic diversity represents the key to breeding success, molecular information is crucial for the optimal selection of crossing parents in breeding programmes. Also, it can help in maintaining or broadening the genetic variation, thus providing the basis for breeders to continuously select improved varieties. Finally, growing interest is nowadays given to the study of epigenetic variation, which has also been considered as a possible cause of heterotic effects.
ii. Genetic dissection of complex traits. Broad genetic resources can now be scanned for marker-trait associations without any limitations in terms of markers' availability, with the final aim of identifying genes and genomic regions that are significantly correlated with quantitative trait variation [37]. Successful examples of Genome Wide Association Scans in small grain cereals have been recently reviewed [37 - 39]. In some cases, they represented the starting point for detecting candidate genes and finally identifying the functional mutation underlying important agronomic traits [31, 40]. In all the other instances, they provide breeders with plenty of marker-traits associations that may be directly exploited for selection of improved breeding populations.
iii. Marker and Genome Assisted Selection. Marker-Assisted Selection (MAS) exploits molecular markers to drive the selection of a relatively small set of loci/genes having major phenotypic effects, to ensure the introgression of

'must-have alleles' into advanced materials. In small grain cereals, MAS was routinely adopted in the private sector, mainly to follow disease resistance alleles [41] and crop-dedicated websites have been developed with specific information (*e.g.* in wheat http://maswheat.ucdavis.edu/; in barley http://www.barleycap.org/). More recently, the advent of high throughput marker technologies has boosted the adoption of Genomic Selection (GS), which takes advantage of all available markers as predictors of the genotypic breeding value. GS was proved as well reliable as more traditional selection approaches; for example, in a comparison between GS, MAS and phenotypic selection for Fusarium head blight resistance in wheat, Rutkoski *et al.* [42] always observed a higher prediction accuracy for GS. In barley and oat, GS is only at the beginning, but rapid progress is foreseen from empirical data and simulation studies, mainly thanks to the reduction of progeny testing for the evaluation of line performance [43].

iv. Genome editing. The cloning and functional annotation of a growing number of genes opened the way to their targeted modification though genome editing technologies or the identification of suitable mutations within mutagenized populations [44, 45]. In this way, new allelic variants can be introduced into the genomes of cultivated species, as an alternative to the standard breeding process based on recombination. Successful examples have already been reported for small grain cereals: for example, in wheat the three MLO genes were knocked out by CRISPR-Cas9 to regenerate powdery mildew resistant plants [32, 46] which were able to induce targeted and stably-inherited mutations in transgene-free barley plants, supporting the potential for downstream biotechnological applications. TILLING (Targeting Induced Local Lesions IN Genome) represents another strategy to identify novel useful alleles.

Taken together, the incorporation of these technologies into breeding and agriculture offers the potential to speed up the selection process and to address the challenges of feeding a rapidly increasing world population. Knowledge-based MAGIC (multiparent advanced generation intercross) and NAM (nested association mapping) populations will help in advancing pre-breeding and breeding programmes of *Triticeae* by pyramiding useful genes in elite germplasm [47]. On the other hand, these achievements highlight new bottlenecks for breeding progress in small grain cereals, such as the manipulation of the recombination frequency and the phenotyping capacity, a trivial step for the identification of the traits to incorporate in the plant for the future.

DESIGNING THE PLANT OF THE FUTURE

Projections of the requirements for global food production at the mid 21st century and beyond require consideration of a series of global change processes that affect production potential as well as demand to significant extents. Reserves of phosphorus and fuel are decreasing [48], and in consequence, related price changes for these resources can potentially play an increasing role in determining the level of intensification in agriculture. The rise in atmospheric CO_2 concentration and other greenhouse gases will change the global climate in the coming decades leading to increased average temperatures, altered patterns of rainfall and increasing frequencies of climatic extremes [49, 50]. The changing climate will lead to geographical shifts of agro-ecological zoning with a trend for poleward displacement of potentially arable lands and increasing drought and heat stress occurrence, especially at low latitudes. These trends are expected to lead to increased irrigation requirements in areas where the needs already at present cannot always be satisfied [51].

The demand for agricultural products is projected to increase substantially, related to quantitative and qualitative changes in the demand structure. Until 2050 the human population is expected to rise to about 9.6 Billion people (UN, 2014) thus leading to an increase of 32% in the number of humans to be nourished relative to the year 2014. At the same time, the increasing wealth of substantial social strata within several big developing economies has already led and is expected to further lead to changes in human diets with increased demand for meat products further increasing the need for plant production. The growing demand adds to the currently still unresolved problem of how to abate mortality due to hunger and malnutrition [52]. Eventual production of bioenergy plants will further increase the demand for agricultural products.

Thus, although deceleration of the growth rates of agricultural production and demand is plausible, the absolute demand for agricultural products will increase substantially till mid-century. The demand for cereals alone in 2050 is estimated to be 3 Billion tonnes, equivalent to an increase by 940 Million tonnes relative to the reference period 2005/2007 [52].

Competing land-use options and requisites of environmental policies constrain the options for meeting the increased demand for agricultural products by extensivation or intensification of the production.

Arable soils are limited and further expansion of the cultivated areas conflicts with nature conservation imperatives [53]. The objectives, to restrain ground and surface water contamination through nutrient leaching and application of high doses of herbicides, pesticides and insecticides, the imperative of an increase in resource use economy and reestablishment of closed material cycles, as well as the need to reduce emissions of CO_2, NO, N_2O, and methane from agricultural systems set limits to the intensification options.

The aim to satisfy the increasing demand for food and concomitantly promote a transition to a sustainable agriculture can be met by combining several approaches: closing yield gaps, increasing agricultural resource use efficiency, reducing waste, and increasing yield potential and stability through the introduction of improved varieties. In many regions of the world agricultural productivity rests behind the level achievable with the most suited varieties and agricultural practices, a phenomenon commonly addressed as the yield gap. These gaps can be closed by *e.g.* improving irrigation, fertilization, and mechanization. Foley *et al.* [54] estimated that bringing yields to 95 (75)% of their potential would increase world food production by 58 (28)%. Also, reducing losses and waste of food along transport and supply chains and by final consumers, estimated to amount to one third to one half of the food produced [54], can contribute to feeding the future world population.

The degree to which the options for closing yield gaps and improving resource and food use efficiencies can be realised will depend on the political, social and economic development paths humanity will pursue in the coming decades. As a full mobilization of the potential inherent in these improvements appears far too optimistic, it is safe to assume that raising yield potential and stability through selection of new varieties will be needed as another key element of the set of solutions.

Growing conditions for cereals are expected to change significantly in the coming decades due to several global change processes. A major ongoing anthropogenic change in the global atmosphere consists in the continuous increase in atmospheric greenhouse gas contents [55]. The emission of greenhouse gases from consumption of fossil fuels in industry, transportation and households, as well as land-use changes and agriculture are at the origin of this modification of the atmospheric composition and are projected to cause further changes in the coming decades.

FACE (Free Air Carbon Dioxide Enrichment) experiments have provided evidence for an increase in cereal grain yields in response to elevation of CO_2 to the levels projected for the middle of the 21st century (about 550 ppm). The crop responses varied between species and also strongly between genotypes. C3 plants on average responded more strongly than C4 plants [56]. The yield increase most often is lower than would be the case if the instantaneous response of photosynthesis to increased levels of the substrate CO_2 could translate proportionately into increased grain production. A meta-analysis of the response of wheat and an ensemble of several crops in response to CO_2 levels expected for mid-Century showed a mean increase in grain yield of +15% (non-significant trend) and +17% respectively [57]. Some first results indicate that between-genotype variability in yield response may be as large as between species variability [56]. Badeck *et al.* [58] reported yield increases between +4% (non-significant trend) and +21% in twelve durum wheat genotypes in a single year of FACE experimentation. Troposphere ozone increases due to higher concentrations of gases promoting its production within the air close to the ground. These changes in ozone concentration strongly vary at regional and local scales and induce reductions in photosynthesis due to stomatal closure and eventually oxidative membrane damage.

Based on the analysis of many studies covering a wide range of regions and crops, Porter *et al.* [56] concluded that negative impacts of climate change on crop yields in recent decades have already been more common than positive impacts and yield decreases are expected to be even more prominent in future especially at low latitudes, while some high latitude growing regions may profit from the rise in temperature. In many, especially low-latitude regions, drought stress episodes are predicted to occur more often, too. Summarising the needs for increased cereal production and the expected effect of global change on productivity, it is likely that increases in yield potential and losses due to higher probabilities of occurrence of abiotic stresses will tend to balance, thus underlining the need to increase yield potential through selection. However, the changing environmental conditions define additional targets for selection. The genetic variability in the yield response to elevated CO_2 and temperatures needs to be explored and the best genotypes for realizing yield increase and minimal negative effects on grain composition identified as bases for further selection.

A major part of the increased yield potential realised through selection during the green revolution was due to the increase in harvest index, allocating a higher fraction of total plant production to harvested organs, *i.e.* grains in cereals. The potential for this kind of improvement is close to being exhausted and the currently slowing breeding progress needs to be accelerated through an increase in biomass production. If the current level of breeding progress can be sustained into the future, wheat yield improvements of 0.9% yr^{-1} [59] would translate into a potential yield that is 57% higher in 2050 compared to the turn of the century. Increases in yield potential during the green revolution were accompanied by increases in the number of grains produced per square metre. Further increases in grains produced per square metre require the definition of a new ideotype for selection that aims to achieve more grains through increased biomass production. At present the unit of land (m^2) potentially can produce 20,000 seeds/m^2 while in 2050 the new varieties would reach 25-30,000 seeds/m^2 in a sustainable manner.

Analyses of the breeding progress in wheat and related changes in photosynthetic and developmental traits led Richards [60] to conclude that "except for a period just before anthesis, photosynthate production is not limiting". With crop photosynthesis responding in a highly plastic manner to changes in sink demand, it is most promising to direct selection efforts towards a modification of sink strength that can sustain high and even elevated photosynthetic rates and increase pre-anthesis canopy photosynthesis in combination with producing a higher number of spikelet primordia and assuring an elevated capacity of the latter to produce viable grains. An analysis of crop RUE (Radiation Use Efficiency) showed that current crop varieties exhibit RUE (for above-ground biomass production) of only up to about 50% of the theoretical maximum RUE and a significant increase in RUE is associated with recent breeding progress in wheat [61] thus indicating a significant lee-way for improvement of RUE.

Along these lines breeding objectives relative to the vegetative phase would be to increase crop duration, which needs to be determined relative to the seasonal course of growth conditions in diverse environments, to increase early vigour through elevated germination speed, size of coleoptiles, specific leaf area (SLA), and leaf expansion rate, with the aim to improve RUE, to optimise tillering rates and to enhance spike growth duration. These elements should all lead to increased biomass production and consequent availability for filling a higher number of grains per square metre combined with an increase in the number of productive tillers. As is well known, the components of final grain number per plant at harvest are determined by fertile spikelets per plant and number of grains per spikelet. Sreenivasulu and Schnurbusch [62] demonstrated the role of genetic factors for floral development and inflorescence architecture in determining these traits in barley and wheat. Higher grain loads also require a selection for increased mechanical stability of the stems in order to keep lodging risks limited. This imperative is more demanding in barley and oats which are more lodging sensitive than wheat.

During grain filling the objective is to sustain green leaf area for longer supported by sustained availability of nutrients, foremost nitrogen, implying an increased NUE. This process is considered strategic for enhancing the water and mineral uptake, transport and the source sink relationship. However, the major challenges that can be defined to reach this target are: (i) Increasing yield potential whilst maintaining quality; (ii) Minimising the gap between yield potential and actual farmers' yield; (iii) Maximising the efficiency of input use for sustainability. In this view, worldwide research efforts have been established to increase wheat productivity. An interesting example is the "20:20 Wheat" Project started in the UK to provide the knowledge base and tools to increase yield potential of wheat to 20 t/ha within the next 20 years. Basically pursued at Rothamsted Research (http://www.rothamsted.ac.uk/our-science/2020-wheat), this challenge is faced through four main programmes: (1) Maximizing yield potential utilising breeding, exploiting novel germplasm, transgenesis and other forms of genome remodelling; (2) Protecting yield potential to mitigate losses due to pests and diseases; (3) Determining soil resource interactions *via* efficient water and nutrients utilisation; (4) Using systems approaches in breeding to account for gene-environment interactions and the adaptation of genotypes in the changing environment.

The allocation of resources to root growth, root growth patterns and symbioses with mycorrhizae and bacteria as well as uptake and transport capacities (including dimensions of vascular bundles) are traits that can be addressed by selection. Intensive work has been recently started to investigate the structure and evolution of the microbiota associated with barley roots. Bulgarelli *et al.* [63], analysing sequences in the metagenome, highlighted a strong positive selection in proteins that are known to directly interact with barley. These interactions drive roots microbiota establishment through specific physiological processes from the surrounding soil biota. This indicates that the identification of the genetic components of the host-microbiome control will be a key for inclusion into breeding programmes.

The rapidly expanding studies on roots, symbiosis and transport will probably mobilize similar tools for application in selection directed towards these traits. New strategies to introduce perennial development of plants in wheat must be carefully considered. The sketched objectives for the selection of the next generations of cereal varieties can be realised by applying the best currently available techniques and will progressively include new emerging tools.

There are a considerable number of mutants related to grain yield to be introduced in elite germplasm in pre-breeding work. It is well known that in barley several mutants, since 1920, have been collected. Most of them have demonstrated their high value for the identification and functionally characterization of key genes important for ideal barley plant architecture [4].

Jaggard *et al.* [64] estimated the yield increase in 2050 that can be expected based on diverse scenarios for future breeding progress, taking into account the effects of elevated CO_2, ozone and changing climates. Their most conservative scenario builds on the assumption that only 70% of the current annual gain realised through breeding can be sustained in the coming decades. Based on this scenario a more than 50% increase in yield potential was predicted for most crop-country combinations. Thus, the ideotype(s) for cereal crops with increased productivity presumably can be realised with the currently developing technologies, assuring a substantial contribution of plant breeding to the endeavour of feeding 9 Billion humans at an improved level of nutrition at mid-century.

Here, we focused on boosting yield potentials and stability of yield. Additional progress can be made by improving biotic stress resistance through breeding which will mainly allow for a switch from intensive biocide applications to less intensive cultivation and thus contribute to meet environmental goals. Furthermore, selection for high yield potential under favourable conditions often also leads to higher yields in less favourable environments subject to abiotic stresses.

The four cereal species, grown in almost every country throughout the globe, are very widespread in the drought-prone lands of the Near East, northern Africa and the highlands of India, in the saline soils of Australia, in Nordic countries, and in areas that, while fertile, are marked by a very short growing season (less than 100 days).

This flexibility is doubtlessly due to cereals' range of genetic variants. However, cereals' capacity to produce biomass even under restrictive conditions implies that they will continue to play a strategic role in food production for the foreseeable future in many areas of the world where drought and other adverse climatic phenomena are inimical to plant growth and development. Due to their phenotypic plasticity, further improvements in "Yield Potential and Stability of Yield" can be expected. Selection specifically directed to create highly tolerant and/or resistant genotypes [65] will contribute to reduce the gap between high yield potential and the actual yield also in very poor small farms (low or zero input) Reliable methods for the detection of copy number variants (CNVs) are nowadays used to perform studies that associate CNVs with agronomic traits (*e.g.*, see [66]), thus preparing the way for fruitful application of CNVs in plant breeding. In fact the interaction between private and public breeding programmes, allowing the introduction of modern varieties very well adapted in fertile as well as in severe stress conditions, represents the modern vision to improve not only grain yield production but another more important goal: the quality of life of all farmers.

CONFLICT OF INTEREST

The authors confirm that they have no conflict of interest to declare for this publication.

ACKNOWLEDGEMENTS

The manuscript has been prepared in the frame of activities funded by MiPAAF project "Risorse genetiche di orzo, avena e triticale per la genomica ed il breeding: conservazione, implementazionee caratterizzazione", D.M.24903 (24/11/2005)

REFERENCES

[1] Cattivelli L, Delogu G, Terzi V, Stanca AM. Progress in Barley Breeding. In: Slafer GA, Ed. Genetic Improvement of Field Crops. New York, USA: Marcel Dekker, Inc. 1993; pp. 95-181.

[2] Pourkheirandish M, Hensel G, Kilian B, *et al.* Evolution of the grain dispersal system in barley. Cell 2015; 162(3): 527-39.
[http://dx.doi.org/10.1016/j.cell.2015.07.002] [PMID: 26232223]

[3] Mayer KF, Waugh R, Brown JW, *et al.* A physical, genetic and functional sequence assembly of the barley genome. Nature 2012; 491(7426): 711-6.
[PMID: 23075845]

[4] Stanca AM, Tumino G, Pagani D, *et al.* The "Italian" barley genetics mutant collection: conservation, development of new mutants and use. Advances in Barley Sciences 2013; pp. 30-5.

[5] von Bothmer R, Komatsuda T. Barley origin and related species. In: Ullrich S, Ed. Barley: Production, Improvement and Uses. Wiley-Blackwell 2011; pp. 14-62.

[6] Ullrich S. Significance, adaptation, production and trade of barley. In: Ullrich S, Ed. Barley: Production, Improvement and Uses. Wiley-Blackwell 2011; pp. 3-13.

[7] Stanca AM, Gianinetti A, Rizza F, Terzi V. Barley: an overview of a versatile cereal grain with many food and feed uses. In: Wringley C, Ed. Encyclopedia of Food Grains. 2nd ed. Oxford Academic Press 2016; pp. 147-52.
[http://dx.doi.org/10.1016/B978-0-12-394437-5.00021-8]

[8] Cattivelli L, Ceccarelli S, Romagosa I, Stanca AM. Abiotic stresses in barley: problems and solutions. Barley: Production, Improvement and Uses. Wiley-Blackwell 2011; pp. 282-306.

[9] Francia E, Tondelli A, Rizza F, *et al.* Determinants of barley grain yield in a wide range of Mediterranean environments. Field Crops Res 2011; 120: 169-78.
[http://dx.doi.org/10.1016/j.fcr.2010.09.010]

[10] Petersen G, Seberg O, Yde M, Berthelsen K. Phylogenetic relationships of Triticum and Aegilops and evidence for the origin of the A, B, and D genomes of common wheat (*Triticum aestivum*). Mol Phylogenet Evol 2006; 39(1): 70-82.
[http://dx.doi.org/10.1016/j.ympev.2006.01.023] [PMID: 16504543]

[11] Heun MR, Schafer-Pregl R, Klawan D, *et al.* Site of einkorn wheat domestication identified by DNA fingerprinting. Science 1997; 278: 1312-4.
[http://dx.doi.org/10.1126/science.278.5341.1312]

[12] Lev-Yadun S, Gopher A, Abbo S. Archaeology. The cradle of agriculture. Science 2000; 288(5471): 1602-3.
[http://dx.doi.org/10.1126/science.288.5471.1602] [PMID: 10858140]

[13] Salamini F, Özkan H, Brandolini A, Schäfer-Pregl R, Martin W. Genetics and geography of wild cereal domestication in the near east. Nat Rev Genet 2002; 3(6): 429-41.
[PMID: 12042770]

[14] Marcussen T, Sandve SR, Heier L, *et al.* Ancient hybridizations among the ancestral genomes of bread wheat. Science 2014; 345(6194): 1250092.
[http://dx.doi.org/10.1126/science.1250092] [PMID: 25035499]

[15] Escobar JS, Scornavacca C, Cenci A, *et al.* Multigenic phylogeny and analysis of tree incongruences in Triticeae (Poaceae). BMC Evol Biol 2011; 11: 181.
[http://dx.doi.org/10.1186/1471-2148-11-181] [PMID: 21702931]

[16] Mayer KF, Rogers J, Doležel J, *et al.* A chromosome-based draft sequence of the hexaploid bread wheat (*Triticum aestivum*) genome. Science 2014; 345(6194): 1251788.
[http://dx.doi.org/10.1126/science.1251788] [PMID: 25035500]

[17] Middleton CP, Senerchia N, Stein N, *et al.* Sequencing of chloroplast genomes from wheat, barley, rye and their relatives provides a detailed insight into the evolution of the Triticeae tribe. PLoS One 2014; 9(3): e85761.
[http://dx.doi.org/10.1371/journal.pone.0085761] [PMID: 24614886]

[18] Slafer GA, Calderini DF. Importance of breeding for further improving durum wheat yield. In: Royo C, Ed. Durum Wheat Breeding: current approaches and future strategies. Food Products Press 2005; Vol. 1: pp. 87-9.

[19] Bozzini A, Corazza L, D'Egidio MG, *et al.* Durum wheat (Triticum turgidum spp. durum). In: Scarascia Mugnozza GT, Pagnotta MA, Eds. An Italian contribution to plant genetics and breeding. Viterbo: Tipolitografia Quattrini 1998; pp. 181-94.

[20] Zohary D, Hopf M, Weiss E. Domestication of plants in the old world Danielle de Carle Added by Danielle de Carle. 4th edition., Oxford, UK: Oxford University Press 2012.
[http://dx.doi.org/10.1093/acprof:osobl/9780199549061.001.0001]

[21] Germeier G, *et al.* AVEQ - A project on Avena Genetic Resources for Nutritional Quality. Bioversity Newsletter for Europe 2011; 42: 5.

[22] Morcia C, Germeier CU, Herrmann M, *et al.* Evaluation of genetic variability for FHB resistance in oat germplasm. EUCARPIA Genetic Resources section meeting; 11-13 June 2013; Alnarp, Sweden. 2013.

[23] Chaffin AS, Huang YF, Smith S, *et al.* A consensus map in cultivated hexaploid oat reveals conserved grass synteny with substantial sub-genome rearrangement. Plant Genome 2016; 9(2): 1-35.
[http://dx.doi.org/10.3835/plantgenome2015.10.0102] [PMID: 27898818]

[24] Kellogg EA, Bennetzen JL. The evolution of nuclear genome structure in seed plants. Am J Bot 2004; 91(10): 1709-25.
[http://dx.doi.org/10.3732/ajb.91.10.1709] [PMID: 21652319]

[25] Zhang J, Chiodini R, Badr A, Zhang G. The impact of next-generation sequencing on genomics. J Genet Genomics 2011; 38(3): 95-109.
[http://dx.doi.org/10.1016/j.jgg.2011.02.003] [PMID: 21477781]

[26] Francia E, Pecchioni N, Policriti A, Scalabrin S. CNV and structural variation in plants: prospects of NGS approaches. In: Sablok G, Ed. Advances in the Understanding of Biological Sciences using Next Generation Sequencing (NGS). Switzerland: Approaches. Springer Int. Pub. 2015; pp. 211-32.
[http://dx.doi.org/10.1007/978-3-319-17157-9_13]

[27] Salse J, Bolot S, Throude M, *et al.* Identification and characterization of shared duplications between rice and wheat provide new insight into grass genome evolution. Plant Cell 2008; 20(1): 11-24.
[http://dx.doi.org/10.1105/tpc.107.056309] [PMID: 18178768]

[28] Ma J, Stiller J, Berkman PJ, *et al.* Sequence-based analysis of translocations and inversions in bread wheat (*Triticum aestivum* L.). PLoS One 2013; 8(11): e79329.
[http://dx.doi.org/10.1371/journal.pone.0079329] [PMID: 24260197]

[29] Spannagl M, Bader K, Pfeifer M, Nussbaumer T, Mayer KF. PGSB/MIPS plant genome information resources and concepts for the analysis of complex grass genomes. Methods Mol Biol 2016; 1374: 165-86.
[http://dx.doi.org/10.1007/978-1-4939-3167-5_8] [PMID: 26519405]

[30] Barabaschi D, Magni F, Volante A. Physical mapping of bread wheat chromosome 5A: an integrated approach. Plant Genome 2015; 8(3): 1-24.
[http://dx.doi.org/10.3835/plantgenome2015.03.0011]

[31] Comadran J, Kilian B, Russell J, *et al.* Natural variation in a homolog of Antirrhinum CENTRORADIALIS contributed to spring growth habit and environmental adaptation in cultivated barley. Nat Genet 2012; 44(12): 1388-92.

[http://dx.doi.org/10.1038/ng.2447] [PMID: 23160098]

[32] Wang S, Wong D, Forrest K, *et al.* Characterization of polyploid wheat genomic diversity using a high-density 90,000 single nucleotide polymorphism array. Plant Biotechnol J 2014; 12(6): 787-96. [http://dx.doi.org/10.1111/pbi.12183] [PMID: 24646323]

[33] Tinker NA, Chao S, Lazo GR, *et al.* A SNP genotyping array for hexaploid oat. Plant Gen 2014; p. 7.

[34] Tondelli A, Xu X, Moragues M, *et al.* Structural and temporal variation in genetic diversity of European spring two-row barley cultivars and association mapping of quantitative traits. Plant Genome 2013; 6: 1-14. [http://dx.doi.org/10.3835/plantgenome2013.03.0007]

[35] Voss-Fels K, Frisch M, Qian L, *et al.* Subgenomic diversity patterns caused by directional selection in bread wheat gene pools. Plant Genome 2015; 8(2): 1-13. [http://dx.doi.org/10.3835/plantgenome2015.03.0013]

[36] Russell J, Dawson IK, Flavell AJ, *et al.* Analysis of >1000 single nucleotide polymorphisms in geographically matched samples of landrace and wild barley indicates secondary contact and chromosome-level differences in diversity around domestication genes. New Phytol 2011; 191(2): 564-78. [http://dx.doi.org/10.1111/j.1469-8137.2011.03704.x] [PMID: 21443695]

[37] Ogura T, Busch W. From phenotypes to causal sequences: using genome wide association studies to dissect the sequence basis for variation of plant development. Curr Opin Plant Biol 2015; 23: 98-108. [http://dx.doi.org/10.1016/j.pbi.2014.11.008] [PMID: 25449733]

[38] Waugh R, Thomas B, Flavell A, *et al.* Genome-Wide Association Scans (GWAS). In: Kumlehn J, Stein N, Eds. Biotechnological Approaches to Barley Improvement. Heidelberg: Springer 2014; pp. 345-66.

[39] Wang M, Wang S, Xia G. From genome to gene: a new epoch for wheat research? Trends Plant Sci 2015; 20(6): 380-7. [http://dx.doi.org/10.1016/j.tplants.2015.03.010] [PMID: 25887708]

[40] Ramsay L, Comadran J, Druka A, *et al.* INTERMEDIUM-C, a modifier of lateral spikelet fertility in barley, is an ortholog of the maize domestication gene TEOSINTE BRANCHED 1. Nat Genet 2011; 43(2): 169-72. [http://dx.doi.org/10.1038/ng.745] [PMID: 21217754]

[41] Thomas WT, Hayes PM, Dahleen LS. Application of molecular genetics and transformation to barley improvement. In: Ullrich SE, Ed. Barley: Production, Improvement, and Uses. Blackwell Publishing Ltd 2011; pp. 122-43. [http://dx.doi.org/10.1002/9780470958636.ch6]

[42] Rutkoski J, Benson J, Jia Y, *et al.* Evaluation of genomic prediction methods for fusarium head blight resistance in wheat. Plant Genome 2012; 5 [http://dx.doi.org/10.3835/plantgenome2012.02.0001]

[43] Schmid KJ, Thorwarth P. Genomic Selection in Barley Breeding. In: Kumlehn J, Stein N, Eds. Biotechnological Approaches to Barley Improvement. Heidelberg: Springer 2014; pp. 367-78.

[44] Tsai H, Howell T, Nitcher R, *et al.* Discovery of rare mutations in populations: TILLING by sequencing. Plant Physiol 2011; 156(3): 1257-68. [http://dx.doi.org/10.1104/pp.110.169748] [PMID: 21531898]

[45] Bortesi L, Fischer R. The CRISPR/Cas9 system for plant genome editing and beyond. Biotechnol Adv 2015; 33(1): 41-52. [http://dx.doi.org/10.1016/j.biotechadv.2014.12.006] [PMID: 25536441]

[46] Lawrenson T, Shorinola O, Stacey N, *et al.* Induction of targeted, heritable mutations in barley and Brassica oleracea using RNA-guided Cas9 nuclease. Genome Biol 2015; 16: 258. [http://dx.doi.org/10.1186/s13059-015-0826-7] [PMID: 26616834]

[47] Mochida K, Shinozaki K. Unlocking Triticeae genomics to sustainably feed the future. Plant Cell Physiol 2013; 54(12): 1931-50. [http://dx.doi.org/10.1093/pcp/pct163] [PMID: 24204022]

[48] Dawson CJ, Hilton J. Fertiliser availability in a resource-limited world: Production and recycling of nitrogen and phosphorus. Food Policy 2011; 36: S14-22. [http://dx.doi.org/10.1016/j.foodpol.2010.11.012]

[49] Easterling WE, Aggarwal PK, Batima P, *et al.* Food, fibre and forest products. In: Parry ML, Canziani OF, Palutikof JP, van der Linden PJ, Hanson CE, Eds. Climate Change 2007: Impacts, Adaptation and Vulnerability Contribution of Working Group II to the Fourth Assessment Report of the Intergovernmental Panel on Climate Change. Cambridge, UK: Cambridge University Press 2007; pp. 273-313.

[50] Coumou D, Rahmstorf S. A decade of weather extremes. Nat Clim Chang 2012; 2: 491-6.

[51] Wallace JS. Increasing agricultural water use efficiency to meet future food production. Agric Ecosyst Environ 2000; 82: 105-19. [http://dx.doi.org/10.1016/S0167-8809(00)00220-6]

[52] Alexandratos N, Bruinsma J. World agriculture towards 2030/2050: the 2012 revision. Agricultural Development Economics Division, Food and Agriculture Organization of the United Nations 2012.

[53] Balmford A, Green RE, Scharlemann JP. Sparing land for nature: exploring the potential impact of changes in agricultural yield on the area needed for crop production. Glob Change Biol 2005; 11: 1594-605. [http://dx.doi.org/10.1111/j.1365-2486.2005.001035.x]

[54] Foley JA, Ramankutty N, Brauman KA, *et al.* Solutions for a cultivated planet. Nature 2011; 478(7369): 337-42. [http://dx.doi.org/10.1038/nature10452] [PMID: 21993620]

[55] Stocker TF, Qin D, Plattner GK, *et al.* Climate Change 2013: The Physical Science Basis. Contribution of Working Group I to the Fifth Assessment Report of the Intergovernmental Panel on Climate Change. Cambridge, UK, and New York, NY, USA: Cambridge University Press 2013.

[56] Porter JR, Xie L, Challinor AJ, *et al.* Food security and food production systems. In: Field CB, Barros VR, Dokken DJ, *et al.*, Eds., Climate Change 2014: Impacts, Adaptation, and Vulnerability Part A: Global and Sectoral Aspects Contribution of Working Group II to the Fifth Assessment Report of the Intergovernmental Panel of Climate Change. Cambridge, United Kingdom and New York, NY, USA: Cambridge University Press 2014; pp. 485-533.

[57] Ainsworth EA, Long SP. What have we learned from 15 years of free-air CO2 enrichment (FACE)? A meta-analytic review of the responses of photosynthesis, canopy properties and plant production to rising CO2. New Phytol 2005; 165(2): 351-71. [http://dx.doi.org/10.1111/j.1469-8137.2004.01224.x] [PMID: 15720649]

[58] Badeck FW, Fares C, Rizza F, *et al.* Durum wheat grain yield and quality under elevated CO2. In: Ventura F, Pieri L, Eds. Atti del XVI convegno nazionale di agrometeorologia. Bologna: Patron Editore 2013; pp. 107-8.

[59] Fischer RA, Edmeades GO. Breeding and cereal yield progress. Crop Sci 2010. [http://dx.doi.org/10.2135/cropsci2009.10.0564]

[60] Richards RA. Selectable traits to increase crop photosynthesis and yield of grain crops. J Exp Bot 2000; 51(Spec No): 447-58. [http://dx.doi.org/10.1093/jexbot/51.suppl_1.447] [PMID: 10938853]

[61] Slattery RA, Ort DR. Photosynthetic energy conversion efficiency: setting a baseline for gauging future improvements in important food and biofuel crops. Plant Physiol 2015; 168(2): 383-92. [http://dx.doi.org/10.1104/pp.15.00066] [PMID: 25829463]

[62] Sreenivasulu N, Schnurbusch T. A genetic playground for enhancing grain number in cereals. Trends Plant Sci 2012; 17(2): 91-101.
[http://dx.doi.org/10.1016/j.tplants.2011.11.003] [PMID: 22197176]

[63] Bulgarelli D, Garrido-Oter R, Münch PC, *et al.* Structure and function of the bacterial root microbiota in wild and domesticated barley. Cell Host Microbe 2015; 17(3): 392-403.
[http://dx.doi.org/10.1016/j.chom.2015.01.011] [PMID: 25732064]

[64] Jaggard KW, Qi A, Ober ES. Possible changes to arable crop yields by 2050. Philos Trans R Soc Lond B Biol Sci 2010; 365(1554): 2835-51.
[http://dx.doi.org/10.1098/rstb.2010.0153] [PMID: 20713388]

[65] Rizza F, Karsai I, Morcia C, *et al.* Association between the allele compositions of major plant developmental genes and frost tolerance in barley (*Hordeum vulgare L.*) germplasm of different origin. Molecular Breeding 2016; 36: 156.
[http://dx.doi.org/10.1007/s11032-016-0571-y]

[66] Francia E, Morcia C, Pasquariello M, *et al.* Copy number variation at the *HvCBF4-HvCBF2* genomic segment is a major component of frost resistance in barley. Plant Mol Biol 2016; 92(1-2): 161-75.
[http://dx.doi.org/10.1007/s11103-016-0505-4] [PMID: 27338258]

SUBJECT INDEX

A

B

Roberto Pilu and Giuseppe Gavazzi (Eds.)

C

D

E

F

H

I

L

M

N

O

P

Q

R

S

T

U

V

W

Y

Z

Made in the USA
Las Vegas, NV
03 November 2026

57425091R00321